SOLID STATE PHYSICS LITERATURE GUIDES
Volume 3

GROUPS IV, V, AND VI
TRANSITION METALS
AND COMPOUNDS
Preparation and Properties

Solid State Physics Literature Guides

Prepared under the auspices of the Research Materials Information Center, Oak Ridge National Laboratory

General Editor: T. F. Connolly

Solid State Division
Oak Ridge National Laboratory*
Oak Ridge, Tennessee

Volume 1: Ferroelectric Materials and Ferroelectricity—1970

Volume 2: Semiconductors—Preparation, Crystal Growth, and Selected Properties—1972

Volume 3: Groups IV, V, and VI Transition Metals and Compounds—Preparation and Properties—1972

Volume 4: Electrical Properties of Solids—Surface Preparation and Methods of Measurement—1972

*Oak Ridge National Laboratory is operated by Union Carbide Corporation for the U.S. Atomic Energy Commission.

SOLID STATE PHYSICS LITERATURE GUIDES
Volume 3

GROUPS IV, V, AND VI TRANSITION METALS AND COMPOUNDS
Preparation and Properties

Edited by
T. F. Connolly
Research Materials Information Center
Solid State Division
Oak Ridge National Laboratory
Oak Ridge, Tennessee

With Commentaries by
D. N. Carlson
Institute for Atomic Research and Development of Metallurgy
Iowa State University
Ames, Iowa
and
Thomas B. Reed
Lincoln Laboratory
Massachusetts Institute of Technology
Lexington, Massachusetts

SPRINGER SCIENCE+BUSINESS MEDIA, LLC · 1972

Library of Congress Catalog Card Number 74-133269

ISBN 978-1-4684-6206-7   ISBN 978-1-4684-6204-3   (eBook)
DOI 10.1007/978-1-4684-6204-3

© Springer Science+Business Media New York 1972
Originally published by IFI/Plenum Data Corporation in 1972
Softcover reprint of the hardcover 1st edition 1972

Introduction

The material in this collection is based mainly on papers actually received by the Research Materials Information Center, although some references are included on specific recommendations. While this might exclude a few relevant papers, it also excludes a much larger number of nonpertinent references that might be chosen on the basis of deceptive titles or inadequate abstracts.

For any collection of this sort the question of what should or should not be included often involves individual bias or differences of opinion of the meaning of terms, and the criticisms of two experts in the same field are often contradictory. For this reason most compilations in this series, in addition to organization under appropriate subject headings, contain one or more sections entitled "Reviews, Bibliographies, and Compilations," in which references to peripheral related subjects are deliberately included. In all other sections the effort is to be as specific as possible, with borderline references kept to a minimum.

At this writing, there are over 70,000 searchable references in the RMIC collection on solid-state inorganic materials science, and the coverage of the field is good back to 1960, although many earlier references are included. Still, there will be omissions and errors in compilations drawn from the collection, and any pointed out to us will be corrected in future editions. (Such corrections should be sent to the Center and not to the publisher.)

The timeliness of these compilations, as well as our ability to answer daily inquiries, depends very largely on the continued receipt by the Center of all papers, reprints, reports, and preprints within our scope. These should be mailed to

T. F. Connolly
Research Materials Information Center
Oak Ridge National Laboratory
P. O. Box X
Oak Ridge, Tennessee 37830

Preface

Since 1963 the Research Materials Information Center has been answering inquiries on the availability, preparation, and properties of ultrapure inorganic research specimens. It has been possible to do this with reasonable efficiency by searching an automated, coded microfilm collection of the report and open literature and of data sheets and questionnaires provided by commercial and research producers of pure materials.

With the growth of the collection to over 70,000 documents and the increase in the demand for more general background information, it has been necessary to compile bibliographies on an increasing variety of subjects. These have been used as indexes to the microfilmed documents for more efficient searching, and in the past distributed in response to individual requests. However, their size and number no longer permit so casual and uneconomic a method of distribution. The "ORNL Solid State Physics Literature Guides" is a practical alternative.

Scope of This Bibliography

The scope of the bibliography is that of the Research Materials Information Center: emphasis is on materials preparation, ultrapurification, crystal growth, characterization, and basic physical properties of research specimens. Engineering and mechanical properties are not included except incidentally in references to handbooks and some serial compilations. The one exception to the coverage indicated by the title is the inclusion of Groups VII and VIII in the section on chalcogenides. The differences between numbers of subdivisions and numbers of references among the major divisions of the bibliography also reflect the coverage of the Center: the greater part of the collection is devoted to the single elements and the oxides.

Organization

The subject organization of the bibliography is given by the Table of Contents, which represents the structures of six separate bibliographies compiled by the RMIC. Sections entitled "General" include listings of reviews, bibliographies, and compilations, or collections of a variety of elements or compounds not subsumable under other headings. Under each subheading the arrangement is chronological, with 1971 first, and alphabetical by first author within each year.

To the rear of this Preface, is an indexed list of other information sources and of new journals or serial publications in in the field of solid-state materials.

Acknowledgment

I am grateful to R. E. Reed, of the ORNL Solid State Division, for his detailed review of the sections on the single elements and the oxides and for useful suggestions on their reorganization. (Errors and omissions, however, are my own responsibility.) To Betty Edwards and Emily Copenhaver my thanks for what must have seemed endless typing, retyping and correcting of these bibliographies over a span of years.

Availability of Documents

U. S. Government contractor reports, usually identified by an alpha-numeric report number, can be purchased from

> National Technical Information Service
> U. S. Department of Commerce
> Springfield, Virginia 22151

and, often, on request from the issuing installation.

USAEC reports are also available from

> International Atomic Energy Agency
> Kaerntnerring A 1010
> Vienna, Austria

> National Lending Library
> Boston Spa
> England

Monographs and reports of the National Bureau of Standards are for sale by

> Superintendent of Documents
> U. S. Government Printing Office
> Washington, D. C. 20402

Theses, listed as Dissertation Abstracts + number, are available in North or South America from

> University Microfilms
> Dissertation Copies
> P. O. Box 1764
> Ann Arbor, Michigan 48106

and elsewhere from

> University Microfilms, Ltd.
> St. John's Road
> Tylers Green
> Penn, Buckinghamshire
> England

Other Information Centers and New Journals

Field	Information centers and other sources	New journals and serials
Ultrapurification and crystal growth	4, 8, 11, 13, 15, 16, 19, 20, 21, 28, 30, 32, 33, 42, 58, 59	9, 11, 15, 24, 31, 32

Field	Information centers and other sources	New journals and serials
Characterization		
Miscellaneous	3, 4, 8, 11, 13, 16, 19, 20, 21, 26, 28, 30, 31, 32, 33, 35, 37, 38, 39, 40, 42, 46, 53, 56, 58, 60, 61, 62	1, 3, 4, 8, 11, 15, 17, 24, 25, 28, 29, 30, 31, 32
Activation analysis	1, 2, 6, 38, 43, 48, 57	12, 22
Neutron diffraction	12, see also Miscellaneous and Activation Analysis	
Mass spectrometry	17, 18, 38, 44, 45	6, 14, 21
Optical methods	5, 27, 39, 54	2, 17, 18, 19, 20, 27, 30
Electron diffraction and microscopy	9, 10, 27, 41, 50	16, 17, 30
X-ray diffraction and spectroscopy	5, 9, 10, 12, 22, 27, 34, 39, 40, 41, 50, 51, 52	17, 30
Electron microprobe analysis	See Electron Diffraction and Microscopy; X-Ray Diffraction and Spectroscopy	16
Measurement of electrical properties	13, 14, 20, 29, 33, 49	25
Thermal methods	4, 7, 23, 24, 25, 30	5, 13, 23
Resonance methods	3, 39, 47, 55	10
Mossbauer effect	3	
Field or ion emission		6

Specialized Information Centers and Other Sources

1. Activation Analysis Documentation Center, Akademiai Kiado, Publishing House of the Hungarian Academy of Sciences, Publicity Department, P.O. Box 24, Budapest 502, Hungary. (5500 punched cards, searchable by elements, classes of materials, discipline, instrumentation; 1935-1968; in English).

2. Activation Analysis Research Laboratory, Texas Agricultural and Mechanical University, College Station, Texas, 77843.

3. Alloy Data Center, Alloy Physics Section, Metallurgy Division, National Bureau of Standards, Washington, D. C. 20234, telephone 301-921-2917. Critically evaluated data on physical properties of metals and alloys.

4. American Society for Metals — Documentation Service, Metals Park, Ohio 44072, telephone 216-ED 8-5151. Includes electronics solid state physics, inorganic chemistry; computer searches; publishes ASM Review of Metal Literature.

5. Atomic Energy Levels Data and Information Center, Institute for Basic Standards, National Bureau of Standards, Washington, D.C. 20234, telephone 301-921-2014. Provides critical appraisal of data on atomic spectra; classification of spectral lines and determination of atomic energy levels; coordination of experimental work in the laboratories working in this field.

6. Zentralstelle fuer Atomkernenergie—Dokumentation, Frankfurt am Main, West Germany. Compiles bibliographies in nuclear science and technology, including several extensive and detailed compilations on activation analyses.

7. Chemical Thermodynamic Data Center, Physical Chemistry Division, Institute for Basic Standards, National Bureau of Standards, Gaithersburg, Maryland, telephone 301-921-2467. Thermochemical data, heats of formation, free energy of formation of chemical substances.

8. Cobalt Information Center, Battelle Memorial Institute, 505 King Ave., Columbus, Ohio 43201, telephone 614-299-3151, extension 2234. (Main office: Centre d'Information du Cobalt, S. A., 35 rue des Colonies, Brussels 1, Belgium; Branch offices: Cobalt Information Centre, 7 Rolls Building, London E. C. 4, England; and Kobalt-Information, Elisabethstrasse 14, Dusseldorf, Germany). Publishes the journal Cobalt; bibliographies; answers specific inquiries and provides technical assistance on cobalt (and alloys and compounds).

9. Commission on Crystallographic Data, International Union of Crystallography. Chairman: Dr. F. W. Matthews, Central Technical Information Unit, Imperial Chemical Industries Ltd., Imperial House, Millbank, London S. W. 1, United Kingdom.

10. Crystal Data Center, Institute for Materials Research, National Bureau of Standards, Washington, D.C. 20234, telephone 301-921-2837. Crystallographic data on all solids; data and related information on unit cell dimension of crystal materials. Purpose is to revise and bring up to date the volume of Crystal Data; to maintain this semicritical compilation; to identify crystalline materials by single crystals; and to collect and maintain data and information on crystalline materials.

11. Centre de Documentation sur les Synthèses Cristallines, Laboratoire de Physique Moléculaire et Cristalline, Faculté des Sciences, Place Eugene-Bataillon 34-Montpellier, France. Identifies European crystal growers by country installation, and material. Up-dated loose-leaf sheets.

12. International Data Center for Work on Crystallography, University Chemical Laboratory, Cambridge, England. Works closely with all groups involved in crysallographic documentation and data evaluation. At present only carbon-containing structures are compiled for data obtained by x-ray and neutron diffraction (for a computer oriented file).

13. Electronic Materials Information Center, Royal Radar Establishment, St. Andrews Road, Gt. Malvern, Worcestershire, England.

14. Electronic Properties Information Center, Hughes Aircraft Company, Centinela and Teale Streets, Culver City, California 90230, telephone 213-391-0711, extension 6596.

15. Bureau d'Information sur les Matériaux ultra Purs, IMP - Inspection Générale, C.E.A. 29, rue de la Fédération, B. P. 510, Paris (XV), France, telephone 273 60-00 poste 55-95. Inquiry service on purification and availability of ultrapure metals.

16. Isotopes Information Center, Oak Ridge National Laboratory, P. O. Box X, Oak Ridge, Tennessee 37830, telephone 615-483-8611, extension 3-1742; Federal Telecommunication System No. 615-483-1742. Information primarily on production and uses of radioisotopes in industry and research. Includes unusual counting methods or new instrumentation for detection of isotopes, instruments that use isotopes in their operation.

17. Mass Spectrometry Data Center, National Bureau of Standards, Washington, D.C. 20234, telephone 301-921-2173.

18. Mass Spectrometry Data Center, Building A 8.1, Atomic Weapons Research Establishment, Aldermaston, Berks., England. (MSDC Mass Spectra Data Sheets)

19. Rare-Earth Information Center, Institute for Atomic Research, Iowa State University, Ames, Iowa 50010, telephone 515-294-2272. Analytical, inorganic, and physical chemistry of the rare-earth elements and compounds; solid state physics and metallurgy. Publishes reviews, compilations, and bibliographies. Inquiry service by mail and telephone.

20. Research Materials Information Center, Oak Ridge National Laboratory, P.O. Box X, Oak Ridge, Tennessee 37830,

telephone 615-483-8611, extension 3-1287; Federal Telecommunication System No. 615-483-1287. Provides references on availability, purification, crystal growth, and properties of all ultrapure inorganic research specimens.

21. The Selenium—Tellurium Information Service, Selenium—Tellurium Development Association, Inc., 345 East 47th Street, New York, New York 10017, telephone 212-688-2632. Publishes periodic selenium and tellurium bibliography; inquiry service.

22. SHARE Library System, a compilation of x-ray diffraction computer programs available on request. IBM Corporation, White Plains, New York.

23. Thermodynamic Properties of Metals and Alloys, Lawrence Radiation Laboratory, University of California, Berkeley, California 94720, telephone 415-843-2740, extension 3817.

24. Thermodynamics Research Center, Research Center of Texas A. and M. University, College Station, Texas 77843. Thermodynamic and other physicochemical properties; infrared, ultraviolet, Raman, mass, and nuclear magnetic resonance spectral data.

25. Thermophysical Properties Research Center, Purdue University, 2595 Yeager Road, West Lafayette, Indiana 47906, telephone 317-743-3827. Provides reference data based on integrated programs of critical evaluation.

26. Analytical Abstracts. Society for Analytical Chemistry, 9-10 Savile Row, London, England.

27. Bulletin Signalétique. Centre de Documentation du C.N.R.S., 15, quai Anatole-France, 75-Paris-VIIe, France. Section 160 - Structure de la matière, I. Physique de l'état condensé. Physique atomique et moléculaire. Spectroscopie; Section 161 - Structure de la matiére, II. Cristallographie; Section 170 - Chimie. Chimie générale et chimie physique. Chimie minerale. Chimie analytique. Chimie organique. Section 761 - Microscopie Electronique. Diffraction Electronique. (All four are monthly abstract publications covering world literature; indexed by subject, author, and installation.)

28. Chemical Abstracts now markets a magnetic tape service. Inquiries should be addressed to Chemical Abstracts, Division of American Chemical Society, Ohio State University, Columbus, Ohio 43210, telephone 614-293-5022.

29. Electrical and Electronics Abstracts; also Current Papers in Electrical and Electronics Engineering. Institute of Electrical and Electronics Engineers, 345 East 47th Street, New York, New York 10012.

30. Metals Abstracts. American Society for Metals, Metals Park, Ohio 44073 or The Institute of Metals, 17 Belgrave Square, London S.W. 1, England. Index terms and complete literature citations available on magnetic tapes for 1968-69. Also 1965-67 for Review of Metal Literature.

31. Nuclear Science Abstracts. Bimonthly; abstracts of world literature, indexed by material, subject, author, and installation. Superintendent of Documents, U.S. Government Printing Office, Washington, D.C. 20402.

32. Physics Abstracts. American Institute of Physics, Office of Publications, 335 East 45th Street, New York, New York 10017.

33. Solid State Abstracts. Abstracts on preparation and properties of high-purity inorganic compounds. Cambridge Communications Corporation, 238 Main Street, Cambridge, Massachusetts 02142.

34. X-Ray Fluorescence Spectrometry Abstracts, Science and Technology Agency, 3 Dyers Building, London E.C. 1, England. New quarterly publication; surveys international literature including conference proceedings and unpublished reports on theory and applications.

35. Periodic bibliographic surveys on most techniques appear in Analytical Chemistry. American Chemical Society, 1155 Sixteenth Street, N.W., Washington, D.C. 20036.

36. Journal of Chemical Documentation. American Chemical Society, 1155 Sixteenth Street, N.W., Washington, D.C. 20036.

37. Current Contents. Institute for Scientific Information, Inc., 325 Chestnut Street, Philadelphia, Pennsylvania 19106. Prints tables of contents of current journals in chemistry, physics, and other fields.

38. Commission for the Establishment of Analytical Methods. Commission d'Etablissement des Methodes d'Analyse (C.E.T.A.M.A.), Commissariat à l'Energie Atomique, 75-Paris, France. Gamma and mass spectrometry, statistical methods, tabulation of radioactive half-lives; reviews.

39. CODATA International Compendium of Numerical Data Projects, Springer Verlag, New York; Heidelberg; Berlin (1969). Identifies, with address, international centers, data projects, and publications concerning (a) optical, x-ray, mass, and magnetic resonance spectra; (b) crystallographic, electrical, and magnetic properties; (c) thermophysical properties; (d) nuclear and isotopic data. Published form and availability of data are noted in each case.

40. Crystal Structures, 2nd edition, R.W.G. Wyckoff (Wiley-Interscience, New York; London; Sydney), Vol. 1: 1963, 467p.; Vol. 2: 1964, 588p.; Vol. 3: 1965 981p.; Vol. 4: 1968, 566p.; Vol. 5, 1966, 785p.

41. A Crystallographic Book List, Commission on Crystallographic Teaching, International Union of Crystallography (1965). (About 800 books on crystallography and related subjects; classification by subjects; many cross references)

42. Pierre de la Breteque, Directeur de Récherches, Société Francaise pour l'Industrie de l'Aluminium, 134 Chemin des Aygalades, 13 Marseille 15e, France. Publishes annual bibliographies on gallium and compounds (including III-V, III-VI, and ternaries) covering preparation and physical properties, containing phase diagrams of new compounds.

43. Gamma-Ray Spectrum Catalogue, National Reactor Testing Station, P.O. Box 1845, Idaho Falls, Idaho 83401, director: R. L. Heath, telephone 208-522-4400, extension 4447. (Gamma-ray reference spectral)

44. Catalog of Mass Spectral Data (loose-leaf data sheets and supplements), Thermodynamics Research Center, Texas A. and M. Research Foundation, F. E. Box 130, College Station, Texas 77843.

45. Index to Mass Spectral Data, American Society for Testing and Materials, 1916 Race Street, Philadelphia, Pennsylvania 19103. Available as ASTM-AMD 10 A-C and 11A (1969), punched cards.

46. For a complete list of Standard Reference Materials available from the National Bureau of Standards, see Catalog of Standard Reference Materials, NBS Spec. Publ. 260 (July 1970 ed.) for sale by the superintendent of Documents, U.S. Government Printing Office, Washington, D.C. for 75 cents. A Standard Reference Material Availability and Price List is issued semiannually and supplied to users on request.

47. NMR, NQR, EPR Current Literature Service issued by the publishers of DMS - Documentation of Molecular Spectroscopy in association with the Institut fur Spektrochemie und Angewandte Spektroskopie, Dortmund and DMS Scientific Advisory Board, London (Butterworth and Company, Publishers, Ltd., London). (Resonance literature from 1963; reference list and card index; 600 journals covered).

48. Index de la Literature Nucleaire Francaise. Monthly computer printout of abstracts; titles in French and English; includes general and analytical chemistry, activation analysis, neutron diffraction, accelerators, neutron sources. Key-word-in-context indexes in French and English; author and installation indexes. (Service Central de Documentation du Commissariat a l'Energie Atomique, Centre d'Etudes Nucleaires de Saclay, B. P. No. 2, 91-Gif-sur-Yvette, France).

49. A Joint Program on Methods of Measurement for Semiconductor Materials, Process Control and Devices has been undertaken (in 1968) by the National Bureau of Standards, Electronic Technology Division, Washington, D.C. [J. C. French is coordinator of the program; NBS-TN-472 (Dec. 1968), MBS-TN-475 (Feb. 1969), NBS-TN-488 (July 1969), NBS-TN-495 (Sept. 1969), and NBS-TN-520 (March 1970) are recent reports.]

50. Powder Diffraction Data File — X-Ray or Electron. Approximately 10,000 patterns, issued in the form of 3 x 5 cards; five types of indexes are available. American Society for Testing and Materials, 1916 Race Street, Philadelphia, Pennsylvania 19104, telephone 215-563-5315. But see Computer detected errors in the ASTM x-ray powder diffraction file, M.C. Nichols (Sandia Corp., Livermore, California), SCL-DR-70-48 (April 1970), 26p.

51. Structure Reports, Vol. 22 (1958-1968), 890p.; Vol. 24 (1960-1968), 795p.; Vol. 25, Cumulative Index for Vols.

15-24 (1969), International Union of Crystallography, Commission on Structure Reports (Oosthoek, Utrecht, Netherlands). (Principally by x-ray but includes results of electron and neutron diffraction and other methods)

52. Advances in X-Ray Analysis. Proceedings of Annual Conference on Applications of X-Ray Analysis, University of Denver (Plenum Press, New York).

53. World Meetings Information Center, CCM Information Corp., 909 Third Avenue, New York, New York 10022. Publishes Current Index to Conference Papers in Chemistry. (Vol. 1, No. 11 in July 1970).

54. Atomic Absorption and Flame Emission Spectroscopy Abstracts. Science and Technology Agency, 3 Dyers Building, London, E.C. 1, England.

55. Nuclear Magnetic Resonance Spectrometry Abstracts. Science and Technology Agency, 3 Dyers Building, London, E.C. 1, England. (First issue January 1971)

56. Diffusion in Metals and Alloys Data Center, Metal Physics Section, Institute for Materials Research, National Bureau of Standards, Washington, D.C. 20234. Diffusion coefficients and activation energies.

57. Activation Analysis Information Center, Analytical Chemistry Division, National Bureau of Standards, Washington, D.C. 20234.

58. Ferroelectric Materials and Ferroelectricity, T.F. Connolly and Errett Turner, eds., Vol. 1 of ORNL Solid State Physics Literature Guides (IFI/Plenum, New York, Washington, London, 1970), 685p.

59. Semiconductors: Preparation, Crystal Growth, and Selected Properties, T.F. Connolly, ed., Vol. 2 of ORNL Solid State Physics Literature Guides (IFI/Plenum, New York, Washington, London, 1971). (Material-organized bibliography of Research Materials Information Center references received thru 1970)

60. Computer-Based Chemical Information Services, Edward M. Arnett, Science 170, 1370-76 (1970). (Review of some services available)

61. Annotated Accession List of Data Compilations of the Office of Standard Reference Data, Herman M. Weisman and G.B. Sherwood, NBS Tech. Note 554 (1970). (About 1300 reference data compilations, critical reviews, bibliographies, and other ancillary reference works)

62. Diffusion Information Center, 22447 Lake Road, Cleveland, Ohio 44116, or P.O. Box 505, CH-4500 Solothurm, Switzerland.

New Journals and Serials

1. Analytical Letters. Marcel Dekker, Inc., 95 Madison Ave., New York, N.Y. 10016. (Vol. 1 in 1967)

2. Applied Spectroscopy Review. Marcel Dekker, Inc., 95 Madison Ave., New York, N.Y. 10016. (Vol. 1 in 1967)

3. Chemical Instrumentation. Marcel Dekker, Inc., 95 Madison Ave., New York, N.Y. 10016. (Vol. 2 in 1969)

4. CRC Critical Reviews in Analytical Chemistry. Chemical Rubber Co., 18901 Cranwood Parkway, Cleveland, Ohio 44128. (Vol. 1 in 1971)

5. High Temperature Science. Academic Press Inc., 111 Fifth Ave., New York, N.Y. 10003. (Vol. 1 in 1969)

6. International Journal of Mass Spectrometry and Ion Physics. Elsevier Publishing Co., P.O. Box 211, Jan Van Galenstraat 335, Amsterdam, Netherlands. (Vol. 2 in 1969)

7. Journal of Chemical Thermodynamics. Academic Press, Inc., 111 Fifth Ave., New York, N.Y. 10003 or 17 Old Queen St., London S.W. 1, England. (Vol. 1 in 1969)

8. Journal of Chromatographic Services. Preston Technical Abstracts, 909 Pitner, Evanston, Ill. 60602.

9. Journal of Crystal Growth. North Holland Publishing Co., P.O. Box 103, Amsterdam, Netherlands. (Vol. 1 in 1967)

10.   Journal of Magnetic Resonance. Academic Press, Inc., 111 Fifth Ave., New York, N. Y. 10003. (Vol. 1 in 1969)

11.   Journal of Materials Science. Chapman and Hall, 11 New Fetter Lane, London E. C. 4, England. (Vol. 1 in 1966)

12.   Journal of Radioanalytical Chemistry. Elsevier Publishing Co., Box 211, Amsterdam, Netherlands. (Vol. 1 in 1968)

13.   Journal of Thermal Analysis. Heyden and Son, Ltd., Spectrum House, Alderton Crescent, London N. W. 4, England. (Vol. 1 in 1969)

14.   Mass Spectrometry Bulletin, Mass Spectrometry Data Centre. Her Majesty's Stationary Office, P. O. Box 569, London, S. E. 1, England.

15.   Materials Research Bulletin. Pergamon Press, Inc., Maxwell House, Fairview Park, Elmsford, N. Y. 10523. (Vol. 1 in 1966)

16.   Micron, The International Quarterly, Journal of Electron Microscopy, Electron Probe Micro-Analysis and Associated Techniques. Structural Publishers, Ltd., Watford, England. (Vol. 1, No. 1, June 1969)

17.   Microstructures. A. Z. Publishing Corp., 647 North Sepulveda Blvd., Bel Air, Los Angeles, Calif. (Vol. 1, No. 1, August-September 1970)

18.   Nouvelle Revue d'Optique Appliquee. Masson et Cie., 120 Bd. Saint Germain, Paris 6, France. (Vol. 1, No. 1, January-February 1970)

19.   Optical Spectra. Optical Publishing Co., Inc., 7 North Pitt St., Pittsfield, Me. 01201. (Vol. 3 in 1969)

20.   Optics Communications. North-Holland Publishing Co., P.O. Box 3489, Amsterdam, Netherlands. (Vol. 1 in 1969)

21.   Organic Mass Spectrometry. Heyden and Sons Ltd., Spectrum House, Alderton Crescent, London N. W. 4, England.

22.   Radiochemical and Raidoanalytical Letters. Elsevier Publishing Co., Laussane, Switzerland. (Vol. 1 in 1969)

23.   Thermochimica Acta. Elsevier Publishing Co., P. O. Box 211, Amsterdam, Netherlands. (Vol. 1, No. 1, March 1970)

24.   Journal of Solid State Chemistry. Academic Press, Inc., 111 Fifth Ave., New York, N. Y. 10003. (Vol. 1 in 1970)

25.   Journal of Non-Crystalline Solids. North-Holland Publishing Co., P. O. Box 3489, Amsterdam, Netherlands. (Vol. 4 in 1970)

26.   Journal of Low Temperature Physics. Plenum Publishing Corp., 227 West 17th St., New York, N. Y. 10011. (Vol. 2 in 1970)

27.   Methodes Physiques d'Analyse. Services du GAMS, L.N.E., 1, rue Gaston-Boissier, Paris XV, France. (Review published quarterly by the Group for Advancement of Spectrographic Methods; Vol. 1 in March 1968)

28.   International Journal of Nondestructive Testing. Gordon and Breach Science Publishers, 8 Bloomsbury Way, London W.C. 1, England. (Vol. 1 in 1970)

29.   Diffusion Data. Diffusion Information Center, 22447 Lake Road, Cleveland, Ohio 44116. (Vol. 4 in 1970)

30.   Crystal Lattice Defects. Gordon and Breach Science Publishers, 8 Bloomsbury Way, London W. C. 1, England. (Vol. 1 in 1970)

31.   CRC Critical Reviews in Solid State Physics. Chemical Rubber Co., 18901 Cranwood Parkway, Cleveland, Ohio 44128. (Vol. 1 in 1971)

32.   Comments on Solid State Physics (A Journal of Critical Discussion of the Current Literature). Gordon and Breach, Science Publishers, Inc., 150 Fifth Ave., New York, N. Y. 10011 or Gordon and Breach, Science Publishers, Ltd., 8 Bloomsbury Way, London W. C. 1, England.

Contents

Contents

Foreword

Any specialized bibliography is at best a poor substitute for the detailed critical review, but in rapidly growing fields it is usually all that is available. In an attempt at a compromise between a simple listing of papers and the desirable exhaustive study, selected sections of the Solid State Physics Literature Guides will be introduced by fairly brief commentaries by experts in the various fields. In most cases the comments should be understood as necessarily tentative, with the authors working against a deadline not usual in more leisurely publications (for example, proof copies of the bibliographies had to be available to them before they could begin). Further, in view of the interim nature of the Guides, the authors of the comments have been asked to feel more free to conjecture than is usual in more formal papers.

We are all in their debt for their willingness to assume such a task under such conditions.

T. F. Connolly

PREPARATION METHODS FOR GROUPS IV, V, AND VI TRANSITION METALS

O. N. Carlson

Institute for Atomic Research and Department of Metallurgy
Iowa State University, Ames, Iowa 50010

Work on preparation methods of the Groups IV, V, and VI transition metals can be classified under four categories: (1) reduction, (2) refining or purification, (3) growth of single crystals, and (4) preparation of thin films. In the ensuing paragraphs an attempt is made to highlight the significant approaches and advances in each of these areas.

Reduction

The extractive metallurgy of all of the metals of these groups has undergone considerable change during the past two decades, as is evidenced by this compilation. Three general preparation methods — aluminothermic, carbothermic, and electrolytic reduction processes — dominate the current literature.

Processes based on the aluminothermic reduction of an oxide have been developed for niobium, tantalum, vanadium, chromium, and molybdenum. While this is a relatively old method dating back to the work of Goldschmidt at the beginning of this century, modern developments in high-temperature and high-vacuum technology have now made it commercially feasible. With the development of the electron-beam furnace, especially, the removal of any retained aluminum and oxygen as well as other residual volatile impurities from the base metal during melting has been greatly facilitated.

Carbothermic reduction processes have been reported for almost all of the metals included in this survey. The ores of the group IV elements are converted to the carbide in a carbon arc furnace, and then chlorinated for use in subsequent and/or reduction steps. The group V elements are prepared by reacting stoichiometric amounts of oxide and carbide *in vacuo* at elevated temperatures to obtain a metallic product usually containing undesirable amounts of carbon or oxygen. Refining of the product is generally required, although by careful control of the reaction conditions a high-quality product may be obtained directly as with niobium and tantalum.

Some effective work has been done on the electrowinning of most of these metals from a fused salt. The salt bath usually consists of a mixture of low-melting alkali halides in which the halide of the metal of interest is dissolved. While all of the transition metals are still produced primarily by metallothermic, carbothermic, or hydrogen reduction processes, electrowinning appears to offer an increasingly attractive alternative. Commercially feasible methods for pro-

---

* Work performed in the Ames Laboratory, U. S. Atomic Energy Commission.

ducing titanium and tantalum by direct electrolysis are known, and methods whereby all of the other metals included in this survey can be electrowon are described in the current literature.

The reduction processes discussed above yield products in solid ingot, sponge, coarse crystal, or powder form. The importance of fine powders in many electronic applications, especially tantalum, molybdenum, and tungsten, is evidenced by the amount of work devoted to this subject.

Refining

A plethora of refining techniques has been employed for these metals including vacuum extraction, zone refining, iodide decomposition, electrorefining, electrotransport, and sublimation.

Vacuum extraction involves the removal of volatile impurities by heating the metal under a low pressure. Most of these metals have relatively low vapor pressures at their melting points and hence can be purified with respect to oxygen and hydrogen as well as those metallic impurities having vapor pressures that are at least a factor of ten higher than that of the base metal. Electron-beam and vacuum-arc melting are widely used in the commercial purification of all of these metals except chromium. Some of the more volatile impurities such as hydrogen, aluminum, magnesium, and, in some cases, oxygen and nitrogen are also removed by vacuum extraction from the solid ingot, usually by induction heating.

Zone refining is used extensively in the purification of all of the metals covered in this survey. Floating zone melting by electron bombardment is widely used, while other methods such as arc-zone and induction zone melting are employed to a lesser extent. Those metals that exhibit a solid phase transformation such as the group IV elements can also be purified by displacement of a transformed zone. Some of the more significant developments in zone refining are the use of multiple zones, continuous and semicontinuous refining, float-zone refining in ultrahigh vacuum, and application of hollow-cathode heat sources.

Purification by the decomposition of a metal iodide on a hot filament or surface has been successfully applied to titanium, zirconium, hafnium, vanadium, niobium, and chromium. Work on the thermodynamic principles and phase equilibria underlying the decomposition of iodides has resulted in a better basic understanding of the decomposition process, especially for the zirconium iodide system.

All of the metals of groups IV and V plus chromium have been purified by electrolysis of the crude metal or scrap from fused salt mixtures, usually chlorides or fluorides. Electrorefining is particularly important in the preparation of high-purity vanadium. Likewise, experiments on ultrapurification by electrotransport have been reported for vanadium and zirconium. This specialized technique depends on the migration of impurities under the influence of an electric field at elevated temperatures. Purification of the bulk metal by sublimation is limited primarily to chromium, although purification of metallic salts, particularly the chlorides, often depends on a sublimation or distillation step prior to their use as a reactant salt in the reduction process.

Single Crystal Preparation

The availability of these metals in high-purity form together with their increased use in the aircraft, space, and nuclear programs has resulted in a demand for single crystals of various sizes, orientations, and purities for use in many different research studies. The groups V and VI elements do not undergo phase transformations and hence are quite readily prepared in single crystal form directly from the melt. Their high melting points and chemical reactivity, however, require the use of high-temperature furnaces and either vacuum or inert gas protective atmospheres. While there are a number of variations in the crystal growing techniques, most of them employ the zone melting principle, in which the liquid zone is maintained by electron-beam, arc, or high-frequency induction heating. Electron-beam heating is used most extensively because of the ease of obtaining the high melting temperatures of these metals under a high vacuum. This method offers the advantages of zone purification and vacuum extraction along with the growth of the monocrystals.

The Czochralski technique of pulling an oriented seed crystal from the melt has been successfully used to grow large molybdenum and niobium single crystals. Likewise molybdenum and tungsten crystals have been grown by decomposition or reduction of their chlorides in a hydrogen atmosphere onto a coarse-grained filament.

The group IV elements exhibit phase transformations which eliminate or at least complicate the use of any of the above methods. Single crystals of these metals are usually prepared by rapid thermal cycling through the $\alpha - \beta$ transformation, high-temperature grain growth, or strain annealing.

Preparation of Thin Films

The importance of these transition elements in electronic components and the superconducting properties of their alloys and compounds have provided much of the impetus for work on their deposition as thin films. The techniques that are most commonly used in the preparation of thin films or coatings of the refractory metals are: (1) sputtering, (2) evaporation, and (3) vapor decomposition.

Sputtering involves the bombardment of a metal by energetic ionized particles, usually inert gas ions, thus dislodging metallic atoms from the surface and enabling them to be deposited elsewhere in the system. The method is applicable to even the highest-melting metals and to the formation of alloy films. The most common methods of producing sputtering are high-voltage d.c. glow discharge and R.F. sputtering. The merits of other techniques, variously described as low-voltage sputtering, tetrode sputtering, and asymmetric a.c. sputtering, are also discussed in the literature.

The evaporation method depends upon vaporization of the metal in an evacuated system and its subsequent condensation upon the desired substrate. As a group these transition metals have some of the highest melting and boiling points of all the metals in the periodic table. Evaporation is, therefore, generally carried out *in vacuo*, usually with electron-beam or vacuum-arc heating. Films of all of the metals including tungsten have been prepared in this way.

Vapor decomposition depends upon the thermal decomposition or chemical reduction of a volatile halide salt such that a metallic film or coating is deposited on a heated surface. Vapors of the iodides of certain metals such as zirconium or vanadium are thermally decomposed and deposited on a hot surface. Other metals such as molybdenum, tungsten, niobium, and tantalum are plated out onto a hot substrate by reduction of their chlorides or fluorides by hydrogen gas.

RECENT DEVELOPMENTS IN THE STUDY OF OXIDES

Thomas B. Reed

Lincoln Laboratory
Massachusetts Institute of Technology
Lexington, Massachusetts 01273

Oxides can be good insulators, metals, or semiconductors. They can conduct electronically or ionically. They include some of the highest-melting compounds and the compounds with highest tensile strength. They can be ferromagnetic or ferroelectric. In fact, if all other compounds were to disappear from the face of the earth, we could still keep an army of scientists busy investigating the properties of the oxides and an army of engineers turning these properties to the use of mankind. One of the richest fields of oxide interest is in the Group IV, V, and VI compounds, as can be seen by the number of references in that section of the bibliography.

Much of our scientific labor proceeds in pedestrian fashion from the opening up of a new field to the filling in of the broad details to the final completion of the picture with fine detail. I will attempt to pick out a few developments of the last decade which have helped us in these labors or made the results more useful.

An old saw states that physicists generally do first-rate work on third-rate materials. Hopefully we can forget this maxim now because during the last decade great progress has been made in the preparation of materials and characterization of these materials to make certain that they were in fact what they were thought to be. For instance, some of the data in the reviews referenced here in the early sixties report phase transitions and magnetic properties of Ti and V oxides which in fact were the properties of a second phase present at too low a concentration to be detected by x-ray. A great deal of time has been wasted in trying to investigate these reported properties and redoing the original work to discover the source of error and correct the results. It is rather pitiful to look back at learned papers which struggle to give reasons for effects that don't exist. Three developments of the last decade will hopefully prevent such errors in the future.

Single crystal growth of compounds has become almost a necessary first step in the investigation (or reinvestigation to correct former errors) of any new compound. In the bibliography about 20% of the papers referenced deal with material preparation and the majority of these are concerned with single crystal growth. Although use of single crystals is not a panacea for all materials problems, the process of crystal growth excludes second phases (or makes them easily visible when they occur) and usually gives at least an order of magnitude reduction in the impurity level. Development of techniques for growth of the oxide crystals can be expected to have the same beneficial effects on oxide research that growth of good germanium and silicon crystals has had in the field of semiconductors.

Oxygen pressure (expressed for convenience as $pO = -\log pO_2$) has the same importance in preparation of oxides that hydrogen ion concentration (expressed as $pH = -\log CH_2$) has in the preparation of compounds from solution. A very important development in oxide chemistry is the development of the zirconia (or thoria) electrode, which serves the same function in measuring pO_2 as the glass electrode does for pH. Stabilized zirconia has a high mobility for oxygen ions and is impervious to all other gases, and a simple electrochemical cell can be set up in which the emf is a direct measure of oxygen pressure, pO. These meters are available commercially and will reliably measure pO in the range from 0 to 30.

The variable-valence oxides show a wide variety of structures. The structure and the deviation from the ideal stoichiometry of these structures are both determined by the oxygen pressure. For instance at least twelve possible vanadium oxides exist, VO, V_2O_3, the Magneli phases V_nO_{2n-1}, VO_2, V_3O_7, V_6O_{13}, and V_2O_3. Synthesis of these compounds requires accurate control of pO in the range $0 < pO < 40$. VO_2 has a semiconducting-to-metal transition at $343^\circ K$, which makes it attractive for both electrical and optical devices. The resistivity ratio (between the semiconducting and metallic phases) varies from 10^4 to 1 as x in VO_x varies from 2.00 to 2.07, a change difficult to detect analytically. Fortunately the pO associated with these two extremes of the VO_2 phase is 4.1 and 2.9 (at $1500^\circ K$), a large change that is easy to control.

Finally, general analysis for bulk samples and thin films (using new techniques such as neutron activation, solid state mass spectroscopy, Auger electron spectroscopy, etc.) has made great strides so that after making a new material one can find the exact stoichiometry and concentration of impurities. Furthermore, elemental chemicals are now easily available relatively inexpensively in much higher purity than they were a decade ago. The suppliers are becoming more sophisticated so that a material labeled "six nines" may actually be as high as "four nines."

Most of the transition metal oxides have several valences. The oxides of Ti, V, and Cr probably have the largest numbers of compounds of any in the periodic table. Structural chemists have studied the way in which variable valence is accommodated, for instance, in the Magneli structures M_nO_{2n-1} (where M is Ti or V), and they find alternate layers of M_2O_3 (edge sharing of octahedra) and MO_2 (corner sharing of octahedra) in the required proportions. The compounds TiO, VO, and NbO are of interest because they have 15—25% vacancies and the oxygen—metal ratio in the first two can vary between 0.8 and 1.3. TiO is metallic throughout this range, while VO is metallic for $x < 1$ and a semiconductor for $x > 1$. Finally, about 20% of the vacancies can be removed by annealing at high pressure.

The oxides of Zr, Hf, and Th are particularly interesting because in the cubic phase they show a high oxygen conductivity and can be used in fuel cells and as oxygen electrodes, as discussed above. Tantalum oxide is of interest because of its use in tantalum capacitors. Various suboxides have interesting structures but are difficult or impossible to prepare. CrO_2 has attracted a lot of attention because of its use in magnetic recording tape. It can be prepared only at high pressure.

From these brief comments it is clear that the oxides of these few metals are a fascinating field of study. It is hoped that this introduction will help in going through the following references.

Addendum

The following were noted since the main body of the bibliography was prepared:

1. Groups IV, V, and VI Transition Metals — Single Elements

Fermi Surface. II. D-Block and f-Block Metals
  A. P. Cracknell
  Advan. Phys. 20:1–141 (1971)

The Metallurgy of Zirconium, Atomic Energy Review, Supplement 1971
  D. L. Douglass
  International Atomic Energy Agency, Vienna, 1971

Electron – Phonon Interaction and Paramagnetic Susceptibilities for Transition Metals
  Nobuo Mori
  J. Phys. Soc. Japan 31:359–367 (1971)

Experimental K-Fluorescence Yields for the Elements Z = 22 to 30 and 32
  P. Suortti
  J. Appl. Phys. 42:5821–5825 (1971)
  Ti, V, Cr, Mn, Fe, Co, Ni, Cu, Zn, and Ge

Selected Values of Chemical Thermodynamic Properties (Tables for Elements 54 through 61 in the Standard Order of Arrangement
  D. D. Wagman, W. H. Evans, V. B. Parker, I. Halow, S. M. Bailey, R. H. Schumm, and K. L. Churney
  Natl. Bur. Stand. Tech. Note 270-5 (March 1971), 49 pp.
  Tables of values for the standard heats and Gibbs (free) energies of formation, entropies, and enthalpies at 298.15°K and heats of formation at 0°K for compounds of V, Nb, Ti, Zr, Hf, Sc, and Y

The Electronic Structure of Pure Metals: Part A. Electron Theory of Pure Metals, Part B. Fermi Surfaces and Physical Properties of Some Real Metals, Progress in Materials Science, Vol. 14
  W. M. Lomer and W. E. Gardner
  Pergamon Press, New York and London (1969)

2. Transition Metal Borides

Structure électronique et propriétés magnetiques des diborures des metaux de transition
  Jacques Castaing

Office National d'Etudes et de Recherches Aerospatiales, Publication No. 140 (1971)
61 pp., 116 references

3. Transition Metal Carbides

Conduction Band of IVa and Va Subgroup Transition Metal Monocarbides. II
  A. S. Borukhovich, P. V. Geld, V. A. Tskhai, L. B. Dubrovskaya, and I. I. Matveenko
  Phys. Stat. Sol. B45:179–187 (1971)

High-Temperature Inorganic Compounds
  G. V. Samsonov
  AEC-tr-6873; TT 67-69068 (1971)
  Translation of "Vysokotemperaturnye Neorganicheskie Soedineniya," Izd. 'Naukova Damka,' Kiev (1965), published for the Atomic Energy Commission and the National Science Foundation, Washington, D. C., by the Indian National Scientific Documentation Center, New Delhi.

Herstellung und Eigenschaften von Karbiden der Übergangsmetalle in ihren Homogenitätsbereichen
  G. V. Samsonov, V. Ya. Naumenko, and L. N. Okhremchuk
  Phys. Stat. Sol. A6:201 (1971)
  Groups IV and V; lattice parameters, electrical resistivity, microhardness, thermal expansion coefficient, and electron work function

4. Binary Transition Metal Nitrides

Review of Thermodynamic Properties of the Chromium – Nitrogen System
  J. P. De Luca and J. M. Leitnaker
  (Metals and Ceramics Division, Oak Ridge National Laboratory, Oak Ridge, Tenn.), ORNL-TM-3618 (Dec. 1971)
  Submitted to J. Am. Ceram. Soc.

About Nitrides and Carbonitrides and Nitride-Based Cemented Hard Alloys
  R. Kieffer, P. Ettmayer, and M. Freudhofmeier
  Modern Developments in Powder Metallurgy, Vol. 5 (Henry H. Hausner, ed.), Plenum Press, New York (1971), pp. 201–214

5. Binary Transition Metal Oxides

Nonstoichiometry, Diffusion and Electrical Conductivity in Binary Metal Oxides
  Per Kofstad
  Wiley-Interscience, New York (1972), 384 pp.

Anomalous Properties of the Vanadium Oxides
  J. B. Goodenough
  Annual Review of Materials Science, Vol. 1, Annual Reviews, Inc., Palo Alto, Calif. (1971), pp. 101-138

Metallic Oxides
  J. B. Goodenough
  Progress in Solid State Chemistry, Vol. 5 (H. Reiss, ed.), Pergamon Press, Oxford, New York, and Toronto (1971), pp. 149-399
  539 references

Electric and Optical Properties of High Quality Crystalline V_2O_4 near the Semiconductor Metal Transition Temperature
  Larry A. Ladd
  (Harvard University, Cambridge, Mass.), AD-727242; HP-26; ARPA-TR-41 (June 1971), 213 pp.
  Experimental measurements and theoretical analysis which permit a tentative explanation of the transition mechanism to be made

Conduction in Non-Crystalline Systems. VIII. The Highly Correlated Electron Gas in Doped Semiconductors and in Vanadium Monoxides
  N. F. Mott
  Phil. Mag. 24:935-958 (1971)

Data Compilation on Vanadium Oxides
  M. Neuberger
  (Electronic Properties Information Center, Hughes Aircraft Company, Culver City, Calif. 90230), IR-79 (Nov. 1971)
  Tables of all available information on the crystal structure, physical, mechanical, optical, thermal, magnetic, and electronic properties of bulk and film samples, as well as graphs

Transition Metal Compounds: Transport and Magnetic Properties
  E. R. Schatz, ed.
  Gordon and Breach, New York (1971), 200 pp.

Electronic Phase Transition
  D. Adler
  Essays in Phys. 1:33-77 (1970)

Proceedings of the Tenth International Conference on the Physics of Semiconductors, Cambridge, Massachusetts, August 17-21, 1970
  S. P. Keller, J. C. Hensel, and Frank Stern, eds.
  CONF-700801 (1970):

  Metal–Insulator Transitions in Transition Metal Oxides, T. M. Rice, D. B. McWhan, and W. F. Brinkman, pp. 293-300

  Variational Approach to the Metal–Semiconductor Transition, T. A. Kaplan and R. A. Bari, pp. 301-303

  Role of the Crystal C/A Ratio in Ti_2O_3 and V_2O_3, J. B. Goodenough, pp. 304-309

  A Raman Study of the Semiconductor–Metal Transition in Ti_2O_3, A. Mooradian and P. M. Raccah, p. 310

  Small and Large Polarons in NiO, D. Adler, pp. 317-320

  Electronic Conduction in V_2O_5 Single Crystals, P. Nagels and M. Denayer, pp. 321-328

6. Binary Transition Metal Chalcogenides

Intercalation Complexes of Lewis Bases and Layered Sulfides: A Large Class of New Superconductors
  F. R. Gambel, J. H. Osiecki, M. Cais, R. Pisharody, F. J. DiSalvo, and T. H. Geballe
  Science 174:493-497 (1971)
  50 materials; T_c all below 5°K; Nb, Ti, and Ta disulfides

Transition Metal Chalcogenides
  G. V. Lashkarev
  Metalloved., Mater. Simp. 1968 (N. V. Ageev, ed.), "Nauka," Moscow (1971), pp. 388-393
  In Russian

Platinum Metal Chalcogenides
  Aaron Wold
  In: Platinum Group Metals and Compounds, Advances in Chemistry Series 98, American Chemical Society, Washington, D. C. (1971).
  Symposium sponsored by the Division of Inorganic Chemistry at the 158th Meeting of the American Chemical Society, New York, N. Y., Sept. 8-9, 1969

1. Groups IV, V, and VI Transition Metals — Single Elements

1.a. Preparation Methods

1.a.1. General

Physico-Chemical Principles of Crystallization Processes Used in the Superpurification of Metals
V. N. Vigdorovich
JPRS 52552 (March 1971)

Some Characteristics of the Purification of Refractory Metals by Electron-Beam Melting
V. E. Efimov
Izv. Akad. Nauk SSSR, Metally, 3:49–52 (1970)

Rare Metals, 3rd ed., rev.
O. A. Songina
Metallurgiya, Moscow (1964)
Translated from Russian by Israel Program for Scientific Translations, Jerusalem, TT70-50021 (1970)

Crystallization of Metals and Alloys in the Suspended State
A. A. Fogel', T. A. Sidorova, Z. A. Guts, M. M. Mezdrogina, and I. V. Korkin
Kristallografiya, 13 (5)883–888 (1968)
Sov. Phys.—Cryst., 13 (5):762–765 (1969)

Preparation of Thin Foils from 0.01 Inch Diameter Wire for Electron Microscopy
J. A. F. Gidley and P. N. Richards
J. Sci. Instr. (J. Phys. E), Ser. 2, Vol. 2, pp. 297–299 (1969)

Carbidothermic Production of Metals of the IVa, Va, and VIa Groups and Their Alloys
R. Kieffer, F. Lihl, and E. Effenberger
In: High Temperature Materials, pp. 276–90 (F. Benesovsky, ed.)
Metallwerk Plansee AG, Reutte/Tirol, Austria (1969)
From 6th International Plansee Seminar on High Temperature Materials, Fundamental and Developments, Reutte/Tirol, Austria. (In German)

Über ein neues, völlig tiegelfreies Schmelzverfahren (Stipp-Verfahren)
J. J. Nickl
Z. Metallk., 60:800–802 (1969)

Purification of a Metal by Electron-Beam Heating
R. E. Reed and J. C. Wilson
(Oak Ridge National Lab., Oak Ridge, Tenn.) Purification of Inorganic and Organic Materials (Morris Zief, ed.), Marcel Dekker, Inc., New York (1969), pp. 235–247

Einkristalle hochschmelzender und seltener Metalle
E. M. Savicky
Nauka, Moscow (1969)

Monocrystals of Refractory and Rare Metals
E. M. Savitskii, ed.
Izdatel'stvo Nauka, Moscow (1969), 192 pp.
Mo, V, W, Nb

Tungsten Bibliography, No. 1, Bulletin of the UNCTAD Committee on Tungsten
TD/TUNGSTEN COM/BIB/L. 1/Rev. 1 (Oct. 1968); GE. 68-21244
Supplement No. 1, TD/TUNGSTEN COM/BIB/L: 1/Rev. 1/Add. 1 (June 1969); GE. 69-12998

Elektronenstrahlschmelzen hochschmelzender und reaktiver Metalle (Ti, Zr, V, Nb, Ta)
H. Bohmeier
Eigenschaften und Anwendung hochschmelzender und reaktiver Metalle, Deutscher Verl. Grundstoffind., Leipzig (1968), pp. 192–210

Production of Monocrystals of High-Melting, Chemically Active Metals and Various Oxides by Zone Melting
N. A. Brilliantov, N. G. Mezentseva, and L. S. Starostina
Rost Kristallov, Akad. Nauk SSSR, Inst. Kristallogr., 6, 297–300 (1965)
Growth of Crystals, Vol. 6B (N. N. Sheftal' and A. V. Shubnikov, eds.), Consultants Bureau, New York (1968), pp. 107–111
W, Nb, Ta, Mo, V, Zr

Techniques of Metals Research, Vol. I: Pts. 1, 2, and 3: Technology of Materials Preparation and Handling; Vol. II: Pts. 1 and 2: Techniques for the Direct Observation of Structure and Its Applications
R. F. Bunshah, ed.
Interscience Publishers, Wiley and Sons, Inc., New York (1968)

The Refractory Metals
F. H. Buttner and R. W. Hale
AD-669 371 (1968)

Crystallization of Levitating Metals
A. A. Fogel', T. A. Sidorova, Z. A. Guts, and M. M. Mezdrogina
Fiz. Tverd. Tela, 9(11):3346-3347 (1967)
Sov. Phys.—Solid State, 9(11):2636-2637 (1968)

Advances in Zone Refining
Robert H. Heil, Jr.
Solid State Tech., 11:21 (1968)

Vapor-Phase Deposition of Refractory Metals and Their Carbides
R. Lorel and B. Pinteau
Galvano, 37:337-341 (1968)

Growth and Imperfections of Metallic Crystals
D. E. Ovsienko, ed.
Consultants Bureau, New York (1968)

Growing of Large Single Crystals of Refractory Metals
A. I. Pekarev, G. N. Shchirenko, and Yu. D. Chistyakov
Izv. Vyssh. Ucheb. Zaved., Tsvet. Met., 11(2):156-157 (1968)

Electron Beam Gun for Evaporation
T. Szuchs and G. Vago
Finommechanika, 7:293-298 (1968)

Production of Pure Metals by Hydrogen Reduction of Their Halides
A. van der Steen and R. Durand
Rev. Met., 65:219-226 (1968)

Dc Sputtering with Rf-Induced Substrate Bias
J. L. Vossen and J. J. O'Neill, Jr.
RCA Rev., 29, 566-81 (1968)
Ti, Ta

The Use of Continuous Zone Refining in Metallurgy
A. S. Zherebovich and V. N. Vigdorovich and A. N. Krestovnikov
Russian Met., No. 2, p. 80 (1968)

Über das Kristallwachstum hochschmelzender Metalle beim Elektronenzonenschmelzen
J. Barthel and R. Scharfenberg
Crystal Growth, Proceedings of an International Conference held in Boston, June 20-24, 1966 (H. Steffen Peiser, ed.), Pergamon Press, New York, London (1967), pp. 133-139

Non-Crucible Zone Melting of Refractory Metals by Means of Point-Focused Electron Beams
E. B. Bas and H. Stevens
Z. Angew. Math. Phys., 18:747-750 (1967)

Levitation Melting of Metals and Alloys
J. W. Downey
Metallurgy Div., Argonne National Lab., Argonne, Ill., ANL-7398 (Dec. 1967), 15 pp.

High-Purity Metals and Alloys: Fabrication, Properties, and Testing
V. S. Emel'yanov and A. I. Evstyukhin, eds.
Consultants Bureau, N. Y. (1967), 175 pp.
Transl. from Russian by G. D. Archard

Levitation Melting. II. Properties of Levitation Coils with Rotational Symmetry and Levi-
tation Experiments
E. Fromm and H. Jehn
Z. Metallk., 58:366 (1967)
Al, Cu, Zr, V, Nb, Mo, Ta, and W

Growth of Metallic Crystals by Thermal Decomposition of the Halides in the Gas Phase
Jacques Gillardeau
CEA-Bibl. 93 (Sept. 1967), 24 pp.

Melting and Casting Refractory or Reactive Metals by Plasma Electron Beam
Seiichiro Kashu and Chikara Hayashi
Nippon Kinzoku Gakkaishi, 31:413-419 (1967)
Ti, Zr, Mo, and Ta

Preparation and Characterization of Ultra-Pure Solids. Annual Report July 1, 1966-June 30, 1967 on ARPA Order 418
C. C. Klick, G. T. Rado, M. R. Achter, and A. I. Schindler
Naval Research Lab., Washington, D. C., NRL-MR-1815; AD-663891 (1967), 79pp.

Some Possibilities of Zone Melting for Purification of Molybdenum and Vanadium
G. P. Kovtun, A. A. Kruglykh, and V. A. Finkel
Ukr. Fiz. Zh., 12:1008-1011 (1967)

Preparation of Unusual Refractory Powders by Flame Processes
J. R. Merriman, S. H. Smiley, and H. L. Kaufman
(Oak Ridge Gaseous Diffusion Plant, Union Carbide Corp., Oak Ridge, Tenn.) Proceedings of the Conference on Chemical Vapor Deposition of Refractory Metals, Alloys, and Compounds, Gatlinburg, Tenn., Sept. 12-14, 1967 (A. C. Schaffhauser, ed.), American Nuclear Society, Inc., Hinsdale, Illinois (1967), pp. 229-241

Evaluation of Metal-Organic Compounds as Materials for Chemical Vapor Deposition
J. A. Papke and R. D. Stevenson
(Ethyl Corp. Research Labs., Ferndale, Mich.) Proceedings of the Conference on Chemical Vapor Deposition of Refractory Metals, Alloys, and Compounds, Gatlinburg, Tenn., Sept. 12-14, 1967 (A. C. Schaffhauser, ed.), American Nuclear Society, Inc., Hinsdale, Ill. (1967), pp. 193-204

Present State-of-the-Art and Prospects for Application of Electron Beam Melting for Obtaining Pure Metals and Improving the Properties of Existing Alloys
B. E. Paton et al.
FTD-HT-23-8-69; AD-693833 (May 1969), 15pp.
Teoriya Formovki. Tr. Soveshch., Moscow (1967), pp. 5-11

Obtaining Single Crystals of Refractory Metals by the Czochralski Method
N. G. Sushkin, O. Budzinskii, and V. E. Efimov
Izv. Akad. Nauk SSSR, Metal., 3:94-100 (May-June 1967)
[Nb and Mo (15 to 20 mm in diameter and 150 to 300 mm long)]

Proceedings Conference on Chemical Vapor Deposition of Refractory Metals, Alloys and Compounds
American Nuclear Society (1967)

Development of Equipment for Single Crystal Growth of High-Melting Point Metals
J. Barthel and R. Scharfenberg
SRS Order No. 4884, ORNL-TR-1403 (Nov. 1966), 14 pp.
Translation from the German

Ziehen von Einkristallen hochschmelzender Metalle (W, Ta, Nb, Mo) nach dem Floating-Zone-Verfahren mit Hilfe von punktfokussierten Elektronenstrahlen
  E. Bas and H. Stevens
  Z. Angew. Math. Phys., 17(3):484–485 (1966)

Literaturbericht über die Herstellung von Schichten aus Molybdän, Wolfram und Rhenium durch thermische Zersetzung ihrer Carbonyle oder durch Reduktion von flüchtigen Halogeniden
  M. Kadner
  (Battelle-Institute. V., Frankfurt/Main, Germany), BMwF-FB K 66-09, Forschungsbericht K 66-09 (April 1966)

A Device for the Zone Refining of High-Melting-Point Metals
  G. P. Kovtun, A. A. Kruglykh, and V. S. Pavlov
  Pribory i Tekhn. Eksperim., 1:211–212 (1966)
  Instr. Exptl. Tech. (USSR), 1:277–228 (1966)
  Mo

High-Purity Metals
  G. W. P. Rengstorff
  (Defense Metals Information Center, Battelle Memorial Institute, Columbus, Ohio), DMIC Rept. 222 (Jan. 1966)

Growing Single Crystals of Refractory Metals and Alloys with a Given Crystallographic Orientation by Electron Beam Zone Melting
  E. M. Savitskii, G. S. Burkhanov, G. L. Tsarev, and N. N. Boka
  Star 4, 1471A (1966)
  JPRS-34181; TT-66-30622

Zone Melting
  Hermann Schildknecht
  (Verlag Chemie, GmbH, Weinheim/Bergstr.; Academic Press, New York and London, 1966)

Electrodeposition of Refractory Metals
  S. Senderoff
  Metals. Rev., 11:97 (1966)

Noncontaminating Crucibles for the Evaporation of Refractory Metals
  H. F. Sterling
  Vide, 21:121–129 (1966)

The Floating Zone Melting of Refractory Metals by Electron Beam Bombardment
  T. Takasi
  Nippon Kinzoku Gakkaishi, 20(11):1017–1021 (1966)

Refractory Metal Crystals
  M. R. Achter et al.
  (Naval Research Lab., Washington, D. C.), Preparation and Characterization of Ultra-Pure Solids (August 1965), pp. 27–46

Evaporation par bombardement electronique de métaux à hauts points de fusion
  M. Aubecz, M. Brabers, M. Henuset, and M. Meulemans
  Mem. Sci. Rev. Met., 62:373 (1965)

The Growth of Pure Single Crystals From Heat-Resistant Metals and the Determination of Their Hall Effect. (Electron) Beam Zone Melting of Tungsten, Molybdenum, Rhenium, Niobium, Tantalum
  N. A. Brilliantov, V. N. Kachinskii, and L. S. Starostina
  Vysokotemp. Neorg. Soedin. Akad. Nauk UkrSSR, Inst. Prob. Materialoved. (1965), pp. 37–40

Floating-Zone Refining by Electron Bombardment
  Marvin S. Brooks, E. Owen Fisk, and Bernard Rubin
  Semicond. Prod., 8:34 (1965)
  Review as a purification and crystal growth technique

Electron Beam Distillation Furnace for Reactive Metals: Design Considerations and Operating Experience
  R. F. Bunshah and R. S. Juntz
  (Lawrence Radiation Lab., Univ. of Calif., Livermore, Calif.), Contract AEC No. W-7405-eng-48, UCRL-12487 (May 1965)

Electromagnetic Forces and Power Absorption in Levitation Melting
  E. Fromm and H. Jehn
  Brit. J. Appl. Phys., 16:653 (1965)

Structures of the Metal Films Produced by Vacuum-Arc Evaporation Method
  Makoto Kikuchi, Sigemaro Nagakura, Hirato Ohmura, and Shigueo Oketani
  Japan. J. Appl. Phys., 4:940 (1965)

Condensation and Evaporation of Solids
  Emile Rutner, Paul Goldfinger, and John P. Hirth, eds.
  Gordon and Breach Science Publishers, New York (1965), 707 pp.

Growing and Plastic Deformation of the Single Crystals of Refractory Metals and Alloys
  E. M. Savitskii, G. S. Burkhanov, Kh. V. Kopetskii, and G. E. Chuprikov
  Rost Kristallov, Akad. Nauk SSSR, Inst. Kristallogr., 6, 308–318 (1965)
  Growth of Crystals, Vol. 6A (N. N. Sheftal' and A. V. Shubnikov, eds.), Consultants Bureau, New York (1968), pp. 117–128

• A New Approach to Depositing Thin Films of Refractory Metals
  H. C. Theuerer
  Chem. Week, April 1965

Vacuum Evaporation of Metals by High Frequency Levitation Heating
  J. Van Audenhove
  Rev. Sci. Instr., 36:383 (1965)

Interaction of Refractory Compounds with Molten Metals
  G. A. Yasinskaya
  Ogneupory, 2:20 (1965)

A Study of Scientific Research Materials
  MRC No. 514 (Materials Research Corporation, Orangeburg, N. Y.), Contract AF-49(638)-1241, May 27, 1963 to May 27, 1965
  Parameters pertinent to the electron beam zone refinement of refractory metals; the effect of ultra-high vacuum electron beam zone refining on the purification of metals; the relation between vacuum and purity in float-zone melting of refractory metals

Superpurification of Metals by Vacuum Distillation: A Theoretical Study
  R. F. Bunshah
  Contract W-7405-eng-48. UCRL-7387 (May 1964). 22 pp.

Preparation of Single Crystals by Arc-Zone Melting (Re, Ta, Nb, Mo, V, Y)
O. N. Carlson, F. A. Schmidt, and W. M. Paulson
(Institute for Atomic Research and Department of Metallurgy, Iowa State University, Ames, Iowa), Trans. ASM Quart., 57:356 (1964)

New Types of Metal Powders
Henry H. Hausner
Met. Soc. Conf., Vol. 23
Gordon and Breach Science Publishers, New York (1964), 116 pp.

Applications of Low Energy Sputtering for Thin Film Deposition
J. W. Nickerson and R. Moseson
Semicond. Prod., 7:33 (1964)

The Importance of Vacuum Techniques for the Preparation and Processing of High-Melting Metals
R. Palme
Vakuum-Tech., 13:229 (1964)

Modes of Operation of an Electron Beam Floating Zone Melting Furnace: Growth of Single Crystals of Metals and Alloys
Heinz G. Sell and Willard M. Grimes
Rev. Sci. Instr., 35:64 (1964)

A Critical Review of Refractories
Edmund K. Storms
(Los Alamos Scientific Lab., N. Mex.), Contracts W-7405-eng-36, LA-2942 (March 1964), 252 pp.

Zone Melting of Metals in Water-Cooled Copper Crucibles
V. G. Yepifanov and A. G. Lesnik
Akad. Nauk Ukr. SSR, Inst. Metallofiziki Sb. Nauchn. Tr., 20: 185-190 (1964)
Cr, Ti, and Mn

Metallurgy of Rare Metals
A. N. Zelikman, O. E. Krein, and G. V. Samsonov
Metallurgiya Redkikh Metallov, second ed., revised and enlarged, Izdatel'stvo Metallurgiya, Moscow (1964), 512 pp.
W, Mo, Ta, Nb, Ti, Zr

Method and Apparatus for Producing Single Crystals of Refractory Material
Centre d'Étude de l'Energie Nucleaire
British Patent 968,785, Sept. 2, 1964

Large Crystals Grown by Recrystallization
K. T. Aust
The Art and Science of Growing Crystals, John Wiley and Sons, Inc., New York (1963), pp. 452-478

Fragilisation des métaux cubiques centres sous l'effet de la récristallisation
M. Herve Bibring
Compt. Rend., 256:3689 (1963)

Preparation and Properties of Sputtered Films
M. H. Francombe
(Philco Scientific Laboratory, Blue Bell, Pennsylvania), CONF-209-12, American Vacuum Society 10th National Meeting, Boston, Mass. (October, 1963)

Techniques for the Fabrication of Ultra-Thin Metallic Foils
F. J. Karasek
(Metallurgy Division, Argonne National Laboratory, Argonne, Illinois), Nucl. Sci. Eng., 17:365-370 (1963)

Method for Vacuum Deposition of Refractory Metal Films
J. L. Nicholson
Rev. Sci. Instr., 34:118 (L) (1963)

A Levitation Zone Melter for Larger Diameter Bars With Positive Process Control
B. F. Oliver
Trans. Met. Soc. AIME, 227:960 (1963)

High Melting-Point Elements
H. W. Schadler
The Art and Science of Growing Crystals, John Wiley and Sons, Inc., New York (1963), pp. 343-364

Relation of Rate and Duration of Evaporation to Background Pressure for the Deposition of Thin Films in Vacuum
H. Schwarz
J. Appl. Phys., 34:2053;2056 (1963)

A Method of Growing Perfect Single Crystals of Refractory Metals
L. S. Starostina, V. N. Kachinskii, and N. A. Brilliantov
Teplofiz. Vysok. Temp., 1(2):310-313 (1963)
High Temperature, 1(2):276-277 (1963)

Production of Single Crystals of Refractory Metals by Czochralski's Method in Conjunction with Electron-Beam Heating
N. G. Sushkin
Teplofiz. Vysok. Temp., 1(2):313-315 (1963)
High Temperature, 1(2):278-279 (1963)

Preparation of Very Thin Suspended Metal Films
G. J. Unterkofler and R. R. Verderber
(Burroughs Corporation, Burroughs Laboratories, Paoli, Pennsylvania), Rev. Sci. Instr., 34:820-821 (1963)

System for Electron-Beam Melting
F. W. Wood, J. L. Hoffman, W. E. Anable, and R. A. Beall
U. S. Dept. of the Interior, Bureau of Mines Rept. of Investigations 6341 (1963)

Current Information On the Refractory Metals
C. E. King
Contract AF33(657)-11214, AD-436333; ERR-FW-049A (Dec. 1, 1962), 80 pp.

High Vacuum Technique
E. Luscher
(Internationales Jahrbuch Chemische Industrie), Vogt-Schild AG, Solothurn, (1962-63), p. 3

Plasmarc Furnace, a New Concept in Melting Metals
R. J. McCullough
J. Metals (Dec. 1962)

On Some Properties of Electron Beam Melted Metals
E. Rexer
Proc. Fourth Symp., pp. 245-275 (1962)

The Growing and Testing of Single Crystals of Refractory Metals, W, Re, Ta, Mo, and Nb
E. M. Savitskii, Kh. V. Kopetskii, A. I. Pekarev, and M. I. Novosadov
Issled. po Zharoproch. Splavam, Akad. Nauk SSSR, Inst. Met., 9:192-194 (1962)

1.a.2. Group IV

Comparaison entre différents modes de purification de hafnium
High Temperature Materials, 6. Plansee Seminar 1968
F. Benesovsky, ed.
Metallwerk Plansee AG, Reutte/Tirol (1969), pp. 237-253

Comment on the Preparation of Thin Foils of Beta Titanium
J. M. Capenos and F. H. Froes
J. Sci. Instr. (J. Phys. E), Ser. 2, 2:735-736 (1969)

Epitaxial Growth of Titanium Films on Mica
E. Grunbaum and R. Schwarz
J. Appl. Phys., 40:3364 (1969)

Electrowinning of Hafnium from Hafnium Tetrachloride
G. M. Martinez, M. M. Wong, and D. E. Couch
Trans. AIME, 245:2237-2242 (1969)

Contamination of Titanium Refined by the Iodide Process in Glass Deposition Units
R. A. J. Shelton
Bull. Inst. Mining Met., 78(751):C 111-C112 (1969)

Epitaxial Growth of Titanium Thin Films
F. E. Wawner, and K. R. Lawless
J. Vac. Sci. Tech., 6:588-590 (1969)

Manufacture Including Purification of Reactive Metals
(To Oregon Metallurgical Corp.), British Patent 1,148,459 (April 10, 1969)
Priority date July 26, 1966, U. S.
Ti and Hf

Purification of Contaminated Group IV: A Refractory Metal Products
(To Oregon Metallurgical Corp.) British Patent 1,148,460 (April 10, 1969). Priority date July 26, 1966, U. S.
Hf, Ti, Zr

Purification of Fe and Zr by "Zone Tranformation" in High Purity Atmosphere
P. Ailloud and J. P. Langeron
Mem. Sci. Rev. Met., 65(15):73 (1968)

Ultrahigh-Vacuum Zone Purification of Zirconium with Analysis of Partial Pressures
D. S. Easton and J. O. Betterton, Jr.
ORNL-4309 (Dec. 1968)
Also in Electron and Ion Beam Science and Technology, Third International Conference (Robert A. Bakish, ed.) The Electrochemical Society, Inc., New York (1969), pp. 469-500

Preparation of Zr from ZrC by Fused Salt Electrolysis
K. Venkatarami et al.
Trans. Indian Inst. Metals, 21:36-41 (1968)

Properties of Iodide-Type Hafnium and Alloys in the Zirconium-Hafnium Systems
A. I. Evstyukhin, I. P. Barinov, and I. I. Korobkov
High-Purity Metals and Alloys, Fabrication, Properties, and Testing (V. S. Emel'yanov, and A. I. Evstyukhin, eds.)
Consultants Bureau, New York (1967), pp. 23-30

Behavior of Ultrathin Zirconium Films upon Exposure to Oxygen
Francis P. Fehlner
J. Appl. Phys., 38:2223 (1967)

The Development of Large Zirconium Crystals by the Alpha-Beta Thermal Cycling Technique
G. T. Higgins
J. Mater. Nucl., Pays-Bas, 22:285-291 (1967)

Sur l'étude de deux critères de pureté du zirconium
Louis Renucci and Jean-Paul Langeron
Compt. Rend., 264:673-676 (1967)

Purification of Iron and Zirconium by Displacement of a Transformed Zone
Pierre Ailloud and Jean-Paul Langeron
Compt. Rend., 262:1656 (1966)

Recrystallization, Grain Growth, and Preferential Orientations in Zone-Melted Zirconium
M. Billion and J.-P. Langeron
Commissariat à Énergie Atomique, 9e Colloque de Metallurgie, Saclay (1965), pp. 97-103, 1966; discussion, pp. 104-105

Study of Grain-Growth Phenomena in Zirconium
J. -C. Colin and P. Lehr
Commissariat à Énergie Atomique, 9e Colloque de Metallurgie, Saclay (1965), pp. 77-94, 1966; discussion, pp. 94-96

Epitaxial Growth of Zirconium on Sodium Chloride and Fluorspar
M. Denoux
Proc. Intern. Symposium Grundprobleme der Physik dünner Schichten, Göttingen (1965), pp. 170-174, 1966

Crystalline Titanium by Sodium Reduction of Titanium Lower Chlorides Dissolved in Sodium Chloride
V. E. Homme and M. M. Wong
U. S. Dept. Interior, Bur. Mines, Rept. Invest. No. 6813 (1966), 27 pp.

Preparation de monocristaux de zircone pure monoclinique
Ann-Marie Anthony and Vutien Loc
Compt. Rend., 260:1383 (1965)

Electrolytic Crystallization of Titanium from Molten Salts
V. S. Balikhin
Izv. Akad. Nauk SSSR, Metally, 7:77-81 (1965)

Sur la récristallisation, la croissance du grain et les orientations préférentielles du zirconium purifié par fusion de zone
M. Billion and J. -P. Langeron
Compt. Rend., 261:2351 (1965)

New Method of Preparing Large Zirconium Crystals
M. Billion and J.-P. Langeron
ORNL-tr-763; Compt. Rend. 260, 152 (1965)

Sur l'oxydation de couches minces orientées de zirconium
M. M. Denoux
Compt. Rend., 260:5003 (1965)

New Improvements of the Zone-Refining Method
J. -P. Langeron
New Physical and Chemical Properties of Metals of Very High Purity, Gordon and Breach Science Publishers, New York (1965), pp. 391-414

Application de la méthode de fusion de zone à la purification du zirconium et étude de certaines propriétés du métal ainsi purifié
J. -P. Langeron
Rev. Hautes. Temp. Refract., 2:137 (1965)

La fusion par bombardement électronique et ses applications en métallurgie
Pierre Lehr and Philippe Albert
Rev. Hautes Temp. Refract., 2:31 (1965)

Zirconium Films from Tungsten Filaments
J. W. Voight
J. Vacuum Sci. Tech., 2:215 (1965)

Sur nouvelles propriétés du zirconium purifié fié par fusion de zone sous ultravide statique
M. Billion and J. -P. Langeron
Compt. Rend., 259:4671 (1964)

Etude du phénomène de croissance du grain dans le cas du zirconium
J. C. Colin
Thèse Doct. 3eme Cycle, Spec. Chim. Metallurg., Paris (1964), 50 pp.

Preparation and Oxidation of Oriented Thin Films of Zr
M. Denoux and J. J. Trillat
Compt. Rend., 258:4683 (1964)

Preparation and Properties of Thin-Film Hard Superconductors
J. Edgecumbe, L. G. Rosner, and D. E. Anderson
(Department of Electrical Engineering, University of Minnesota, Minneapolis, Minnesota), J. Appl. Phys., 35:2198 (1964)

Sur la réduction sous vide de la zircone
Gerard Gosse, Louis Renucci, Phillippe Albert, and Pierre Lehr
(Laboratoire de Vitry du C. N. R. S.), C. R. Acad. Sc. Paris, 258:589-591 (1964)

Ion Exchange Procedures II, Separation of Zirconium, Neptunium and Niobium
J. H. Holloway and F. Nelson
J. Chromatog., 14:255 (1964)

Significance of Refractory Metals for the Production of High-Field Superconductors
D. Koch
Metall., 18:714-718 (1964)

Application de la méthode de fusion de zone à la purification du zirconium et étude de certaines propriétés de ce métal
J. -P. Langeron
Centre d'Etudes de Chimie Metallurgique du CNRS, thesis presented to Faculty of Sciences, University of Paris, France (1964)

Sur les propriétés du zirconium de zone fondue
Jean-Paul Langeron
C. R. Acad. Sc. Paris, 258:1801-1804 (1964)

Purification du zirconium par fusion de zone sous ultra-vide statique
J. P. Langeron
(Centre d'Etudes de Chimie Metallurgique, Vitry-sur-Seine, France), Mem. Sci. Rev. Met., 61:637 (1964)

Préparation d'un monocristal de Ti à partir de TiH_2
A. Oka, T. Kubo, and N. Oshima
J. Chem. Soc. Japan, Pure Chem. Sect., 85:394 (1964)

Electrolytic Titanium from $TiCl_4$, I. Operation of a Reliable Laboratory Cell
Myron J. Rand and Lawrence J. Reimert
(New Jersey Zinc Company, Palmerton, Pennsylvania), J. Electrochem. Soc., 111:429-434 (1964)

Variable-Gradient, Electron-Beam Heating Methods for Growing Single Crystals of Zirconium
J. C. Wilson and M. L. Picklesimer
Proc. 1st Intern. Conf. Electron Ion Beam Sci. Technol., Toronto (1964), pp. 502-520

Large Crystals Grown by Recrystallization
K. T. Aust
The Art and Science of Growing Crystals, John Wiley and Sons, Inc., New York (1963), pp. 452-478

The Physical Metallurgy of Zirconium
D. L. Douglass
(General Electric Co., Pleasanton, California), At. Energy Rev., 1:71-237 (1963)

The Iodide Method of Refining Zirconium. The Transfer of Nonmetallic Additions During Refining
V. S. Emel'yanov, A. I. Evstyukhin, and D. D. Abanin
OTS-64-21263, pp. 1-7; JPRS-22697, p. 1-7
Met. i Metalloved. Chistykh Metal., Sb. Nauchn. Rabot, 4, (May 31, 1963)

Purification du zirconium par la méthode de la zone fondue verticale sous ultra-vide
J. -P. Langeron
Colloque sur la Fusion de Zones et la Cristallisation en Colonnes, Kernforschungszentrum Karlsruhe, Germany (June 5-7, 1963), pp. 215

Application de la fusion au four à bombardement électronique à l'étude de la purification du zirconium. Elimination de l'oxygène
L. Renucci, Ph. Albert, and P. Lehr
Mem. Sci. Rev. Met., 60:829 (1963)

Preparation of Single Crystals of Zirconium and Zirconium Alloys
J. C. Wilson
(Oak Ridge National Laboratory, Oak Ridge, Tenn.), ORNL-3470 (1963), p. 59

A System for Electron-Beam Melting
F. W. Wood, J. L. Hoffman, W. E. Anable, and R. A. Beall
U. S. Dept. of the Interior, Bureau of Mines Rept. of Investigations 6341 (1963)

Electron-Beam Purification of Hafnium
W. E. Anable and R. A. Beall
(Bureau of Mines, Albany Metallurgy Research Center, Albany, Oregon), USBM-U-929 (June 1962)

Preparation of Small Samples of Ductile Titanium and Zirconium from the Isotopic Oxides by Iodide Refining
N. D. Veigel and J. M. Blocher, Jr.
(Battelle Memorial Institute, Columbus, Ohio), J. Electrochem. Soc., 109:647 (1962)

Purification of Hafnium Tetrachloride by Al-
kali-Chloride Fusion
  A. Adams and H. O. Poppleton
  (Bureau of Mines, Albany Metallurgy Research Center, Ore.),
    BM-RI-5870 (Oct. 1960)

Procedures for the Metallographic Prepara-
tion of Beryllium, Titanium and Refractory
Metals
  R. D. Buchheit, C. H. Brady, and G. A. Wheeler
  (Defense Metals Information Center, Battelle Memorial
    Institute, Columbus, Ohio). DMIC-Memo-37 (1959), 33 pp.

Preparation of High Purity W, Mo, Ta, Nb, and
Zr
  George A. Moore and L. L. Wyman
  National Bureau of Standards, Washington, D. C. (July 1959)
    13pp. Project 7360
  Contract AF33 (616)-58-11. (AD-232137). WADC-TR-59-314
    OTS
  Period covered: February 1958 to February 1959

Floating Zone Purification of Zirconium
  G. D. Kneip, Jr., and J. O. Betterton, Jr.
  (Metallurgy Division, Oak Ridge National Laboratory, Oak
    Ridge, Tennessee), J. Electrochem. Soc., 103:684 (1956)

Preparation of Single Crystals of Titanium and
Their Mode of Deformation
  A. T. Churchman
  Nature, 171:706 (1953)
  Strain anneal method

Electrolytic Separation of Zirconium
  V. A. Plotnikov and E. B. Gutman
  NP-2232, Zh. Prikl. Khim., 19:826-832 (1946)

1.a.3. Group V

A Foil-Wrapping Technique to Prevent the Con-
tamination of Vanadium Annealed in Silica in the
Temperature Range 1000°-1250°C
  K. Abe, K. Toma, H. Yoshinaga, and S. Morozumi
  J. Less-Common Metals, 23:213-216 (1971)

Deposition of High-Purity Nb Films by Tetrode
Sputtering
  S. Ogawa, K. Takiguchi, and T. Hasegawa
  J. Vacuum Sci. Tech., 8:192-194 (1971)

R. F. Sputtered Tantalum Films Deposited in an
Oxygen-Doped Atmosphere
  P. N. Baker
  Thin Solid Films, 6:R57-R60 (1970)

Some Notes on the Control of Floating-Zone Re-
fining
  G. D. King
  J. Phys. E: Sci. Instr., 3:730-732 (1970)
  Nb and Pt

Preparation of Vanadium Isotope Foils
  L. I. Kovalenko, A. A. Rozen, and L. N. Reshetova
  Pribory Tekhn. Eksper. 1:239 (1970)
  Inst. and Exper. Tech., 1:280 (1970)

Phase Forming Processes in Tantalum Films
through Sputtering
  M. Nakamura, M. Fujimori, and Y. Nishimura
  Japan. J. Appl. Phys., 9:557-571 (1970)

Getter-Bias Sputtering of High Purity Metal
Films in a High Current Vacuum Discharge in
the 10^{-4} Torr Range
  R. Pinto and B. M. Shaha
  Japan. J. Appl. Phys., 9:174 (1970)

Electron-Beam Float Zone and Vacuum Purifica-
tion of Vanadium
  R. E. Reed
  J. Vacuum Sci. Tech., 7:5105-5112 (1970)

Structure and Properties of R. F. Sputtered, Su-
perconducting Tantalum Films
  F. Schrey, R. D. Mathis, R. T. Payne, and L. E. Murr
  Thin Solid Films, 5:29-40 (1970)

Vanadium Purification by Transformation to
the Tetrachloride and Reduction by Hydrogen.
Study of the Reduction Mechanism
  Christian Val
  Commissariat à l'Energie Atomique, Centre d'Etudes Nu-
    cléaires, Saclay, France, CEA-R-3939 (Jan. 1970); thesis,
    Univ. Paris, Orsay, France

Tantalum Metal by Bomb Reduction of Ta_2O_5
  H. A. Wilhelm, R. M. Bergman, and F. A. Schmidt
  J. Metals, pp. 1-5 (January 1970)

Design and Operation of Stable Rf-Biased Super-
conducting Point-Contact Quantum Devices, and
a Note on the Properties of Perfectly Clean
Metal Contacts
  J. E. Zimmerman, P. Thiene, and J. T. Harding
  J. Appl. Phys., 41:1572-1580 (1970)
  Nb

The Influence of Conducting Underlays on the
Properties of Sputtered Tantalum Films
  Thin Solid Films, 6:307-320 (1970)

Electrolytic Refining of Tantalum in Molten
Salts
  V. S. Balikhin
  Izv. Akad. Nauk SSSR, Metal. No. 6, 79-84 (Nov.-Dec. 1969)

Préparation des couches minces de niobium
sous vide nitrapousse, leur traitement thermique et leurs propriétés électriques
  K. I. Burmenko, G. F. Ivanovskaja, A. F. Orlov, and K. A.
    Osipov
  Fiz. Khim. Obrabot. Mater., SSSR, No. 5, 62-67 (1969)

Purification of Vanadium by Electrotransport
and Preparation of Vanadium and Vanadium
Base Alloys
  O. N. Carlson and F. A. Schmidt
  Annual Summary Research Report Ceramic and Mechanical
    Engineering, Chemical Engineering, Chemistry, Mathe-
    matics and Computer Science, Metallurgy, Physics, and
    Reactor Divisions, July 1, 1968 - June 30, 1969, Ames
    Laboratory, Ames, Iowa, Contract W-7405-eng-82, IS-
    2100 (July 1969)

Tantalum Crystal Grown by the Electron Beam Zone Melt. I. Horizontal Hearth Method
S. Dohi and K. Kudo
Mem. Defense Acad. (Japan), 9:369–376 (1969)

Tantalum Crystals Grown by the Electron Beam Zone Melt. II. Floating Zone Melting
S. Dohi and T. Matsuyama
Mem. Defense Acad. (Japan), 9:377–384 (1969)

Contribution to the Behavior of Nitrogen at Electron Beam Melting of Vanadium, and to the Determination of Nitrogen
H. J. Gottwald, K. Krone, and J. Krueger
Metall., 23:1284–1289 (1969)
Nitrogen content approached a constant value of 0.3 wt.%

Preparation of Metallic Niobium and Tantalum by Reduction of Their Pentachlorides Using a Large Excess of Zinc
G. Jangg, R. Kieffer, and P. Topic
Monatsh. Chem., 100:379–384 (1969)

Recherche et élaboration d'une technologie permettant d'obtenir des couches mincés de niobium supraconductrices en vue de leur application à l'électronique et à l'électrotechnique
J. Maldy and C. Guenel
D. G. R. S. T. Final Report, Contract 6801298 (1969), 43 pp., Centre Rech. C.G.E., 81-Marcoussis, France

Eine Bemerkung zur Wasserstoffentgasung von Niob
G. Melchior
J. Less-Common Metals, 18:430–431 (1969)

High Purity Niobium Films Formed by Glow Discharge Cathode Sputtering
John R. Rairden III and James T. Furey
General Electric Co., U. S. Patent 3,432,416 (March 11, 1969)

Electron Beam Floating Zone Refining of Niobium
R. E. Reed
(Oak Ridge National Lab., Oak Ridge, Tenn.), Electron and Ion Beam Science and Technology, Vol. 1, Proceedings of Second International Conference held in New York, April 17–20, 1966; Metallurgical Society Conferences, Vol. 51 (Robert Bakish, ed.), Gordon and Breach, New York (1969), pp. 225–258

Inhomogeneous Distribution of Tantalum and Tungsten in Zone-Refined Niobium
R. E. Reed
ORNL-4334 (Feb. 1969)

Technologie de fabrication de fils de Ta
S. Stolarz and A. Rausz
Rudy Metale Niezelazne, Polska, 14:38–44 (1969)

Vanadium Purification, Final Report
J. C. La Vake, C. T. Wang, et al.
Combustion Engineering, Inc., CEND-3742-356 (Nov. 1969)

Texture de cristaux basaltiques et orientation cristallographique des dendrites dans des lingots de molybdène
Dokl. Akad. Nauk SSSR, 184:1088–1090 (1969)

Growth of Oriented Niobium Bicrystals by Arc-Zone Melting
W. F. Brehm, Jr., and J. L. Gregg
J. Less-Common Metals, 14:463 (1968)

Preparation of Vanadium
O. N. Carlson, and F. A. Schmidt
Annual Summary Research Report: Ceramic and Mechanical Engineering, Chemical Engineering, Chemistry, Mathematics and Computer Science, Metallurgy, Physics, and Reactor Divisions, July 1, 1967 - June 30, 1968, Ames Laboratory (Iowa State University of Science and Technology, Ames, Iowa), Contract W-7405-eng-82, IS-1900 (July 1968), p. M-1

The Refining Effects on the Hardness of Tantalum by Electron-Beam Melting
S. Dohi and A. Inoue
Mem. Def. Acad., Japan, 8(1):423–428 (1968)

High-Purity Vanadium by Metallothermic Reduction of Vanadium Trichloride
M. J. Ferrante, F. E. Block, and J. L. Schaller
BM-RI-7145 (July 1968)

High-Purity Niobium Through Aluminothermic Reduction of Niobium Pentoxide
C. K. Gupta and P. K. Jena
(Bhabha Atomic Research Centre, Bombay, India), BARC-328 (1968), 12 pp.

Production of Tantalum Metal by Alumino-Thermic Reduction of Its Pentoxide
C. K. Gupta and P. K. Jean
J. Metals, 20:25–28 (1968)

Über die Entgasung von Vanadium
Gerhard Horz
Z. Metallk., 59:832–834 (1968)

Preparation of Thin Films of V, Nb and Ta for Nuclear Studies
V. N. Karev, A. P. Klyucharev, L. I. Kovalenko, and A. A. Rozen
Izv. Akad. Nauk SSSR, Ser. Fiz., 32:328–331 (1968)

Effect of the Conditions of Growth on the Orientation, Purity and Structural Perfection of Tantalum Single Crystals Obtained by a Crucibleless Zone Refining Method
A. A. Kralina, T. V. Zholtikova, V. O. Esin, and V. A. Sazonova
Fiz. Metallov Metalloved., 25:333–340 (1968)

Metal Films Sputtered at Low Voltages
William W. Lee
J. Appl. Phys., 39:5366 (1968)
Ta - abnormally high resistivity

High-Purity Vanadium
K. P. V. Lei and T. A. Sullivan
J. Less-Common Metals, 14:145–147 (1968)

Formation of F. C. C., B. C. C. and β-Tantalum Films by Evaporation
R. B. Marcus and S. Quigley
Thin Solid Films, 2:467–477 (1968)

Zone refined Nb, Ta and Fe
K. T. Marshall
Rev. Met., 65:165–169 (1968)

Purification of Vanadium, Elimination of Oxygen, Carbon and Nitrogen
A. Pattoret, J. Trouve, and A. Accary
Rev. Intern. Hautes Temp. Refract., 5:205-211 (1968)

Zum Rekristallisationsverhalten von Sintertantal
E. Pink and H. Karle
Planseeber. Pulvermet., 16:104-107 (1968)

Reduction of Tantalum Pentachloride by Hydrogen on a Heated Surface
Ya. M. Polyakov, I. F. Lisonvik, and A. N. Krestovnikov
Joint Publications Research on Treatment of Metals (July 1968), pp. 1-9

Electron Beam Float Zone Refining and Vacuum Degassing of Niobium Single Crystals
R. E. Reed
Presented at Electrochem. Soc. Meeting, Montreal, Canada, Oct. 6-11, 1968, on Preparation and Purification of Ultrapure Metals

Evaluation of Niobium Source Materials Suitable for Electron Beam Float Zone Melting
R. E. Reed, R. P. Tucker, and C. L. Brooks
ORNL-4246 (April 1968)

Sources of Contamination During Electron-Beam Melting
R. E. Reed, C. W. Dean, R. E. McDonald, and J. F. Emery
Contract W-7405-eng-26, ORNL-TM-2208 (July 1968), 20 pp.
Also published in Electron and Ion Beam Science and Technology, Third International Conference, Robert A. Bakish, ed., Electrothermics and Metallurgy Div., The Electrochemical Soc., Inc., New York, N. Y., pp. 516

Purification of Vanadium by Electrotransport
F. A. Schmidt
Annual Summary Research Report: Ceramic and Mechanical Engineering, Chemical Engineering, Chemistry, Mathematics and Computer Science, Metallurgy, Physics, and Reactor Divisions, July 1, 1967 - June 30, 1968, Ames Laboratory (Iowa State University of Science and Technology, Ames, Iowa), Contract W-7405-eng-82, IS-1900 (July 1968), p. M-6

Preparation of Nb and Ta Metal Powders
J. C. Sehra, D. K. Base, and P. K. Jena
Trans. Ind. Inst. Met., 21:21-23 (1968)

The Deposition of Niobium Thin Films by dc Diode and Substrate Bias Sputtering
J. Sosniak
J. Appl. Phys., 39:4157-4163 (1968)

Rôle des métaux des terres rares dans la purification du niobium
F. Trombe and G. Male
Rev. Int. Hautes Temp., 5:295-298 (1968)
WAPD-Trans-126, translation, 8 pp.

Preparation of Very High Purity Vanadium Tetrachloride
C. Val and A. Accary
Bull. Soc. Chim. France No. 11, 4327-4334 (1968)
CEA-TP-7279

Vanadium Purification, Quarterly Progress Report No. 5, October 1 - December 31, 1967
C. T. Wang, Wah Chang, E. F. Barock, J. C. La Vake, and S. S. Christopher
(Combustion Engineering, Inc., Windsor, Conn.), Contract AT(30-1)-3742, CEND-3742-329 (Jan. 1968), 14 pp.

Vanadium Purification, Quarterly Progress Report No. 6, January 1 - March 31, 1968
C. T. Wang and Wah Chang
(Combustion Engineering, Inc., Windsor, Conn.), Contract AT(30-1)-3742, CEND-3742-336 (April 1968), 43 pp.

Vanadium Metal by the Carbon Reduction of Vanadium Oxides
H. A. Wilhelm and J. K. McClusky
Annual Summary Research Report: Ceramic and Mechanical Engineering, Chemical Engineering, Chemistry, Mathematics, and Computer Science, Metallurgy, Physics, and Reactor Divisions, July 1, 1967 - June 30, 1968, Ames Laboratory (Iowa State University of Science and Technology, Ames, Iowa), Contract W-7405-eng-82, IS-1900 (July 1968), p. M-5

Electrowinning of Niobium
M. M. Wong and D. E. Kirby
Electrochem. Tech., 6:119 (1968)

Electron-Beam Purification of V
W. E. Anable
BM-RI-7014 (1967)

Preparation of Vanadium Metal
O. N. Carlson and F. A. Schmidt
(Ames Lab., IS-1600 (1967)

High Vacuum Purification of Niobium and Tantalum and Its Effect on the Electrical Resistance
Reiner-Joachim Dinter
ORNL-tr-2039, translated by R. P. Tucker and R. Gregg Mansfield, Z. Metallk., 58:70-75 (1967)

Iodide-Type Vanadium
V. S. Emel'yanov, A. I. Evstyukhin, V. I. Statsenko, and A. I. Dashkovskii
High-Purity Metals and Alloys. Fabrication, Properties, and Testing (V. S. Emel'yanov, and A. I. Evstyukhin, eds.) New York Consultants Bureau, New York (1967), pp. 1-5

Substrate-Condensate Chemical Interaction and the Vapor Deposition of Epitaxial Niobium Films
T. E. Hutchinson and K. H. Olsen
J. Appl. Phys., 38:4933 (1967)

Thin Films of Niobium for Cryotron Ground Planes
R. E. Joynson, C. A. Neugebauer, and J. R. Rairden
J. Vacuum Sci. Tech., 4:171-178 (1967)

The Properties of Niobium Crystals Produced by Zone-Melting
H. C. Kim and P. L. Pratt
Mater. Res. Bull., 2:323-335 (1967)

Structure of the Ultra-Thin Films Obtained by the Evaporation of Tantalum and Iridium in Vacuum
D. I. Lainer and V. A. Kholmyanski
Kristallografiya, 12(6):1051-1057 (1967)
Sov. Phys. — Cryst., 12(6):913-918 (1968)

Preparation of High-Purity Nb by Thermal Decomposition of NbCl$_5$
L. V. McCarty and W. T. Bachmann
Electrochem. Tech., 5:299-301 (1967)

Niobium: Purification and Perfection
R. E. Reed
ORNL-4097 (April 1967)

The Growth of Niobium Single Crystals with Good Crystalline Perfection
R. E. Reed, H. D. Guberman, and T. O. Baldwin
(Solid State Division, Oak Ridge National Lab., Oak Ridge, Tenn.), Proc. Intern. Conf. Crystal Growth, June 20-24, 1966, Suppl. to J. Phys. Chem. Solids (H. Steffen Peiser, ed.), Pergamon Press, New York (London, 1967), pp. 829-832

Vanadium Purification, Quarterly Progress Report No. 4, May 1 - September 30, 1967
Y. S. Shen, S. Worcester, J. C. La Vake, and S. S. Christopher
(Combustion Engineering, Inc., Windsor, Conn.), Contract AT(30-1)-3742, CEND-3742-325 (Nov. 1967), 32 pp.

Vanadium Purification, Quarterly Progress Report Feb. 1 - April 30, 1967
V. S. Shen, S. Worcester, D. E. Mahagin, and S. S. Christopher
(Combustion Engineering, Inc., Windsor, Conn.), Contract AT(30-1)-3742, CEND-3742-311 (May 1967), 37 pp.

EB Melting Furnace for Nb, Ta
H. Bourgeois
J. Four. Electr., 7:193-197 (1966)

A Process for Preparing High-Purity Vanadium
O. N. Carlson, F. A. Schmidt, and W. E. Krupp
J. Metals, 18:320 (1966)

Vanadium Purification, Quarterly Progress Report July 1 - Sept. 30, 1966
S. S. Christopher and S. Worcester
(Combustion Engineering, Inc., Windsor, Conn.), Contract AT(30-1)-3742, CEND-3742-288; QPR-1 (Nov. 1966), 39 pp.

Research and Development of Superconducting Thin Films
R. S. Collier, R. A. Kamper, L. O. Mullen, and R. J. Duff
(National Bureau of Standards, Boulder, Colo.), N67-18689, Dec. 31, 1966

Solvent Extraction of Niobium and Tantalum, III. Extraction Mechanism in Oxalic Solutions with Long-Chain Tertiary Amines
C. Djordjevic, H. Gorican, and S. L. Tan
J. Less-Common Metals, 11:342 (1966)

On Producing Single Crystal Rods of Tantalum by the Strain-Anneal Method Using Electron Beam Heating
S. Dohi
Mem. Defense Acad., Math., Phys., Chem. Eng. (Yokosuka, Japan), 5(4):409-413 (1966)

Production of Iodide-Type Vanadium
V. S. Emel'yanov, A. I. Evstyukhin, V. I. Statsenko, and A. I. Dashkovskii
Met. i Metalloved. Chistykh Metal., Sb. Nauchn. Rabot, 5:5-11 (1966)

On the Electron Beam Refining of Niobium
H. Kimura, Y. Sasaki, and S. Uehara
Trans. Natl. Res. Inst. Metals (Tokyo), 8:15 (1966)

Deposition of Tantalum, Tantalum Oxide, and Tantalum Nitride with Controlled Electrical Characteristics
E. Krikorian and R. J. Sneed
J. Appl. Phys., 37:3674 (1966)

Electron Beam Melting and Refining of V
J. Kruger and H. Winterhager
Metall., 20:430 (1966)

Growth of Vanadium on Silicon Substrates
K. J. Miller, M. J. Grieco, and S. M. Sze
J. Electrochem. Soc., 113:902 (1966)

The Electrodeposition of Coherent Deposits of Refractory Metals. IV. The Electrode Reactions in the Deposition of Niobium
S. Senderoff and G. W. Mellors
J. Electrochem. Soc., 113:66 (1966)

Properties of Zone-Refined Nb and Ta
M. K. Thomas, E. S. Tankins, and J. F. Erthal
Met. Soc. Can., 34:451-468 (1966)

Columbium Metal by the Aluminothermic Reduction of Cb$_2$O$_5$
H. A. Wilhelm, F. A. Schmidt, and T. G. Ellis
J. Metals, 18:1 (1966)

Possibility of Evaporation-Depositing Thin Layers of Tantalum from the Liquid Phase by Means of Electron Bombardment
Bartlomiej Zelechowski
Przeglad Elektron., 7:539-540 (1966)

Production of Niobium by Vapor Phase Reduction of Niobium Pentachloride
J. A. Brothers and W. F. Pesold
(to Wyandotte Chemical Corp.), U. S. Patent 3,216,822 (Nov. 1965)

A Process for Preparing High Purity Vanadium
O. N. Carlson, F. A. Schmidt, and W. E. Krupp
(Ames Lab., Ames, Iowa), Contract W-7405-eng-42, IS-1239 (Sept. 20, 1965), 15 pp.

Growing Large Single Crystals of Niobium (Columbium) by the Strain-Anneal Method
T. G. Digges, Jr. and M. R. Achter
Trans. AIME, 230:1737 (1965)

Epitaxial Growth on MgO of Niobium Films Investigated by Electron Microscopy
T. E. Hutchinson
J. Appl. Phys., 36:270 (1965)

The Effect of Yttrium on the Recrystallization and Grain Growth of Tantalum
L. D. Kirkbride, J. A. Basmajian, D. R. Stoller, W. E. Ferguson, R. H. Perkins, and D. N. Dunning
J. Less-Common Metals, 9:393 (1965)

Vacuum Melting and Refining of Impure V
J. Krueger and H. Winterhager
Compt. Rend. Coll. Int. Met. Vide, Bruxelles (1965), p. 108

Electrodeposition of Coherent Deposits Refractory Metals I. Niobium
G. W. Mellors and S. Senderoff
J. Electrochem. Soc., 112:266 (1965)

A New Method for the Preparation of Strain-Free Single Crystal Tensile Specimens for High Melting Point Metals and Some Results on the Plastic Deformation on Niobium Single Crystals
E. Votava
ORNL-tr-794; Phys. Stat. Sol., 5:421 (1965)

Tantalum Films Deposited by Asymmetric A-C Sputtering
F. Vratny and D. J. Harrington
J. Electrochem. Soc., 112:484 (1965)

Deposition of Tantalum Films with an Open-Ended Vacuum System
J. W. Balde, S. S. Charschan, and J. J. Dineen
Bell System Tech. J., 43:127 (1964)

Preparation of Single Crystals by Arc-Zone Melting (Re, Ta, Nb, Mo, V, Y)
O. N. Carlson, F. A. Schmidt, and W. M. Paulson
(Institute for Atomic Research and Department of Metallurgy, Iowa State University, Ames, Iowa), Trans. ASM Quart., 57:356 (1964)

Preparation and Properties of Thin-Film Hard Superconductors
J. Edgecumbe, L. G. Rosner, and D. E. Anderson
(Department of Electrical Engineering, University of Minnesota, Minneapolis, Minnesota), J. Appl. Phys., 35:2198 (1964)

Significance of Refractory Metals for the Production of High-Field Superconductors
D. Koch
Metall., 18:714-718 (1964)

Preparation of High-Purity Tantalum
O. P. Kolcin and I. K. Berlin
At. Energ. (USSR), 17:400 (1964)

Recrystallization of Vanadium
E. A. Loria, W. D. Ludemann, and E. A. Rowe
(Bureau of Mines, Metallurgy Res. Ctr., Reno, Nevada), BM-RI-6547 (1964)

Tantalum-Film Technology
D. A. McLean, N. Schwartz, and E. D. Tidd
Proc. IEEE, 52:1450 (1964)

Restwiderstandsmessungen an Elektronenstrahlgeschmolzenem Tantal zur Untersuchung des Zonenreinigungseffektes
Barry L. Mordike
Z. Metallk., 55:304-306 (1964)

Preparation of Niobium by Reduction of $NbCl_5$ by Hydrogen
L. A. Nielson, Ya. M. Polyakov, and A. N. Krestovnikov
Zh. Prikl. Khim., 37:669-672 (Mar. 1964)

Deposition of Tantalum from the Vapor Phase
Ya. M. Ployakov and G. Z. Zamesova
UCRL-Trans-1211 (L); transl. Izv. Vysshikh Uchebn. Zaveden., Tsvetn. Met., 4:130 (1964), 9pp.

Préparation et propriétés de superconducteurs synthétiques pour courant intense et champ intense
E. Saur
Phys. Verh. D. P. G., Dtsch., 8:318 (1964)
Nb and Nb_3Sn

Substructure and Impurities in Niobium Single Crystals Grown by the Electron-Beam, Floating-Zone Technique
C. S. Tedmon, Jr., and R. M. Rose
(Massachusetts Institute of Technology, Cambridge), Proc. 1st Intern. Conf. Electron Ion Beam Sci. Tech., Toronto (1964), pp. 521-540

A New Method for Preparation of Strain-Free Single-Crystal Tensile Samples of High-Melting Metals and Some Results on the Plastic Deformation of Niobium Single Crystals
E. Votava
(Union Carbide European Research Associates, Brussels), Phys. Stat. Sol., 5:421-434 (1964)

Large Crystals Grown by Recrystallization
K. T. Aust
The Art and Science of Growing Crystals, John Wiley and Sons, Inc., New York (1963), pp. 452-478

Investigation of the Distribution of Additions in Niobium after Zone Refining
A. I. Evstyukhin, V. V. Nikishanov, and I. V. Milov
OTS-64-21263, pp. 46-63
JPRS-22697, pp. 46-63, Met. i Metalloved. Chistykh Metal., Sb. Nauchn. Rabot, 4 (May 31, 1963)

Properties of Superconducting Niobium Films Made by Asymmetric AC Sputtering
R. Frerichs and C. J. Kircher
J. Appl. Phys., 34:3541 (1963)

On the Vacuum Reduction Process for the Production of Niobium Metal
Hirozo Kimura and Yasuo Sasaki
Trans. Natl. Res. Inst. Metals (Tokyo), 5:213-218 (1963)

Über das Rekristallisationsverhalten von Elektronenstrahlgeschmolzenem Niob
Ingard Kvernes
Z. Metallk., 54:449 (1963)

Work-Hardening in Niobium Single Crystals
T. E. Mitchell, R. A. Foxall, and P. B. Hirsch
Phil. Mag., 8:1895-1896 (1963)
Single crystals by electron-beam zone melting

Effect of Oxygen on the Lattice Constant, Hardness and Ductility of Vanadium
S. A. Bradford and O. N. Carlson
Trans. Quart. Am. Soc. Metals, 55:169 (1962)

Superconductive Films Made by Protected Sputtering of Tantalum or Niobium
R. Frerichs
J. Appl. Phys., 33:1898-1899 (L) (1962)

High Vacuum Technique
E. Luscher
(Internationales Jahrbuch Chemische Industrie), Vogt-Schild AG, Solothurn, (1962-63), p. 3

Electrodeposition Studies of Molybdenum, Tungsten, and Vanadium in Organic Solvents
Robert E. Meredith and Thomas T. Campbell
(Bureau of Mines, Albany Metallurgy Research Center, Ore.),
BM-RI-6303 (Nov. 1962), 17 pp.

Growth of Refractory Crystals Using the Induction Plasma Torch
Thomas B. Reed
(Lincoln Laboratory, Massachusetts Institute of Technology, Lexington 73, Massachusetts), J. Appl. Phys., 32:2534 (1962)

Quelques cas d'application du vide dans la métallurgie des métaux non ferreux
B. Tougarinoff
(Société générale métallurgique de Hoboken, Belgique), Schweizer Archiv., 28:491 (1962)

Zone Refining of Niobium by Arc Melting
A. I. Yevstyukhin, V. V. Nikishanov, and I. V. Milov
Issled. po Zharoproch. Splavam, Akad. Nauk SSR, Inst. Met., 9: 218–226 (1962)

The Role of Dilute Binary Transition Element Additions on the Recrystallization of Columbium
E. P. Abrahamson, II
(Watertown Arsenal, Watertown 72, Mass.), AIME Trans., 221:1196–1198 (1961)

The Recrystallization and Ductile-Brittle Transition Behavior of Tungsten. Effect of Impurities on Polycrystals Prepared from Single Crystals
B. C. Allen, D. J. Maykuth, and R. I. Jaffee
J. Inst. Metals, 90:120–128 (1961)

Preparation of High-Purity Vanadium Metal by the Iodide Refining Process
O. N. Carlson and C. V. Owen
(Institute for Atomic Research and Department of Chemistry, Iowa State University, Ames, Iowa), J. Electrochem. Soc., 108:88 (1961)

Elastic Moduli of Vanadium
G. A. Alers
Phys. Rev., 119: 1532–1533 (1960)
Single crystals by electron beam floating zone

Investigation of Electrolytic Processes for Preparation of High Purity Niobium Metal
A. J. Kolk, Jr., M. E. Sibert, and M. A. Steinberg
(Horizons, Inc., Cleveland), Contract AT(30-1)-1894, TID-6101 (1960), 26 pp.

Single Crystal Growth and Purification of Tantalum
D. P. Seraphim, J. I. Budnik, and W. B. Ittner III
Trans. AIME, 218:527–533 (1960)
Strain anneal

Research on the Production of Ultra Pure Refractory Metals Final Rept. July 1, 1959 Through June 30, 1960
Contract NOas 59-6248-c, Alloyd Corp., Cambridge, Mass.
AD-246722 (Sept. 27, 1960)

Procedures for the Metallographic Preparation of Beryllium, Titanium and Refractory Metals
R. D. Buchheit, C. H. Brady, and G. A. Wheeler
(Defense Metals Information Center, Battelle Memorial Institute, Columbus, Ohio). DMIC-Memo-37 (1959), 33 pp.

Refining of Niobium by the Iodide Method
D. M. Chizhikov and A. M. Grin'ko
Zh. Neorg. Khim., 4(5):982–984 (1959)

Preparation of High Purity W, Mo, Ta, Nb, and Zr
George A. Moore and L. L. Wyman
National Bureau of Standards, Washington, D. C. (July 1959), 13pp. Project 7360
Contract AF33(616)-58-11. (AD-232137). WADC-TR-59-314 OTS
Period covered: February 1958 to February 1959

The Preparation and Purification of Tantalum Single Crystals and Measurements of the Critical Field Curve for Single and Polycrystalline Specimens
J. I. Budnik, W. B. Ittner, and D. P. Seraphim
Physica, 24(S), S 151 (A) (1958)
Strain-anneal

The Growth of Molybdenum, Tungsten, and Columbium Crystals by Floating Zone Melting in Vacuum
E. Buehler
(Bell Telephone Laboratories, Inc., Murray Hill, N. J.), Trans. AIME, 212:694–698 (1958)

The Electrodeposition of Coherent Deposits of Refractory Metals. II. The Electrode Reactions in the Deposition of Tantalum
S. Senderoff, G. W. Mellors, and W. J. Reinhardt
J. Electrochem. Soc., 112:840 (1956)

The Separation of Niobium and Tantalum by the Fusion-Extraction Method (Studies on the seperation of elements of similar properties IV)
Kozo Nagashima and Shizno Fujiwara
AEC-tr-1790 (Translated by Toshio Nakai), Nippon Kagaku Zasshi 74:383–385 (1953)

Concerning the Electrolytic Deposition of Metallic Niobium, II
N. Izgaruishev (Isgarischew) and G. E. Kaplan
AEC-tr-3626, translated by H. P. Raaen (Oak Ridge National Lab.)
Z. Electrochem., 40:33–36 (1934)

Concerning the Electrolytic Deposition of Metallic Niobium and Its Separation From Tantalum
N. Izgaruishev and A. F. Prede
AEC-tr-3635, translated by H. P. Raaen (Oak Ridge National Lab.)
Z. Elektrochem., 39:283–288 (1933)

1.a.4. Group VI

Growth and Structure of Thin Chromium Films Condensed on Ultra-High Vacuum Cleaved NaCl and KCl Crystals
J. Forssell and B. Persson
J. Phys. Soc. Japan, 29:1532–1545 (1970)

Dendritic Growth of Evaporated Chromium
Y. Fukano and C. M. Wayman
J. Crystal Growth, 7:163–176 (1970)

New Method for Preparing Unusually Oriented
Tungsten Tips and Their Growth Mechanism
   F. Okuyama and T. Hibi
   Japan. J. Appl. Phys., 9:15 (1970)

Morphologie und Realstruktur von aus der
Gasphase durch thermische Zersetzung von
Molybdänpentachlorid hergestellten Molyb-
däneinkristallen
   G. Weise and R. Gunther
   J. Cryst. Growth, 6:167-183 (1970)

Wachstum von Wolframeinkristallen aus der
Gasphase über chemische Reaktionen
   G. Weise and G. Owsian
   J. Less-Common Metals, 22:99-116 (1970)

Die Textur von Molybdänabscheidungen aus der
Gasphase
   G. Weise and H. Wadewitz
   J. Crystal Growth, 7:313-316 (1970)

Growth Spirals on Electrolytically Polished Sur-
faces of Mo Single Crystals
   V. D. Yaroshevich and G. I. Shakhalova
   Fiz. Tverd. Tela, 11: 3582 (1969)
   Sov. Phys. – Solid State, 11:2998-2999 (1970)

Electro- and Thermotransport in Single-Crys-
talline Molybdenum and Surface Change of High-
Melting Metals during Directional Diffusion
   I. V. Zakurdaev
   Fiz. Tverd. Tela, 11(12):3463-3473 (1969)
   Sov. Phys. –Solid State, 11(12):2905-2913 (1970)

On the Effect of the Laminar Inclusion of For-
eign Elements in Zone Melting on the Effec-
tive Distribution Coefficients
   J. Berthel and K. Eichler
   Krist. Tech., 2:205-215 (1967)
   ORNL-TR-1978 (April 1969), 12 pp.
   W and Mo

Mo-Einkristalle in Argon-Plasma
   A. J. Borisov et al.
   in: Einkristalle hochschmelzender und seltener Metalle
   E. M. Savitskii, ed. Nauka, Moscow (1969), pp. 165-167

Zonenschmelzen von Einkristallen (Mo)
   I. A. Brodskii et al.
   in: Einkristalle hochschmelzender und seltener Metalle
   E. M. Savitskii, ed. Nauka, Moscow (1969), pp. 14-21

Herstellung von Blechteilen aus Mo-Einkristallen
   E. V. Chibisova et al.
   in: Einkristalle hochschmelzender und seltener Metalle
   E. M. Savitskii, ed. Nauka, Moscow (1969), pp. 168-172

Mo-Einkristalle für kalte Kathoden
   G. I. Esarev et al.
   in: Einkristalle hochschmelzender und seltener Metalle
   E. M. Savitskii, ed., Nauka, Moscow (1969), pp. 179-181

Growth of Tungsten Single Crystals on Glowing
Tips in a Mixture of $N_2 + H_2$
   K. H. von Grote
   Optik, 28:360-373 (1969)

Epitaxial Chromium Deposition
   David W. Hardesty
   J. Electrochem. Soc., 116:1194-1197 (1969)

Shape of Crystallites Grown in the Case of
Oblique Incidence of a Molecular Beam of
Chromium on a Substrate
   V. P. Kostyuk and I. N. Shklyarevskii
   Fiz. Tverd. Tela, 11:539-541 (Feb. 1969)
   Sov. Phys. – Solid State, 11:436 (1969)

Herstellung von Mo-Pulver aus Hexacarbonyl
   O. D. Kritschevskaja et al.
   Poroshkovaya Met., 9(8):17-20 (1969)

Dendrites of W Grown on the Hot Filament
   Y. Kumamoto
   Nippon Tungsten Rev. (Sept. 1969), pp. 88-90

Untersuchung von deuterierten W-Einkristallen
   V. A. Kuznetsov et. al.
   in: Einkristalle hochschmelzender und seltener Metalle
   E. M. Savitskii, ed., Nauka, Moscow (1969), pp. 73-79

Preparation of Ultrafine Mo-Powder
   R. Matsuzaki, Y. Saeki, and K. Funaki
   Denki Kagaku Oyobi Kogyo Butsuri Kagaku, 37:412-417 (1969)

Croissance anormale des grains et formation
de monocristaux lors du recuit du molybdène
file hydrostatiquement
   G. A. Mochalov, E. D. Martynov, B. I. Beresnev, A. I.
      Evstyukhin, D. K. Bulychev, A. A. Rusakov, K. P. Rodio-
      nov, and Yu. N. Ryabinin
   Fiz. Metal. Metalloved. SSSR, 27(5):870-872 (1969)

Herstellung von Mo-Einkristallen im festen
Zustand
   G. A. Mochalov et al.
   in: Einkristalle hochschmelzender und seltener Metalle
   E. M. Savitskii, ed., Nauka, Moscow (1969), pp. 11-14

Verformung von Mo-Einkristallen, hergestellt
aus der Gasphase
   G. A. Mochalov et al.
   in: Einkristalle hochschmelzender und seltener Metalle
   E. M. Savitskii, ed., Nauka, Moscow (1969), pp. 127-132

Textur von verformten und rekristallisierten
Mo-Einkristallen
   G. A. Mochalov et al.
   in: Einkristalle hochschmelzender und seltener Metalle
   E. M. Savitskii, ed., Nauka, Moscow (1969), pp. 154-159

Étude de l'orientation préférentielle des cris-
tallites dans les lingots obtenus par fusion
au faisceau électronique
   B. A. Movchan, T. A. Molodkina, and V. N. Statkevich
   Fiz. Khim. Obrabot. Mater., SSSR, 6:43-46 (1969)

Einkristalle hochsmelzender und seltener
Metalle
   E. M. Savitskii, ed.
   Nauka, Moscow (1969)

Einkristalle von Mo und C
   E. M. Savitskii et al.
   in: Einkristalle hochschmelzender und seltener Metalle
   E. M. Savitskii, ed., Nauka, Moscow (1969), pp. 58-65

Zonengeschmolzene W-Einkristalle nach Tem-
peraturzyklen
   E. M. Savitskii et al.
   in: Einkristalle hochschmelzender und seltener Metalle
   E. M. Savitskii, ed., Nauka, Moscow (1969), pp. 89-96

Sur la chimisorption du carbone et de l'oxygène à la surface de molybdène
 Jean-Paul Touboul, Pierre Ailloud, Lea Minel, and Jean-Paul Langeron
 Compt. Rend. 269, Serie C, 116-19 (1969)

The Influence of Annealing on the Recrystallization Behaviour and the Mechanical Properties of Doped and Undoped Molybdenum Wire
 F. A. M. M. van Meel, A. C. van Maaren, and G. J. van Weezel
 In: High Temperature Materials (F. Benesovsky, ed.), Metallwerk Plansee AG., Reutte/Tirol, Austria, 1969, pp. 172-181
 From 6th International Plansee Seminar on High Temperature Materials, Fundamental and Developments, Reutte/Tirol, Austria, CONF-680603.

Chromium Coatings Prepared by Chemical Vapor Deposition
 Gene F. Wakefield
 J. Electrochem. Soc., 116:5-9 (1969)

The Direct Current Effect in Zone Melted Tungsten Crystals
 Peter Adam
 Ph.D. thesis, Technische Univ. Berlin (1968), 92 pp.
 Surface or volume diffusion in crystal formation

Electron Beam Melting of Tungsten
 J. C. E. Amos and P. H. Morton
 Rev. Int. Hautes Temp. Refract., 5:141-144 (1968)

Chromium and Nickel Powders Made by Reduction of Their Oxides with Magnesium, Lithium, or Sodium Vapors
 Alan Arias
 NASA-TN-D-4714; E-4386 (Aug. 1968), 23 pp.

Mechanism of Electrolytic Deposition of Chromium
 K. P. Batashev and I. A. Belozerova
 J. Appl. Chem. USSR, 41:1834-1839 (1968)

Preparation of Anhydrous Chromous Chloride
 Robert L. de Beauchamp and Thomas A. Sullivan
 BM-RI-7194 (1968)

Preparation of High-Purity Metals and Semiconductor Materials
 A. I. Belyaev, E. A. Zhemchuzhina, V. V. Krapukhin, L. A. Firsanova, and Yu. D. Chistyakov
 Sb., Mosk. Inst. Stali Splavov, No. 52, 300-14 (1968)
 Electron-beam melting of W and Mo

Sur la purification du chrome par la méthode de la zone fondue
 J. Bigot
 Mem. Sci. Rev. Metal., France, 65:61-65 (1968)

Purification du chrome par la méthode de la zone fondue et par traitement thermique sous hydrogène. Etude de la temperature de transition ductile-fragile du métal obtenu
 M. J. Bigot
 (Ph.D. thesis, Univ. Paris), 1969. Arch. Orig. Centre Document C.N.R.S. No. 2739 (25 Oct. 1968)

Tungsten Hexafluoride (WF$_6$)
 Maurice Carles
 (Commissariat a l'Energie Atomique, Pierrelatte, France), CEA-Bib-125 (Sept. 1968), 30 pp.

Chemical Vapor Deposition of Tungsten
 J. Chin and J. Horsley
 (Gulf General Atomic, Inc., San Diego, Cal.), Contract

AT(04-3)-167, GA-8772, CONF-681059-1 (EEE Thermionic Conversion Specialist Conf., Boston, Mass.), Oct. 1968, 14 pp.

Heteroepitaxy of Molybdenum on Tungsten from the Liquid State
 Yu. D. Chistyakov, A. I. Pekarev, and M. V. Gartman
 Kristallografiya, 12(4):688:693 (1968)
 Sov. Phys. – Cryst., 12(4):597-600 (1968)

Texture of W Formed by Deposition from Vapor
 R. K. Chuzko et al.
 Proceedings of International Conference on Crystal Growth, Birmingham (1968), pp. 219-224

Texture of Tungsten Formed by Deposition from the Vapour
 R. K. Chuzhko, I. V. Kirillov, Yu. N. Golovanov, and A. P. Zakharov
 J. Crystal Growth, 3, 4, pp. 219-224 (1968)

Structure of Electro-Deposited Chromium
 W. H. Cleghorn, D. H. Warrington, and J. M. West
 Electrochim. Acta, 13:331-334 (1968)

Electron-Beam Zone Purification and Analysis of Tungsten
 D. R. Hay, R. K. Skogerboe, and E. Scala
 J. Less-Common Metals, 15:121-127 (1968)

Preliminary Study of Pyrolytic Tungsten Deposition
 D. O. Hobson
 Contract W-7405-eng-26, ORNL-TM-2074 (Jan. 1968), 37 pp.

Vapor-Phase Deposition of Tungsten from Tungsten Hexafluoride and Hydrogen
 F. W. Hoertel
 High Temperature Refractory Metals (W. A. Krivsky, ed.), Met. Soc. Conf., Vol. 34, Pt. 1, Gordon and Breach, New York (1968), pp. 519-527

Chemical Vapor Deposition of High Performance Tungsten Thermionic Emitter
 R. G. Hudson, T. Tagami, and L. Yang
 J. Metals, 20:157A (1968)

Pseudomorphic Deposits of Chromium on Nickel
 W. A. Jesser and J. W. Matthews
 Phil. Mag., 17:475-479 (1968)

Influence of Annealing on the Recrystallization of Doped and Undoped Mo Wires
 F. A. M. van Meel
 Iron Steel Inst. London Publ. No. 113, pp. 49-51 (1968)

Preparation of Vapor-Deposited Tungsten at Atmospheric Pressure
 Emil J. Mehalchick and Martin B. MacInnis
 Electrochem. Tech., 6:66 (1968)

Preparation and Properties of Chromium Films
 A. A. Milgram and Chin-Shun Lu
 J. Appl. Phys., 39:2851 (1968)
 Vacuum deposition

A Low-Energy Ion Source for the Deposition of Chromium
 B. A. Probyn
 British J. Appl. Phys. (J. Phys. D), 1:457-466 (1968)

Tungsten Whiskers by Vapor-Phase Growth
 A. G. Starliper and H. Kenworthy
 (Bureau of Mines, Rolla, Maryland), BM-RI-7118 (April 1968), 17pp.

Chemical Deposition of Mo on Si
Takuo Sugano, Hsun-Kwei Chou, Minoru Yoshida, and Tateki Nishi
Japan J. Appl. Phys., 7:1028-1038 (1968)

Production and Properties of Ultradisperse Tungsten Powders
Yu. V. Tsvetkov, S. S. Deineka and D. M. Chizhikov, I. K. Tagirov, and I. Ya. Basieva
In: Powder Metallurgy in Modern Technology
Nauka, Moscow (M. K. Rybaltschenko, ed.) (1968), pp. 143-15

Chemische Transportreaktionen als Raffinationsverfahren zur Darstellung von Reinstchrom
H. Winterhager und U. Holtkamp
Metall, 22:33-38 (1968)

Properties of Ultrafine Tungsten Powder
H. Yamamoto, K. Ueki, and N. Ichiyama
J. Japan. Soc. Powder Met., 15:174-183 (1968)
Also: ORNL-tr-2326

Electron-Beam Melting of Molybdenum in a Controlled Atmosphere
T. N. Zagorskaya, G. F. Ivanovskiy, and T. K. Lyakhovich
Russian Met., No. 2, p. 87 (1968)

Some Properties of Tungsten and Molybdenum Wire Made from Single Crystals
R. T. Andreeva and N. S. Rozinova
Metallurgy of Tungsten, Molybdenum, and Niobium (V. A. Reznitschenko, ed.), Nauka, Moscow (1967), pp. 149-154

Forging and Drawing Textures of Molybdenum Wire Obtained from Single Crystals of Different Orientations
A. A. Babareko
Izv. Akad. Nauk SSSR, Metally, No. 1, pp. 140-144 (1967)

Evaluation of Solid State Purification of Molybdenum
W. S. Bennett and F. A. Smidt, Jr.
(Metallurgy Research Section, Metallurgy Department, Pacific Northwest Laboratory, Richland, Washington), BNWL-379 (Feb. 1967)

Vapor Deposition of Tungsten by Hydrogen Reduction of Tungsten Hexafluoride. Process Variables and Properties of the Deposit
Jean F. Berkeley, Abner Brenner, and Walter E. Reid, Jr.
J. Electrochem. Soc., 114:561 (1967)

Chemical Vapor Deposition of Mo onto Si
J. J. Casey, R. R. Verderber, and R. R. Garnache
J. Electrochem. Soc., 114:201 (1967)

Epitaxy of Chromium on Cleavage Faces of Rocksalt in U.H.V.
W. A. Crossland and C. A. Marr
Japan. J. Appl. Phys., 6:544-546 (1967)

Effect of Deposition Conditions on the Orientation of Chemical Vapor-Deposited Tungsten and Molybdenum
J. I. Federer, W. C. Robinson, and R. M. Steele
Proceedings of the 1967 Thermionic Conversion Specialist Conference, Palo Alto, California (Oct. 30 - Nov. 2, 1967), the Inst. of Electrical and Electronics Engineers, Inc., New York, New York

Production and Control of Purity of W and Mo Single Crystals
V. D. Kapustin, Yu. S. Belomyttsev, V. N. Bykov, N. N. Mikhailov and V. S. Tsoi
Izv. Akad. Nauk SSSR, Metal., No. 2, 71-78 (March-April 1967)

The Preparation of Fine Recrystallized Tungsten Wire for Transmission Electron Microscopy
E. F. Koch and J. L. Walter
GE-67-C-151 (Sept. 1967), 9 pp.

Grain Growth During Sintering of Tungsten
N. C. Kothari
J. Appl. Phys., 38:2395 (1967)

Single Crystals of Molybdenum with Very Small Residual Resistance Ratios
H. Opperman, G. Weise, J. Barthel, and K. H. Berthel
Phys. Stat. Sol., 22:K151-153 (1967)

Electrodeposition of Coherent Deposits of Refractory Metals. V. Mechanism for the Deposition of Molybdenum from a Chloride Melt
S. Senderoff and G. W. Mellors
J. Electrochem. Soc., 114:556-560 (1967)

Electrodeposition of Coherent Deposits of Refractory Metals. VI. Mechanism of Depositions of Molybdenum and Tungsten from Fluoride Melts
S. Senderoff and G. W. Mellors
J. Electrochem. Soc., 114:586 (1967)

Purification of Tungsten Hexachloride
F. A. Skirvin, T. T. Campbell, and F. E. Block
BM-RI-6984 (July 1967), 21 pp.

Method for Preparing Thin Chromium Foil
J. Stephen and D. N. Osborne
(Atomic Energy Research Establishment, Harwell, England), AERE-R-5430 (March 1967), 10 pp.

Growth and Preferred Orientations of Large Elongated Grains in Doped W Sheet
J. L. Walter
Trans. Met. Soc. AIME, 239:272-286 (1967)

Die Wachstumsformen von Einkristallhalbkugeln bei der thermischen Zersetzung von Molybdän und Wolframhalogeniden
G. Weise
Kristall u. Tech., Dtsch., 2:339-358 (1967)

Characterization of Tungsten Single Crystals Grown from the Gas Phase
G. Weise
In: Second International Symposium on Hyperpure Products in Science and Technology, Dresden (J. Kunze, B. Pegel, K. Schlaubitz, and D. Schulze, eds.), Akademie-Verlag, Berlin, 1967, pp. 305-312
In German

Electron beam zone melting of tungsten and molybdenum
J. Barthel and R. Petri
(Deut. Akad. Wiss. Berlin, Inst. Metallphysik und Reinstmetalle Dresden), Reinststoffdarstellung, 2. Internationales Symposium Reinststoffe in Wissenschaft und Technik, Dresden (Sept. 28 - Oct. 2, 1965, 1966); ORNL-tr-1402, 21 pp.
Survey of electron beam zone melting of tungsten and molybdenum

a. Preparation Methods

The Recrystallization of 99.997% Chromium
C. R. Brinkman and C. H. Pitt
J. Less-Common Metals, 11:323 (1966)

Face-Centered Cubic Tungsten Films Obtained
by Ion Beam Sputtering
K. L. Chopra, M. R. Randlett, and R. H. Duff
(Legemont Laboratory, Kennecott Copper Corporation, Lexington, Mass.), Appl. Phys. Letters, 9:402 (1966)

Herstellung und Zusammensetzung von Beta-Wolfram
L. C. Dufour
Bull. Soc. Chim. France, S. 2107-2108 (1966)

Orientation, Purity and Perfection of Molybdenum Single Crystals Growth by Electron-Beam Zone Melting
V. O. Esin, V. I. Levit, E. P. Romanov, and L. V. Smirnov
Fiz. Metallov Metalloved., 22:415-419 (1966)

Effect of Lanthanum and Yttrium on the Recrystallization of Molybdenum
V. P. Goltsev
AD-659666; FTD-HT-67-29, 8pp.
Dokl. Akad. Nauk BSSR, 10:196-198 (1966)

Zone Refining of Mo
G. P. Kovtun, A. D. Kruglykh, and I. G. Kyakov
Izv. Akad. Nauk SSSR, Metally, 1:71-72 (1966)

The Impurities in Aluminothermal Cr and Its Purification by Zone Refining
M. Mancaux
Ann. Chim. (Paris), No. 1, p. 461 (1966)

. Tungsten Single Crystal Sheet
L. Raymond and F. R. Charvat
Am. Inst. Aero. Astron. J., 4:895 (1966)

Spark Erosion Fabrication of Single Crystal Tungsten Ribbon Filaments
F. L. Reynolds
(Lawrence Radiation Lab., University of California, Berkeley, California), Rev. Sci. Instr., 37:1730-1731 (1966)

Effect of Fluorine Impurities on the Grain Stability of Thermochemically Deposited Tungsten
A. C. Schaffhauser and R. L. Heestand
ORNL Preprint P12-04-6-14; published in the Proceedings of the 1966 Thermionic Conversion Specialist Conterence

Deposition of Tungsten and Rhenium by Solid-Gas Interaction
A. M. Shroff
Prace Przemysl. Inst. Elektroniki (Poland), No. 3 (1967), pp. 447-462, Franco Polish Colloquium on Vacuum Technique, Warsaw, Poland, June 27-29 (1966)

Application of Electron Bombardment to the Zone Refining of W and Mo
A. M. Shroff and C. Barromee
Rev. Hautes Temp. Refract., 3:385-393 (1966)

The Growth of Needle-like Crystals on Tungsten in Argon
Ernest H. S. van Someren
Z. Metallk., 57:633 (1966)

The Electron Beam Melting of Tungsten
T. Takaai
Nippon Kinzoku Gakkaishi, 20:1027-1031 (1966)

Structure of Epitaxial Films of Cobalt and Chromium Prepared by Sublimation
N. Takahashi and H. Martina
Compt. Rend., 263:203-206 (1966)

Preparation of Submicron Tungsten Powder by Hydrogen Reduction of Tungsten Hexachloride
J. E. Tress, T. T. Campbell, and F. E. Block
BM-RI-6835 (1966)

New Electron Beam Optics for the Preparation of Insulating Crystals by the Floating Zone Method
C. E. Tyan, D. P. Considine, J. J. Hawley, and R. C. Marshall
AFCRL-66-539 (Aug. 1966), 28 pp.

Effect of Alloying on Grain Refinement of Electron-Beam-Melted Tungsten
Walter R. Witzke
NASA-TN-D-3531 (July 1966), 19 pp.

Sur la purification du chrome par zone fondue
Jean Bigot and Simone Talbot-Besnard
Compt. Rend., 261:121 (1965)

A Program of Basic Research on Mechanical Properties of Reactor Materials
R. H. Chambers and T. A. Trozera
(General Atomic, San Diego, Calif.), Contract AT(04-3)-167, GA-6413, (June 18, 1965), 30 pp.

Epitaxy of Iron and Chromium on Silver
R. Cinti, J. Devenyi, P. Escudier, R. Montmo, and A. Yelon
Compt. Rend., 260:6849 (1965)

Fabrication and Properties of Tungsten Single Crystals. Part I. Growth of Single Crystal Tungsten to Be Used for Rolling into Single Crystal Sheet. Part II. Rolled Single Crystal Tungsten Sheet
G. W. Edwards and J. W. Arthur
(Linde Division, Union Carbide Corp., Indianapolis, Indiana) Contract NOw-64-0055, AD-464 350 (May 14, 1965)

Thermochemical Deposition of Refractory Metals, Alloys, and Compounds for Application in Thermionic Devices
J. I. Federer, R. L. Heestand, F. H. Patterson, and C. F. Leitten
Presented at Intern. Conf. on Thermionic Elec. Power Generation, London, Sept. 1965, Contract W-7405-ENG-26, ORNL-P-1446-1; CONF-650908-9, 13 pp.

Bias Sputtering of Molybdenum Films
R. Glang, R. A. Holmwood, and P. C. Furois
(IBM East Fishkill, Systems Manufacturing Div., Hopewell Junction, N. Y.), TR 22-173, June 7, 1965; presented at 3rd Intern. Vacuum Cong., Stuttgart, Germany (June 28 to July 2, 1965)

Zone-Refining Tungsten in the Presence of a Superimposed Direct Current
D. R. Hay and E. Scala
Trans. AIME, 233:1153 (1965)

Vacuum Deposited Molybdenum Films
R. A. Holmwood and R. Glang
J. Electrochem. Soc., 112:827 (1965)

High Vacuum Zone Refining of Molybdenum
G. F. Ivanovskii and T. N. Zagorskaya
Izv. Akad. Nauk SSSR, Metally, 3:65-69 (1965)

Dendritisches Wachstum von Wolframkristallen
aus der Dampfphase
  Kuno Kirner
  Bosch Techn. Berichte, 1:2 (1964)
  Z. Metallk., 56:179 (1965)

Plasma Jet Atomization of Molybdenum
  A. N. Krasnov
  (Air Force Systems Command, Wright Patterson AFB, Ohio,
    Foreign Tech. Division), FTD-TT-65-1381/1+2+4; AD-
    626961 (Jan. 1966), 12pp.
  Poroshkovaya Met., Akad. Nauk Ukr. SSR (Kiev), No. 3, pp.
    1-5 (1965)

Vapor Deposited Tungsten for Thermionic
Emitters
  J. R. Lindgren, R. G. Mills, and A. F. Weinberg
  GA-6473 (1965)

Experimental Investigation of the Opacity of
Small Particles
  P. J. Marteney
  (NASA, Washington Contract NASw-847, NASA-CR-211 (April
    1965), 52 pp.
  W-producing dispersions of submicron-radius solid particles

Analyse statistique des directions de croissance preférentielle des monocristaux de molybdène obtenus pare récristallisation zonale électronique, sans creuset
  A. I. Pekarev, Yu. D. Chistyakov, and G. N. Shchirenko
  Izv. Vysshikh Uchebn. Zavedenii, Svetn. Metal., 8:65-72 (1965)

Croissance de monocristaux de métaux réfractaires et de leurs alliages avec une orientation cristallographique donnée par fusion
zonale à faisceau électronique
  E. M. Savitskii, G. S. Burkhanov, G. L. Tsarev, and N. N.
    Bokareva
  Pribory i Tekhn. Eksper., No. 4, pp. 248-250 (1965)
  Inst. and Exper. Tech., No. 4, pp. 1004-1005 (1965)

Epitaxially Deposited Molybdenum Films on
Silicon
  R. N. Tucker
  J. Electrochem. Soc., 112:184C (1965)

Preparation of Single Crystals by Arc-Zone
Melting (Re, Ta, Nb, Mo, V, Y)
  O. N. Carlson, F. A. Schmidt, and W. M. Paulson
  (Institute for Atomic Research and Department of Metallurgy,
    Iowa State University, Ames, Iowa), Trans. ASM Quart.,
    57:356 (1964)

Fabrication and Properties of Tungsten Single
Crystals. Growth of Single Crystal Tungsten
to Be Used for Rolling into Single Crystal Sheet
  F. R. Charvat, G. W. Edwards, and J. W. Arthur
  QPR-3; AD-456521, Contract NOw-64-0055-c (1964), 26 pp.

Epitaxie du molybdène obtenu par décomposi-
tion partielle de la molybdenite
  Eveline Gillet and Marcel Gillet
  Compt. Rend., 258:6382 (1964)

Field-Aided Electron Beam Zone Refining of
Tungsten
  D. R. Hay and E. Scala
  (Department of Engineering Physics and Materials Science,
    Cornell University, Ithaca, N. Y.), First World Conf. Elec-
    tron and Ion Beam Technology (May 4-7, 1964), Toronto,
    Canada

Growth of Tungsten Monocrystals on Glowed
Tungsten Wires
  T. W. Hoffmann and J. Nikloborc
  (Univ. of Wroclaw, Poland), Acta Phys. Polon., 25:633-634
    (1964)

Preparation of Tungsten Metal Powder
  S. K. Kantan, R. V. Roghavan, and N. K. Rao
  Trans. Indian Inst. Metals, 17:191 (1964)

Effects of Purity and Structure of Recrystalli-
zation, Grain Growth, Ductility, Tensile, and
Creep Properties of Arc-Melted Tungsten
  William D. Klopp and Peter L. Raffo
  NASA-TN-D-2503, N64-33085 (1964), 49 pp.

Significance of Refractory Metals for the Pro-
duction of High-Field Superconductors
  D. Koch
  Metall., 18:714-718 (1964)

Growing Large Tungsten Crystallites by Re-
crystallization During Sintering of Tungsten
Samples
  O. V. Mitrofanov
  N64-11019 OTS: Fiz. Metallov i Metalloved. (Moscow), 13(5):
    760-762, (1962)
  STAR, 2:206(A), (Jan. 23, 1964)

Modes of Operation of an Electron Beam Float-
ing Zone Melting Furnace: Growth of Single
Crystals of Metals and Alloys
  Heinz G. Sell and Willard M. Grimes
  (Lamp Parts Engineering Department, Westinghouse Lamp
    Division, Bloomfield, New Jersey), Rev. Sci. Instr., 35:64
    (1964)

Growth of Tungsten and Molybdenum Crystals
from Cryolite-Oxide Melts (by Zone Refining)
  V. N. Vigdorovich and V. V. Marychev
  Dokl. Chem. Tech. Proc. Acad. Sci. USSR Chem. Tech. Sect.,
    159:157-160 (1964)

Study of Filiform Crystals of Beryllium and
Chromium
  V. M. Amonenko, B. M. Vasyutinskii, G. N. Kartmazov,
    I. I. Papirov, G. F. Tikhinskii, and V. A. Finkel
  UCRL-Trans.-1152(L), Issled. po Vysokoprochnym Splavam
    i Nitevidnym Kritallam, Akad. Nauk SSSR, Inst. Met., Dokl.
    Nauchn. Sessii, 1962:34-36 (1963), 6 pp.

Large Crystals Grown by Recrystallization
  K. T. Aust
  The Art and Science of Growing Crystals, John Wiley and
    Sons, Inc., New York (1963), pp. 452-478

Obtaining Molybdenum Single Crystals and Their
Properties
  V. S. Emel'yanov, A. I. Evstyukhin, G. A. Leont'ev, and A. N.
    Semenikhin
  JPRS-22697, pp. 39-45
  Met. i Metalloved. Chistykh Metal., Sb. Nauchn. Rabot, No. 4
    (May 31, 1963)
  OTS-64-21263, pp. 39-45

Techniques for the Fabrication of Ultra-Thin
Metallic Foils
  F. J. Karasek
  (Metallurgy Division, Argonne National Laboratory, Argonne,
    Illinois), Nucl. Sci. Eng., 17:365-370 (1963)

b. Assay Methods

The Effect of Temperature and Heating Rate on the Secondary Recrystallization of Doped Tungsten Wires
M. Mannerkoski
(Oy Fiskars AB, Research Laboratory, Aminnefors, Finland), J. Inst. Metals, 92:149 (1963-64)

Preparation of Molybdenum Single Crystals by Electron Beam Floating Zone Melting
N. P. Ming, T. Y. Fan, C. Li, Y. S. Hsu, and T. Feng
Rept. Emm-65-3; TT66-61212; AD 602-629, AD 632-405 (Feb. 1966), 14 pp.
Wu Li Hsueh Pao (Chinese People's Republic), 19:160-164 (1963)

Growth of Tungsten Crystals from Vapor
O. V. Mitrofanov
Kristallografiya, 7(5):780-783 (1962)
Sov. Phys. – Cryst., 7(5):630-632 (1963)

Refractory Metals Phase Diagrams
E. J. Rapperport
NMI 9249, Contract AF 33(616)7157, (Feb. 8, 1963), 7 pp.
AD 402 630; U. S. Govt. Res. Rept., 38:51 (A) (1963)

High Vacuum Technique
E. Luscher
(Internationales Jahrbuch Chemische Industrie), Vogt-Schild AG, Solothurn (1962-63), p. 3

Electrodeposition Studies of Molybdenum, Tungsten, and Vanadium in Organic Solvents
Robert E. Meredith and Thomas T. Campbell
(Bureau of Mines, Albany Metallurgy Research Center, Ore.), BM-RI-6303 (Nov. 1962), 17 pp.

Electron Beam Melting of Molybdenum
Kazuo Tsuya and Noriyoshi Aritomi
Trans. Natl. Res. Inst. Metals, 4:70 (1962)

Use of a Precious Metal Anode in the Electrowinning of High-Purity Chromium from a Fluoride Bath
J. A. Whittaker
(Fulmer Research Institute, Stoke Poges, Bucks, England), J. Electrochem. Soc., 109:986 (1962)

Fabrication and Properties of Tungsten and Tungsten Alloy Single Crystals
Linde Company, Contract NOw 61-1671, Mar. 31, 1962, 74 pp.
AD 273 951
U. S. Govt. Res. Rept., 37:45 (A) (1962)

The Recrystallization and Ductile-Brittle Transition Behavior of Tungsten. Effect of Impurities on Polycrystals Prepared from Single Crystals
B. C. Allen, D. J. Maykuth, and R. I. Jaffee
J. Inst. Metals, 90:120-128 (1961)

Preparing Monocrystals of Molybdenum and Tungsten by Crucibleless Zone Melting
N. A. Brilliantov, L. S. Starostina, and O. P. Fedorov
Kristallografiya, 6(2):261-264 (1961)
Sov. Phys. – Cryst., 6(2):202-204 (1961)

Observation Concerning Zone Refining and Thermal Treatment of Molybdenum from Low Temperature Resistance Measurement
E. Buehler and J. E. Kunzler
AIME, Trans., 221:957-961 (1961)

•Production of Monocrystal Tungsten Emitters
O. V. Mitrofanov
Instr. and Experimental Tech. (1961), pp. 1007-1009
Recrystallization

Preparation of Chromium
N. W. Silcox, A. F. Armington, and G. F. Dillon
Conf. Ultrapurifi. Semiconduct. Mat. (A) (Apr. 1961)

Iodide Method of Refining Chromium
V. S. Emel'yanov, A. I. Evstyukhin, D. D. Abanin, and V. I. Statsenko
Met. i Metalloved. Chistykh Metal., Sb. Nauchn. Rabot, 2:14-26 (1960)

Research on the Production of Ultra Pure Refractory Metals Final Rept. July 1, 1959 Through June 30, 1960
Contract NOas 59-6248-c, Alloyd Corp., Cambridge, Mass. AD-246722 (Sept. 27, 1960)

Vacuum Distillation of Chromium
V. M. Amonenko, A. A. Kruglykh, and G. F. Tikhinskii
Fiz. Metallov Metalloved., 7:868-874 (1959)

Procedures for the Metallographic Preparation of Beryllium, Titanium and Refractory Metals
R. D. Buchheit, C. H. Brady, and G. A. Wheeler
(Defense Metals Information Center, Battelle Memorial Institute, Columbus, Ohio). DMIC-Memo-37 (1959), 33 pp.

Preparation of High Purity W, Mo, Ta, Nb, and Zr
George A. Moore and L. L. Wyman
National Bureau of Standards, Washington, D. C. (July 1959), 13 pp. Project 7360
Contract AF33(616)-58-11. (AD-232137). WADC-TR-59-314 OTS
Period covered: February 1958 to February 1959

The Purification of Tungsten by Electron-Bombardment Floating-Zone Melting
Walter R. Witzke
(Lewis Research Center, National Aeronautics and Space Administration, Cleveland, Ohio), Transactions of the Vacuum Metallurgy Conf. (1959), p. 140

An Improved Method of Producing Iodide Chromium
V. S. Emel'yanov, A. I. Evstyukhin, D. D. Abanin, and V. I. Statsenko
Met. i Metalloved. Chistykh Metal., Sb. Nauchn. Rabot, 1:44-62 (1959)

Method and Apparatus for the Production of Single Crystals
S. S. Brenner and C. R. Morelack
General Electric, U. S. Patent 2,836,524 (May 27, 1958), 2 pp.
Rod-like single crystals of W by vapor deposition

The Growth of Molybdenum, Tungsten, and Columbium Crystals by Floating Zone Melting in Vacuum
E. Buehler
(Bell Telephone Laboratories, Inc., Murray Hill, N. J.), Trans. AIME, 212:694-698 (1958)

Recrystallization Diagram of Molybdenum
E. M. Savitskii, V. V. Baron, and K. N. Ivanova
Dokl. Akad. Nauk SSSR, 113(5):1070-1072 (1957)
Dokl. – Chem., 113(5):355-358 (1957)

Innovations in Electron-Beam Zone Refining of Molybdenum
  H. L. Prekel and A. Lawley
  (Franklin Inst. Research Labs., Philadelphia, Pa.), CONF-660414-4. ORAU. Gmelin, AED-CONF-66-119-4
  From 2nd International Conf. on Electron and Ion Beam Science and Technology, New York

Experimental Program
  (General Dynamics Corp., San Diego, California, General Atomic Div.), GA-7978, pp. 11-46
  High-purity tungsten single crystals were produced from vapor-deposited tungsten by electron-beam floating-zone refining

1.b. Assay Methods

1.b.1. General

Nonferrous Metallurgy. II. Zirconium, Hafnium, Vanadium, Niobium, Tantalum, Chromium, Molybdenum, and Tungsten
  R. Z. Bachman and C. V. Banks
  Anal. Chem. Ann. Rev., 43:120R-145R (1971)
  712 refs.

Sources of Information on Ultrapurification and Characterization
  T. F. Connolly
  To be published in Fractional Solidification, Vol. III, Techniques of Ultrapurity (Morris Zief, ed.), Marcel Dekker, New York (1971)
  Lists national and international information services, information centers, books, journals, reviews, bibliographies, and data compilations dealing with ultrapurification, activation analysis, neutron, x-ray and electron diffraction, mass and optical spectometry, electron microprobe analysis, measurements of electrical properties, thermal-analysis, resonance methods, the Mössbauer effect, and field and ion emission microscopy—all with reference to the production and characterization of ultrapure inorganic solid research specimens.

Modern Analytical Techniques for Metals and Alloys
  Parts 1 and 2, Vol. 3, Techniques of Metals Research
  R. F. Bunshah, ed.
  Wiley Interscience, New York (1970), 946 pp.

Some Problems in the Analysis of Auger Electron Spectra
  T. W. Haas, J. T. Grant, and G. J. Dooley, III
  J. Vacuum Sci. Tech., 7:43-45 (1970)
  16th National Vacuum Symposium, Seattle, Wash., CONF-691006

Auger-Electron Spectroscopy of Transition Metals
  T. W. Haas, J. T. Grant, and G. J. Dooley
  Phys. Rev., 1:1449-1458 (1970)

Die Bestimmung von Kohlenstoffspuren in hochschmelzenden Metallen
  Erik Lassner
  Mikrochimica Acta (Wien), pp. 820-830 (1970)

Determination of the Light Elements in Metals: A Bibliography of Activation Analysis Papers
  G. J. Lutz, ed.
  NBS-TN-524 (May 1970), 71 pp.
  Indexed by element determined—boron, carbon, nitrogen, oxygen, phosphorus, silicon and sulfur

Mass Spectrometry: Residual Gas Analysis during Vacuum Melting
  W. E. Anable and E. D. Calvert
  (Bureau of Mines, Washington, D. C.), BM-RI-7293 (PB-186148) (Sept. 1969), 15 pp.
  Mo, Hf, Ti, and Fe

Nonferrous Metallurgy. II. Zirconium, Hafnium, Vanadium, Niobium, Tantalum, Chromium, Molybdenum, and Tungsten
  R. Z. Bachman and C. V. Banks
  Anal. Chem., 41:112R-140R (1969)
  Determination of, 821 refs.

Contribution to Activation Analysis by Charged Particles: Determination of Carbon and Oxygen in Pure Metals, Possibilities of Sulphur Determination
  Jean-Luc Debrun, Jean-Noel Barrandon, and Ph. Albert
  Natl. Bur. Std. Mod. Trends in Activation Analysis, 2:774-784 (June 1969)

Determination of Cobalt and Zinc in High-Purity Niobium, Tantalum, Molybdenum and Tungsten Metals by Atomic-Absorption Spectrophotometry after Separation by Extraction
  E. M. Donaldson, D. J. Charette, and V. H. E. Rolko
  Talanta, 16:1305-1310 (1969)

Zur Bestimmung von Sauerstoff, Stickstoff und Wasserstoff in hochschmelzenden Metallen. III. Ermittlung optimaler Analysenbedingungen für die Stickstoff- und Sauerstoffbestimmung in Niob und Tantal nach dem Vakuumextraktionsverfahren und durch Vergleich mit anderen Methoden
  K. Friedrich, E. Lassner, T. Kraus, G. Paesold, and F. Schlat
  Hochtemperatur-Werkstoffe, 6. Plansee-Seminar, June 24-28, 1968, Reutee/Tirol (1969), pp. 1016-1027

Electron-Microprobe Analyser for Very Light Element. Application to Study of Nitrogen and Oxygen Distributions in Various (Metal) Compositions
  Pierre Lanusse, Claude Le Pennec, Francoise Pichoir, and Guy Roy
  Récherche Aerospatiale, 132:33-48 (1969)
  Cr and Nb

Determination of Oxygen in High Melting Point Materials by Fast Neutron Activation
  Haruo Muto and Yohichi Gohshi
  Japan Analyst, 18:600-603 (1969)

Activation Analysis with 14 MeV Neutrons
  A. F. Voigt, R. G. Clark, and W. A. Stensland
  Annual Summary Research Report Ceramic and Mechanical Engineering, Chemical Engineering, Chemistry, Mathematics and Computer Science, Metallurgy, Physics, and Reactor Divisions, Ames Laboratory, Ames, Iowa, Contract W-7405-eng-82, July 1, 1968-June 30, 1969, IS-2100 (July 1969)

Determination of Oxygen in Titanium and Refractory Metals by Activation Analysis
  G. H. Anderson
  Nucleonics in Aerospace (Paul Polishuk, ed.), Plenum Press, New York (1968), pp. 317-322
  (see CONF-670714, 2nd Intern. Symp. on Nucleonics in Aerospace, Columbus, Ohio)

Columbium Chromatographic Separation of Niobium, Tantalum, Molybdenum, and Tungsten
  J. S. Fritz and L. H. Dahmer
  Anal. Chem., 40:20-23 (1968)

Determination of Traces of Nitrogen in Refractory Metals and Alloys by Hydrofluoric Acid-Phosphoric Acid-Potassium Dichromate Decomposition and Indophenol Photometry
  Silve Kallmann, Everett W. Hobart, Hans K. Oberthin, and Walter C. Bienza, Jr.
  Anal. Chem., 40:332 (1968)

The Direct Determination of Fluoride in Molybdenum, Rhenium, and Tungsten
  Bruce A. Raby
  (Lawrence Radiation Lab., Univ. California, Livermore), UCRL-50522 (Oct. 1968), 18 pp.

Evaluation of the Vacuum Fusion Method for the Determination of Oxygen and Nitrogen in the Refractory Metals
  E. S. Tankins
  (Naval Air Warminster Development Ctr., Johnsville, Pa.), Met. Soc. Conf., 34, Pt. 1, 325-34 (1968)
  From Conf. on High Refractory Metals, New York. See CONF-650227-(Pt. 1)

Determination of Oxygen in Niobium and Titanium by Fast Neutron Activation
  A. V. Andreev, I. Ya. Barit, and I. M. Pronman
  Zavod. Lab., 33:1105-07 (1967)

Activation Analysis: Cockcroft-Walton Generator Nuclear Reactor, LINAC, June 1966-June 1967
  James R. Devoe, ed.
  NBS Tech. Note 428, Nov. 1967

Zur Bestimmung von Sauerstoff, Wasserstoff und Stickstoff in hochschmelzenden Metallen I. Literaturübersicht
  K. Friedrich and E. Lassner
  J. Less-Common Metals, 13:156-170 (1967)
  Comparison of methods; 106 refs.

Trace Characterization (Chemical and Physical)
  W. W. Meinke and B. F. Scribner, eds.
  NBS Monograph 100 (April 1967), 570 pp.
  Based on lectures and discussions of the 1st Materials Research Symposium held at Gaithersburg, Md., Oct. 3-7, 1966. Electrochemical methods; optical and x-ray spectroscopy; x-ray diffraction; optical methods; chemical spectrophotometry; nuclear methods; mass spectroscopy; preconcentration; sampling and reagent; and electron and optical microscopy

Indexes to the Oak Ridge National Laboratory Master Analytical Manual (1953-1966)
  Helen P. Raaen and Ann Klein Haas, eds.
  (Oak Ridge National Lab., Oak Ridge, Tenn.), Contract W-7405-eng-26, TID-7015 (Indexes) Rev. 4 (1967)
  Loose leaf supplemental sheets issued annually
  Key word index composed from method titles, a bibliographic index that is equivalent to a table of contents of the Manual, an author index, and method-number cross indexes; methods: ionic, radiochemical, spectrographic, mass spectrometric, activation analysis

Possibilities of Oxygen Determination in Zr, Mo, Hf and W by Irradiation in He-3 and He-4 Particles
  G. Revel and P. Albert
  Practical Aspects of Activation Analysis with Charged Particles (M. G. Ebert), European Communities, Brussels (1968), pp. 261-265
  (2nd Conf. Practical Aspects of Activation Analysis with Charged Particles, Liège (Sept. 21-22, 1967)
  Zr, Mo, Hf, W

Absorption Spectra of Zr, Nb, Mo, Pd and Ag in the Ultrasoft X-Ray Region
  I. I. Zhukova, V. A. Fomichev, and T. M. Zimkina
  Izv. Akad. Nauk SSSR, Fiz., 31:952-956 (1967)

Analysis of the New Metals Titanium, Zirconium, Hafnium, Niobium, Tantalum, Tungsten and Their Alloys
  W. T. Elwell and D. F. Wood
  Pergamon Press, New York (1966)

X-Ray Mass Absorption Coefficients
  G. Hugues and J. B. Woodhouse
  Opt. RX Microanal. IVeme Congr. Intern., Orsay, 1965, Hermann, Paris (1966)
  Cu, Ni, Ti, Zr, Pd, Ag, Ta, Au, Al

Spectrophotometric Determination of Molybdenum in Steel, Tantalum, Niobium, or Tungsten
  C. L. Luke
  Anal. Chim. Acta, 34:302 (1966)

Reference Methods for Refractory Metals Analysis
  G. L. Miller, ed.
  Advisory Group for Aerospace Research and Development, Paris, France, NP-17627 (1966), 80 pp.
  Mo, Nb, Ta, and W

Comparaison de quelques méthodes modernes d'analyses de traces d'impuretés, spectrometrie de masse, dilution isotopique, activation
  Étienne Roth, A. Cornu, and Ph. Albert
  Z. Anal. Chem., 218:24 (1966)

Determination of Nitrogen in Metals by an Isotope-Dilution Method
  K. B. Orlova and E. N. Vitol
  Zh. Analit. Khim., 20:694 (1965)

Simultaneous Determination of Oxygen and Nitrogen in Refractory Metals by the Direct Current Carbon-Arc Gas Chromatographic Technique
  Royce K. Winge and Velmer A. Fassel
  Anal. Chem., 37:67 (1965)

The Determination of Low Energy X-Ray Mass Absorption Coefficients (Thin Foil Cu, Mn, Ge, Ni, Sc, Ti, V, Fe, Co, and Zn)
  Randall W. Carter and Robert H. Rohrer
  Contract AT(40-1) 2953, TID-21542 (1964), 48 pp.

The Application of a Flash Discharge Lamp to the Determination of Impurities in Thin Films
  W. G. Guldner
  IEEE Trans. CP-11, 9 (1964)

Emission Spectrometry
Bourdon F. Scribner and Marvin Margoshes
(National Bureau of Standards, Washington, D.C.), Anal. Chem., 36:329R–343R (April 1964)

Critical Review of the Analytical Methods of the Titanium, Vanadium, and Chromium Transition Element Groups
R. Z. Bachman and C. V. Banks
Progress in Nuclear Energy. Series IX. Analytical Chemistry, Vol. 3, Part 4, MacMillan, New York (1963), 67 pp.
574 refs.

The Effect of Imperfect Vacua Upon the Purity of Evaporated Materials
G. L. Krieger
(Sandia Corp., Albuquerque, N. Mex.), Contract AT(29-1)-789, SCTM-249-61(14) (Sept. 1961, reprinted Jan. 1962), 8 pp.

Description of a New Additional Device for Castaing Microanalyser — New Application for Elementary and Structural Microanalysis
J. Lemaitre and R. Theisen
(ORGEL Program, Joint Nuclear Research Center, Ispra Establishment (Italy)—Metallurgy and Ceramics Service, Brussels), Eur 109.f (1962), 13 pp.

Ultra-High Purity Metals
Papers presented at a seminar of American Society of Metals (October 21 and 22, 1961), American Society of Metals, Metals Park, Ohio (1962)

Extension of Sensitivity in Analysis of Impurities in Solids by Mass Spectrometry
C. M. Stevens
(Argonne National Lab., Ill.), Symposium on Extension of Sensitivity for Determining Various Constituents in Metals, Special Technical Publication No. 308, Philadelphia, American Society for Testing and Materials (1961), pp. 58–68

Spectrochemical Analysis of High-Purity Metals: A Review and Bibliography of Recent Literature
Maurice J. Peterson
(Bureau of Mines, College Park, Md.), BM-IC-8039 (Jan. 1960), 52 pp.

Low-Temperature Resistance Measurements as a Means of Studying Impurity Distributions in Zone-Refined Ingots of Metals
J. E. Kunzler and J. H. Wernick
Trans. Met. Soc. AIME, 212:856 (1958)

1.b.2. Group IV

Studies on the Analytical Chemistry of Hafnium and Zirconium. Part I. A Review of Methods for the Determination of Hafnium and Zirconium in Admixture
A. Brookes and A. Townshend
Analyst, 95:529–534 (1970)

Determination of Impurities in Titanium and Titanium Dioxide by Neutron Activation Analysis. Part V. Destructive and Non-Destructive Determination of Manganese, Indium and Uranium
R. Neirinckx, F. Adams, and J. Hoste
Anal. Chim. Acta, 50:31–38 (1970)

Analytical Chemistry of Zirconium and Hafnium
S. V. Elinson and K. I. Petrov
(Ann Arbor-Humphrey Science Publishers, Inc., Ann Arbor, Mich., 1969), 251 pp.
841 refs.

Spectrochemical Analysis of Titanium and Its Alloys. I. Spectrographic Determination of Impurities in Titanium by the Point-to-Plane Low Voltage Spark Technique
Tomomi Mizuno and Tetsuo Matsumura
J. Japan. Inst. Metals, 33(8):944–948 (1969)

Neutron Activation Analysis of Trace Impurities in Titanium and Titanium Dioxide
Rudi Dominique Neirinckx
Thesis, State University of Ghent, Ghent, Belgium (1969)

Nonferrous Metallurgy. I. Light Metals: Aluminum, Beryllium, Titanium, and Magnesium
R. T. Oliver and E. P. Cox
Anal. Chem., 41:101R–11R (1969)
Metallurgical analyses reviewed; 266 refs.

Neutron Capture Gamma-Ray Studies of Silicon, Titanium and Erbium Isotopes
K. C. Tripathi
(Inst. of Physics, Goteborg Univ., Sweden), PB-188813 (1969), 16 pp.

Analysis by Nuclear Methods of Impurities Introduced in Metals During the Preparation of Surfaces; Application to Zirconium
J. Nucl. Mater., 29:144–153 (1969)

Analysis of Zirconium and Its Alloys
JAERI Report No. 4050 (1969), p. 212
Japan Atomic Energy Research Institute, Tokaimura, Nakagun, Ibaraki-ken, Japan

X-Ray Determination of Traces of Hafnium in Zirconium Metal or Traces of Zirconium in Hafnium Metal after Separation by Ion Exchange
C. L. Luke
Anal. Chim. Acta, 41:453–458 (1968)

Quantitative Determination of Oxygen in Titanium Hydride and Titanium Metal
V. R. Negina, E. A. Kozyreva, and Z. S. Bogatova
Zh. Anal. Khim., 23:563–567 (1968)

Determination of Impurities in Titanium and Titanium Dioxide by Neutron Activation Analysis I. Simultaneous Determination of 16 Trace Elements in Titanium
R. Neirinckx, F. Adams, and J. Hoste
Anal. Chim. Acta, 43:369–380 (1968)

Determination of Small Amounts of Manganese in Metallic Titanium
V. N. Tikhanov
Ind. Lab., 34:1744 (1968)

Separation of Zirconium and Hafnium, and the Preparation of Hafnium Dioxide by the Methods of Selective Extraction of the Thiocyanate from Ethyl Acetate
I. V. Vinarov, E. S. Gertsenshtein, G. I. Beek, and I. E. Kovaleva
Ukr. Khim. Zh., 34:153–155 (1968), LIB-Trans-269, 10 pp. (V. Gavriloff, translator)
Zr and Hf

[180]Hf for Activation Analysis of Hafnium
A. Boudin and F. Hanappe
Radiochim. Acta, 8:188–191 (1967)

Internal Standard Techniques for Determination of Oxygen in Mg, Steel, and Ti by Activation Analysis
B. L. Twitty and K. M. Fritz
Anal. Chem., 39:527–529 (1967)

Modern Methods of Separating Zirconium and
Hafnium
   I. V. Vinarov
   Russian Chem. Rev., 36:522-536 (1967)

Application of Photoactivation to the Determination of Germanium in Titanium
   Shigeki Abe
   Anal. Chem., 38:1622 (1966)

Analytical Chemistry of Zirconium and Hafnium.
Series: Analytical Chemistry of the Elements
   S. V. Elinson and K. I. Petrov
   IPST-2124; TT-67-27074 (1965), 206 pp.

The Analysis of Pure Zirconium by Neutron
Radioactivation
   E. P. Mignonsin and P. Albert
   Bull. Soc. Chim. France, 2:553 (1965)

Determination of Microgram Quantities of Chloride in High Purity Titanium by X-Ray Spectrochemical Analysis
   J. S. Rudolph and R. J. Nadalin
      (Westinghouse Research Laboratories, Beulah Road, Churchill
      Boro, Pittsburgh 35, Pa.), Anal. Chem., 36:1815 (1964)

L'analyse systematique du zirconium après irradiation dans les neutrons
   L. Fournet
   Ann. Chim., 7:763-784 (1962)

Precipitation of Zirconium Hydride in Alpha
Zirconium Crystals
   D. G. Westlake and E. S. Fisher
      (Argonne National Laboratory, Argonne, Illinois), Trans. AIME,
      224:254 (1962)

Fluorescence Analysis of Trace Amount of Hafnium in Zirconium Using a Silicon Crystal
   J. C. Parks, Jr., D. G. Plackmann, and G. H. Beyer
      Advances in X-ray Analysis, Vol. 4 (William M. Mueller, ed.),
      Plenum Press, New York (1961), pp. 488-494

Etude de la résistivité électrique du zirconium.
Application à l'étude de la récristallisation du
métal après écrouissage. Influence de la pureté
   L. Renucci, J. P. Langeron, and P. Lehr
   Mem. Sci. Rev. Met., 58:699 (1961)

Methods of Determining Impurities in Zirconium
   Yu. A. Chernikhov and B. M. Dobkina
   TT-837 (Translated by D. A. Sinclair) Zavodskaya Lab., 22:
      1019-1024 (1956), 18 pp.

Trace Element Determination in Niobium and Zirconium Metal by Radioactivation Analysis
   J. F. Emery, W. T. Mullins, L. C. Bate, and G. W. Leddicotte
   (Oak Ridge National Lab., Tenn.), TID-7629, pp. 239-244

1.b.3. Group V

Zur Bestimmung von Sauerstoff, Wasser, und
Stickstoff in hochschmelzenden Metallen. V. Vergleichende Untersuchungen zur Sauerstoff- und
Stickstoffbestimmung in Niob und Tantal
   K. Friedrich, E. Lassner, and G. Paesold
   J. Less-Common Metals, 22:429-438 (1970)
   Nb and Ta

Zur Bestimmung von Sauerstoff, Wasserstoff und
Stickstoff in hochschmelzenden Metallen. IV.
Ursache der besonderen Eignung von Platinschmelzen fur die Sauerstoffbestimmung in Niob-
und Tantal und Abhängigkeit der Analysenwerte
von der Art der Probenvorbereitung
   E. Lassner and G. Kraft
   J. Less-Common Metals, 22:83-90 (1970)
   Nb and Ta

Spectrophotometric Determination of a Trace of
Molybdenum in Tantalum with Zn-Dithiol
   T. Yamane, K. Iida, T. Mukoyama, and T. Fukasawa
   Japan. Anal., 19:808-812 (1970)

Determination of Niobium by Photon-Activation
Analysis Based on Internal-Reference Method
   Yoshinaga Oka, Tong-chuin Pung, and Tatsuya Saito
   Bull. Chem. Soc. Japan, 43:1083-1087 (1970)

Zur Bestimmung von Sauerstoff, Stickstoff und
Wasserstoff in hochschmelzenden Metallen. III.
Ermittlung optimaler Analysenbedingungen für
die Stickstoff- und Sauerstoffbestimmung in Niob
und Tantal nach dem Vakuumextraktionsverfahren
und durch Vergleich mit anderen Methoden
   K. Friedrich, E. Lassner, T. Kraus, G. Paesold, and F. Schlat
   Hochtemperatur-Werkstoffe. 6. Plansee-Seminar, June 24-28,
      1968, Reutte/Tirol (1969), pp. 1016-1027
   Nb and Ta

The Destructive and Nondestructive Analysis of
Vanadium for Carbon by Gamma Activation
   J. S. Hislop and D. A. Wood
      (Atomic Energy Research Establishment, Harwell, England),
      AERE-R-6165 (Aug. 1969), 18 pp.

Inhomogeneous Distribution of Tantalum and Tungsten in Zone-Refined Niobium
   R. E. Reed
      Oak Ridge National Laboratory's Radiation Met. Section, Solid
      State Division (Feb. 1969), pp. 3-9
      Autoradiographs, using β-emission, neutron activated

Nondestructive Neutron Activation Analysis for
Trace Impurities in Niobium
   F. A. Rodriguez, S. J. Gage, and K. M. Ralls
      Proc. of Seventh Symposium on Nondestructive Evaluation of
      Components and Materials in Aerospace, Weapons Systems,
      and Nuclear Applications, April 23-25, 1969, San Antonio,
      Texas, Western Periodicals Co., North Hollywood, Calif.
      (1969), pp. 373-378

Spectrochemical Determination of Yttrium in
Tantalum Metal and Tantalum-Tungsten Alloys
   Oliver R. Simi and Robert T. Phelps
      (Los Alamos Scientific Lab., New Mexico), Contract
      W-7405-Eng-36, LA-4127 (May 19, 1969), 10 pp.
      1 to 100 ppm

Separation of Niobium and Tantalum by Reaction
of the Metal Pentachlorides with Calcium Oxide
and Calcium Fluoride
   J. E. Conway and F. D. Stevenson
   J. Less-Common Metals, 14:303-313 (1968)

Thermodynamic Analysis of the Solid Solutions of
Carbon, Nitrogen, and Oxygen in Niobium and
Tantalum
   E. Fromm
      ORNL-tr-1942; translated from J. Less-Common Metals, 14:
      113-125 (1968), 16 pp.

Neutron-Activation Determination of Tantalum in Niobium and of Niobium in Tantalum
E. N. Gil'bert, V. A. Pronin, and V. G. Torgov
Radiokhimiya, 10:500-501 (1968)

Standard tests for Gas Determination in Nb
K. M. Gusev and L. V. Volkova
In: Methods of Gas Determination in Metals, Nauka, Moscow (1968), pp. 61-64

A Spectral Analysis Method for the Detection of Calcium and Magnesium in Tantalum
K.-H. Herbst
Spectrochim. Acta, Part B, 23:489-493 (1968)

Spectrophotometric Determination of Traces of Chromium in Pure Tantalum with Diphenyl-carbazide after Prior Isolation as Acetylacetonate
Andreas Hofer and Rudolf Heidinger
Z. Anal. Chem., 233:415-418 (1968)

Improvements of the Sensitivity of Molybdenum and Tungsten Determination in Niobium and Tantalum
S. Kallmann, E. W. Hobart, and H. K. Oberthin
Talanta, 15:982-985 (1968)

Determination of Impurities in Nb
N. N. Kuznecova and L. S. Kraus
Metalloved. Term. Obr. Metallov, No. 5, pp. 66-72 (1968)

Evaluation of Sampling Reproducibility of Nitrogen and Oxygen in Niobium Rods
K. A. Snesarev, I. M. Blokh, and Yu. A. Karpov
In: Methods of Gas Determination in Metals, Nauka, Moscow (1968), pp. 64-67

Determination of Nitrogen in Pure Niobium by Neutron Activation at 14 MeV
I. Ya. Barit, B. S. Kudinov, R. M. Musselyan, and I. M. Pronman
Zavod. Lab., 33:1108-1109 (1967)

Zur Bestimmung von Sauerstoff, Wasserstoff und Stickstoff in hochschmelzenden Metallen. II. Ermittlung optimaler Analysenbedingungen für die Sauerstoffbestimmung in Molybdän nach dem Vakuumextraktionsverfahren
K. Friedrich and E. Lassner
J. Less-Common Metals, 13:171-178 (1967)

Determination of Impurities in High-Purity Niobium
N. N. Kuznetsova and L. S. Krauz
High-Purity Metals and Alloys. Fabrication, Properties, and Testing (V. S. Emel'yanov and A. I. Evstyukhin, eds.), Consultants Bureau, New York (1967), pp. 51-57

Determination of Tantalum in Metallic Niobium
L. N. Makogonova and V. I. Kurbatova
Tr. Vses. Nauch.-Issled. Inst. Stand. Obraztsov Spektral. Etalonov., 3:50-55 (1967)

Determination of Niobium and Tantalum
V. Mayer
Hutn. Listy, 22:122-126 (1967)

Interstitial Impurities in Ta and Nb
S. Miura, J. Takamura and M. Yamshita
Suiyo-kai-Shi, 16:144-156 (1967)

Examen au microscope électonique du niobium contaminé à haute température par des éléments d'insertion
H. Bibring, B. Jouffrey, and F. Girard
Mem. Sci. Rev. Met., 63:59 (1966)

Segregation of Tantalum at Very Low Concentrations in Niobium by Controlled Solidification: A Neutron-Activation Study
F. H. Cocks, R. M. Rose, and J. Wulff
J. Less-Common Metals, 10:157 (1966)

Determination of Small Amounts of Molybdenum in Niobium and Tantalum by Atomic-Absorption Spectroscopy in a Nitrous Oxide-Acetylene Flame
G. F. Kirkbright, M. K. Peters, and T. S. West
Analyst, 91:705-708 (1966)

Determining Impurities in High-Purity Niobium
N. N. Kuznetsova and L. S. Krauz
Metallovedenie i Termicheskaya Obrabotka Metallov
Atomizdat, Moscow (1966), pp. 66-72

Determination of Impurities in Tantalum and Niobium by Neutron Radioactivation
G. Aubouin, F. Dugain, and J. Laverlochere
Bull. Soc. Chim. France, 2:547 (1965)

Determination of Oxygen in Niobium, Tantalum, and Beryllium
J. Bril and F. Dugain
Bull. Soc. Chim. France, 2:562 (1965)

Determination of Common Impurities in Niobium Metal and Oxide by the Powder-D.C. Arc Technique
Leroy S. Brooks
Spectrochim. Acta, 21:1023 (1965)

Determination of Common Impurities in Niobium Metal by the Fusion-D.C. Arc Technique
Leroy S. Brooks
Spectrochim. Acta, 21:1029 (1965)

Determination of Micro Amounts of Tantalum in Niobium by Using Neutron Activation and Gamma Spectrometry
F. Dugain and J. Laverlochere
Anal. Chem., 37:998 (1965)

The Spectrophotometric Determination of Trace Amounts of Nickel in Tantalum Metal
Ross D. Gardner, Carter H. Ward, and William H. Ashley
LA-3152, Contract W-7405-eng-36 (1964), 10 pp.

Interference-Free Determination of Impurities in Pure Niobium by Neutron Activation and Gamma Spectrometry
J. Grosel and H. Sorantin
(Osterreichische Studiengesellschaft für Atomenergie GmbH, Seibersdorf, Austria), SGAE-CH-11/1964 (1964), 20 pp.

The Spectrophotometric Determination of Microgram Quantities of Cobalt in Tantalum and Plutonium
T. K. Marshall, J. W. Dahlby, and G. R. Waterbury
Contract W-7405-eng-36, LA-3124 (1964), 16 pp.

An Electron Microscopic Investigation of Phenomena Associated with Solid Solution of Oxygen in Niobium
J. Van Landuyt
Phys. Stat. Sol., 6:957 (1964)

b. Assay Methods

Determination of Hydrogen in Niobium
R. J. Walter and H. G. Offner
Anal. Chem., 36:1779 (1964)

The Determination of Tantalum and Niobium
M. H. Cockbill
(London and Scandinavian Metallurgical Co. Ltd., London),
Analyst, 87:611-629 (1962)

Study of Effects of Impurities on the Mechanical
Properties of Niobium. Quarterly Progress Report No. 1, Jan. 1, 1962 through March 31, 1962
Torleif Lindtveit, Paul Storvik, and Ingard Kvernes
(Sentralinstitutt for Industriell Forskning, Blindern, Norway),
NP-12634 (1962), 8 pp.

Extraction and Determination of Iron As the
Bathophenanthroline Complex in High-Purity Niobium, Tantalum, Molybdenum, and Tungsten Metals
Elsie M. Penner and W. R. Inman (Department of Mines and
Technical Surveys, Ottawa 1, Ontario, Canada), Talanta, 9:
1027-1036 (1962)

Radioactivation Analysis of Niobium in Tantalum
by Irradiation with Protons in a Cyclotron
I. Tomita and H. Saisho
Nature, 195:1189-1190 (1962)

Spectrographic Determination of Impurities in
Niobium
A. A. Baskin, E. I. Zakharov, K. I. Petrov, and E. I. Rzhekhina
Zh. Anal. Khim., 16:627-630 (1961)

Methods for the Analysis of Columbium and Its
Alloys
E. W. Hobart and D. E. Fornwalt
(Pratt and Whitney Aircraft Div., United Aircraft Corp., Connecticut Aircraft Nuclear Engine Lab., Middletown), PWAC-340
(1961), 56 pp.

Determination of Hydrogen in Tantalum, Niobium,
and Uranium
A. N. Zaidel and K. I. Petrov
CEA-tr-R-640
Translated into French from Zavodskaya Lab., 24:1000-1 (1958),
5 pp.

The Determination of Oxygen, Nitrogen, and Hydrogen in Molybdenum, Tungsten, and Niobium
G. V. Mikhailova, Z. M. Turovtseva, and R. Sh. Khalitov
AEC-tr-3184
Translated by S. J. Rothman from Zh. Anal. Khim., 12:338-341
(1957), 6 pp.

Trace Element Determination in Niobium and Zirconium Metal by Radioactivation Analysis
J. F. Emery, W. T. Mullins, L. C. Bate, and G. W. Leddicotte
(Oak Ridge National Lab., Tenn.), TID-7629, pp. 239-244

The X-Ray K-Absorption Fine Structure of Niobium and Copper at Cryogenic Temperatures
Thomas Arthur Boster
Dissertation Abstr., 66-11,896

Dosage de l'oxygène et de l'azote dans le niobium par spectrographic de mass à étincelles
Georges Vidal, Pierre Galmard and Pierre Lanusse
ONERA TP 537

Round-Robin Samples for Vanadium Working Group
F. A. Schmidt and O. N. Carlson
(Ames Lab., Ames, Iowa), nd.
A summary of the analytic data obtained on the "Round-Robin"
samples for the Vanadium Working Group. The participating organizations include Ames Laboratory, Argonne National Laboratory, Battelle Memorial Institute, U. S. Bureau of Mines (Albany), Oak Ridge National Laboratory, Oregon Metallurgical Corp., Wah Chang Albany and Westinghouse ARD

1.b.4. Group VI

Microdetermination of Fluorine in High-Purity
Tungsten Metal by Photoactivation Analysis
G. Buzzelli and K. McDonald
Gulf General Atomic, San Diego, Calif., GA-10125 (Nov. 1970),
9 pp.

Determinations of Manganese, Copper, Nickel,
Tungsten and Sodium in Molybdenum with a Single
Neutron Irradiation
M. Fedoroff
Bull. Soc. Chim. France, 3:1233-1236 (1970)

Dosage du palladium dans le molybdène par activation neutronique
Michel Federoff
Compt. Rend., C 270, 486-87 (1970)

Sulfur Determination in Molybdenum by Activation
Analysis in Thermal Neutrons
M. Federoff
Compt. Rend., C 271(6):399-402 (1970)

Determination of Carbon in High Purity Metals
by Photon Activation Analysis
G. J. Lutz and L. W. Masters
Anal. Chem., 42:948-950 (1970)

Determination of the Internal and Superficial
Oxygen Content of Molybdenum Obtained Using 54
MeV Alpha Particles and Low Energy Deuterons
Lyliane Mazagol, Jean Tousset, and Michele Boissier
LYCEN-7076 (October 1970), 14 pp.

Activation Analysis of Gallium in Tungsten
A. Salamon
J. Radioanal. Chem., 4:81-86 (1970)

Radioactivation Determination of Impurities in
Yttrium and Molybdenum by Means of Extraction
Displacement Chromatography
O. V. Stepanets, Yu. V. Yakovlev, and I. P. Alimarin
Zh. Anal. Khim., 25:1906-1911 (1970)

Etude de la résistivité électrique à basse température comme critère de pureté du chrome
J. Bigot
Compt. Rend., Ser. C, 268:1035-1038 (1969)

Analytical Chemistry of Molybdenum
A. I. Busev
Analytical Chemistry of Elements, Ann Arbor-Humphrey
Science Publishers, Inc., Ann Arbor, Mich. (1969), 290 pp.

Activation Analysis Using [4]He and [3]He Ions. Determination of Trace Oxygen in High Purity Iron,
Nickel, and Chromium. Studies on Carbon Determination
J.-L. Debrun, J.-N. Barrandon, and Ph. Albert
Bull. Soc. Chim. Fr., 3:1011-1016 (1969)

Determination of Titanium in High-Purity Molybdenum and Tungsten Metals with Diantipyrylmethane after Separation by Extraction of Its Cupferron Complex
E. M. Donaldson
Talanta, 16:1505-1512 (1969)

Rapid Determination of Tungsten and Hafnium in Pure Molybdenum
 G. V. Leushkina, E. M. Lobanov, A. G. Dutov, and N. P. Matveeva
 Zadvod. Lab., 35:715-717 (1969)

Determination of Molybdenum by Atomic Absorption Spectroscopy
 T. V. Ramakrishna, P. W. West, and J. W. Robinson
 Anal. Chim. Acta, 44:437-439 (1969)

Determination of Oxygen in Molybdenum Samples by Activation with Neutrons with Energies of 14 MeV
 S. R. Abdurakhmanova, V. A. Kireev, L. V. Navalikhin, and Yu. N. Talanin
 Zh. Anal. Khim., 23:1188-1191 (1968)

Chemical Vapor Deposition of Tungsten
 J. Chin and J. Horsley
 (Gulf General Atomic, San Diego, California), Contract AT (04-3)-167, GA-8772; CONF-681059-1. Presented at IEEC Thermionic Conversion Specialist Conference, Boston, October 21-24, 1968 (Oct. 1968), 15 pp.
 Principal problems with activation analysis of fluorine are caused by gamma attenuation correction factors and a lack of suitable standards

Determination of Impurities in Molybdenum and Tungsten by Activation Analysis
 H. G. Doege and H. Grosse-Ruyken
 Isotopenpraxis, 4:262-268 (1968)

Analysis of Chromium by γ-Spectrometry After Separation of the Radioelements by Ion Exchange
 F. Dugain, C. Castre, and B. Beyssier
 Anal. Chim. Acta, 42:39-50 (1968)

Determination of Aluminium in Molybdenum by Neutron Activation
 M. Fedoroff
 Compt. Rend., 267(19):1227-1229 (1968)

Spectrographic Determination of Tungsten in Molybdenum After Chemical Concentration
 A. D. Khlystova
 Zh. Anal. Khim., 23:211-213 (1968)

Quantitative Spectrographic Determination of Hafnium in Tungsten by a Vacuum Cup Solution Method
 Edgar S. Peck
 Anal. Chem., 40:324 (1968)

Bureau of Mines Research on the Analysis of High-Purity Tungsten
 Staff (Bureau of Mines, Pittsburgh, Pa.), BM-IC-8397, Oct. 1968

Spectral and Spectrochemical Methods for the Determination of Impurities in Tungsten
 N. I. Tarasevich, A. A. Zheleznova, L. V. Moreiskaya, and O. Kh. Kogan
 Vestn. Mosk., Ser. II. Khim., No. 1, 85-89 (1968)

Determination of Combined Nitrogen in Tungsten Metal Powder
 S. P. Awasthi, S. Sahasranaman, and M. Sundaresan
 Analyst, 92:650-652 (1967)

Zur Bestimmung von Sauerstoff, Wasserstoff und Stickstoff in hochschmelzenden Metallen. I. Literaturübersicht
 K. Friedrich and E. Lassner
 J. Less-Common Metals, 13:156-170 (1967)

Comparison of Analytical Methods for Oxygen in Chromium Metal
 E. Lassner
 Z. Erzberg u. Metallhüttenwesen, 20(10):466-468 (1967)

Chemico-Spectral Method of Determining 10^{-4} Percent Metal Impurities in Pure Chromium, Using Ion-Exchange Chromatography
 P. V. Marchenko, A. I. Voronina, and M. L. Kaplan
 Ukr. Khim. Zh. 33(8):838-842 (1967)

Structure des films de tungstène
 Edouard Miska and Marcel Gillet
 Compt. Rend., Ser. B, 264:1267-1270 (1967)

Spectrophotometric Determination of Tungsten in Molybdenum
 B. Neef and H. G. Doge
 Talanta, 14:967-972 (1967)

Effect of the Successive Appearance of the Spectral Lines of Molybdenum with Different Excitation Potentials
 A. G. Orlov
 Zh. Prikl. Spektrosk., Akad. Nauk USSR, 7:416-418 (Sept. 1967)

Application de l'analyse par activation neutronique à l'étude des impuretés dans le molybdène, le tungstène et de la graphite nucléaire
 Guy Pinte
 (Commissariat a l'Energie Atomique, Saclay, France. Centre d'Études Nucléaires), CEA-R-3267 (Oct. 1967), 76 pp.

Determination of Aluminum and Sodium in Tungsten by Nondestructive Activation Analysis
 P. Quittner, A. Simonits, and A. Elck
 Talanta, 14:417-420 (1967)

Determination of Nitrogen in Chromium Metal by Wet Distillation
 H. Rothman
 Z. Erzberg u. Metallhüttenwesen, 20(10):469-470 (1967)

Determination of Oxygen and Hydrogen in Tungsten and Other Metals by a Vacuum-Fusion Gas-Chromatographic Method
 D. F. Wood and G. Wolfenden
 Anal. Chim. Acta, 38:385-402 (1967)

Electron Microscopic and Microdiffraction Investigation of Nonmetallic Impurities in Molybdenum and Its Alloys
 N. V. Ageev, D. V. Ignatov, and M. M. Kantor
 FTD-HT-66-695, AD-647-729
 Dokl. Akad. Nauk SSSR 167(3):635-636 (1966)
 Dokl. — Chem. Tech., 167(3):46-48 (1966)

Determination of Impurities in Molybdenum by Emission Spectroscopy
 E. E. Albin and W. F. Morris
 (Lawrence Radiation Lab., Univ. of California, Livermore), Gen. Chem. Tech. Note 245, UCID-15039 (Oct. 5, 1966)

Determination of Oxygen in Mo by Neutron Activation
 A. V. Andreev, I. Ya. Barit, and R. M. Musaelyan and others
 Zh. Anal. Khim., 21(12):1453-1456 (1966)

Spektrophotometrische und aktivierungsanalytische Bestimmung von Verunreinigungen in Molybdän und Wolfram
 H. G. Doge, G. Ehrlich, H. Grosse-Ruyken, O. Grossmann, and B. Neef

c. Structure

(Deut. Akad. Wiss. Berlin, Inst. Metallphysik und Reinstmetalle, Dresden; Tech. Univ. Dresden, Inst. Anorganische und Anorganisch-Technische Chemie), Reinststoffanalytik, 2. Internationales Symposium Reinststoffe in Wissenschaft und Technik, Dresden, Sept. 28–Oct. 2, 1965 (1966)

A Study of Spectrochemical Detection Limits of Selected Elements in Tungsten
R. Dyck
Anal. Chem., 37:1046 (1965)

Preparing Infiltrated Tungsten for Examination
Robert E. Matt and Isamu Yoshioka
Metal Progr., 85(4):95–96 (Apr. 1964)

Determination of Trace Amounts of Molybdenum and Niobium in Tungsten Using a Small Anion Exchange Column
Ken F. Sugawara
(Air Force Materials Lab., Wright-Patterson AFB, Ohio), AD–608074; ML–TDR–64–268 (1964), 16 pp.

Extraction and Determination of Iron as the Bathophenanthroline Complex in High-Purity Niobium, Tantalum, Molybdenum, and Tungsten Metals
Elsie M. Penner and W. R. Inman
(Department of Mines and Technical Surveys, Ottawa 1, Ontario, Canada), Talanta 9:1027–1036 (1962)

Spectrochemical Analysis of High-Purity Tungsten
R. W. Lewis, C. F. Earl, J. L. Potter, and J. R. Wells
(Bureau of Mines, Boulder City Metallurgy Research Lab., Nev. and Bureau of Mines, Reno Metallurgical Research Center, Nev.), BM–RI–5814, (Jan. 1961), 14 pp.

Semiquantitative Spectrographic Analysis of Tungsten
M. J. Peterson and C. L. Chaney
(Bureau of Mines, College Park, Md.), BM–RI–5903 (Feb. 1961), 20 pp.

Analyse spectrographique de la teneur en silicium, en aluminium et en fer dans le chrome métallique
(Spectrographic Analysis of the Content of Silicon, Aluminum, and Iron in Metallic Chromium)
I. A. M. Kalinskii (Kalinsky)
CEA–tr–R–602
(Translated into French by B. de Trezvinsky) Zavodsk. Lab., 24:755 (1958)

The Detection of Impurities and Cold-Working in Sintered Tungsten by the Measurement of Residual Resistance
Erich Krautz and Herman Schultz
AEC–tr–5315
Z. Metallk., 49:399–403 (1958)

The Determination of Oxygen, Nitrogen, and Hydrogen In Molybdenum, Tungsten, and Niobium
G. V. Mikhailova, Z. M. Turovtseva, and R. Sh. Khalitov
AEC–tr–3184
Translated by S. J. Rothman from Zh. Anal. Khim., 12:338–341 (1957), 6 pp.

Determination of Trace Element in W and Mo by Atomic Absorption Spectrometry
J. H. Muntz and L. L. Roush
AFML–TR–66339

Determination of Trace Elements in Tungsten and Molybdenum by Atomic Absorption Spectrophotometry
J. H. Muntz and L. L. Roush
CONF–660611–7; ORAU, Gmelin, AED–CONF–66–172–9 (from Society of Applied Spectroscopy Meeting, Chicago, Ill.), 13 pp.

Intersite Exchange of Tungsten Standard Samples
R. S. Wood, Jr.
(E. I. Du Pont de Nemours and Company, Savannah River Plant, Aiken, S. C.), DP–904, p. 29

Determination of Trace Impurities in Molybdenum by Spark-Source Mass Spectrography
Georges Vidal, Pierre Galmard, and Pierre Lanusse
ONERA TP 563

1.c. Structure

1.c.1. General

Electron Structure and Crystal Structure of Transition Metals
V. K. Grigorovich
In: Structure and Properties of Heat-Resistant Metals and Alloys, Israel Program for Scientific Translations, Jerusalem (1970), pp. 61–70, N70–2597512–17

Formation de la texture dans les métaux déposé sous vide
L. M. Gert and A. A. Barad-Zakhryapin
Fiz. Metallov Metalloved., SSR 28(2):375–377 (1969)
Nb, Zr, Ti

Interplanar Angles and Standard Projections for H.C.P. Metals (Cd, Zn, Mg, Re, Ti, Zr, Tl, Ru, Hf, Os, Yt, and Be)
R. K. Govila
TMS Paper Selection No. A69–12, Metallurgical Society of AIME, 345 E. 47th St., New York, N. Y. 10017 (1969), 27 pp.

Vacancies and Interstitials in Body-Centred Cubic Transition Metals
J. Nihoul
Interstitials and Vacancies in Metals, North-Holland, Amsterdam (1969), pp. 839–888

Twinning and Low Temperature Properties of Nb and Mo
B. J. Shaw
AD–681 864 (1969)

Computer Program for Calculating Interplanar Angles and Indexing Back-Reflection Laue Films of Hexagonal Crystals
B. A. Cheadle and F. E. Lane
Trans. AIME, 242:1488 (1968)

Substruktur in Mo, W, Ta
V. O. Esin and T. V. Ushkova
Wachstum und Fehler in Kristallen (1968), pp. 220–227

Effects of Elastic Anisotropy on Dislocations in Hcp Metals
E. S. Fisher and L. C. R. Alfred
Trans. AIME, 242:1575–1586 (1968)

Variation de la structure cristalline dans les couches superficielles écrouies de W, Nb, et Mo lors du recuit stationnaire sous vide
O. I. Kozyrskii and L. V. Tikhonov
Fiz. Metallov Metalloved., SSSR, 26(6):1084–1090 (1968)

Structure Defects in Metal Single Crystals Grown from Melts
A. A. Kralina
Kristallografiya, 12(5):954–955 (1967)
Sov. Phys. — Cryst., 12(5):834–836 (1968)

Crystal Structure and Thermal Expansion of Scandium, Titanium, Vanadium, and Manganese
Norbert Schmitz-Pranghe and Philipp Duenner
Z. Metallk., 59:377–382 (1968)

Determination of Patterns of Macromosaic Substructure in Metal Single Crystals by X-ray Diffraction Microscopy
V. O. Esin and T. V. Ushkova
Kristallografiya, 12(1):136–137 (1967)
Sov. Phys. — Cryst., 12(1):108–110 (1967)

A Handbook of Lattice Spacings and Structures of Metals and Alloys, Vol. 2
W. B. Pearson
Vol. 8 of International Series of Monographs on Metal Physics and Physical Metallurgy (G. V. Raynor, ed.), Pergamon Press, New York; London; Paris (1967), 1446 pp.

The 4f and 5f Elements
M. Haissinsky and C. K. Jorgensen
ORNL-tr-1636
J. Chem. Phys., 63:1135–1138 (1966)

The Lattice Stability of Metals. V. Vanadium, Chromium, Cobalt, and Nickel
Larry Kaufman
(ManLabs, Inc., Cambridge, Mass.), Contract Nonr-2600(00), TR-12; AD-628985, (Feb. 1966), 25 pp.

X-Ray Diffraction
Lynne L. Merritt, Jr., and William E. Streib
(Indiana Univ., Bloomington), Anal. Chem., 36:399R–404R (April 1964)

A Critical Review of Refractories
Edmund K. Storms
(Los Alamos Scientific Lab., N. Mex.), Contract W-7405-eng-36, LA-2942 (March 1964), 252 pp.

The Electrical Resistivity of Dislocations
Z. S. Basinski, J. S. Dugdale, and A. Howie
(Division of Pure Physics, National Research Council, Ottawa, Canada), Phil. Mag., 8:1989 (1963)

Absorption Contrast Effect in BCC Metals
S. E. Bronisz and D. L. Douglas
(University of California, Los Alamos Scientific Laboratory, Los Alamos, New Mexico), J. Appl. Phys., 34:713 (1963)

Thermally Activated Dislocation Motion in FCC and Refractory BCC Metals
R. H. Chambers
Appl. Phys. Letters, 2:165 (1963)

$\{11\bar{2}1\}$ Twinning (Zr)
A. G. Crocker
(Physics Department, Battersea College of Technology, London, S. W. 11), Phil. Mag., 8:1077 (1963)

On the Dynamical Theory of Electron Transmission Microscope Images of Dislocations and Stacking Faults
R. Gevers
(Solid State Physics Department, S. C. K. — C. E. N., Mol), Phys. Stat. Sol., 3:415 (1963)

The Study of Crystal Imperfections in Thermal Conductivity Measurements
K. Mendelssohn
Proceedings of the International Conference on Crystal Lattice Defects, 1962, Conference
J. Phys. Soc. Japan, 18, Suppl. II (1963)

Field Ion Microscopy of the Defect Structure of Metal Crystals
E. W. Müller
(Pennsylvania State University, Pennsylvania, U.S.A.), Proc. Intern. Conf. on Crystal Lattice Defects, 1962, J. Phys. Soc. Japan, 18, Suppl. II, 1 (1963)

Défauts ponctuels dans les solides ferromagnétiques et ordre directionnel
L. Neel
(Faculté des Sciences, Grenoble), J. Phys., 24:513 (1963)

Electron Microscopy of Thin Foils
G. Thomas
(California. Univ., Berkeley; Lawrence Radiation Lab.), Contract W-7405-eng-48
UCRL-11009 (Oct. 18, 1963)
CONF-356-9
From 1963 American Society of Metals/Metals and Materials Show, Cleveland, Oct. 1963, 23rd National Convention of the Society for Nondestructive Testing, 73 pp.

Precipitation of Vacancies in Metals
G. Thomas and J. Washburn
(Department of Mineral Technology and Inorganic Materials Research Division, Lawrence Radiation Laboratory, University of California, Berkeley, California), Rev. Mod. Phys., 35:992 (1963)

Structure and Properties of Evaporated Metal Films
J. R. Anderson, B. G. Baker, and J. V. Sanders
J. Catalysis, 1:443 (1962)

The Influence of Defects on the Lattice Thermal Conductivity at Low Temperatures
H. Bross
NP-tr-963 (translated by L. F. Secretan), Phys. Stat. Sol., 2:481–516 (1962), 72 pp.

Selected Applications of High-Temperature X-Ray Studies in the Metallurgical Field
H. J. Goldschmidt
(The B.S.A. Group Research Centre, Birmingham, England), Advances X-Ray Analysis, Vol. 5, Plenum Press, New York, 1962, pp. 191–212
Review

Description of a New Additional Device for Castaing Microanalyser—New Applications for Elementary and Structural Microanalysis
J. Lemaitre and R. Theisen
(ORGEL Program, Joint Nuclear Research Center, Ispra Establishment (Italy)—Metallurgy and Ceramics Service, Brussels), Eur 109.f (1962), 13 pp.

The Effects of Chemical Impurities and Physical Imperfections on the Thermoelectricity of Metals
W. B. Pearson
Chap. 10 of Ultra-High-Purity Metals, Seminar of Am. Soc. for Metals (Oct. 21 and 22, 1961), ASM, Metals Park, Ohio (1962)

Direct Observation of Crystal Imperfections by Field Ion Microscopy
E. W. Müller
(The Pennsylvania State University, University Park, Pennsylvania), Imperfections in Crystals (eds. J. B. Newkirk and J. H. Wernick), Interscience Publishers, New York (1961), p. 77

Growth from the Melt. I. Influence of Surface Intersections in Pure Metals
G. F. Bolling and W. A. Tiller
(Westinghouse Research Laboratories, Churchill Borough, Beulah Road, Pittsburgh 35, Pennsylvania), J. Appl. Phys., 31:1345 (1960)

Electrical and Thermal Resistivity of Dislocations
Peter Carruthers
Phys. Rev. Letters, 2:336 (1959)

The Growth of Metal Single Crystals
R. W. K. Honeycombe
(Department of Metallurgy, University of Sheffield), Met. Rev., 4:1 (1959)

Moire Patterns on Electron Micrographs, and Their Application to the Study of Dislocations in Metals
G. A. Bassett, J. W. Menter, and D. W. Pashley
(Tube Investments Research Laboratories, Hinxton Hall, Cambridge), Proc. Roy. Soc. A, 246:345-368 (1958)

X-Ray Structural Analysis of Certain Systems of Transition Metals
Ya. Ya. Cherkashin, Ya. I. Gladyskevskii, P. I. Kripyakevich, and Yu. B. Kuz'ma
NP-tr-286, pp. 166-173
Zh. Neorg. Khim., 3:650-653 (1958), 8 pp.

Spin-Disorder Effects in the Electrical Resistivities of Metals and Alloys
B. R. Coles
Phil. Mag. Suppl., 7(25):40 (1958)

A Handbook of Lattice Spacings and Structures of Metals and Alloys
W. B. Pearson
Vol. 4 of International Series of Monographs on Metal Physics and Physical Metallurgy (G. V. Raynor, ed.), Pergamon Press, New York; London; Paris (1958), 1044 pp.

Grain Boundary Migration Motivated by Substructure in High Purity Metal Crystals
J. W. Rutter and K. T. Aust
(General Electric Research Laboratory, Schenectady, New York), Acta Met., 6:375 (1958)

Stereometric Metallography
S. A. Saltykov
NP-tr-823, Parts I and II
Stereometricheskaya Metallografiya, a publication of the State Scientific-Technical Publishing House of Literature on Ferrous and Non-Ferrous Metallurgy, Moscow (1958), 604 pp.
Second Revised and Supplemented Edition

A Method of Growing Single Crystals of Metals With Prescribed Spatial Orientation and Natural Crystallographic Faces
D. M. Chigvinadze
AERE-Lib/Trans-658 (translated by R. C. Murray), Zh. Tekh. Fiz., 25:805-811 (1955), 9 pp.

The Effect of Impurities on Mechanical Twinning and Dislocation Behavior in Body-Centered Cubic Metals
J. O. Stiegler and C. J. McHargue
ORNL-TM-542

1.c.2. Group IV

Sur la comparaison des résultats fournis par les mesures de dureté et de résistivité lors de l'étude de la restauration et de la récristallisation du zirconium pur
Marc Billion, Louis Renucci, and Jean-Paul Langeron
Compt. Rend., 270:895-898 (1970)

Crystallographic Techniques and Data for Transmission Electron Microscopy of Zirconium
A. Jostsons and J. G. Napier
Australian Atomic Energy Commission Research Establishment, Lucas Heights), AAEC/E-204 (Feb. 1970), 47 pp.

Structure and Electrical Properties of Sputtered Films of Hafnium and Hafnium Compounds
F. T. J. Smith
J. Appl. Phys., 41:4227-4231 (1970)

Effect of Ultrasonic Waves on Phase Recrystallization of Titanium
I. A. Gindin and G. N. Malik
Fiz. Metal. Metalloved., 28:1091-1094 (1969)

Interstitial Order-Disorder Transformation in the Titanium-Oxygen System
M. Hirabayashi, M. Koiwa, and S. Yamaguchi
The Mechanism of Phase Transformations in Crystalline Solids, Proceedings of an International Symposium Organized by the Institute of Metals and Held in the University of Manchester from July 3-5, 1968, Monograph and Report Series No. 33, The Institute of Metals, London (1969), pp. 207-211
Electron-diffraction and calorimetric measurements, a series of titanium-oxygen specimens with compositions from $O/Ti = 0.09$ to 0.40

Etude radiocristallographique du titane finement divisé, à différents états
A. I. Razhchenko, I. F. Martynova, V. V. Kononenko, N. F. Geletii, and L.-Yu. Tuzyak
Fiz. Metallov Metalloved. SSSR, 27(5):833-836 (1969)

Defects in Quenched Zirconium
M. L. Swanson, G. R. Piercy, G. V. Kidson, and A. F. Quenneneville
(Solid State Science Branch, Chalk River Nuclear Laboratories, Atomic Energy of Canada, Ltd., Chalk River, Ontario, Canada), AECL-3428 (Sept. 1969)

Hafnium. I. Microstructure
F. W. Vahldiek
J. Less-Common Metals, 19:83-92 (1969)

Indexing of Electron Diffraction Spots Obtained from Zirconium Lattice
A. Gardini
(Centro Informazioni Studi Esperienze, Milan, Italy), CISE-116 (June 1968), 53 pp.

Titanium Etching in Hydrochloric Acid
E. S. Zhiganova and K. P. Batashev
Zh. Prikl. Khim., 41(1):206-207 (1968)

Etude des caractéristiques structurales du zirconium polycrystallin et de la croissance granu-
laire
M. Dechamps
Thèse Doct. 3eme Cycle, Metall. spec., Paris (1967), 59 pp.

X-ray Study of Cold Work and Recovery in Some
HCP Metals and Alloys
Shrikant Lele and T. R. Anantharaman
Z. Metallk., 58:461-464 (1967)
Ti, Zr, Hf

Recrystallization, Grain Growth, and Preferen-
tial Orientation in Zone Melted Zirconium
M. Billion and J. P. Langeron
Etude sur la corrosion et la protection du zirconium et de ses
alliages (M. Saleese and G. Chaudron, eds.), Centre d'études
Nucléaires de Saclay, Gif-sur-Yvette, France, 1966 and
Presses Universitaires de France, Paris, (1966), pp. 97-105

Etching Dislocations in Zirconium
D. Mills and G. B. Craig
Electrochem. Tech., 4:300-303 (1966)

Sur un aspect superficiel présenté par le zir-
conium recuit en phase β
Louis Renucci and Jean-Paul Langeron
Compt. Rend., Ser. C, 263:39-41 (1966)

The Preparation of Titanium for Transmission
Electron Microscopy
L. A. Rice and R. N. Orava
Trans. AIME, 236:1502 (1966)

Properties of Titanium and Rutile (TiO_2) Thin
Films
F. O. Zrntz
AFML-TR-66-122 (March 1966)

A Study of Cold Rolling Titanium by Transmis-
sion Electron Microscopy
Toshimi Yamane and Jitsuhiko Ueda
(Technical Research Laboratory, Hitachi Shipbuilding and
Engineering Co., Ltd., Sakurajima, Osaka, Japan), Trans.
Japan Inst. Metals, 5:43 (1964)

The Physical Metallurgy of Zirconium
D. L. Douglass
(General Electric Co., Pleasanton, California), At. Energy
Rev., 1:71-237 (1963)

Crystal Structures of Titanium, Zirconium, and
Hafnium at High Pressures
John C. Jamieson
(Univ. of Chicago, Chicago, Illinois) Science, 140:72-73 (1963)

Solid-Solid Transitions in Titanium and Zirco-
nium at High Pressures
A. Jayaraman, K. Klement, Jr., and G. C. Kennedy
(Institute of Geophysics and Planetary Physics, University of
California, Los Angeles, California), Phys. Rev., 131:644
(1963)

Zig-Zag Twins in Zirconium
R. E. Reed-Hill and E. R. Buchanan
Acta Met., 11:73-75 (L) (1963)

Preferred Orientation and Anisotropy in Titanium
W. T. Roberts
(Research Department, I.C.I. Metals Division, Witton, Birming-
ham, Great Britain), J. Less-Common Metals, 4:345 (1962)

Deformation Modes of Zirconium at 77°K, 575°K,
and 1075°K
E. J. Rapperport and C. S. Hartley
(Nuclear Metals, Inc., Concord, Mass.), Contract AT(30-1)-
1565, NMI-1243 (Oct. 6, 1960), 37 pp.

Exposure Diagrams for X-Ray Film Work on Titanium and Zirconium
K. Sagel
AEC-tr-3128 (translated by K. S. Bevis) Metall., 11:769 (1957),
2 pp.

1.c.3. Group V

X-ray Determination of Phonon Dispersion in
Vanadium
R. Colella and B. W. Batterman
Phys. Rev., B 1:3913-3921 (1970)

Dislocation Dynamics in Niobium Single Crystals
K. R. Evans and E. B. Schwenk
Acta Met., 18:1-8 (1970)

A Phase Transition in Vanadium
V. A. Finkel', V. I. Glamazda, and G. P. Kovtun
Zh. Éksper. Teor. Fiz., SSSR, 57(4):1065-1068 (1969)
Sov. Phys. — JETP, 30(4):581-583 (1970)

Effects of Low-Temperature Rolling and Anneal-
ing on the Structure of Niobium Single Crystals
I. A. Gindin et al.
Fiz. Metal. Metalloved., 30(2):426-431 (1970)

Oxidation of Thin Niobium Films at Low Temper-
atures
V. V. Klechkovskaya
Kristallografiya, 15(2):358-361 (1970)
Sov. Phys. — Cryst., 15(2):299-301 (1970)

Clustering of Zr and N in Nb
D. Mosher, C. Dollins, and C. Wert
Acta Met., 18:797-803 (1970)

Interstitial Ordering in Tantalum
P. Rao and G. Thomas
Scripta Metallurgica, 4(4):243-244 (1970)

Die Textur von Molybdänabscheidungen aus der
Gasphase
G. Weise and H. Wadewitz
J. Crystal Growth, 7:313-316 (1970)

Inelastic Scattering of Neutrons by Localized
Vibrations of Interstitial Hydrogen in Metal
Lattices: Case of a Vanadium Lattice
G. Blaesser, J. Peretti and G. Toth
(European Atomic Energy Community, Ispra, Italy), EUR-4216
(1969), 41 pp.

X-Ray Study of the Precipitation of Hydrogen in
Vanadium
A. Bonfiglioli and B. W. Batterman
(Cornell Univ., Materials Science Ctr., Ithaca, N. Y.)
NYO-3789-8 (Aug. 1969), 14 pp.

Preparation of Thin Vanadium Foils for Transmission Electron Microscopy
J. Bressers and P. Helbach
J. Sci. Instr. (J. Phys. E), Series 2, Vol. 2:828-830 (1969)

A Detailed Study of the Deformation of High Purity Niobium Single Crystals
M. S. Duesbery and R. A. Foxall
Phil. Mag., 20:719-751 (1969)

A New Etchant for Thin Films of Tantalum and Tantalum Compounds
J. Grossman and D. S. Herman
J. Electrochem. Soc., 116:674 (1969)

Gitterdynamik bei Nb, Ta
E. E. Lachteenkova
Ann. Acad. Sci. Fenn., A6, 321-346 (1969)

Discussion of "The Effect of Hydrogen on the Structure and Properties of Vanadium"
Edward A. Loria
Trans. Met. Soc. AIME, 245:1366-1367 (1969)

Revelation of Dislocations in Niobium by Means of Etch Patterns
Jose Manuel da Cunha Monteiro
(Junta de Energia Nuclear, Sacavem, Portugal, Laboratorio de Fisica e Engenharia Nucleares), LFEN-99-a (1969), 29 pp.

Point Defects in Vanadium
J. H. Perepezko, R. F. Murphy, and A. A. Johnson
Phil. Mag., 19:1-6 (1969)

Point Defects Studies in Niobium
R. Pichon, F. Vandni, P. Bichon, G. De Keating-Hart, and P. Moser
(Centre d'Études Nucleaires de Grenoble, Commissariat à l'Energie Atomique), CEA-CONF-1513 (1969), 39 pp.

Metallographic and X-Ray Studies of Vanadium Single Crystals
M. V. Pikunov et al.
In: Single Crystals of Refractory and Rare Metals (E. M. Savitskii, ed.), Nauka, Moscow (1969), pp. 69-73.

Dislocation Damping and Interstitial Pinning in High Purity Single Crystal Tantalum
R. C. Sanders, S. R. Painter, and T. J. Turner
Scripta Met., 3:855-888 (1969)

Lattice Dynamics of Niobium. I. Measurements of the Phonon Frequencies. II. Kohn Anomalies in Niobium
R. I. Sharp
Proc. Phys. Soc. (Solid State), J. Phys., C 2:421-443 (1969)

Mössbauer Effect of ^{57}Fe Impurities Bound in Vanadium and Vanadium Hydrides
A. Simopoulos and I. Pelah
J. Chem. Phys., 51:5691-5695 (1969)

Investigation of Ordering in Interstitial Impurities in Tantalum Using an Electron Diffraction Microscope
M. P. Usikov and A. G. Khachaturyan
Kristallografiya, 13(6):1045-1055 (1968)
Sov. Phys. — Cryst., 13(6):910-918 (1969)

Niobium: Lattice Parameter and Density
R. L. Barns
J. Appl. Phys., 39:4044-4045 (1968)

Studies of the Lattice Properties of High Field Superconductors and Vanadium
B. W. Batterman
(Cornell University, Ithaca, New York), Contract AT(30-1)-3789, NYO-3789-5 (Sept. 20, 1968), 2 pp.

Imperfect Structure in Tantalum Single Crystal Rods
S. Dohi and H. Akiyama
Mem. Defense Acad. (Math., Phys., Chem., Eng., Japan), 8: 437-440 (1968)

Spark-Machine Damage in Niobium Single Crystals as Indicated by Etch Figures
H. D. Guberman
J. Appl. Phys., 39:2975-2976 (1968)

Structure of Ultrathin Films Produced by Vacuum Evaporation of Tantalum and Iridium
D. I. Lainer and V. A. Kholmyanskii
Kristallografiya, 12(6):1051-1057 (1967)
Sov. Phys. — Cryst., 12(6):913-918 (1968)

Dislocation-Interstitial Interactions in Single Crystal Tantalum
Richard C. Sanders and Thomas J. Turner
Kernforschungsanlage Intern. Conf. on Vacancies and Interstitials in Metals, Vol. 2 (Sept. 1968), pp. 531-545
See N69-12994 03-17

The Structure of Thin and Very Thin Tantalum Films
U. S. Urazaliev
Fiz. Metallov Metallovedenie (USSR), 26(1):97-100 (1968)

Dislocation Structures in Niobium (Columbium) Single Crystals Observed by Optical Microscopy
R. G. Vardiman and M. R. Achter
Trans. AIME, 242:196 (1968)

Dislocation Relaxation in Tantalum at Low Temperatures
L. Verdini and L. A. Vienneau
Can. J. Phys., 46:2715-2717 (1968)

Texture Development and Crystal Perfection in Nb Annealed at 2000°C
J. W. Scandelin and L. S. Birks
Trans. Met. Soc. AIME, 239:1268 (1967)

Etching Phenomena in the {111} X-Ray Diffraction
E. Zedler
J. Appl. Phys., 38:2046 (1967)

X-ray Diffraction Study of the Perfection of Niobium (Columbium) Single Crystals
M. R. Achter, C. L. Vold, and T. G. Digges, Jr.
Trans. Met. Soc. AIME, 236:1597 (1966)

X-ray Study of Wire-Drawn Niobium and Tantalum
R. P. I. Adler and H. M. Otte
Proc. Ann. Conf. Advances in X-Ray Analysis, 1965, 9:115-129 (1966)

Structure of Sputtered Tantalum
D. Mills
J. Can. Ceram. Soc., 35:48-52 (1966)

Crystal Structure of Tantalum, Niobium, and Vanadium between 110-400°K
Yu. N. Smirnov and V. A. Finkel'
Zh. Éksperim. Teor. Fiz., 49:1077-1082 (1965)
Sov. Phys. — JETP, 22(4):750-753 (1966)

Dislocation Arrangements in Ta Single Crystals
W. A. Sptizig and T. E. Mitchell
Acta Met., 14:1311-1323 (1966)

Ordering of Interstitial Impurities in Niobium
J. Van Landuyt, R. Gevers, and S. Amelinckx
Phys. Stat. Sol., 13:467 (1966)

Dislocation Structures in Deformed and Recovered Tantalum
Troy W. Barbee, Jr., and R. A. Huggins
J. Less-Common Metals, 8:306 (1965)

Dislocation-Interstitial Interactions in Single-Crystal Tantalum
S. H. Carpenter and G. S. Baker
J. Appl. Phys., 36:1733 (1965)

A New Structure in Tantalum Thin Films
M. H. Read and C. Altman
Appl. Phys. Letters, 7:51 (1965)

Low-Temperature Deformation of Body-Centered Cubic Metals. I. Yield and Flow Stress Measurements
J. W. Christian and B. C. Masters
Proc. Roy. Soc., A 281:223-239 (1964)

Etude de l'influence du recuit de récristallisation sur la structure et sur les propriétés du tantale et de ses alliages
M. I. Gavrilyuk, V. B. Milova, and V. A. Khostantinov
Fiz. Metal. i Metalloved., 18:389-395 (1964)

Effects of Nitrogen, Methane, and Oxygen on Structure and Electrical Properties of Thin Tantalum Films
D. Gerstenberg and C. J. Calbrick
J. Appl. Phys., 35:402 (1964)

Substructure and Mechanical Properties of Refractory Metal Wires
Roy Kaplow, John F. Peck, Frank T. J. Smith, and David A. Thomas
(Massachusetts Inst. of Tech., Cambridge), Contract AF33 (657-11238, ML-TDR-64-56) (Feb. 1964), 52 pp.

The Effect of Interstitials on Mechanical Twinning in Body-Centered Cubic Metals
J. O. Steigler, C. K. H. DuBose, and C. J. McHargue
(Oak Ridge National Laboratory, Oak Ridge, Tenn.), Acta Met., 12:263-264 (1964)

A New Method for Preparation of Strain-Free Single-Crystal Tensile Samples of High-Melting Metals and Some Results on the Plastic Deformation of Niobium Single Crystals
E. Votava
(Union Carbide European Research Associates, Brussels)
Phys. Stat. Sol., 5:421-434 (1964)

Niobium Hardness at 20-2120°C
V. A. Borisenko
"Sbnornik, Voprosy Vysokotemperaturnoi Prochnosti v Mashinostroenii" (Kiev, Academy of Sciences (1963), pp. 31-35

Twinning at 77°K of "Flash-Annealed" Niobium
A. Fourdeux and A. Wronski
(Union Carbide European Research Associates, s.a., 95, rue Gatti de Gamond, Bruxelles 18), Nature, 199:1284-1285 (1963)

Work-Hardening Mechanisms in Body-Centered Cubic Metals
Donald P. Gregory
(Pratt and Whitney Aircraft Div., United Aircraft Corp., Connecticut Aircraft Nuclear Engine Lab., Middletown), Contract AF33 (616)-7855, AD-411124, ASD-TDR-62-354, Pt. II (Mar. 1963), 48 pp.

Effects of Lattice Structural Defects on the Electric Resistance and Hall Effect of Tungsten and Tantalum
Erich Krautz and Hermann Schultz
Z. Angew. Phys., 15:1-4 (1963)

The Engineering Properties of Tantalum and Tantalum Alloys
F. F. Schmidt and H. R. Ogden
(Battelle Memorial Institute, Defense Metals Information Center, Columbus, Ohio), DMIC-Rept. 189 (September 13, 1963) Contract AF33 (616)-7747, 112 pp.

Dislocations in Deformed and Annealed Niobium Single Crystals
J. O. Stiegler, C. K. H. Dubose, R. E. Reed, Sr., and C. J. McHargue
(Oak Ridge National Laboratory, Oak Ridge, Tenn.), Acta Met., 11:851-860 (1963)

Solid Solubility Limits of Y and Sc in the Elements W, Ta, Mo, Nb, and Cr
A. Taylor
(Westinghouse Electric Corp., Research Laboratories, Beulah Road, Churchill Borough, Pittsburgh 35, Pa.), Contract No. AF33(616)-8315, ASD-TDR-63-204 (1963)

Effect of Oxygen on the Lattice Constant, Hardness and Ductility of Vanadium
S. A. Bradford and O. N. Carlson
Trans. Quart. Am. Soc. Metals, 55:169 (1962)

The Effects of Carbon on the Hardness, Microstructure, and Cold Working Properties of High-Purity Niobium
F. R. Cortes and A. L. Feild, Jr.
J. Less-Common Metals, 4:169-180 (Apr. 1962)

Deformation of Tantalum Single Crystals
D. P. Ferriss, R. M. Rose, and J. Wulff
Trans. Met. Soc. AIME, 224:975 (1962)

Transmission Electron Microscopy Studies of Deformed and Annealed Niobium
J. O. Stiegler
(Oak Ridge National Laboratory, Oak Ridge, Tenn.), ORNL-3313 (1962), p. 159

Hardness Anisotropy of Columbium
D. L. Douglass
Trans. Am. Soc. Metals, 54:322-330 (1961)

Substructure and Mechanical Properties of Refractory Metals
B. S. Lement, D. A. Thomas, S. Weissmann, W. S. Owen, and P. B. Hirsch
(Manufacturing Labs., Inc., Cambridge, Mass.), Contract AF33 (616)-6838, WADD-TR-61-181 (Apr. 1961), 255 pp.

Determination of Interstitial Solid-Solubility Limit in Tantalum and Identification of the Precipitate Phases
D. A. Vaughan, O. M. Stewart, and C. M. Schwartz
AIME Trans., 221:937-946 (1961)

Metallochemical Properties of Niobium
I. I. Kornilov
NP-tr-704
Doklady Akad. Nauk SSSR, 135:1399-1401 (1960), 6 pp.

Twinning in Vanadium
C. J. McHargue
(Oak Ridge National Laboratory, Oak Ridge, Tenn.), Acta
Met., 8:900 (1960)

The Clustering of Oxygen in Solid Solution in
Niobium
Ronald Gibala
Dissertation Abstr., 65-7104, available from University Micro-
films, Ann Arbor, Mich., 129 pp.

1.c.4. Group VI

The Absence of Polymorphism in Chromium
A. K. Butylenko et al.
Structure and Properties of Heat Resistant Metals and Al-
loys, in: Israel Program for Scientific Translations,
Jerusalem (1970), N70-25975 12-17, pp. 33-40

A Study of Vacancies in Tungsten Wires Quenched
in Superfluid Helium
R. J. Gripshover, M. Khoshnevisan, J. S. Zells, and J. Bass
Phil. Mag., 22:757-777 (1970)

A Neutron Elastic Scattering Study of Chromium,
Iron, and Nickel in the Energy Region 1.77 to
2.76 MeV
B. Holmqvist, S. G. Johansson, G. Lodin, M. Salama, and T.
Wiedling
(Aktiebolaget Atomenergi, Studsvik, Sweden), AE-385 (1970),
26 pp.

Dislocation Structures in Single-Crystal Tung-
sten and Tungsten Alloys
J. R. Stephens
Met. Trans., 1:1293-1301 (1970)

Fine Structure of Tungsten Single Crystals
V. I. Tiraspolski et al.
Fiz. Metallov Metalloved., 29:175-179 (1970)

Dislocation Helices in Molybdenum
C. Becker and B. Pegnel
Phys. Stat. Sol., 32:443-446 (1969)

Structure et propriétés des condensats minces,
déposés sous vide, de molybdène
Yu. S. Belomytcev, P. Yu. Yaroshevich, L. N. Saratovskii, and
T. M. Pavlovich
Fiz. Metallov Metalloved. SSSR, 27(1):77-80 (1969)

Dislocations in Mo and W Single Crystals
T. J. Benieva, T. V. Golub, and V. S. Skopin
Fiz. Metallov Metalloved., 28:700-704 (1969)

Restauration et récristallisation de monocristaux
de Mo déformés par laminage
N. V. Dubovyc'ka, L. N. Larykov, and Yu. D. Yakovenko
Dop. Akad. Nauk URSR, A 31(11):1039-1042 (1969)

Interpretation of Diffraction Patterns from a
New Modification of Chromium
J. Forssell and B. Persson
J. Phys. Soc. Japan, 27:1368 (1969)

Autoionic Microscope Investigation of the Orien-
tation Relations in Polycrystalline Tungsten
R. I. Garber, Zh. I. Dranova, and I. M. Mikhailovskii
Fiz. Metallov Metalloved., 27:573-576 (1969)
Preferred grain orientation

Temperature Dependence of Chromium X-Ray Interference in the Temperature Range of 293-
1100°K
E. I. Geshko
Ukr. Fiz. Zh., 14:335-336 (1969)

Effect of Vacuum Conditions in the Apparatus
and the Rate of Condensation on the Preferred
Orientation of Chromium Films
V. S. Kogan and A. L. Seryugin
Fiz. Metallov Metalloved., 28(4):621-628 (1969)

Dislocation Structure in Heat-Treated Mo Single
Crystals
V. A. Krachmalev and G. A. Klein
In: Single Crystals of Refractory and Rare Metals (E. M.
Savitskii, ed.), Nauka, Moscow (1969), pp. 86-89

Grain Boundary Dislocations in Tungsten Wire
B. Loberg and H. Norden
Arkiv Fysik, 40(28):413-419 (1969)
Electron and field-ion microscopy

Observation of Dislocations in a Tungsten Grain
Boundary by Combined Electron and Field-Ion
Microscopy
B. Loberg, H. Norden, and D. A. Smith
Arkiv Fysik, 40(38):513-519 (1969)

The Effect of Grinding on the Dislocation Struc-
ture of As-Grown Molybdenum Single Crystals
A. Luft and L. Kaun
Phys. Stat. Sol., 34:135 (1969)

Der Einfluss der Umformungstexturen auf die
Eigenschaften hochschmelzender Metalle
R. Palme
Hochtemperatur-Werkstoffe. 6. Plansee-Seminar, June 24-28,
1968, Reutte/Tirol (1969), pp. 159-163

Dislocation Helices in Molybdenum. II. Plastic
Deformation by Moving Helices
B. Pegel and C. Becker
Phys. Stat. Sol., 35:157 (1969)

Analyse du profil des raies X des dépôts électro-
lytiques de Cr
Y. Sakamoto
J. Soc. Mater. Sci. Japan, 18(194):997-1003 (1969)

The Recovery of Cold-Worked Molybdenum
L. Stals and J. Nihoul
Physica, 42:165-178 (1969)

Lattice Anisotropy in Antiferromagnetic Chromi-
um
M. O. Steinitz, L. H. Schwarz, J. A. Marcus, E. Fawcett, and
W. A. Reed
Phys. Rev. Letters, 23:979 (1969)

Field-Ion Microscope Study of Grain Boundary
Structure and Textures of Polycrystalline Sub-
stances
A. L. Suvorov
Fiz. Metal. Metalloved., 28:813-817 (1969)

Dislocation Velocity and Macro-Deformation in
Molybdenum
 T. Vreeland and A. P. L. Turner
 Scripta Met., 3:193-194 (1969)

Cleavage Cracks on Chromium Crystals due to
Spark Machining
 J. M. Anderson and A. D. Stewart
 J. Inst. Metals-Met. Rev., 96:96 (1968)

Structure and Morphology of Electrodeposited
Molybdenum Dendrites
 C. J. Bechtoldt, Fielding Ogburn, and J. Smit
 J. Electrochem. Soc., 115:813-816 (1968)

An Investigation of the Crystal Structure of Anti-
ferromagnetic Chromium
 F. H. Combley
 Acta Cryst., 24 B:142 (1968)

Some Metallographic Techniques Applied to Molyb-
denum Single Crystals. I. A New Electrolytic
Etchant for Dislocations in Molybdenum
 G. C. Das
 Prakt. Metallogr., Dtsch., 5(4):199-205 (1968)

Metallography of Molybdenum in Color
 R. Hasson
 Microscope, 16:329-334 (1968)

Surface Structure of Negative Crystals (Pores)
in Molybdenum
 G. A. Klein and V. A. Krakhmalev
 Kristallografiya, 12(6):1096 (1967)
 Sov. Phys. – Cryst., 12(6):964 (1968)

Anisotropy of Chromium Films Prepared by the
Oblique Incidence of a Molecular Beam on a Sub-
strate
 Fiz. Tverd. Tela, 10(1):315-316 (1968)
 Sov. Phys. – Solid State, 10(1):250-252 (1968)

Dislocation Etch Pitting Studies on Tungsten
Single Crystals under Potentiostatic Conditions
 S. R. Maloof
 Corrosion, 24:283-290 (1968)

Influence of Annealing on the Recrystallization
of Doped and Undoped Mo Wires
 F. A. M. van Meel
 Iron Steel Inst. London Publ., No. 113, 49-51 (1968)

Metallographische Untersuchungen der Primarre-
kristallisation von Molybdän und Wolfram
 E. Pink
 Planseeber. Pulvermet., 16:277-282 (1968)

Dislocation Velocity Measurements in High Purity
Molybdenum
 H. L. Prekel, A. Lawley, and H. Conrad
 Acta Met., 16:337-345 (1968)

Investigation of the Moessbauer Effect of Ultra-
fine Tungsten Particles
 S. Roth and E. M. Hoerl
 Acta Phys. Austr., 27:264-270 (1968)

Fine Structure of Tungsten Single Crystals
 E. M. Savitskii and G. L. Tsarev
 Growth and Imperfections of Metallic Crystals (D. E. Ovsienko,
  ed.) Consultants Bureau, New York (1968), 268 pp.

Multiple Diffraction Origin of Low Energy Electron Diffraction Intensities (II)
 R. M. Stern, A. Gervais, and M. Menes
 Acta Cryst., A 25:393-394 (1969), AFOSR-1263-67 (Aug. 1968),
  4 pp.

Lattice Constants, Thermal Expansion Coefficients, and Densities of Molybdenum and the Solubility of Sulfur, Selenium, and Tellurium in It
at 1100°
 Martin E. Straumanis and Ramesh P. Shodhan
 Z. Metallk., 59(6):492-495 (1968)

X-Ray Topographic Pictures of Mo Single Crystals
with the Lang Method
 C. Becker and D. Schulze
 Monatsber. Deut. Akad. Wiss. Berlin, 9:92-93 (1967)

Etch Pitting of Dislocations in Molybdenum Single
Crystals
 G. C. Das and P. L. Pratt
 Mater. Res. Bull., 2:667 (1967)

Stacking Faults in Tungsten
 J. Demny
 Phys. Stat. Sol., 22:K 1 (1967)

Influence of Spark Discharges on the True Struc-
ture of Tungsten Crystals
 G. V. Dudko, O. V. Mitrofanov, and O. A. Ovechin
 Elektronnaya Obrabot. Mat., No. 2, 3-7 (1967)

High Angle Substructure in Electron-Beam Zone
Melted W
 D. R. Hay and E. Scala
 Trans. AIME, 239:1272 (1967)

Electron Optical Investigation of the Secondary
Recrystallization of Drawn Tungsten Wires Based
on Thermal Electron Emission
 O. Horacsek and T. Millner
 Z. Metallk., 58(5):345-348 (1967)

An Electron Diffraction Study of the Crystal
Structure of a New Modification of Chromium
 Kazuo Kimoto and Isao Nishida
 J. Phys. Soc. Japan, 22:744 (1967)

Revelation of the Substructure of Tungsten,
Molybdenum, and Rhenium Single Crystals
 N. P. Koroleva and M. V. Pikunov
 Zavod. Lab., No. 6, 733-35 (1967)

Bestimmung der Gitterparameter von Kobalt und
Chrom im Temperaturbereich von 20° C bis -180° C
 S. Muller, Ph. Dunner, and N. Schmitz-Pranghe
 Z. Angew. Phys., 22:403 (1967)

Orientation, Deformation Texture and Indentation
Anisotropy in W and Mo Single Crystals
 C. J. Sandwith
 Dissertation, Oregon State University, Corvallis, Oregon (1967)

Microstructure of Tungsten Single Crystals
 E. M. Savitskii and G. L. Tsarev
 Rost Kristallov (Growth of Crystals) (Naukova Dumka, Kiev,
  1967), pp. 280-296 (in Russian)

Etching Behavior of Small-Angle Boundaries in
Tungsten
 H. Wadewitz
 Realstruktur und Eigenschaften von Reinststoffen, Teil 3 (J.
  Kunze, B. Pegel, K. Schlaubitz, and D. Schulze, eds.), Akade-
  mie-Verlag, Berlin (1967), pp. 287-294

Structure of Very Thin Tantalum and Molybdenum Films
P. N. Denbigh and R. B. Marcus
J. Appl. Phys., 37:4325-4330 (1966)

Resistivity and Structure of Sputtered Molybdenum Films
F. M. d'Heurle
Trans. AIME, 236:321 (1966)

Effect of Orientation on Ductile to Brittle Transition of Tungsten Single Crystals
Ralph G. Garlick
NASA-TM- X-1252 (July 1966), 8 pp.

Physical Characterization of Molybdenum Single Crystals for Irradiation Experiments
H. E. Kissinger, J. L. Brimhall, and B. Mastel
(Battelle-Northwest Laboratory, Richland, Washington), Contract AT(45-1-)-1830, BNWL-SA-777; CONF-661126-1, November 1966 (Intern. Conf. on Characterization of Materials, University Park, Pa.)

Transmission Electron Microscopic Studies of Undoped W Wire (0.38 mm) Annealed Isothermally at 1180°C
G. Mima and K. Nishida
Nippon Kinzoku Gakkaishi, 20(11):1047-1051 (1966)

The Morphology and Size Analysis of Fine Metal Powders
Minoru Ozasa and Hiroshi Fukuma
Proc. Intern. Conf. Modern Developments in Powder Metallurgy, New York, 1965 (1966), pp. 144-152
W, Mo

The Direct Observation of Etch-Pits at Dislocations in Molybdenum
H. L. Prekel and A. Lawley
Phil. Mag., 14:545 (1966)

Observations of Dislocations in High Angle Boundaries in Tungsten
H. F. Ryan and J. Suiter
Acta Met., 14:847 (1966)

Precise Determination of Lattice Parameters (Chromium) by Step-Scanning X-Ray Diffractometry
M. A. Sanborn
J. Mater., 1:481-490 (1966)

Structure des couches de Mo et W obtenues par décomposition thermique de carbonyles sous vide
Struktura i Svoistva Metallicheskikh Plenok, Naukova Dumka, Kiev, (1966), pp. 84-87

Dislocation Interstitial Interactions in Single Crystal Tungsten
S. H. Carpenter and G. S. Baker
Acta Met., 13:917 (1965)

A Study of Etch-Figures of Dislocation Lines in Molybdenum Crystals
Fong Duan, Ming Nay Ben, and Li Chi
Sci. Sinica (Peking), 14:1130-1144 (1965)

Direct Observation of Dislocations in Refractory Metals
Tuan Feng and Nai-pen Min
Nan Ching Ta Hsueh Pao, Tzu Jan K'o Hsueh-9:165-174 (1965)
Mo and W monocrystals

Deformation Structures in Zone-Melted Molybdenum Above Room Temperature
A. Lawley
(Franklin Inst. Res. Labs., Philadelphia, Pa.), Contract Nonr-4434(00), 1-B219502 (Sept. 1965)

The Detection of Dislocations in Tungsten and Molybdenum by Electrolytic Etching
Erwin Pink
(Metallwerk Plansee Ag, Reutte/Tirol, Austria), Prakt. Metallog., 2:58 (1965)

Investigation of the Fine Structure of Tungsten Single Crystals by Microdiffraction of Ultra-Fast Electrons and Surface Tension and Structure of Molten Iron, Cobalt, and Nickel Borides, Studies in Structure of Tungsten and Molten Iron, Cobalt, and Nickel Borides
N. M. Popov, E. M. Savitskii, G. L. Tsarev, F. N. Tavadze, and I. A. Bairamashvili
JPRS-31225, TT-65-31722 (July 1965), 13 pp., 27 refs.
Dokl. Akad. Nauk SSSR, 162:64-69 (1965)

Etch-Pitting Characteristics of High-Purity Molybdenum
H. L. Prekel and A. Lawley
(Franklin Inst. Res. Labs., Philadelphia, Pa.), Contract Nonr-4434(00), 1-B2195-1, 1965

Dislocation Structure in Chromium, Chromium-Rhenium, and Chromium-Iron Alloys
C. N. Reid and A. Gilbert
J. Less-Common Metals, 10:77 (1965)

The Occurrence of Cylindrical Texture in Tungsten Wires
G. D. Rieck and A. S. Koster
Trans. AIME, 233:770 (1965)

X-ray Diffraction Study of Deformation by Filing in B.C.C. Refractory Metals
E. N. Aqua and C. N. J. Wagner
Phil. Mag., 9:565 (1964)

Rolled Molybdenum Single Crystals: Deformation Textures, Recrystallization, and Transition Temperature
R. Blickensderfer, R. Siemens, G. Asai, and H. Kato
Bureau of Mines report of investigations 6539 (1964)

Spiral Displacements and Coaxial Prismatic Slips in Molybdenum Single Crystals Made by Electron Beam Zone Melting
T. Feng, N. P. Ming, C. Li, C. Y. Soong, and S. M. Pan
Sci. Sinica, 13:339-341 (1964)

Investigation of Room-Temperature Slip In Zone-Melted Tungsten Single Crystals
R. G. Garlick and H. B. Probst
(Research Metallurgist, Refractory Metals Section, and Head, Refractory Compounds Section, respectively, Lewis Research Center, National Aeronautics and Space Administration, Cleveland, Ohio), Trans. Met. Soc. AIME, 230:1120 (1964)

Cleaning Methods, Gas Adsorption and Corrosion Characteristics of a (100) Chromium Single Crystal Using Low Energy Electron Diffraction
C. A. Haque and H. E. Farnsworth
Surface Sci., 1:378 (1964)

Growth of Tungsten Monocrystals on Glowed Tungsten Wires
T. W. Hoffmann and J. Nikloborc
(Univ. of Wroclaw, Poland), Acta Phys. Polon., 25:633-634 (1964)

Observations on Prismatic and Grown-In Dislocations in Zone-Melted Molybdenum
A. Lawley and H. L. Gaigher
Phil. Mag., 8:1713 (1963)

Deformation Structures in Zone-Melted Molybdenum
A. Lawley and H. L. Gaigher
(The Franklin Institute Laboratories, Philadelphia 3, Pennsylvania), Phil. Mag., 10:103 (1964)

Identification of Prismatic Loops in Zone-Melted Molybdenum
A. Lawley and J. D. Makin
Phil. Mag., 10:737 (1964)

Grain Boundary Topography in Tungsten
H. F. Ryan and J. Suiter
Phil. Mag., 10:727 (1964)

Controlled Addition of Small Amounts of Oxygen to Niobium (Columbium)
C. S. Tedmon, Jr., R. M. Rose, and J. Wulff
Trans. AIME, 230:1732 (1964)

Slip-Induced Cleavage in Polycrystalline Tungsten
A. Wronski and A. Fourdeaux
J. Less-Common Metals, 6:413 (1964)

Plastic Deformation of Single Crystals of Molybdenum
Pang-hsin Chou
Wu Li Hsueh Pao, 19:285-296 (1963)

Obtaining Molybdenum Single Crystals and Their Properties
V. S. Emel'yanov, A. I. Evstyukhin, G. A. Leont'ev, and A. N. Semenikhin
OTS-64-21263, p. 39-45
JPRS-22697, pp. 39-45, Met. i Metalloved. Chistykh Metal., Sb. Nauchn. Rabot 4, (May 31, 1963)

Work-Hardening Mechanisms in Body-Centered Cubic Metals
Donald P. Gregory
(Pratt and Whitney Aircraft Div., United Aircraft Corp., Connecticut Aircraft Nuclear Engine Lab., Middletown), Contract AF33 (616)-7855, AD-411124, ASD-TDR-62-354, Pt. II (Mar. 1963), 48 pp.

The Influence of Lattice Defects on the Electrical Resistance and the Hall-Effect in Tungsten and Tantalum
E. Krautz and H. Schultz
Z. Angew. Phys., 15:1-4 (1963)

Dislocation Uncertainty in Relation to Observations on BCC Molybdenum
A. Lawley and H. L. Gaigher
Appl. Phys. Letters, 2:123 (1963)

Solid Solubility Limits of Y and Sc in the Elements W, Ta, Mo, Nb, and Cr
A. Taylor
(Westinghouse Electric Corp., Research Laboratories, Beulah Road, Churchill Borough, Pittsburgh 35, Pa.), Contract No. AF33 (616)-8315, ASD-TDR-63-204 (1963)

Dislocations at Isolated Grains in a Single-Crystal Matrix of Molybdenum
A. Lawley and H. L. Gaigher
(The Franklin Institute Laboratories, Philadelphia 3, Pennsylvania), J. Appl. Phys., 33, 3594 (1962)

X-Ray Diffraction Analysis of Crystallite Size and Lattice Strain in Tungsten Wire
A. J. Opinsky, J. L. Orehotsky, and C. W. W. Hoffman
J. Appl. Phys., 33:708 (1962)

Dislocation Substructures in Deformed and Recovered Molybdenum
R. Benson and G. Thomas
Prog. 19th Ann. Meeting Electron Microscope Soc., (A) (1961), p. 4

The Direct Observation of Lattice Defects by Field Ion Microscopy
D. G. Brandon and M. Wald
(Univ. of Cambridge, England), Phil. Mag., 6:1035-1044 (1961)

Substructure and Mechanical Properties of Refractory Metals
B. S. Lement, D. A. Thomas, S. Weissmann, W. S. Owen, and P. B. Hirsch
(Manufacturing Labs., Inc., Cambridge, Mass.), Contract AF33 (616)-6838, WADD-TR-61-181 (Apr. 1961), 255 pp.

Production of Monocrystal Tungsten Emitters
O. V. Mitrofanov
Instr. Exp. Tech., pp. 1007-1009 (1961) (USSR)

Subcrystals in Large Vapour-Grown Crystals in Tungsten Wires
G. D. Rieck
Acta Met., 6:360 (1958)

The Texture of Drawn Tungsten Wires
G. D. Rieck
Philips Res. Rept., 12:423 (1957)

A Set of Effective Coordination Number (12) Radii for the β-Wolfram Structure Elements
S. Geller
Acta Cryst., 9:885 (1956)

Impurity Atom-Dislocation Interactions and Subsequent Effects on Mechanical Properties of Refractory Metals
M. A. Adams and H. Nesor
(Materials Research Corp., Orangeburg, New York), Contract AF 33 (616)-7596, ASD-TDR-62-11

Dislocation Interstitial Interactions in Single Crystal Tungsten and Tantalum
Steve H. Carpenter
Dissertation Abstr. 65, 1786, 86 pp., University Microfilms, Ann Arbor, Mich.

Investigation of the Effect of Interstitial Impurities on the Structure and Properties of Single Crystals of Tungsten
E. M. Savitskii and G. L. Tsarev
TT-66-11244, 16 pp.

1.d. Surface Structure (field and ion emission, LEED, chemisorption)

1.d.1. General

Modern Methods of Surface Analysis Symposium, Bell Telephone Laboratories, Murray Hill, N. J., May 1970
Surface Sci., 25 (1) (1971) (entire issue)

Surface Studies by Electron Diffraction
P. J. Estrup and E. G. McRae
Surface Sci., 25:1-52 (1971)
139 refs.

Contact Potential Measurements of the Adsorption of Alkali Metals on Ta(110) and W(100) Crystals
D. L. Fehrs and R. E. Stickney
Surface Sci., 24:309-331 (1971)

Emission of W, Mo, and Nb with Adsorbed Cs and Ba
V. D. Bondarenko et al.
Teplofiz. Vysok. Temp., 8(1):211-213 (1970)
High Temperature, 8(1):201-203 (1970)

The Surface Energies of Solid Molybdenum, Niobium, Tantalum and Tungsten
E. N. Hodkin, M. G. Nicholas, and D. M. Poole
J. Less-Common Metals, 20:93-103 (1970)

Influence of High-Temperature Heat Treatment on Surface Order and Faceting of Some Mo, W, Ta and Re Crystal Surfaces
H. E. Farnsworth, C. A. Haque, D. M. Zehner, and G. Barton
Surface Sci., 17:1-6 (1969)

Determination of Heats of Desorption for Potassium and Cesium Ions and Atoms from Tungsten, Molybdenum and Tantalum
N. I. Ionov, E. N. Lebedeva, and Ts. S. Marionova
Zh. Tekhn. Fiz., 39:1323-1324 (1969)

Field-Ion Microscopy and Electronic Structure of Metal Surfaces
E. W. Muller
Quart. Rev. Chem. Soc. London, 23:177-186 (1969)

The Incorporation of Chemisorbed Species
M. W. Roberts
Recent Progress in Surface Science, Vol. 3 (J. F. Danielli, A. C. Riddiford, and M. D. Rosenberg, eds.), Academic Press, New York and London (1970), pp. 1-22

Chemisorption Reactions on W, Mo and Ta Films
M. W. Roberts and L. Whalley
Trans. Faraday Soc., 65:1377-1385 (1969)

Surface Phenomena of Thermionic Emitters
Round Table Conference held at Julich, Germany (June 10, 1969), Kernforschungsanlage Julich, Jul-Conf-3 (Nov. 1969), 70 pp.
Zr, W, Mo

Emission von Mo, Ta, Nb und Zr
V. D. Dmitriev and G. K. Cholpov
Teplofiz. Vys. Temp., 6(3):550-551 (1968)
High Temperature, 6(3):527-528 (1968)

Emission Parameters of Tantalum and Molybdenum Single Crystals

O. D. Protopopov, E. V. Mikheeva, B. N. Sheinberg, and G. N. Shuppe
Fiz. Tverd. Tela, 8(4):1140-1146 (1968)
Sov. Phys. — Solid State, 8(4):909-914 (1968)

The Connection Between Work Function and Some Physical Characteristics of Transition Metals
G. V. Samsonov and I. Ya. Kondratov
Izv. VUZ Fiz., No. 6, 103-105 (1968)
Sc, Ti, V, Cr, Y, Zr, Nb, Mo, La, Hf, Ta, and W

Anisotropy of Metal Work Functions
M. Kaplit
(Univ. of Penn., Philadelphia), Research in the Conversion of Various Forms of Energy by Unconventional Techniques (March 1967), 11 pp.

Zur Feldemission von Chrom, Kobalt und Hafnium
Gunter Krause und Dietrich Stark
Z. Phys., 201:69-74 (1967)

On Field Evaporation
D. G. Brandon
Phil. Mag., 14:803-820 (1966)

Photoelectron Emission in the Extreme Ultraviolet Region
L. Heroux et al.
AFCRL-66-718 (October 1966)

Laser-Induced Spontaneous Electron Emission from Rear Side of Metal Foils
Walter L. Knecht
Appl. Phys. Letters, 8:254 (1966)

Parameters of Emission from Faces of W, Mo, and Ta Monocrystals
G. N. Shuppe
Izv. Akad. Nauk SSSR, Ser. Fiz., 30:1935-1941 (1966)

Investigations of Electron Emission Characteristics of Low Work Function Surfaces
L. W. Swanson, A. E. Bell, and L. C. Crouser
(Field Emission Corp., McMinnville, Oregon), Report No. 2, Jan. 1-Mar. 31, 1966, Contract NASw-1082, NASA-CR-75630 (May 1966), 38 pp.

Literature Search of Physical Property Data and of Composite Surface Work Function Models, Semiannual report May 1-Nov. 1, 1966
L. W. Swanson, A. E. Bell, C. H. Hinrichs, L. C. Crouser, and B. E. Evans
(Field Emission Corp., McMinnville, Oregon), Contract NAS3-8910, NASA-CR-85562 (December 15, 1966), 132 pp.

Vacuum Thermionic Work Functions of Polycrystalline Be, Ti, Cr, Fe, Ni, Cu, Pt, and Type 304 Stainless Steel
R. G. Wilson
J. Appl. Phys., 37:2261 (1966)

Vacuum Thermionic Work Functions of Polycrystalline Nb, Mo, Ta, W, Re, Os, and Ir
R. G. Wilson
J. Appl. Phys., 37:3170 (1966)

Electron and Ion Emission from Polycrystalline Surfaces of Nb, Mo, Ta, W, Re, Os, and Ir in Cesium Vapor
R. G. Wilson
J. Appl. Phys., 37:4125 (1966)

Emission Spectrometry
Bourdon F. Scribner and Marvin Margoshes
(National Bureau of Standards, Washington, D. C.), Anal. Chem.,
36:329R-343R (April 1964)

A Critical Review of Refractories
Edmund K. Storms
(Los Alamos Scientific Lab., N. Mex.), Contract W-7405-eng-
36, LA-2942 (March 1964), 252 pp.

Large Crystals Grown by Recrystallization
K. T. Aust
The Art and Science of Growing Crystals, John Wiley and Sons,
Inc., New York (1963), pp. 452-478

Work-Function Measurements and Lattice Defects
J. A. Dillon, Jr., and R. M. Oman
(Department of Physics, Brown University, Providence, Rhode
Island), reprinted from 1963 Transactions, The Tenth Na-
tional Vacuum Symposium, American Vacuum Society

Specimen Preparation by Spark-Erosion Cutting
H. H. Ehlers and D. F. Kolesar
(Lincoln Laboratory), MIT Technical Rept., 303 (March 14,
1963)

Preparation and Properties of Sputtered Films
M. H. Francombe
(Philco Scientific Laboratory, Blue Bell, Pennsylvania), CONF-
209-12, American Vacuum Society 10th National Meeting,
Boston, Mass. (October, 1963)

Investigation of Surface Energy States of Single
Crystal Metals
R. C. Menard and A. A. Anderson
(General Mills, Inc. Electronics Division, Aerospace Research,
St. Paul, Minnesota), Contract AF33(657)-8038, Project 7022,
Task 7022-01, ARL 63-139 (August 1963)

Field Emission Microscopy of Clean Surfaces
With Electrons and Positive Ions
E. W. Müller
(Pennsylvania State University, University Park, Pa.), Annals
N. Y. Acad. Sci., 101, Article 3, 585-598 (1963)

Generation of Clean Surfaces in High Vacuum
R. W. Roberts
Paper presented at the Conference on Sorption Properties of
Vacuum Deposited Metal Films, Liverpool (April 1963)
Brit. J. Appl. Phys., 14:537 (1963)

Precipitation of Vacancies in Metals
G. Thomas and J. Washburn
(Department of Mineral Technology and Inorganic Materials Research Division, Lawrence Radiation Laboratory, University
of California, Berkeley, California), Rev. Mod. Phys., 35:992
(1963)

Thermionic Emission from Metal Crystals in
Alkali Metal Vapors
H. F. Webster and P. L. Read
(General Electric Research Laboratory, Schenectady, New York)
No. 63-RL-3347E (May 1963), 54 pp.
Emission density from W, Mo, Ta, Nb, Ni, Re, and Nb carbide
in Cs, Rb, and K vapors as a function of alkali vapor pres-
sure, emitter temperature, and emitter crystal face

Electron Microscopy
M. C. Botty and A. M. Thomas
(American Cyanamid Co., Stamford, Conn.), Anal. Chem., 34:
127 (1962)
Review paper — 218 references

Electron Microscopy
Fifth International Congress for Electron Microscopy held in
Philadelphia, Pa. (August 29 to Sept. 5, 1962), Volume 1,
Non-Biology (Sydney S. Breese, Jr., ed.), Academic Press,
New York (1962)

Crystalline Structure and Surface Reactivity
H. C. Gatos
Science, 137:311 (1962)

Metallographic Etching by Monoenergetic Ion
Bombardment
P. S. Maiya and J. B. Newkirk
(Department of Metallurgical Engineering, Cornell University,
Ithaca, New York), Trans. ASM, 55:474 (1962)

A Versatile Cathodic Etcher
J. W. Ward
(Los Alamos Scientific Laboratory of the University of Califor-
nia, Los Alamos, New Mexico), Contract W-7405-ENG. 36,
LAMS-2946 (April 1962)

Extension of Sensitivity in Analysis of Impurities
in Solids by Mass Spectrometry
C. M. Stevens
(Argonne National Lab., Ill.), Symposium on Extension of Sensitivity for Determining Various Constituents in Metals,
Special Technical Publication No. 308, Philadelphia, American Society for Testing and Materials (1961), pp. 58-68

Beobachtung von Umwandlungs- und Oxydations-
Vorgängen im Elektronen-Emissions-Mikroskop
H. Duker
Proc. Eur. Reg. Conf. on Electron Microscopy, Delft, Vol. 1:
456 (1960)

Emissionsmikroskopie mit ionenausgelösten Elek-
tronen
H. Duker
Schweiz. Archiv. Angew. Wiss. Tech., 26:199 (1960)

Beobachtung und Registrierung veränderlicher
Vorgänge im Emissions-Elektronenmikroskop
H. Duker
(Max-Planck-Institut für Metallforschung, Abt. Sondermetalle,
Stuttgart, und Institut für experimentelle und angewandte
Phsyik der Universität Tübingen), Radex Rundschau, Heft 6
(1960)

Propriétés des métaux et alliages aux basses
températures (Properties of metals and alloys
at low temperatures)
E. Justi
CEA-tr-A-789
Z. Metallk., 51(1):1-17 (1960)

Soviet Research in Field Electron and Ion Emis-
sion, 1955-1959; An Annotated Bibliography
T. W. Marton and R. Klein
(National Bureau of Standards, Washington, D. C.), National
Bureau of Standards Tech. Note 75 (Oct. 1960)

Electron Microscopy
T. G. Rochow, Ann M. Thomas, and M. C. Botty
(American Cyanamid Company, Stamford, Connecticut), Anal.
Chem., 32:92 (1960)
Review paper-198 references

Improved Method of Etching by Ion Bombardment
T. K. Bierlein and B. Mastel
(General Electric Company, Hanford Atomic Products Opera-
tion, Richland, Washington), Rev. Sci. Instr., 30:832 (1959)

Etching Reagents for Dislocations in Metal Crystals
L. C. Lovell, F. L. Vogel, and J. H. Wernick
(Bell Telephone Laboratories, Inc., Murray Hill, N. J.), Metal Progr., 75:96 (1959)

Etching
B. D. Cuming and A. J. W. Moore
(Division of Tribophysics, C.S.I.R.O., University of Melbourne), J. Australian Inst. Metals, 3:124–142 (August 1958)

Electron Microscopy
T. G. Rochow and M. C. Botty
(Research Division, American Cyanamid Co., Stamford, Conn.) Anal. Chem., 30:640 (1958)
Review paper - 327 references

Observations Made During Electrolytic Polishing of Metals
H. Mohlberger
Metall., 11:756–761 (1957)
Translation AEC-TR-4036

Cathodic Vacuum Etching
T. R. Padden and F. M. Cain, Jr.
(Westinghouse Atomic Power Division, Bettis Field, Pittsburgh), Contract AT-11-1-Gen-14 with the U. S. Atomic Energy Commission, Metal Progr., 66:108 (1954)

Secondary Electron Emission from Single Crystals of Titanium, Germanium, Tungsten, and Beryllium
George G. Goetz
Dissertation Abstr., 64-4063, 121 pp.

1.d.2. Group IV

Auger Spectroscopy of Titanium
H. E. Bishop, J. C. Riviere, and J. P. Coad
Surface Sci., 24:1–17 (1971)

Sur la chimisorption du carbone à la surface du zirconium
Pierre Ailloud, Jean-Paul Touboul, and Jean-Paul Langeron
Compt Rend, Ser. C, 270:184–186 (1970)

Dissociative Adsorption of Carbon Dioxide on Titanium Film
K. Kawasaki, N. Hayashi, S. Ebisawa, and T. Sugita
Surface Sci., 20:209–212 (1970)

Field Ion Microscopy of Hafnium
T. Reisner, O. Nishikawa, and E. W. Muller
Surface Sci., 20:163–173 (1970)

Zur Feldemission der Metalle Titan, Zirkon, und Hafnium
Jurgen Hospodarsky and Dietrich Stark
Z. Physik, 229:102–108 (1969)

Determination de l'énergie superficielle du titane en phase solide
V. N. Kostikov, A. V. Kharitonov, and V. I. Savenko
Fiz. Metal. Metallov., 26:947–948 (1968)

The Structure and Recrystallization Behaviour of Titanium Oxidation Films
D. L. Douglass and J. Van Landuyt
A.T.B. Metall, 7(4) (1967)

Adsorption, Surface Reaction and Mutual Displacement of CO, CO_2, and O_2 on Titanium Film
K. Kawasaki, T. Sugita, and S. Ebisawa
Surface Sci., 7:502–506 (1967)

Adsorption of Oxygen on Ultra-Thin Titanium Films
F. P. Fehlner
Nature, 210:1035 (1966)

Structure of an Adsorbed Layer on a Titanium (0001) Surface
Allan M. Russell
Appl. Phys. Letters, 8:177 (1966)

The Physical Metallurgy of Zirconium
D. L. Douglass
(General Electric Co., Pleasanton, California), At. Energy Rev., 1:71–237 (1963)

Study of Electrical and Physical Characteristics of Secondary Emitting Surfaces
W. T. Peria, ed.
(Minnesota, Univ., Minneapolis), Contract AF33(657)-8040 (Jan. 23, 1963), 126 pp.

Observations on Regular Arrays of Etch Pits in a Titanium-10% Molybdenum Alloy
T. H. Schofield and A. E. Bacon
(National Physical Laboratory, Teddington, England), Acta Met., 7:403 (1959)

1.d.3. Group V

Adsorption of Oxygen and Nitrogen on Thin Films of Tantalum
K. Fischer and E. Fromm
Vakuum-Technik, 19:134–138 (1970)

Adsorption of Oxygen and Nitrogen on Tantalum Films
Konrad Fischer and Eckehard Fromm
Z. Metallkunde, Vol. 61 (1970):
Part I. Experimental Methods, pp. 710–717;
Part II. Investigations with Oxygen, pp. 805–810;
Part III. Investigations with Nitrogen, pp. 811–814;
Part IV. Temperature Dependence of the Rate of Chemisorption and Physical Adsorption of Nitrogen, pp. 815–818

Chemisorption and Absorption of Nitrogen with Tantalum
Takeo Oguri
Japan. J. Appl. Phys., 9:1453–1460 (1970)

Studies on Chemisorption of Nitrogen on Tantalum with the Field Emission Microscope
Takeo Oguri
Japan. J. Appl. Phys., 9:1461–1466 (1970)

Oxidation of Metals (Ta) by Atomic O
P. G. Dickens, R. Heckinbottom, and J. W. Linnett
Trans. Faraday Soc., 65:2235–2247 (1969)

Chemical Shifts in Auger Electron Spectroscopy from the Initial Oxidation of Ta(110)
T. W. Haas and J. T. Grant
Phys. Letters, 30A:272 (1969)
Less than one monolayer

Influence of Oxygen on the Work Function of Ta
T. Krivachy and J. Klugel
Surface Sci., 15:358–360 (1969)

Study of Surface Self-Diffusion of Tantalum by Field-Emission Microscopy
D. M. Pautov and I. L. Sokol'skaya
Fiz. Tverd. Tela, 10(8):2473-2479 (1968)
Sov. Phys. — Solid State, 10(8):1941-1945 (1969)

Determination of Anode-Oxide Film Thickness on Tantalum by Capacitance Measurement
A. F. Bogoyavlenskii and G. I. Zakhvatov
Ind. Lab., 34:1800 (1968)

Auger Electron Studies of Surface of Nb
B. D. Campbell
LA-4010 (1968)

Low-Energy Electron Scattering from Clean and Hydrogen-Covered Nb (110) Surfaces
T. W. Haas
J. Appl. Phys., 39:5854 (1968)

Studies of Physical and Chemical Properties of Solid Surfaces
A. G. Jackson
(Systems Research Labs., Dayton, Ohio), Contract F33615-67-C-1041, ARL-68-0194 (Nov. 1968)

Investigation of Surface Self-Diffusion of Niobium in a Strong Electric Field by Means of a Field Emission Microscope
G. A. Odishariya
Fiz. Tverd. Tela, 10(5):1425-1432 (1968)
Sov. Phys. — Solid State, 10(5):1130-1135 (1968)

Emission Parameters of Faces of a Niobium Single Crystal
O. D. Protopopov and I. V. Strigushchenko
Fiz. Tverd. Tela, 10(3):943-945 (1968)
Sov. Phys. — Solid State, 10(3):747-748 (1968)

Field Electron Microscope Study of Niobium Surfaces
William R. Savage
Phys. Stat. Sol., 25:131-138 (1968)

Etude de la migration superficielle du Ta par le microscope à émission de champ de Muller
N. Drandarov, K. Bobev and V. Kanev
Izv. Inst. Elektron., Sofia, 3:27-38 (1967)

Adsorption of Cs and O on Ta
D. F. Fehrs and R. R. Stickney
27th Annual Conf. Physical Electronics, American Physical Society, held at Massachusetts Institute of Technology, Cambridge, Mass. (March 20-22, 1967), pp. 77-84

Study of the Niobium (110) Surface Using Low-Energy Electron Diffraction Techniques
T. W. Haas
Surface Sci., 5:345-358 (1966)

Low Energy Electron Diffraction Study of the Formation of TaO (111) on Ta (110)
J. E. Boggio and H. E. Farnsworth
Surface Sci., 3:62 (1965)

Das Gesamtemissionsvermögen von Niob
G. Horz, W. Durrschnabel, and E. Gebhardt
J. Nucl. Mater., 17:277 (1965)

Field-Evaporation End Form of Tantalum
S. Nakamura and E. W. Muller
J. Appl. Phys., 36:2535 (1965)

Interaction of Niobium with Active Gases at Very Low Pressures. I. Adsorption of Nitrogen by Niobium
R. A. Pasternak and R. Gibson
Acta Met., 13:1031 (1965)

Low Energy Electron Diffraction and Photoelectric Study of (110) Tantalum as a Function of Ion Bombardment and Heat Treatment
J. E. Boggio and H. E. Farnsworth
Surface Sci., 1:399 (1964)

A New Feature in the Metallographic Etching of Niobium
R. S. Eary and R. D. Johnson
Metallurgia, 69:43-49 (1964)

Determination and Application of Thermophysical Properties of Refractory Metals
G. D. Rieck
The Science and Technology of Tungsten, Tantalum, Molybdenum, Niobium, and Their Alloys, Pergamon Press, Oxford (1964), pp. 205-217

Work-Function Measurements and Lattice Defects
J. A. Dillon, Jr., and R. M. Oman
(Department of Physics, Brown University, Providence, Rhode Island), reprinted from 1963 Transactions, the Tenth National Vacuum Symposium, American Vacuum Society

Study of Electrical and Physical Characteristics of Secondary Emitting Surfaces
W. T. Peria, ed.
(Minnesota Univ., Minneapolis), Contract AF33(657)-8040 (Jan. 23, 1963), 126 pp.

Contributions to Emission Microscopy
E. Guyenot and E.-A. Soa
Cesk. Casopis Fys., 12:608-610 (April 1962)

The Emittance of Chromium, Columbium, Molybdenum, Tantalum, and Tungsten
W. D. Wood, H. W. Deem, and C. F. Lucks
(Defense Metals Information Center, Battelle Memorial Institute, Columbus, Ohio), DMIC-Memo-141, (December 10, 1961), 48 pp.

Spectral Emissivity, Total Emissivity, and Thermal Conductivity of Molybdenum, Tantalum, and Tungsten Above 2300°K
R. D. Allen, L. F. Glasier, and P. L. Jordon
J. Appl. Phys., 31:1382-1387 (1960)

Mechanical and Physical Properties of the Refractory Metals, Tungsten, Tantalum, and Molybdenum Above 4000°F
L. F. Glasier, Jr., R. D. Allen, and I. L. Saldinger
(Aerojet-General Corp., Azusa, Cal.), Rept. M-1826 (April 1959), 89 pp.

1.d.4. Group VI

Adsorption of Barium on a (110) Face of a Molybdenum Single Crystal
B. V. Bondarenko and V. I. Makhov
Sov. Phys. — Solid State, 12:1522-1524 (1971)

Adsorption of Uranium on Tungsten Single Crystal Planes
R. A. Collins and B. H. Blott
J. Phys. D: Appl. Phys., 4:102-113 (1971)

Adsorption of Zirconium on Tungsten Single Crystal Planes
R. A. Collins and B. H. Blott
J. Phys. D: Appl. Phys., 4:114-117 (1971)

Some Applications of Ion-Electron Image Conversion in Field Ion Microscope Studies of Surface Reactions
G. K. L. Cranstoun and D. R. Pyke
Appl. Phys. Letters, 18:341-344 (1971)
Oxygen and carbon monoxide adsorption on tungsten

Crystallographic Anisotropy of the Reflection of Electrons from a Tungsten Single Crystal
M. V. Gomoyunova and B. Z. Aliev
Sov. Phys. – Solid State, 12:1981-1983 (1971)

Simultaneous LEED and RHEED Studies of the Growth of Zirconium on the Tungsten (100) Surface
G. E. Hill, I. Marklund, and J. Martinson
Surface Sci., 24:435-450 (1971)

The Sharpening of Field Emitter Tips by Ion Sputtering
A. P. Janssen and J. P. Jones
J. Phys. D: Appl. Phys., 4:118-123 (1971)

The Adsorption of Co on the (110) Plane of Tungsten
Carl Kohrt and Robert Gomer
Surface Sci., 24:77-103 (1971)

Coadsorption of Yttrium and Oxygen Atoms on Faces of a Tungsten Crystal
O. K. Kultashev and A. P. Makarov
Sov. Phys. – Solid State, 12:1850-1853 (1971)

A Study of Hydrogen Adsorption on a (100) Tungsten Surface Using a Simple HEED System
M. R. Leggett and R. A. Armstrong
Surface Sci., 24:404-416 (1971)

Radiotracer and Photoemission Studies of Co Chemisorbed on Mo (100)
L. D. Matthews
Surface Sci., 24:248-254 (1971)

Adsorption of Lanthanum on Individual Faces of Tungsten Single Crystals
B. M. Palyukh and A. I. Yakivchuk
Sov. Phys. – Solid State, 12:2189-2190 (1971)

Study of the Interaction of Ammonia with Tungsten Surfaces by Thermal Desorption Spectrometry
Y. K. Peng and P. T. Dawson
J. Chem. Phys., 54:950-961 (1971)

Successive Adsorption of Nitrogen, Carbon Monoxide, and Oxygen on Tungsten
V. N. Ageev and N. I. Ionov
Fiz. Tverd. Tela, 12(5):1573-1574 (1970)
Sov. Phys. – Solid State, 12(5):1245 (1970)

Adsorption of CO on the Surface of a Tungsten Single Crystal Partially Covered with Tb
V. S. Ageikin, Yu. G. Ptushinskii, and B. P. Polozov
Fiz. Tverd. Tela, 12(1):221-226 (1970)
Sov. Phys. – Solid State, 12(1):176-180 (1970)

Field Ion Microscope Studies of Iridium Adatom Clusters on (110) Tungsten Surfaces
D. W. Bassett
Surface Sci., 21:181-185 (1970)

Field Ion Microscope Studies of Transition Metal Adatom Diffusion on (110), (211), and (321) Tungsten Surfaces
D. W. Bassett and M. J. Parsley
J. Phys. D: Appl. Phys., 3:707-716 (1970)

Mechanism of Work-Function Reduction by Oxygen Adsorption
J. M. Chen
J. Appl. Phys., 41(12):5008 (1970)
Na-covered W(112) surface

Na-Induced Facetting of W(112)
J. M. Chen
Appl. Phys. Letters, 16:449-452 (1970)

Interaction of N_2 with (100) W
L. R. Clavenna and L. D. Schmidt
Surface Sci., 22:365-391 (1970)

Scattering of Low-Energy Electrons from Hydrogen-Covered Mo (100) Surfaces
George J. Dooley, III, and T. W. Haas
J. Chem. Phys., 52:993-996 (1970)

Chemisorption on Single Crystal Molybdenum (112) Surfaces
G. J. Dooley, III, and T. W. Haas
J. Vacuum Sci. Tech., 7:49-52 (1970)

Determination of Various Mo (110) and W (110) Surface Structures from LEED Results Using a Multiple-Scattering Method
G. Gafner
Surface Sci., 19:9-20 (1970)

The Surface Potential of Hydrogen on Tungsten (100), (110), and (112) Single Crystal Surfaces
B. J. Hopkins and S. Usami
Surface Sci., 23:423-426 (1970)

Oxygen Adsorption, Reconstruction, and Thin Oxide Film Formation on Clean Metal Surfaces: Ni, Fe, W, and Mo
A. M. Horgan and D. A. King
Surface Sci., 23:259-282 (1970)

Electron Diffraction at Crystal Surfaces. IV. Computation of LEED Intensities for "Muffin-Tin" Models with Applications to Tungsten (001)
P. J. Jennings and E. G. McRae
Surface Sci., 23:363-388 (1970)

Work Function Changes Due to the Adsorption of Chlorine, Bromine and Iodine on Tungsten Single Crystal Surfaces
C. W. Jowett and B. J. Hopkins
Surface Sci., 22:392-410 (1970)

Studies on Adsorption and Nucleation of Ge on Tungsten Surface by FEM
H. Kim, H. Araki, and E. Sugata
Japan. J. Appl. Phys., 9:1445-1452 (1970)

Sauerstoffdesorption von W durch Elektronenbeschuss
A. Klopfer
Vakuum-Technik, 19:1-8 (1970)

Entrapment of Helium Ions at (100) and (110) Tungsten Surfaces
E. V. Kornelsen
Can. J. Phys., 48:2812-2823 (1970)

Adsorption and Work Function of Ytterbium and Neodymium on Tungsten
Ts. S. Marinova and Yu. V. Zubenko
Fiz. Tverd. Tela, 12:516–519 (1970)
Sov. Phys. — Solid State, 12:398–400 (1970)

Migration and Evaporation of Ytterbium and Neodymium on Tungsten
Ts. S. Marinova and Yu. V. Zubenko
Fiz. Tverd. Tela, 1:520–525 (1970)
Sov. Phys. — Solid State, 12:401–403 (1970)

Experimental Study of the Surface Structure and Electronic Properties of Single Crystal Molybdenum and Tungsten Ribbons, Final Report, Nov. 1, 1964–Oct. 31, 1969
D. R. Morgan
(Saint Mary's Coll., Winona, Minn.), Contract AT(11-1-1417, COO-1417-4 (Jan 1970)), 37 pp.

Room-Temperature Adsorption of Oxygen on Tungsten Surfaces: A Review
R. G. Musket
J. Less-Common Metals, 22:175–191 (1970)

Auger Electron Spectroscopy Study of Oxygen Adsorption on W(110)
R. G. Musket and John Ferrante
J. Vacuum Sci. Tech., 7:14–17 (1970)

Electron-Impact Desorption of Ions from Polycrystalline Tungsten
M. Nishijima and F. M. Propst
Phys. Rev., B2:2368–2383 (1970)

Desorption and Migration of Cesium Adsorbed on Tungsten in the Presence of an Electric Field
S. Nomura and E. Sugata
Japan. J. Appl. Phys., 9:1229–1235 (1970)

Angular Dependence of the Apparent Work Function of Carbon-Adsorbed Tungsten Tip
F. Okuyama and T. Hibi
Japan. J. Appl. Phys., 9(9):1091–1094 (1970)

Surface and Bulk Carburization of Tungsten Single Crystals
D. F. Ollis and M. Boudart
Surface Sci., 23:320–346 (1970)

Field-Induced Surface Rearrangement in a Field-Ion Microscope
A. van Oostrom
Appl. Phys. Letters, 17(5):206–208 (1970)
N on W

A Comparison Between FIM and LEED Studies of Surface Reconstruction
K. D. Rendulic
Surface Sci., 21(2):401–412 (1970)
Oxygen-covered tungsten (100), (110), and (112) planes

A Relationship Between Surface Potential and Electronegativity for Adsorption on Tungsten Single Crystals
A. J. Sargood, C. W. Jowett, and B. J. Hopkins
Surface Sci., 22:343–356 (1970)

High Resolution FEM Image of Tungsten
M. Sasaki and T. Hibi
Japan. J. Appl. Phys., 9:405 (1970)

Beyond the Local-Density Approximation: Surface Properties of (110) W
J. R. Smith
Phys. Rev. Letters, 25:1023–1026 (1970)

Residence Time Measurements for the Surface Ionization of K on W: The Effects of Surface Contaminants
J. N. Smith, Jr., J. Wolleswinkel, and J. Los
Surface Sci., 22:411–425 (1970)

A New Method of Measuring the Tip Temperature in a Field Electron Microscope
Z. Sisa, L. Eckertova, and J. Babor
Czech. J. Phys., 20B:241–246 (1970)

Adsorption and Nucleation of Silver on Tungsten
E. Sugata and K. Takeda
Phys. Stat. Sol., 38:549–557 (1970)

Interaction of H_2 with (100) W. II. Condensation
P. W. Tamm and L. D. Schmidt
J. Chem. Phys., 52:1150–1160 (1970)

Field Evaporation Rates of Tungsten
T. T. Tsong and E. W. Muller
Phys. Stat. Sol., (a) 1:513 (1970)

Migration of Ti on Tungsten in Strong Electric Fields
G. G. Vladimirov, B. K. Medvedev, and I. L. Sokol'skaya
Fiz. Tverd. Tela, 12:539–544 (1970)
Sov. Phys. — Solid State, 12:413–416 (1970)

Low-Energy Electron Diffraction Study of Tungsten (100) Surface
P. S. P. Wei
J. Chem. Phys., 53:2939–2945 (1970)

Calculation of Migration Energies and Binding Energies for Tungsten Adatoms on Tungsten Surfaces
P. Wynblatt and N. A. Gjosten
Surface Sci., 22:125–136 (1970)

Molecular Beam Study of the Scattering of Rare Gases from the (110) Face of a Tungsten Crystal
S. Yamamoto and R. E. Stickney
J. Chem. Phys., 53:1594–1604 (1970)

The Change in Work Function of the (100) Surface of Tungsten as O_2, N_2, CO_2, and H_2 Are Adsorbed
A. E. Abey
J. Appl. Phys., 40:284 (1969)

Desorption of Hydrogen from Tungsten
V. N. Ageev and N. I. Ionov
Zh. Tekh. Fiz., 34:1523–1529 (Aug. 1969)

Desorption of Ions and Neutral Molecules from Adsorbed Gas on Tungsten Surface by Low-Voltage Electrons
V. N. Ageev and Z. N. Kutsenko
Zh. Tekh. Fiz., 39:1275 (1969)

Reaction of Water Vapors with Tungsten Surface
V. N. Ageev, N. I. Ionov, and Yu. K. Ustinov
Zh. Tekh. Fiz., 39:1337 (1969)

Adsorption of Carbon Monoxide on (110) and (113) Surfaces of Tungsten Single Crystals Partly Covered by Oxygen
V. S. Ageĭkin and Yu. G. Ptushinskiĭ
Fiz. Tverd. Tela, 10(7):2168-2176 (1968)
Sov. Phys. — Solid State, 10(7):1698-1705 (1969)

The Surface Free Energy of Solid Molybdenum
B. C. Allen
J. Less-Common Metals, 17:403-410 (1969)

Surface Characterization by HEED
R. A. Armstrong
Bull. Radio Electr. Engng. Div., Nation. Res. Council Canada, 19(2):1-3 (1969)
Co on W

The Effect of an Electric Field on the Surface Diffusion of Rhenium Adsorbed on Tungsten
D. W. Bassett and M. J. Parsley
Brit. J. Appl. Phys., 2:13 (1969)

Theoretical and Experimental Investigation of the Physics of Crystalline Surfaces, Annual Report Feb. 1, 1968-Jan. 31, 1969
E. Bauer
(Naval Weapons Center, California), NASA-CR-101279 (Jan. 31, 1969), 42 pp.
Quantitative studies of the elastic and inelastic interactions of slow electrons with tungsten single crystal surfaces

Structure and Work Function of Chromium Films on Tungsten
B. V. Bondarenko, V. I. Makhov, and A. M. Kozlov
Fiz. Tverd. Tela, 11:3574-3576 (1969)
Sov. Phys. — Solid State, 11:2991-2993 (1970)

Ellipsometry-LEED Study of the Adsorption of Oxygen on (011) Tungsten
J. J. Carroll and A. J. Melmed
Surface Sci., 16:251-264 (1969)

Mean Adsorption Lifetimes and Activation Energies of Silver and Gold on Clean, Oxygenated, and Carburized Tungsten Surfaces
A. Y. Cho and C. D. Hendricks
J. Appl. Phys., 40:3339 (1969)

Surface Studies of Uranium and Zirconium on Tungsten
R. A. Collins
Thesis, Univ. Southampton, England (1969)

Hydrogen Adsorption of the (110), (112), (100), and (111) Tungsten Single-Crystal Faces
R. A. Collins and B. H. Blott
J. Appl. Phys., 40:5390-5391 (1969)

Ion-Induced Electron Ejection from a Gas-Covered Tungsten Single Crystal
D. L. V. Couchman and R. B. Burtt
Brit. J. Appl. Phys. (J. Phys. D), Ser. 2, Vol. 2, pp. 1177-1179 (1969)

Low-Temperature Surface Migration of Tungsten Activated by Ion Bombardment
Zh. I. Dranova and I. M. Mikhaĭlovskiĭ
Fiz. Tverd. Tela, 12(1):132-137 (1969)
Sov. Phys. — Solid State, 12(1):104-108 (1970)

Influence de la température sur l'émission de champ par certaines faces du tungstène dans le domaine des faibles densités de courant
M. Drechsler, Jean-Marie Bermond, and Jean-Paul Prulhiere
Compt. Rend., 269B:1267-1270 (1969)

Sorption of O at Very Low Pressures by Mo
N. Endow and R. A. Pasternak
J. Vacuum Sci. Technol., 6:340-342 (1969)

Adsorption of CO on Tungsten: Field Emission from Single Planes
T. Engel and R. Gomer
J. Chem. Phys., 50:2428-2437 (1969)

Electronic and Adsorption Properties of Barium on the (100) Face of Tungsten
A. G. Fedorus, Yu. M. Konoplev, and A. G. Naumovets
Fiz. Tverd. Tela, 11:207-209, Jan. 1969
Sov. Phys. — Solid State, 11:160-162 (1969)

Some Studies of the Cr (100) and Cr (110) Surfaces
J. T. Grant and T. W. Haas
Surf. Sci., 17(2):484-485 (1969)

Distribution of Crystal Orientation and Work Function in Tungsten Ribbons
G. A. Haas and R. E. Thomas
J. Appl. Phys., 40:3919-3924 (1969)

Surface Potentials of the Common Gases on Oriented W-Surfaces
B. J. Hopkins, S. Usami, and B. Williams
Vide, 24:26-27 (1969)

Determination of Kinetic Characteristics of Surface Ionization of Cesium Atoms in Tungsten by the Scintillation Method
N. I. Ionov, E. N. Lebedeva, and M. A. Mittsev
Zh. Tekh. Fiz., 39:1905-1910 (Oct. 1969)

Low-Energy Electron-Diffraction Studies of the Interaction of Oxygen with a Molybdenum (100) Surface
H. K. A. Kan and S. Feuerstein
J. Chem. Phys., 50:3618-3623 (1969)

Experimental Study of Positive Ion Emission from W
I. O. Kaposi and M. Riedel
Acta Chim. Acad. Sci. Hungary, 61(4):349-366 (1969)

Formation of a Barium Oxide Film on a Tungsten Filament
T. S. Kirsanova, T. A. Tumareva, and B. M. Shaikhim
Fiz. Tverd. Tela, 11:1331-1335 (1969)
Sov. Phys. — Solid State, 11:1077-1088 (1969)

Investigation of Sodium Adsorption on the Surface of a Tungsten Single Crystal in a Field-Emission Microscope
E. V. Klimenko and V. K. Medvedev
Fiz. Tverd. Tela, 10(7):1986-1990 (1968)
Sov. Phys. — Solid State, 10(7):1562-1565 (1969)

Adsorption of Hydrogen by Tungsten and the Field-Emission Pattern of Tungsten
A. P. Komar, Yu. K. Ustinov, and V. D. Belov
Fiz. Tverd. Tela, 11:142-148, Jan. 1969
Sov. Phys. — Solid State, 11:104-108 (1969)

Ionic Entrapment in Tungsten Monocrystals: A Survey of Effects Observed in Thermal Desorption
E. V. Kornelsen
J. Vacuum Sci. Technol., 6(1):173-174 (1969)
25th National Vacuum Symp., Pittsburgh (Oct. 30-Nov. 1, 1968)

Electronic Field-Emission from a Tungsten Knife Edge
  G. V. Krivoshchekov and E. G. Shirokov
  Radio Eng. Electron Phys., 14:987-989 (1969)

Chemisorption of Iodine on Tungsten (100)
  J. J. Lander and J. Morrison
  Surf. Sci., 17(2):469-473 (1969)

Effect of Temperature on the Work-Function Minimum of Cesiated Tungsten Surfaces
  T. J. Lee, B. J. Hopkins, and B. H. Blott
  J. Appl. Phys., 40:3825 (1969)

Observations of the Field-Evaporation End Form of Tungsten
  B. Loberg and H. Norden
  Arkiv Fys., 39(4):383-395 (1969)

Cesium Diffusion at a Tungsten Surface
  H. M. Lova and H. D. Widerick
  Can. J. Phys., 47:657-663 (1969)

Electron Diffraction Study of Cesium Adsorption on Tungsten
  A. U. MacRae, K. Muller, J. J. Lander, et al.
  Surface Sci., 15:483-497 (1969)

Electrochemical Polishing of Chromium
  G. P. Maitak and I. N. Yudenkova
  J. Appl. Chem. USSR, 42:776-781 (Apr. 1969)

Adsorption and Decomposition of Ammonia on Tungsten
  Kun-ichi Matsushita and Robert S. Hansen
  J. Chem. Phys., 51:472-475 (1969)

Adsorption of Barium on the (110) Face of a Tungsten Single Crystal
  V. K. Medvedev
  Fiz. Tverd. Tela, 10(11):3469-3470 (1968)
  Sov. Phys. — Solid State, 10(11):2752-2753 (1969)

Single-Specimen FEM-LEED Studies: Carbon on Tungsten
  Allan J. Melmed
  J. Appl. Phys., 40:2330 (1969)

Field Evaporation of Tungsten Atoms
  A. J. W. Moore and J. A. Spink
  Surface Sci., 17:262-266 (1969)

The Adsorption of Ionic Salts on a W (110) Surface
  J. Morrison and J. J. Lander
  Surface Sci., 18(2):428-430 (1969)

Measurement of the Surface Free Energy Anisotropy of Clean Tungsten with a Field Ion Microscope
  A. Muller and M. Drechsler
  Surface Sci., 13(2):471-490 (1969)

Calculation of Band-Structure Effects in Field-Emission Tunneling from Tungsten
  Douglas Nagy and P. H. Cutler
  (Osmond Lab., Penn. State Univ., University Park, Penn.), Phys. Rev., 186:651-656 (1969)
  Also AFOSR-70-1629TR (April, 1969), 8 pp.

Field-Ion and Electron Microscope Studies of Lattice Defects in Tungsten and Gold
  Hans Norden
  Thesis, Chalmers University of Technology, Goeteborg (1969), 19 pp.

Adsorption, Migration, and Evaporation of Yttrium on a Tungsten Single Crystal
  P. M. Palyukh and L. L. Sivers
  Fiz. Tverd. Tela, 10(7):2018-2021 (1968)
  Sov. Phys. — Solid State, 10(7):1585-1588 (1969)

The Effect of a Field on the Equilibrium Concentration of Yttrium Atoms Adsorbed on a Tungsten Point
  B. M. Palyukh and L. L. Sivers
  Fiz. Tverd. Tela, 10(12):3723-3725 (1968)
  Sov. Phys. — Solid State, 10(12):2962-2964 (1969)

Desorption of Yttrium from Tungsten in a Strong Electric Field
  B. M. Palyukh and L. L. Sivers
  Fiz. Tverd. Tela, 11(4):1000-1003 (1969)
  Sov. Phys. — Solid State, 11(4):814-816 (1969)

The Cohesive Energy of Surface Atoms of Tungsten on Various Crystal Planes
  Marc Herbert Richman
  Japan. J. Appl. Phys., 8:1273-1274 (1969)

Oxidation of Tungsten at Room Temperature
  V. F. Rybalko, V. Ya. Kolot, and Ya. M. Fogel'
  Fiz. Tverd. Tela, 10(10):3176-3177 (1968)
  Sov. Phys. — Solid State, 10(10):2518-2519 (1969)

Adsorption von O an W
  V. F. Rybalko and others
  Zh. Tekh. Fiz., 39:1717-1719 (1969)

Influence of Oxygen on the Thermonic Emission of a (110) Orientated Tungsten Surface
  H. E. J. Schins and E. Van Andel
  (EURATOM, Joint Nuclear Research Center, Ispra, Italy), EUR-4224-e (April 1969)

Adsorption of Cs on Tungsten: Measurements on Single-Crystal Planes
  Z. Sidorski, I. Pelly, and R. Gomer
  J. Chem. Phys., 50:2382-2391 (1969)

Investigation of Adsorption of Yttrium on Individual Faces of a Tungsten Single Crystal Using an Electron Field-Emission Projector
  L. L. Sivers
  Fiz. Tverd. Tela, 11(4):1035-1036 (1968)
  Sov. Phys. — Solid State, 11(4):843-844 (1969)

Temperature Dependence of Mean Free Path in Secondary Electron Emission (Tungsten)
  H. Taub, R. M. Stern, and V. F. Dvoryankin
  Phys. Stat. Sol., 33:573-577 (1969)

A Study of Facetting of Tungsten Single Crystal Surfaces
  J. C. Tracy and J. M. Blakeley
  Surface Sci., 13(2):313-336 (1969)

Electric-Transport of Adsorbed Thorium Atoms and Barium Oxide Molecules on the (113) Face of a Tungsten Single Crystal (and a Method of Studying the Migration of Adsorbed Atoms and Molecules)
  Yu. S. Vedula
  Ukr. Fiz. Zh., 14:132-137 (1969)

Image Intensity of the Field-Ion Microscopy of W
  V. G. Weizer and A. F. Forestieri
  NASA TN D-5125 (1969)

Temperature Dependence of Field Emission of
Single Crystal Planes of Tungsten
  J. Wysocki
  Acta Phys. Polon, 35:195-198 (1969)

Observations of Reactions in Solids and Formation of Surface Layers with the Emission Electron Microscope
  Ch. Zaminer, R. Graber, and L. Wegmann
  J. Vacuum Sci. and Tech., 6(1):269-272 (1969)

Mechanism of Thermal Desorption of Oxygen on
the (100) Face of a Tungsten Crystal
  Ya. P. Zingerman
  Fiz. Tverd. Tela, 11:3621-3624 (1969)
  Sov. Phys. — Solid State, 11:3038-3040 (1970)

Double-Layer Adsorption of Oxygen on the (100)
Face of a Tungsten Crystal at Room Temperature
  Ya. P. Zingerman and V. A. Ishchuk
  Fiz. Tverd. Tela, 10(12):3720-3722 (1969)
  Sov. Phys. — Solid State, 10(12):2960-2961 (1969)

Angular Dependence of Energy Spectra of Secon-
dary Ions Scattered from Single-Crystal Surfaces
(Molybdenum)
  U. A. Arifov and A. A. Aliev
  Zh. Éksper. Teor. Fiz. USSR, 54(2):354-361 (1968)
  Sov. Phys. — JETP, 27(2):190-193 (1968)

Potential and Kinetic Electron Emission from the
(110) and (111) Faces of a Mo Single Crystal
  U. A. Arifov, R. R. Rakhimov, and Kh. Dzhurakulov
  Fiz. Tverd. Tela, 10(4):1166-1172 (1968)
  Sov. Phys. — Solid State, 10(4):925-929 (1968)

CO Adsorption on W
  R. A. Armstrong
  Can. J. Phys., 46:949-958 (1968)

Investigation of the Adsorption of Potassium Ions
on the Surface of Molybdenum
  N. G. Ban'kovskii, E. M. Stefanovskaya, and V. P. Fedorinov
  Fiz. Tverd. Tela, 10(1):153-158 (1968)
  Sov. Phys. — Solid State, 10(1):113-117 (1968)

Inner Sources in Low Energy Electron Diffrac-
tion: Tungsten (110)
  R. Baudoing, R. M. Stern, and H. Taub
  Surface Sci., 11:255-264 (1968)

Effect of Fluorine and Oxygen on the Work Func-
tion of W
  J. G. Bergman and R. E. Stickney
  NASA-CR-85-160, (1968)

Effects of Hydrogen Contamination on the Work
Function of W
  R. A. Blott et al.
  Surface Sci., 11:149-152 (1968)

Emissions from Tungsten
  V. S. Borodin and F. G. Rutberg
  Teplofiz. Vys. Temp., 6(3):566-568 (1968)
  High Temperature, 6(3):546-548 (1968)

Field Evaporation of a Screw Dislocation in
Tungsten
  Robert W. Carroll and Clarence C. Schubert
  J. Appl. Phys., 39:2339 (1968)

Mean Adsorption Lifetimes and Binding Energies
of Silver and Gold on Clean, Oxygenated and Car-
burized Tungsten Surfaces
  A. Cho
  Thesis, University of Illinois (1968), 109 pp.
  University Microfilms, Ann Arbor, Mich.

Field Emission Through Single Strontium Atoms
Adsorbed on a Tungsten Surface
  H. E. Clark and R. D. Young
  Surface Sci., 12:383-389 (1968)

Adsorption Studies Based on Thermionic Emission
Measurements. I. Cesium on Single-Crystal Tung-
sten
  J. L. Coggins and R. E. Stickney
  Surface Sci., 11:355-369 (1968)

Adsorption of O on W
  J. Czyzewski and C. Workowski
  Acta Phys. Pol., 33:913-921 (1968)

Chemisorption Studies by LEED, final report
  W. E. Danforth
  (Bartol Research Foundation, Swarthmore, Pa.), Contract AT
    AT(30-1)-3830, NYO-3830-5 (March 29, 1968), 79 pp.

Adsorption of O on W
  W. Engelmaier and R. E. Stickney
  Surface Sci., 11:335-369, 370-394 (1968)

Adsorption of Cesium Fluoride on Tungsten
  B. E. Evans, L. W. Swanson, and A. E. Bell
  Surface Sci., 11:1-18 (1968)

Low-Energy Electron Diffraction Study of Oxygen
Adsorption on Molybdenum (111) Surface
  John Ferrante and Gilbert C. Barton
  (National Aeronautics and Space Administration, Lewis Re-
    search Center, Washington), NASA-TN-D-4735 (Aug. 1968),
    19 pp.

Migration of Gases on W
  M. Folman and R. Klein
  Surface Sci., 11:430-442 (1968)

A Study of Grain-Boundary Structure in W with
the Field-Ion Microscope
  R. I. Garber, Zh. I. Dranova, and I. M. Mikhailovskii
  Zh. Eksp. Teor. Fiz., 54(3):714-720 (1968)
  Sov. Phys. — JETP, 27(3):381-383 (1968)

Chemisorption of Nitrous Oxide by Tungsten
  R. P. H. Gasser and C. P. Lawrence
  Surface Sci., 10(1):91-101 (1968)

Behavior of Cesium on the Faces of a Large Tung-
sten Single Crystal
  N. A. Gorbatyi, L. V. Reshetnikova, and V. M. Sultanov
  Fiz. Tverd. Tela, 10(4):1185-1192 (1968)
  Sov. Phys. — Solid State, 10(4):940-946 (1968)

Interaction of Oxygen, Carbon Monoxide and Nitro-
gen with (001) and (110) Faces of Molybdenum
  K. Hayek, H. E. Farnsworth, and R. L. Park
  Surface Sci., 10:429-445 (1968)

Sticking Probabilities of H on Mo-Filaments
  D. O. Hayward and N. Taylor
  Trans. Faraday Soc., 64:1904-1914 (1968)

Absolute Photoelectric Yield of Tungsten Between
31.6 and 304 Å
  L. J. Herous and M. Cohen
  (Air Force Cambridge Research Labs., L. G. Hanscom Field,
    Mass.), AFCRL-69-0231; AD-689 535 (1968), 14 pp.

Calibration methods in the ultraviolet and X-ray regions of the spectrum, Munich Symposium (May 1968)

Adsorption of Barium onto Oriented Tungsten Surfaces
B. J. Hopkins and B. J. Smith
J. Chem. Phys., 49:2136-2140 (1968)

Adsorption of Gold on Tungsten
J. P. Jones
SCI Mono., No. 28, 263-90 (1968)

Low-Energy Electron Diffraction Studies of the Interaction of Oxygen with a Molybdenum (100) Surface
H. K. A. Kan and S. Feuerstein
(Aerospace Corp., El Segundo, Calif.), Contract F04701-68-C-0200, Rept. TR-0200(4250-20)-3 SAMSO-TR-68-429; AD-679976 (Sept. 1968)

Electron Emission from W
M. Kaplit et al.
Adv. Energy Conversion, 7:177-189 (1968)

Field-Emission Patterns of a Tungsten Single Crystal in the Case of Hydrogen Adsorption
A. P. Komar, Yu. K. Ustinov, and V. D. Belov
Dokl. Akad. Nauk SSSR, 179:568, 1968
Sov. Phys. — Dokl., 13:253-255 (1968)

Cesium Ion Emission from a Small Hole in Single-Crystal Tungsten
Guntis Kuskevics
J. Appl. Phys., 39:4076-4079 (1968)

Adsorption and Electron Emission of Molybdenum Films on a Tungsten Single Crystal
V. A. Kuznetsov and B. M. Tsarev
Fiz. Tverd. Tela, 9(9):2524-2528 (1967)
Sov. Phys. — Solid State, 9(9):1987-1991 (1968)

Field Evaporation from Tungsten and the Bonding of Surface Atoms
A. J. W. Moore and J. A. Spink
Surface Sci., 12:479-496 (1968)

Some Properties of Thin W-Films
H. M. Montagu-Pollack and others
Surface Sci., 12:1-18 (1968)

Comparison of Brightness of Field-Ion Image for Different Metals
Shogo Nakamura, Tsukasa Kuroda, and Toshiyuki Adachi
Japan J. Appl. Phys., 7:1127 (1968)
W and Mo

Investigation of the Structure and Work Function of Platinum Films on Tungsten and of Adsorption of Cesium on Such Films
A. G. Naumovets and A. G. Fedorus
Fiz. Tverd. Tela, 10(3):801-808 (1968)
Sov. Phys. — Solid State, 10(3):627-633 (1968)

Adsorption of Aluminium on Tungsten in the Field Electron Microscope
H. Neumann
Ann. Phys., 21:414-416 (1968)

Dependence of the Critical Field Strength for Vacuum Breakdown of Tungsten on the Field Emitting Area
A. Van Oostrom

Paper No. 4167, Proc. Intern. Symp. on Discharges and Electrical Insulation in Vacuum, Paris, September 1968 (1968) p. 174

Field-Emission Study of the Adsorption of Sodium on Faces of Tungsten and Rhenium Single Crystals
A. P. Ovchinnikov and B. M. Tsarev
Fiz. Tverd. Tela, 9(7):1927-1938 (1967)
Sov. Phys. — Solid State, 9(7):1519-1524 (1968)

Binding of Transition Elements on W-Surface
E. W. Plummer
Dissertation, Cornell University, 1968

Atomic Binding of Transition Metals on Clean Single-Crystal Tungsten Surfaces
E. W. Plummer and T. N. Rhodin
J. Chem. Phys., 49:3479-3496 (1968)

Mass-Spectrometric Investigation of the Interaction of Oxygen with (110) and (100) Faces of a Tungsten Single Crystal
Yu. G. Ptushinskii and B. A. Chuikov
Fiz. Tverd. Tela, 10(3):722-728 (1968)
Sov. Phys. — Solid State, 10(3):565-570 (1968)

Field-Ion Microscope Studies of Planar Faults
S. Ranganathan
Field-Ion Microscopy (John J. Hren, and S. Ranganathan, eds.) Plenum Press, New York (1968), pp. 120-136

Emission from W
W. R. Rueth and D. Lichtman
Surface Sci., 12:96-101 (1968)

Reactions of Hydrocarbons on W
R. R. Rye
Dissertation, Iowa State Univ., Ames (1968)

Simultaneous Adsorption of Hydrogen and Nitrogen on Tungsten at 300°K
J. H. Singleton
J. Vacuum Sci. Tech., 5:109-117 (1968)

The Analysis of Field-Ion Micrographs: Stacking Faults in Tungsten
D. A. Smith and K. M. Bowkett
Phil. Mag., 18:1219 (1968)

Secondary Emission from W
R. M. Stern and H. Taub
Phys. Rev. Letters, 20:1340-1343 (1968)

Field-Evaporation of Tungsten in Field Ion Microscope
Shozo Tamaki and Tsukasa Kuroda
Japan. J. Appl. Phys., 7:1202-1209 (1968)

A Study of Facetting of W Single-Crystal Surfaces
J. Charles Tracy and J. M. Blakely
Bull. Am. Phys. Soc., 13:1647 (1968)

Oxygen Adsorption on Mo
M. P. Vasko and J. G. Pluschinski
Ukr. Fiz. Zur., 13:344-347 (1968)

Desorption of Titanium from Tungsten in a Strong Electric Field
G. G. Vladimirov
Fiz. Tverd. Tela, 10(4):1207-1213 (1968)
Sov. Phys. — Solid State, 10(4):957-962 (1968)

Mechanism and Kinetics of the Process of Sorption of Oxygen on the (111) Face of a Tungsten Single Crystal
Ya. P. Zingerman and V. A. Ishchuk
Fiz. Tverd. Tela, 9(9):2529-2538 (1967)
Sov. Phys. – Solid State, 9(9):1992-2000 (1968)

Mass-Spectrometric Investigation of the Sorption of Oxygen on the Surface of Tungsten
V. S. Ageikin and Yu. G. Ptushinskii
Ukr. Fiz. Zh., 12(9):1483-1488 (1967)

Adsorption of CO on a W(100) Surface
J. Anderson and P. J. Estrup
J. Chem. Phys., 46:563-567 (1967)

Adsorption of Hydrogen on Tungsten Single Crystal Surfaces
R. A. Armstrong
Surface Sci., 6:478 (1967)

Desorption Kinetics of Ga on W
J. R. Arthur
J. Chem. Phys., 46:188-192 (1967)

Oxidation of the (112) Face of Tungsten
C. C. Chang and L. H. Germer
Surface Sci., 8:115-129 (1967)
Electron diffraction

Chemisorption of Carbon Monoxide on Molybdenum
A. D. Crowell and L. D. Matthews
Surface Sci., 7:79-89 (1967)

Electron Emission from W
A. A. Dorozkin
Tr. Leningrad. Politechn. Inst., 277:89-92 (1967)

Tungsten Outgassing
J. B. Eby and M. P. Schrank
J. Chem. Phys., 46:272-276 (1967)

Characterization of Chemisorption by LEED
P. J. Estrup and J. Anderson
Surface Sci., 8:101-114 (1967)
W

LEED Studies of the Adsorption Systems W(100) + N₂ and W(100) + N₂ + CO
P. J. Estrup and J. Anderson
J. Chem. Phys., 46:567-570 (1967)

Oxygen Adsorption on W
F. J. Estrup and J. R. Anderson
27th Annual Conference Physical Electronics, American Physical Society, held at Massachusetts Institute of Technology, Cambridge, Mass. (March 20-22, 1967), pp. 47-53

Investigation of Surface Bombardment Damage by LEED
H. E. Farnsworth and K. Hayek
Surface Sci., 8:35-56 (1967)

Field Ion Microscope Investigation of the Diffusion of Carbon in Tungsten
Robert Dexter French
Thesis from Brown University, Providence, Rhode Island (1967), 96 pp.

The Structure and Properties of Metal Surfaces
N. A. Gjostein and W. L. Winterbottom
Fundamentals of Gas-Surface Interactions, Academic Press, Inc., New York (1967), pp. 42-74

Sticking Probabilities, Heats of Adsorption and Redistribution Processes of Nitrogen on Tungsten Films at 195 and 290°K
D. O. Hayward, D. A. King, and F. C. Tompkins
Proc. Roy. Soc., 297:305 (1967)

Surface Self-Diffusion on Tungsten by the Mass Transfer Method
Lai Ho-yi
Jernkontorets Ann., 151:801-821 (1967)

A LEED Study of CO and CO₂ Adsorption on Mo(110)
A. G. Jackson and M. P. Hooker
Surface Sci., 6:297-308 (1967)

Proton Scattering on a W Single Crystal
V. S. Kulikauskas, M. M. Malov, and A. F. Tulinov
Zh. Éksp. Teor. Fiz., 53(2):487-489 (1967)
Sov. Phys. – JETP, 26(2):321-322 (1968)

Oxygen Chemisorption on W Ribbons
A. E. Lee and B. A. Pethica
J. Chem. Phys., 46:141-155 (1967)

The Work Functions of Cesium on the (100) and (110) Oriented Faces of Tungsten Single Crystals
T. J. Lee, B. H. Blott, and B. J. Hopkins
Appl. Phys. Letters, 11:3_1 (1967)

Work Function of Tungsten Single Crystals
H. M. Love and J. R. Wilson
Can. J. Phys., 45:225 (1967)

A Field-Ion Microscope Investigation of Oxygen in Tungsten
Eugene S. Machlin
Trans. Quart., 60:260 (1967)

Experimental Study of the Surface Structure and Electronic Properties of Single Crystal Molybdenum and Tungsten Ribbons, progress report Nov. 1, 1966-Aug. 1, 1967
Donald R. Morgan and William E. Blass
Contract AT(11-1)-1417, COO-1417-3 (August 7, 1967), 13 pp.
LEED

Field-Effect Emission – A Method for the Study of Adsorption and Its Effects on Electron Emission
J. Nikliborc
Prace Przemysl. Inst. Elektroniki (Poland), No. 3, 339-346 (1967)
Franco-Polish Colloquium on vacuum technique, Warsaw, Poland (June 27-29, 1966)
K, Sr and Ge on W

Adsorption and Electron Emission of Potassium Films on Faces of a Tungsten Single Crystal
A. P. Ovchinnikov
Fiz. Tverd. Tela, 9(2):628-633 (1967)
Sov. Phys. – Solid State, 9(2):483-487 (1967)

The Surface Physics of Simple and Complex Electron Emitters, final report Nov. 1, 1954-July 31, 1967

Alex A. Petrauskas and Edward A. Coomes
Contract Nonr-1623(01), AD-660211 (June 1967), 74 pp.
The kinetics of adsorbed layers of KCl on W, Cl on W, Sr on
W, O on W, and O on Mo have been investigated in the tem-
perature range 300°K to 1900°K

Atomic Perfection and Field Emission from
Tungsten (110) Surface
E. W. Plummer and T. N. Rhodin
Appl. Phys. Letters, 11:194-196 (1967)

Chemisorbed β Phase of N_2 on W
J. L. Robins, W. K. Warburton, and T. N. Rhodin
J. Chem. Phys., 46:665-673 (1967)

Anisotropic Work Functions of Molybdenum
Single Crystals
E. M. Savitskii, I. V. Burov, L. N. Litvak, G. S. Burkhanov, and
N. N. Bokareva
Sov. Phys. — Tech. Phys., 11:974-976 (1967)

Inelastic Scattering of Low Energy Electrons
from Surfaces
E. J. Scheibner and L. N. Tharp
Surface Sci., 8:247-265 (1967)

Adsorption of O on W Surfaces
L. W. Swanson et al.
J. Chem. Phys., 46:54-62 (1967)

Energy Spectra of Inelastically Scattered Elec-
trons and LEED Studies of Tungsten
L. N. Tharp and E. J. Scheibner
J. Appl. Phys., 38:3320-3330 (1967)

Adsorption on Single Crystal Planes, final report
Russell D. Young and Howard E. Clark
(National Bureau of Standards, Washington), ARL-67-0198; AD-
663 427 (Sept. 1967), 37 pp.

Investigation of the Process of Sorption of Oxy-
gen on (100) and (110) Faces of a Tungsten Single
Crystal Using the Electron-Stimulated Desorp-
tion Effect
Ya. P. Zingerman and V. A. Ishchuk
Fiz. Tverd. Tela, 9:797-804 (1967)
Sov. Phys. — Solid State, 9:623 (1967)

Mechanism of the Sorption of Oxygen on the (100)
Face of a Tungsten Single Crystal
Ya. P. Zingerman and V. A. Ishchuk
Fiz. Tverd. Tela, 8(10):2999-3003 (1966)
Sov. Phys. — Solid State, 8(10):2394-2397 (1967)

Low-Energy Electron Diffraction Study of the
(100) Plane of a Tungsten Single Crystal
R. O. Adams
RFP-643 (April 1966)

Field Ionization Studies of Tungsten
R. W. Carroll
Dissertation Abstr. 26, 3645 (1966)

Trapping and Thermal Re-emission of He from
Polycrystalline W
K. Erents and G. Carter
Vacuum, 16:523-528 (1966)

Chemisorption of Hydrogen on W(100)
P. J. Estrup and J. Anderson
J. Chem. Phys., 45:2254 (1966)

LEED Studies of Thorium Adsorption on Tung-
sten
P. J. Estrup, J. Anderson, and W. E. Danforth
Surface Sci. Neth., 4:286-298 (1966)

Adsorption of Nitrogen on the (113) Plane of
Tungsten Single Crystals
V. M. Gavrilyuk, V. K. Medvedev, and T. P. Smereka
Fiz. Tverd. Tela, 8:2482-2484 (1966)
Sov. Phys. — Solid State 8:1983 (1967)

Diffraction Study of Oxygen Adsorption on a (110)
Tungsten Face
L. H. Germer and J. W. May
Surface Sci. Neth., 4:452-470 (1966)

Investigation of the Adsorption of Lithium on the
Surface of a Tungsten Single Crystal in a Field-
Emission Projector
V. M. Gavrilyuk and V. K. Medvedev
Fiz. Tverd. Tela, 8(6):1811-1818 (1966)
Sov. Phys. — Solid State, 8(6):1439-1444 (1966)

Emission and Adsorption Properties of the W-La
System
N. A. Gorbatyi and S. Khashimova
Fiz. Tverd. Tela, 8(5):1441-1448 (1966)
Sov. Phys. — Solid State, 8(5):1150-1155 (1966)

Low-Energy Electron-Diffraction Study of Molyb-
denum (110) Surfaces
T. W. Haas and A. G. Jackson
J. Chem. Phys., 44:2921 (1966)

The Adsorption of Oxygen on the Surface of (110)
and (100) Oriented Tungsten Single Crystals
B. J. Hopkins and K. R. Pender
Surface Sci. Neth., 5:155-159 (1966)

Adsorption of H on (110), (100), (111), and (113)
Oriented W Single Crystal Surfaces
B. J. Hopkins and K. R. Pender
Surface Sci., 5:316-324 (1966)

Surface Ionization of Cesium and Thermionic
Emission from Planar Single Crystals of Tung-
sten
D. R. Koenig (thesis)
UCRL-11857 (June 1966), 305 pp.

Thermal Desorption of Inert Gases from Tung-
sten Monocrystals
E. V. Kornelsen and M. K. Sinha
Appl. Phys. Letters, 9:112 (1966)

Electron Probe Surface Mass Spectrometry Study
of Oxygen on Molybdenum
D. Lichtman and T. R. Kirst
Phys. Letters, 20:7 (1966)

Adsorption and Electron Emission of Alkaline-
Earth Metal Films on Tungsten, Iridium, and
Rhodium
V. I. Makukha and B. M. Tsarev
Fiz. Tverd. Tela, 8(5):1417-1427 (1966)
Sov. Phys. — Solid State, 8(5):1130-1138 (1966)

Adsorption of Carbon Monoxide on a Tungsten
(110) Surface
J. W. May and L. H. Germer
J. Chem. Phys., 44:2895 (1966)

Coadsorption of O and Co upon a (100) W Surface
J. W. May, L. H. Germer, and C. C. Chang
J. Chem. Phys., 45:2383 (1966)

Flash Desorption and Isotopic Mixing of Hydrogen and Deuterium Adsorbed on W, Ir, and Rh
V. J. Mimeault and R. S. Hansen
J. Chem. Phys., 45:2240 (1966)

Adsorption of Silicon on Tungsten
Hans Neumann
Ann. Phys., 7:144-158 (1966)

Adsorption of Silicon on Tungsten Single Crystal Surfaces in the Field Electron Microscope
H. Neumann
Z. Naturforsch., 21:2122-2124 (1966)

Anisotropy of Work Function of Mo Monocrystals
E. M. Savitskii, I. V. Burov, L. N. Litvak, G. S. Burkhanov, and N. N. Bokareva
Zh. Tekhn. Fiz., 36:1310 (1966)

Adsorption of Potassium on Tungsten: Measurements on Single-Crystal Planes
L. D. Schmidt and R. Gomer
J. Chem. Phys., 45:1605-1623 (1966)

Evaporative Lifetimes of Copper, Chromium, Beryllium, Nickel, Iron, and Titanium on Tungsten and Oxygenated Tungsten
H. Shelton and A. Y. H. Cho
J. Appl. Phys., 37:3544 (1966)

Adsorption of Au on W by Electron Microscopy
I. L. Sokolskaya and Yu. V. Zubenko
Izv. Akad. Nauk. SSSR, Ser. Fiz., 30(12):1921-1928 (1966)

Interaction of Boron with Tungsten Single-Crystal Substrates
Charles W. Tucker, Jr.
Surface Sci., 5:179-186 (1966)

Field Emission Microscopy of 112-, 100-Oriented W Tip
H. Utsugi, T. Suwa, and H. Yokoyama
Nippon Kagaku Zasshi, 87:1289-1293 (1966)

Field Emission Microscopy of the Residual Gases Adsorbed on W
H. Utsugi, H. Yokoyama, M. Maeda, and T. Suwa
Nippon Kagaku Zasshi, 87:1284-1288 (1966)

Experimental Evidence Supporting Chemisorption Without Surface Rearrangement for Nitrogen on Tungsten at and Below Room Temperatures
A. van Oostrom
Phys. Letters, 22:137 (1966)

Anomalous Work Function of the Tungsten (110) Plane
R. D. Young and H. E. Clark
Appl. Phys. Letters, 9:265 (1966)

Adsorption of Oxygen on the (110) Face of a Tungsten Single Crystal
Ya. P. Zingerman and V. A. Ishchuk
Fiz. Tverd. Tela, 8(3):912-919 (1966)
Sov. Phys. — Solid State, 8(3):728-733 (1966)

Emission Properties of Mo Single Crystals
U. V. Azizov, U. V. Vakhidov, V. M. Sultanov, B. N. Sheinberg, and G. N. Shuppe
Fiz. Tverd. Tela, 7(9):2759-2762 (1965)
Sov. Phys. — Solid State, 7(9):2232-2234 (1966)

The Thermal Stability and Rearrangement of Field-Evaporated Tungsten Surfaces
D. W. Bassett
Proc. Roy. Soc., A286:191 (1965)

Über die Beobachtung einer Korngrenzenwanderung auf der Wolframkathode im Feldelektronenmikroskop
H. Hesse
Phys. Stat. Sol., 8:695 (1965)

Adsorption and Surface Diffusion of Copper on Tungsten
A. J. Melmed
J. Chem. Phys., 43:3057 (1965)

High Temperature Observation of Tungsten Cathode by Field Emission Microscope
Fumio Okuyama and Tadatoshi Hibi
Japan J. Appl. Phys., 4:337 (1965)

Field-Microscopic Investigations of Tungsten-Molybdenum Mixed-Crystal Systems
E. K. Caspary and E. Krautz
(OSRAM-Studiengesellschaft, Augsburg, Ger.), Z. Naturforsch., 19a: 591-595 (May 1964)

A Study of Etch-Figures of Dislocation Lines in Molybdenum Crystals
Duan Fong, Nay-ben Ming, and Chi Li
(Nanking Univ., China), Wu Li Hsueh Pao, 20:337-351 (Apr. 1964)

Surface Potentials of Nitrogen on Individual Crystal Faces of Tungsten
A. A. Holscher
(Koninklijke/Shell-Laboratorium, Amsterdam), J. Chem. Phys., 41:579-580 (1964)

Field Ion Microscopic Studies of Tungsten and Platinum
E. S. Machlin
NASA-CR-58711, Grant NsG-294-63 (July 1, 1963 to June 30, 1964), 9 pp.

Preparing Infiltrated Tungsten for Examination
Robert E. Matt and Isamu Yoshioka
Metal Progr., 85(4):95-96 (Apr. 1964)

Determination and Application of Thermophysical Properties of Refractory Metals
G. D. Rieck
The Science and Technology of Tungsten, Tantalum, Molybdenum, Niobium, and Their Alloys, Pergamon Press, Oxford (1964), pp. 205-217

Mobility of Edge Dislocations on {110} Planes in Tungsten Single Crystals
H. W. Schadler
(General Electric Research Laboratory, Schenectady, New York), Acta Met., 2:861 (1964)

Mass Spectrometric Study of the Oxidation of Tungsten
P. O. Schissel and O. C. Trulson
Union Carbide Research Institute Tech. Rept. No. C-26 (October 1964), Contract DA-30-069-ORD-2787

Tungsten (110) Surface Characteristics in Low-Energy Electron Diffraction
R. M. Stern
(Department of Physics, Polytechnic Institute of Brooklyn), Appl. Phys. Letters, 5(11):218 (1 Dec. 1964)

A Low Energy Electron Diffraction Study of the Structural Effect of Oxygen on the (111) Face of a Tungsten Crystal
N. J. Taylor
Surface Sci., 2:544 (1964)

Reflection and Diffraction of Slow Electrons from Single Crystals of Tungsten
I. H. Khan, J. P. Hobson, and R. A. Armstrong
(Radio and Electrical Engineering Division, National Research Council, Ottawa, Canada), Phys. Rev., 129:1513-1523 (1963)

The Effect of Temperature and Heating Rate on the Secondary Recrystallization of Doped Tungsten Wires
M. Mannerkoski
(Oy Fiskars AB, Research Laboratory, Aminnefors, Finland), J. Inst. Metals, 92:149 (1963-64)

Preparation of Molybdenum Single Crystals by Electron-Beam Floating-Zone Melting
Nai-pen Ming, Ts'un-yi Fan, Ch'i Li, Yu-shan Hsu, and Tuan Feng
FTD-TT-64-350
Wu Li Hsueh Pao, 19:160-164 (1963)

Corrosion of Tungsten and Iridium by Field Desorption of Nitrogen and Carbon Monoxide
J. F. Mulson and E. W. Muller
(Department of Physics, Pennsylvania State University, Pennsylvania), J. Chem. Phys., 38:2615-2619 (1963)

Temperature Dependence of the Work Function of Single Crystal Planes of Tungsten in the Range 78°-293°K
A. van Oostrom
Phys. Letters, 4:34 (1963)

Absolute Measurements of the Ionization Yield of Alkali-Atoms on Metal Surfaces. Part I. Potassium Atoms on Tungsten and Platinum Surfaces
Walter Schoen
SC-T-70-4019
Translated by M. I. Weinreich, Sandia Labs., Albuquerque, N. Mexico, from Z. Phys., 176:237-252 (1963)

On the Nature of Dislocation Etch Pits in Tungsten
I. Berlec
J. Appl. Phys., 33:197-202 (1962)

Contributions to Emission Microscopy
E. Guyenot and E.-A. Soa
Cesk. Casopis Fys., 12:608-610 (April 1962)

The Emittance of Chromium, Columbium, Molybdenum, Tantalum, and Tungsten
W. D. Wood, H. W. Deem, and C. F. Lucks
(Defense Metals Information Center, Battelle Memorial Institute, Columbus, Ohio), DMIC-Memo-141, (December 10, 1961), 48 pp.

Spectral Emissivity, Total Emissivity, and Thermal Conductivity of Molybdenum, Tantalum, and Tungsten Above 2300°K
R. D. Allen, L. F. Glasier, and P. L. Jordon
J. Appl. Phys., 31:1382-1387 (1960)

Mechanical and Physical Properties of the Refractory Metals, Tungsten, Tantalum, and Molybdenum Above 4000°F
L. F. Glasier, Jr., R. D. Allen, and I. L. Saldinger
(Aerojet-General Corp., Azusa, Cal.), Rept. M-1826 (April 1959), 89 pp.

Observations on Regular Arrays of Etch Pits in a Titanium-10% Molybdenum Alloy
T. H. Schofield and A. E. Bacon
(National Physical Laboratory, Teddington, England), Acta Met., 7:403 (1959)

Dislocation Etch Pits in Tungsten
Ursula E. Wolff
(Refractory Metals Laboratory, Lamp Wire & Phosphors Department, General Electric Company, Cleveland, Ohio), Acta Met., 6:559-561 (1958)

Ion Reflection and Secondary Electron Emission During Impact of Alkali Ions on Pure Molybdenum Surfaces
Curt Brunnee
ORNL-tr-467, Z. Physik, 147:161(1957)

Molybdenum Catalyst Bibliography 1964-1967
E. N. Losey and D. K. Means
Climax Molybdenum Company, New York, NP-17809 Suppl. 1, 189 pp.

Initial Oxidation of Tungsten and Tantalum
Doyle W. Rausch
Dissertation Abstr., 65, 13, 273, 125 pp.

1.e. Electrical Properties

1.e.1. General

Electrical Resistance, Thermal Conductivity and Thermoelectric Power of Transition Metals at High Temperatures
T. Aisaka and M. Shimizu
J. Phys. Soc. Japan, 28:646 (1970)
Mo, W, Pd, Pt, Ir, Rh

Thermopower of Cubic Transition Metals
R. Carter, A. Davidson, and P. A. Schroeder
J. Phys. Chem. Solids, 31:2374-2377 (1970)
V, Nb, Ta, Mo, W, Re, Rh, Ir

Investigation of Structure and Electrical Conductivity of Molybdenum and Titanium Thin Films Deposited on NaCl Single Crystals
R. S. Panchishin, Z. V. Stasyuk, and D. M. Freik
Fiz. Tverd. Tela, 10(9):2731-2737 (1968)
Sov. Phys. — Solid State, 10(9):2149-2153 (1969)

Thermoelectric Powers of Transition Metals at High Temperatures
M. V. Vedernikov
Adv. Phys., 18:337-370 (1969)

Low-Temperature Electrical Properties of Some Transition Metals and Transition-Metal Carbides
F. W. Clinard, Jr., and C. P. Kempter
J. Less-Common Metals, 15:59 (1968)
Ti, Zr, Hf, Nb, Ta, Cr, Mo, W

Survey of Electrical Resistivity Measurements on 16 Pure Metals in the Temperature Range 0 to 273°K
L. A. Hall
NBS-TN-365 (Feb. 1968), 113 pp.

Ion-Bombardment-Induced Resistivity Changes in Thin Films of Silver, Gold, Titanium, and Tungsten
B. Navinsek and G. Carter
Can. J. Phys., 46:719-723 (1968)
From Conf. on Atomic Collisions and Penetration Studies with Energetic (keV) Ion Beams, Chalk River, Ontario

The Development of Knowledge and Understanding of the Anomalous Resistivity of Diluted Metallic Solutions of Transition Metals
G. J. Van Den Berg and C. J. Gorter
Helv. Phys. Acta, 41:1230-1233 (1968)

Ionentransport von C in Ta, Ti and W
D. F. Kalinovitsch, I. I. Kovenski, and M. D. Smolin
Ukr. Fiz. Zh., 12:688 (1967)

Hall Effect in the Transition Metals
N. V. Volkenshtein and E. V. Galoshina
Fiz. Metallov i Metalloved., 20:475-478 (1965)

Electronic Structure and Superconductivity of Transition Metals and Their Alloys
B. R. Coles
Rev. Mod. Phys., 36:139 (1964)

Superconductivity in the Transition Metals
T. H. Geballe
Rev. Mod. Phys., 36:134 (1964)

A Critical Review of Refractories
Edmund K. Storms
(Los Alamos Scientific Lab., N. Mex.), Contract W-7405-eng-36, LA-2942 (March 1964), 252 pp.

Galvanomagnetic Properties of Single Crystals of Transition Metals
N. E. Alekseevskii, V. S. Egorov, G. E. Karstens, and B. N. Kazak
Soviet Phys. — JETP 16, 519-520 (1963)

The Electrical Resistivity of Dislocations
Z. S. Basinski, J. S. Dugdale, and A. Howie
(Division of Pure Physics, National Research Council, Ottawa, Canada), Phil. Mag., 8:1989 (1963)

Electrical Resistivity of Certain Transition Metals at High Temperatures
I. I. Kovenskii and G. V. Samsonov
Fizika Metallov i Metallovedenie, 15(6):940-941 (1963)
The Physics of Metals and Metallography, 15(6):124-125 (1963)

Spin-Dependent Scattering and Resistivity of Magnetic Metals and Alloys
T. van Peski-Tinbergen and A. J. Dekker
(Instituut voor Kristalfysica der Rijksuniversiteit, Groningen, Nederland), Physica, 29:917-937 (1963)

On the Measurement of Resistivity of Metal Bars by Eddy Current Decay
R. Stern, M. Levy, R. Kagiwada, and I. Rudnick
Appl. Phys. Letters, 2:80 (1963)

Preparation of Very Thin Suspended Metal Films
G. J. Unterkofler and R. R. Verderber
(Burroughs Corporation, Burroughs Laboratories, Paoli, Pennsylvania), Rev. Sci. Instr., 34:820-821 (1963)

Variation of the Elastic Moduli at the Superconducting Transition
G. A. Alers and D. L. Waldorf
IBM J. Res. Develop., 6:89-93 (1962)

Effect of Plastic Deformation and Annealing Temperature on Superconductive Properties
J. H. Hauser and E. Buehler
Phys. Rev., 125:142-148 (1962)

Diffusion Studies of Vacancies and Impurities
David Lazarus
J. Phys. Rad., 23:772 (1962)

Thermo-Electricity at Low Temperatures. IX. The Transition Metals as Solute and Solvent
D. K. C. MacDonald, W. B. Pearson, and I. M. Templeton
Proc. Roy. Soc., A 266:161-184 (1962)

The Effects of Chemical Impurities and Physical Imperfections on the Thermoelectricity of Metals
W. B. Pearson
Chap. 10 of Ultra-High-Purity Metals, Seminar of Am. Soc. for Metals (Oct. 21 and 22, 1961), ASM, Metals Park, Ohio (1962)

A Deformation (Strain) Apparatus for the Measurement of the Electrical Resistance of Metals at Lowest Temperatures, and Investigations into the Plastic Behavior of Copper Monocrystals
Wolfgang Schule, Otto Buck, and Eberhard Koster
SCL-T-434 (1962)
Z. Metallk., 53:172-177 (1962)

Ultra-High-Purity Metals
Papers presented at a seminar of American Society of Metals (October 21 and 22, 1961), American Society of Metals, Metals Park, Ohio (1962)

Changes in the Electrical Resistance of Several Metals up to Pressures of 250,000 kg/cm^2
L. F. Vereshchagin, A. A. Semerchan, N. N. Kuzin, and S. V. Popova
Soviet Phys. — Dokl. 6, 391-392 (1961)

Propriétés des métaux et alliages aux basses températures (Properties of Metals and Alloys at Low Temperatures)
E. Justi
CEA-tr-A-789
Z. Metallk., 51(1):1-17 (1960)

Messung des Temperaturkoeffizienten von Kontaktpotentialen
K. Andes
Helv. Phys. Acta, 32 (1959)

Electrical and Thermal Resistivity of Dislocations
Peter Carruthers
Phys. Rev. Letter, 2:336 (1959)

Measurements of Thermoelectricity Below 1°K —III
D. K. C. MacDonald, W. B. Pearson, and I. M. Templeton
Phil. Mag., 4:380 (1959)

Electrical and Thermal Resistivity of the
Transition Elements at Low Temperatures
  G. K. White and S. B. Woods
  Phil. Trans. Roy. Soc., 251A:273 (1959)

Spin-Disorder Effects in the Electrical Resistivities of Metals and Alloys
  B. R. Coles
  Phil. Mag. Suppl., 7(25):40 (1958)

Low-Temperature Resistance Measurements as
a Means of Studying Impurity Distributions in
Zone-Refined Ingots of Metals
  J. E. Kunzler and J. H. Wernick
  Trans. Met. Soc. AIME, 212:856 (1958)

Research on the Electrochemical Behavior of
Metallic Single Crystals, Parts 1, 2, and 3
  I. R. Piontelli, U. Bertocci, L. Bicelli, G. Poli, B. Rivolta,
    G. Sternheim, and C. Tamplenizza
  AEC-tr-3583 translated from Rend. ist. Lombardo sci Pt.
    I. Classe sci. mat. e. nat., 91, 347-54 (1957); II. Roberto
    Piontelli, Guido Poli and Lucia Paganini, Ibid. 355-70;
    III. Roberto Piontelli, Ugo Betocci, and Claudio Tamplenizza, Ibid., 378-85. 59 pp.

Electrodeless Measurements of Electrical Conductivity by the Rotating Field Method
  A. Roll, H. Fleger, and H. Motz
  CEA-tr-A-263
  Z. Metallk., 47:708-713 (1956)

Measurement of the Electrical Resistance of
Metals and Alloys at High Temperatures
  P. Chiotti
  Rev. Sci. Instr., 25:876-883 (1954)

Electrodeless Measurement of the Electrical
Resistivity of Metals and Alloys at High Temperatures
  H. Speidel
  CEA-tr-A-252
  Thesis D93, T. H. Stuttgart (1940), 25 pp.

Temperature Dependence of Thermoelectromotive Force and Specific Electrical Resistance
of Titanium, Vanadium, Chromium, and Their
Borides, Carbides, and Nitrides
  S. N. L'vov and V. F. Nemchenko
  Kherson Pedagogical Institute LA-TR-67-17

1.a.2. Group IV

Electrical Properties of Titanium, Zirconium,
and Hafnium Films from 300°K to 1.3°K
  P. E. Friebertshauser and J. W. McCamont
  J. Vacuum Sci. Tech., 6:184-187 (1969)

Thermoelectric Effects in Anisotropic Titanium
Foils
  L. A. Kalinets, I. G. Mel'nik, and A. E. Bryukhanov
  Fiz. Tverd. Tela, 11:3335-3337 (1969)
  Sov. Phys. — Solid State, 11:2702-2704 (1970)

Electrical Properties of Thin Titanium Films
  Ramesh Chander, R. E. Howard, and S. C. Jain
  Indian J. Pure Appl. Phys., 5:397-400 (1967)

Variation of the Electrical Resistance of Titanium Films under Argon Ion Bombardment
  G. F. Ivanovskii and T. D. Radzhabov
  Fiz. Tverd. Tela, 8(4):1271-1272 (1966)
  Sov. Phys. Solid State, 8(4):1013-1014 (1966)

Electrical Resistivity Data and Bibliography
on Titanium and Titanium Alloys
  John T. Milek
  EPIC Interim Report No. IR-27 (March 1966)

The Hall Effect in Single Crystals of Titanium
  Georgiana W. Scovil
  Appl. Phys. Letters, 9:247 (1966)

Significance of Refractory Metals for the Production of High-Field Superconductors
  D. Koch
  Metall., 18:714-718 (1964)

Sur les propriétés du zirconium de zone fondue
  Jean-Paul Langeron
  C. R. Acad. Sc. Paris, 258:1801-1804 (1964)

The Hall Effect and Paramagnetic Susceptibility
of Hafnium
  N. V. Volkenshtein and E. V. Galoshina
  Fiz. Metal. Metalloved., 18:784 (1964)

Structure and Electrical Properties of Evaporated and Sputtered Titanium Films
  D. Gerstenberg
  Ann. Physik, 11:354-364 (1963)

Observation on Electrical Resistivity of Titanium
  R. J. Wasilewski
  AIME Trans., 224:5-8 (1962)

Etude de la résistivité électrique du zirconium. Application à l'étude de la récristallisation du métal après écrouissage. Influence de
la pureté
  L. Renucci, J. P. Langeron, and P. Lehr
  Mem. Sci. Rev. Met., 58:699 (1961)

Hall Effect, Resistivity, and Magnetoresistivity
of Th, U, Zr, Ti, and Nb
  T. G. Berlincourt
  Phys. Rev., 114:969 (1959)

The Electrical Conductivity of Zone Melted Zr
at Low Temperatures
  L. Renucci, J. P. Langeron, and P. Lehr
  Compt. Rend., 249:113 (1959)

Hall Effect in Titanium, Chromium, Vanadium,
and Manganese
  Simon Foner
  Phys. Rev., 107:1513-1516 (1957)

Low Temperature Resistivity of Transition
Elements: Vanadium, Niobium, and Hafnium
  G. K. White and S. B. Woods
  (Contribution from the Division of Pure Physics, National
    Research Council, Ottawa, Canada), Can. J. Phys., 35:892
    (1957)

e. Electrical Properties

1.e.3. Group V

Concerning the Fermi Surface of Niobium
N. E. Alekseevskii, K. Kh. Bertel, and A. V. Dubrovin
Zh. Éksper. Teor. Fiz. Pis'ma Redaktsiyu, 10:74–76 (1969)
JETP Letters (USSR), 10:74–76 (Aug. 5, 1969)

Influence of Small Additions of Yttrium on the
High-Temperature Resistance of Niobium
F. Bollenrath and G. Happek
Met. Ital., 61:473–480 (1969)

Effect of Thermal Treatment on the Hall Mo-
bility in Thin Vanadium Films
S. C. Jain and Ramesh Chander
Thin Solid Films, 4:R11–R13 (1969)

Recovery of Electrical Resistance in Worked
Niobium
G. S. Martkoplishvili, I. A. Naskidashvili, and N. A.
Maisuradze
Fiz. Metal. Metalloved., 28(3):569–571 (1969)

Transverse Magnetoresistance in Vanadium and
in a Chromium-Rich Chromium — Vanadium Alloy
Kerry Scott Nelson
(Iowa State University of Science and Technology, Ames,
Iowa), Contract W-7405-eng-82, IS-T-295 (Feb. 1969),
54 pp.

Nernst Effect in Superconducting Lead and
Niobium
A. Seher
Kernforschungungsanlage Julich, Germany, JUEL-598-FN
(June 1969), 79 pp.

Contact Potential Difference of Double-Layer
Ba and Au Films on Nb
E. A. Tishin and B. M. Tsarev
Fiz. Tverd. Tela, 10(7):2196–2197 (1968)
Sov. Phys. — Solid State, 10(7):1719–1720 (1969)

Low Temperature Electrical Resistivity of Pure
Niobium
George W. Webb
Ph.D. thesis, Univ. Calif., San Diego (1967), 85 pp.
University Microfilms, Inc., Ann Arbor, Mich., No. 68-807
Phys. Rev., 181:1127–1135 (1969)

The Effect of Tungsten Content on the Residual
Resistivity of Zone Refined Molybdenum
J. H. Evans and B. L. Eyre
Phys. Stat. Sol., 25:K39 (1968)

Resistivity and Hall Effect in Superconducting
Nb
A. T. Fiory and B. Serin
Phys. Rev. Letters, 21:359–361 (1968)

Resistance of Ta Capacitors
B. Goudswaard
Electrochem. Tech., 6:178–182 (1968)

Surface Impedance of Nb
W. D. Hibler
Phys. Rev. Letters, 21:742:745 (1968)

Temperature Coefficient of Resistance in Thin
Vanadium Films
S. C. Jain and Ramesh Chander
J. Appl. Phys., 39:5343 (1968)

Integral Hemispherical Radiating Power and
Specific Electrical Resistance of Tantalum in
the Temperature Interval 1200°–2800°K
V. A. Petrov, V. Ya. Chekhovskoi, and A. E. Sheindlin
Teplofiz. Vys. Temp., 6(3):548–549 (1968)
High Temperature, 6(3):525–526 (1968)

Microwave Properties of Nb
J. P. Turneauru
Adv. Cryogenic Eng., 13:109–115 (1968)

Structure and Electrical Resistance of Tantalum
Films Obtained by Cathode Sputtering
U. S. Urazaliev, B. D. Galkin, and R. D. Ivanov
Fiz. Metallov. Metalloved., 25:1025–1028 (1968)

Low-Density Tantalum
B. H. Vromen
Bell Labs. Records, 46:327–32 (1968)
Very high resistance

Hall Effect and Resistivity in Nb and V
C. H. Weijsenfeld
Phys. Letters, 28 A:362–363 (1968)

Resistance Stability of Electron-Beam Vapor-
ized Tantalum Layers
Gerhard Wolf
Wiss. Z. tech. Hochsch. Karl-Marx-Stadt, 10:181–183 (1968)
In German

Experimental Investigation of the Surface Con-
ductivity of Nb
S. Sh. Akhmedov and V. R. Karasik
Zh. Éksp. Teor. Fiz., Pis'ma Redakt., 5(12):442–444 (1967)
JETP Letters, 5(12):358–359 (1967)

Galvanomagnetic Properties of Niobium Single
Crystals
N. E. Alekseevskii, K. H. Bertel, A. V. Dubrovin, and G. E.
Karstens
Zh. Éksp. Teor. Fiz., Pis'ma Redakt., 6:637–642 (1967)
JETP Letters, 6:132–135 (1967)

Hall Effect and Electrical Resistivity of Vana-
dium at 20 – 300°K
E. B. Amitin, Yu. A. Kovalevskaya, and Yu. Z. Kovdrya
Fiz. Tverd. Tela, 9(3):905–908 (1967)
Sov. Phys. — Solid State, 9(3):704–706 (1967)

Electrical Conductivity and Hall Effect in Thin
Vanadium Films
Ramesh Chander, R. E. Howard, and S. C. Jain
J. Appl. Phys., 38:4092 (1967)

Influence of High Vacuum Refining on the Elec-
trical Resistance of Nb and Ta
R. J. Dinter
Z. Metallk., 58:70–75 (1967)

Schottky Deviations from Tantalum Single Crystals
Michael Stephen Dubas
Ph.D Thesis from St. Louis University, St. Louis, Missouri
(1967), 219 pp.

Tantalum Thin Film Single-Sideband
R. K. P. Galpin and others
J. Solid-State Circuits, 2:26–31 (1967)

Conductivité des couches ultra-mincés de Ta
obtenues par vaporisation sous vide
D. I. Lajner and V. A. Kholmjanskij
Izv. Akad. Nauk SSSR, Neorg. Mater., 3:1673-1675 (1967)
Inorg. Mater., 3(9):1459-1461 (1967)

High Temperature Capacitors
R. E. Stapleton
WAED, 67:24E (1967)

Continuous Contactless Resistivity Measurement
J. M. Clarke and B. L. Mordike
Appl. Mater. Res., 5:181-186 (1966)
Ta

Thermoelectric Properties of Vanadium
I. P. Druzhinina, T. M. Vladimirskaya, and A. A. Fraktovnikova
Izmeritel. Tekhn., 8:41-42 (1966)

Emissivity and Electrical Resistivity of Tantalum at High Temperatures
Gerhard Hoerz
Z. Metallk., 57:871-883 (Dec. 1966)

Electrical and Structural Properties of Epitaxial bcc Tantalum Films
R. B. Marcus
J. Appl. Phys., 37:3121 (1966)

A Bibliography on Tantalum Metal Films for Electronic Applications and Related Information
John T. Milek
(Electronic Properties Information Center, Hughes Aircraft Co., Culver City, Calif.), EPIC Interim Report IR-18 (Jan. 1966)

Effect of Plastic Deformation on Electrical Resistivity of Nb
D. P. Gregory
Acta Met., 13:135 (1965)

Pouvoir émissif et résistivite électrique du vanadium et des solutions solides vanadium-azote et vanadium-oxygène
G. Horz, E. Gebhardt, and W. Durrschnabel
Z. Metallk., 56:554-560 (1965)

Critical Alternating Currents in Superconducting Niobium Strips
R. M. F. Linford
Phys. Letters, 17:18 (1965)

The Effect of Dissolved Nitrogen on the Electrical Resistance of Niobium
R. A. Pasternak and B. Evans
Trans. AIME, 233:1194 (1965)

Effect of Hydrogen on Electrical Resistance of Niobium
V. M. Pletenev
Izv. Vysshikh Uchebn. Zavednii, Cvetn. Metal., SSSR, 8:113-116 (1965)
(Wright-Patterson AFB, Ohio, Foreign Technology Division), FTD-HT-66-556; AD-665701 (Aug. 1967), 10 pp.

Thermoelectric Properties of Niobium in the Temperature Range 300° - 1200°K
V. Raag and H. V. Kowger
J. Appl. Phys., 36:2045 (1965)

Specific Electrical Restivity and Interstitials in Tantalum Crystals
L. I. van Torne
J. Less Common Metals, 9:152 (1965)

Electrical Resistance Measurements on Hydrogen-Charged Tantalum and Niobium
M. Verani Borgucci and L. Verdini
Phys. Stat. Sol., 9:243 (1965)

X-Ray Diffraction Study of Deformation by Filing in B. C. C. Refractory Metals
E. N. Aqua and C. N. J. Wagner
Phil. Mag., 9:565 (1964)

Effects of Nitrogen, Methane, and Oxygen on Structure and Electrical Properties of Thin Tantalum Films
D. Gerstenberg and C. J. Calbrick
J. Appl. Phys., 35:402 (1964)

Superconducting Thin Films of Niobium, Tantalum, Tantalum Nitride, Tantalum Carbide, and Niobium Nitride
D. Gerstenberg and P. M. Hall
J. Electrochem. Soc., 111:936 (1964)

Significance of Refractory Metals for the Production of High-Field Superconductors
D. Koch
Metall., 18:714-718 (1964)

Restwiderstandsmessungen an elektronenstrahlgeschmolzenem Tantal zur Untersuchung des Zonenreinigungseffektes
Barry L. Mordike
Z. Metallk., 55:304-306 (1964)

Some Electrical and Chemical Properties of the (111) Niobium Surface
R. M. Oman and J. A. Dillon, Jr.
Surface Sci., 2:227 (1964)

The Specific Heats and Resistivities of Molybdenum, Tantalum, and Rhenium
R. E. Taylor and R. A. Finch
J. Less-Common Metals, 6:283-294 (1964)

The Influence of Lattice Defects on the Electrical Resistance and the Hall-Effect in Tungsten and Tantalum
E. Krautz and H. Schultz
Z. Angew. Phys., 15:1-4 (1963)

The Thermoelectric Power in Chromium and Vanadium
A. R. Mackintosh and L. Sill
J. Phys. Chem. Solids, 24:501-506 (1963)

The Electrical Resistivity of Vanadium and Vanadium-Chromium Solid Solutions
M. A. Taylor and C. H. Llewellyn Smith
Physica, 28:453-460 (1962)

Low Temperature Tensile Properties of Zone-Refined Niobium (Columbium)
J. F. Wellings and R. Maddin
(University of Pennsylvania, School of Metallurgical Engineering, Philadelphia), NP-11795 (1962), 19 pp.

Effect of Dissolved Gases on Some Superconducting Properties of Niobium
   W. DeSorbo and G. E. Nichols
   Bull. Am. Phys. Soc. Ser. II, 6:267 (A) (1961)

The Specific Heats and Resistivities of Molybdenum, Tantalum, and Rhenium from Low to Very High Temperatures
   R. E. Taylor and R. A. Finch
   (Atomic International, Div. of North American Aviation, Inc., Canoga Park, Calif.), Contract AT(11-1)-GEN-8, NAA-SR-6034 (Sept. 15, 1961), 37 pp.

Single-Crystal Growth and Purification of Tantalum
   D. P. Seraphim, J. I. Budnick, and W. B. Ittner III
   Cryogenics Dept. of the International Business Machines Research Laboratory in Poughkeepsie, N. Y.
   Trans AIME, 218:527 (1960)

Hall Effect, Resistivity, and Magnetoresistivity of Th, U, Zr, Ti, and Nb
   T. G. Berlincourt
   Phys. Rev., 114:969 (1959)

The Preparation and Purification of Tantalum Single Crystals and Measurements of the Critical Field for Single and Polycrystalline Specimens
   J. I. Budnick, W. B. Ittner, and D. P. Seraphim
   Cryogenics Department, International Business Machines Research Laboratory, Poughkeepsie, New York
   Physica, 24 (S), S151 (A) (1958)

Hall Effect in Titanium, Chromium, Vanadium, and Manganese
   Simon Foner
   Phys. Rev., 107:1513-1516 (1957)

Low Temperature Resistivity of Transition Elements: Vanadium, Niobium, and Hafnium
   G. K. White and S. B. Woods
   (Contribution from the Division of Pure Physics, National Research Council, Ottawa, Canada), Can. J. Phys., 35:892 (1957)

1.e.4. Group VI

High Temperature Resistivity Measurements on Chromium
   J. M. Anderson, A. D. Stewart, and I. Ramsay
   Phys. Stat. Sol,, 37:325-328 (1970)

Mean Free Path of the Current Carriers in Molybdenum
   V. V. Boiko and V. A. Gasparov
   Fiz. Tverd. Tela, 12(1):310-312 (1970)
   Sov. Phys. − Solid State, 12(1):254-255 (1970)

The Effect of Deposition Parameters on the Structure and Resistivity of Molybdenum Films
   J. R. Bosnell and U. C. Voisey
   Thin Solid Films, 6:107-111 (1970)

Evolution, sous vide, de la résistance des couches mincés de chrome préparées par evaporation thermique sur des supports en verre refroidis à − 120°C, en relation avec leur structure
   Gerard Chassaing and Maurice Sigrist
   Compt. Rend., 270B:227-230 (1970)

High-Speed (Subsecond) Measurement of Heat Capacity, Electrical Resistivity, and Thermal Radiation Properties of Molybdenum in the Range 1900 to 2800°K
   A. Cezairliyan, M. S. Morse, H. A. Berman, and C. W. Beckett
   J. Res. Natl. Bur. Std., 74A:65-92 (1970)

Electrical Properties and Superconductivity of Rhenium and Molybdenum Films
   P. E. Frieberthauser and H. A. Notarys
   J. Vacuum Sci. Tech., 7:485-488 (1970)

Lorenz Number of Chromium
   J. F. Goff
   Phys. Rev. B, 1(4):1351;1362 (1970)

Transport Properties of Chromium through the Neel Point
   M. J. Laubitz and T. Matsumura
   Phys. Rev. Letters, 24:727-730 (1970)

Comments on 'Lorenz-Function Enhancement Due to Inelastic Processes near the Neel Point of Chromium'
   J. P. Moore, R. K. Williams, and D. L. McElroy
   Phy. Rev. Letters, 24:587-588 (1970)

Special Features in the Electrical Resistance of Chromium at 0.36-320°K
   E. E. Semenenko and V. I. Tutov
   Fiz. Metal. Metalloved., 29(5):952-956 (1970)

Quenching and Annealing of Tungsten
   Ronald J. Gripshover and Jack Bass
   Bull. Am. Phys. Soc., 14:82 (1969)
   Quenched-in resistivities

Anomaly in the Electrical Resistivity of Chromium Metal in the Vicinity of Spin-Flip Temperature
   Takehiko Matsumoto, Takashi Sambongi, and Tadayasu Mitsui
   J. Phys. Soc. Japan, 26:209 (1969)

Electrical Resistivity of Tungsten Films Prepared by WF_6 Reduction
   A. F. Mayadas, J. J. Cuomo, and R. Rosenberg
   J. Electrochem. Soc., 116:1742-1745 (1969)

Lorenz-Function Enhancement Due to Inelastic Processes near the Neel Point of Chromium
   G. T. Meaden, K. V. Rao, and H. Y. Loo
   Phys. Rev. Letters, 23:475-477 (1969)

Critical Exponents and Electrical Resistivity near the Neel Point of Chromium
   G. T. Meaden and N. H. Sze
   Phys. Letters, 29A:162 (1969)

Effect of the Spin-Flip Transition on the Thermal and Electrical Resistivities of Chromium
   G. T. Meaden, K. V. Rao, H. Y. Loo, and N. H. Sze
   J. Phys. Soc. Japan, 27:1073 (1969)

Galvanomagnetic Phenomena in Thin Films of Some Transition Metals
   O. A. Panchenko, P. P. Lutsishin, and Yu. G. Ptushinskii
   Zh. Eksper. Teor. Fiz. (USSR), 56(1):134-138 (1969)
   Sov. Phys. − JETP, 29(1):76-78 (1969)
   Cr, W, Pd, Pt

Effect of Magnetic Field on the Minimum of the Electrical Resistance of Chromium
E. E. Semenenko and V. I. Tutov
Fiz. Met. Metallov., 27:343 (1969)

Hall Effect in Chromium
E. B. Amitin and Yu. A. Kovalevskaya
Fiz. Tverd. Tela, 10:1884–1886 (1968)
Soviet Phys. – Solid State, 10:1483 (1968)

Electrical Resistance of Chromium Near the Neel Point
E. B. Amitin and Yu. A. Kovalevskaya
Fiz. Tverd. Tela, 9(9):2731–2734 (1967)
Sov. Phys. – Solid State, 9(9):2145–2147 (1968)

Resistance of Mo-Films
T. H. Ansbacher
AD-690 121 (1968)

Electrical Conductance of Thin Ultrahigh-Vacuum-Evaporated Films of Tungsten on Glass
P. J. Dobson and B. J. Hopkins
J. Appl. Phys., 39:3074 (1968)

The Structure of, and the Contact Potential Difference Between, Polycrystalline Tungsten Foil and Vapour-Deposited Films of Tungsten on Glass
P. J. Dobson and B. J. Hopkins
British J. Appl. Phys., 1:1241–1244 (1968)

Studies of the Residual Resistance of W with Varying C Content
E. Krautz and H. Schultz
Z. Metallk., 59:133 (1968)

Influence of Dielectric Films on the Resistance of Vacuum-Deposited Chromium Films
Alvin A. Milgram and Chih-Shun Lu
J. Appl. Phys., 39:1624 (1968)

Physical Properties of Chromium from 77 to 400°K
J. P. Moore, R. K. Williams, and D. L. McElroy
(Oak Ridge National Lab., Oak Ridge, Tenn.) Proceedings of the 7th Conf. on Thermal Conductivity, Gaithersburg, Maryland, Nov. 13–16, 1967
National Bureau of Standards, Wash., D. C. (1968), p. 297–310
Electrical resistivity, thermal conductivity, and absolute Seebeck coefficient

Second Minimum in the Electrical Resistivity Versus Temperature Curve for Chromium
T. M. Sabine and A. C. Svenson
Phys. Letters, 28A:443–444 (1968)

Electrical Resistivity of Antiferromagnetic Chromium
Takashi Sambongi and Tadayasu Mitsui
J. Phys. Soc. Japan, 24:1168 (1968)

The Temperature Dependence of the Electrical Resistance of Molybdenum
V. I. Makarov and T. A. Sverbilova
(Akad. Nauk URSR, Kiev) in its Investigating the Electron. Properties of Metals and Alloys (1967), pp. 124–129
See N70-17576 06-17

Electrotransport in W
G. M. Neumann and W. Hirschwald
Z. Naturforsch., 22A:388–392 (1967)

The Hall Coefficient of Chromium at 4.2 and 77°K and at Room Temperature in Magnetic Fields up to 147 kOe
Klaus Schroder and Barrie S. Shabel
Z. Metallk., 58:727–729 (1967)

Anomalous Decrease of the Electrical Resistivity of Metallic Chromium Due to Alloying with Vanadium and Titanium
M. V. Vedernikov
Fiz. Tverd. Tela, 9(10):3018–3021 (1967)
Sov. Phys. – Solid State, 9(10):2381–2383 (1968)

Magnetoresistance of Molybdenum and Tungsten (Deviation from the Kohler Rule)
N. V. Volkenshtein, V. A. Novoselov, and V. E. Startsev
Fiz. Metal. Metalloved., 24:677–682 (1967)

The Effects of Deformation on the Electrical Resistivity of Molybdenum Single Crystals
L. D. Whitmire and F. R. Brotzen
Trans. Met. Soc. AIME, 239:824 (1967)

"Reply to Comments on 'The Low-Temperature Electrical Resistivity of Lattice Defects in Deformed Tungsten Single Crystals'"
H. B. Shukovsky, R. M. Rose, and J. Wulff
Acta Met., 15:391–393 (1967)
By A. S. Wronski and A. A. Johnson

"Comments on 'The Low-Temperature Electrical Resistivity of Lattice Defects in Deformed Tungsten Single Crystals'"
A. S. Wronski and A. A. Johnson
Acta Met., 15:389–391 (1967)
By H. B. Shukovsky

Effect of Glow Discharge on Chromium Thin-Film Resistance
S. M. Dembicka-Jellonkowa and E. Murawski
Electron. Letters (GB), 2:292–293 (1966)

Thermoelectric Properties of a Pure and Ba(111) - Covered Film Lattice of W Single Crystals
L. N. Dikova, M. I. Smorodinova, and E. P. Sytaya
Dokl. Akad. Nauk Uz.SSR, 23:19 (March 1966)

The Seebeck Coefficients of W
W. Fulkerson and J. P. Moore
High Temperature Thermometry (1966)

The Permeation and Diffusion of Hydrogen in Molybdenum
P. M. S. Jones, R. Gibson, and J. A. Evans
AWRE Report 0-16/66 (February 1966)

Galvanomagnetic and Thermomagnetic Transport Coefficients in Tungsten
J. R. Long
Bull. Am. Phys. Soc., 11:848 (1966)

The Effects of Deformation on the Electrical Resistivity of Molybdenum Single Crystals
L. D. Whitmire and F. R. Brotzen
(Rice University, Houston, Texas) NASA-CR-77029 (1966)
26 pp. Grant NsG-6-59

Electrical Resistivity and Transverse Electrical Magnetoresistivity of Chromium
S. Araja and G. R. Dunmyre
J. Appl. Phys., 36:3555 (1965)

The Resistivity and Structure of Chromium
Thin Films
 P. A. Gould
 British J. Appl. Phys., 16:1481 (1965)

Resistivity and Temperature Coefficient of
Pure Molybdenum
 R. A. Holmwood and R. Glang
 J. Chem. Eng. Data, 10:162 (1965)

Magnetoresistance transversale de rubans de
chrome de haute pureté
 T. J. Bastow and R. Street
 Phil. Mag., 10:269-276 (1964)

The Specific Electric Resistance of Electron-
Beam-Zone-Melted Tungsten Between 1.4 and
27°K
 K. H. Berthel
 (Institut für Angewandte Physik der Reinststoffe, Dresden)
 Phys. Stat. Sol., 5:399-404 (1964)

Rest-Resistance Behavior, Diameter Effect,
and Current Dependence of the Resistance Ratio of Zone-Melted Tungsten at Low Temperatures
 K. H. Berthel
 (Institut für Angewandte Physik der Reinststoffe, Dresden)
 Phys. Stat. Sol., 5:159-168 (1964)

Electrical Resistivity of Electron-Beam Zone-
Refined Tungsten Single Crystals Between 1.4
and 27°K
 K. H. Berthel
 Physics and Techniques of Low Temperatures
 Publishing House of the Czech. Acad. of Sciences, Prague,
 Czechoslovakia (1964), pp. 116-126

Significance of Refractory Metals for the Pro-
duction of High-Field Superconductors
 D. Koch
 Metall., 18:714-718 (1964)

The Recovery of the Electrical Resistivity of
Cold-Worked Tungsten
 H. Schultz
 (OSRAM-Studiengesellschaft, Augsburg, Germany)
 Acta Met., 12:649-664 (1964)

The Specific Heats and Resistivities of Molyb-
denum, Tantalum, and Rhenium
 R. E. Taylor and R. A. Finch
 J. Less-Common Metals, 6:283-294 (1964)

The Influence of Lattice Defects on the Electrical Resistance and the Hall-Effect in Tungsten and Tantalum
 E. Krautz and H. Schultz
 Z. Angew. Phys., 15: 1-4 (1963)

Resistivity and Electron Microstructure of
Quenched Tungsten
 F. W. Kuether
 Bull. Am. Phys. Soc. Ser. II, 8:66 (A) (1963)

The Thermoelectric Power in Chromium and
Vanadium
 A. R. Mackintosh and L. Sill
 J. Phys. Chem. Solids, 24:501-506 (1963)

Initial Study of Electrical Resistivity of a
Chromium Single Crystal (4 to 330°K)
 Sigurds Arajs, R. N. Colvin, and M. J. Marcinkowski
 J. Less-Common Metals, 4:46-51 (1962)

Superconductivity in Molybdenum
 T. H. Geballe and others
 Phys. Rev. Letters, 8:313 (1962)

Transitions In Chromium
 Jean-Francois Marin
 U. S. Atomic Energy Commission Contract No. AT(04-3)221
 and Office of Naval Research, Contract No. Nonr-220(30),
 AD-14740 (Jan. 1962)

Superconductivity of Chromium Alloys
 B. T. Matthias, T. H. Geballe, V. B. Compton, E. Corenzwit,
 and G. W. Hull, Jr.
 Phys. Rev., 128:588-590 (1962)

The Electrical Resistivity of Vanadium and
Vanadium-Chromium Solid Solutions
 M. A. Taylor and C. H. Llewellyn Smith
 Physica, 28:453-460 (1962)

Observation Concerning Zone Refining and
Thermal Treatment of Molybdenum from Low
Temperature Resistance Measurement
 E. Buehler and J. E. Kunzler
 AIME Trans., 221:957-961 (1961)

Electrical Resistivity of Chromium in the Vi-
cinity of the Néel Temperature
 M. J. Marcinkowski and H. A. Lipsitt
 J. Appl. Phys., 32:1238-1240 (1961)

The Specific Heats and Resistivities of Molyb-
denum, Tantalum, and Rhenium from Low to
Very High Temperatures
 R. E. Taylor and R. A. Finch
 (Atomic International, Div. of North American Aviation, Inc.,
 Canoga Park, Calif.) Contract AT(11-1)-GEN-8, NAA-
 SR-6034 (Sept. 15, 1961), 37 pp.

The Detection of Impurities and Cold-Working
in Sintered Tungsten by the Measurement of
Residual Resistance
 Erich Krautz and Herman Schultz
 AEC-tr-5315
 Z. Metallk., 49:399-403 (1958)

Hall Effect in Titanium, Chromium, Vanadium,
and Manganese
 Simon Foner
 Phys. Rev., 107:1513-1516 (1957)

Low Temperature Resistivity of the Transition
Elements: Cobalt, Tungsten, and Rhenium
 G. K. White and S. B. Woods
 Can. J. Phys., 35:656-665 (1957)

Photoelectric and Thermionic Schottky Devia-
tions for Tungsten Single Crystals
 David F. Stafford
 Dissertation Abstr., 64-3772, 111 pp.

1.f. Electronic Structure

1.f.1. General

Structure électronique des défauts dans les
métaux de transition
Guy Allan
Ann. Phys., 5:169–202 (1970)
59 refs.

Emission of W, Mo and Nb with Absorbed Cs
and Ba
V. D. Bondarenko
Teplofiz. Vysok. Temp., 8:211–212 (1970)
High Temperature, 8(1):201–203 (1970)

Photoelectric Work Functions of Transition,
Rare-Earth, and Noble Metals
D. E. Eastman
Phys. Rev., B 2(1):1-2 (1970)
Polycrystalline films of Sc, Ti, V, Cr, Mn, Fe, Co, Ni, Cu,
Y, Zr, Nb, Mo, Pd, Ag, La, Ce, Nd, Sm, Eu, Gd, Hf, Pt,
and Au

Electron Interaction in Transition Metal X-Ray
Emission Spectra
B. Ekstig, E. Kaeline, E. Noreland, and R. Manne
UUIP-702 (April 1970), 28 pp.

Electronic Structure of the Metals of the Vana-
dium and Chromium Groups
C. Reale
Phys. Letters, 32A:197–198 (1970)

Positive and Negative Ion Sublimation from
Transition Metal Surfaces: A Review of Some
Recent Results
M. D. Scheer
J. Res. Natl. Bur. Stand., 74A:37–43 (1970)

Self-Consistent Band Calculations for Vanadium
and Chromium
M. Yasui, E. Hayashi, and M. Shimizu
J. Phys. Soc. Japan, 29:1446–1455 (1970)

New Shortwave Satellites in Ka Spectra of Iron
Transition Group Elements
A. Z. Zhmudskii, L. F. Chestnykh, and A. S. Erniyazov
Fiz. Tverd. Tela, 12(5):1527–1529 (1969)
Sov. Phys. — Solid State 12(5):1198–1199 (1970)
Ti, V, Cr, Fe, and Ni

On the Structure of the Transition Metals II.
Computed Densities of States
N. W. Dalton and R. A. Deegan
J. Phys. C. (Solid State Phys.) Ser. 2, Vol. 2:2369–2372
(1969)

Photoemission Studies of d-Band Structure in
Sc, Y, Gd, Ti, Zr, Hf, V, Nb, Cr and Mo
D. E. Eastman
Solid State Commun., 7:1697–1700 (1969)

The Physics of Metals. I. Electrons
J. M. Ziman, ed.
University Press, Cambridge, England, (1969), 433 pp.

Structure Dependence of D Bands in Transition
Metals
R. A. Deegan
Phys. Rev., 171:659 (1968)

Zur Zustandsdichte der Übergangsmetalle. I.
Isochromatenspektroskopische Untersuchungen
an kubisch-raumzentrierten Übergangsmetallen
H. Merz and K. Ulmer

Z. Phys., 210:92–110 (1968)
Ti, V, Cr, Zr, Nb, Mo, Hf, Ta, W

Density of States for Transition Metals. II. Iso-
chromat Spectroscopy of hcp IIIA and IVA
Metals
H. Merz and K. Ulmer
Z. Phys., 212:435–448 (1968)

Positron Annihilation as a Probe for Many
Body Effects in Metals
Ted Warren Mihalisin
Thesis from University of Rochester, Rochester, New York
(1968), 205 pp.

X-Ray Spectra and Electronic Structure of Elements of the First and Second Transition Series
V. V. Nemoshkalenko
Nekot. Vop. Elektron. Strukt. Perekhodnykh Metal. (1968),
pp. 121–170

Lokalisierte d-Elektronen in Übergangsmetallen
H. Stachowiak
Acta Phys. Polon., 33:325–336 (1968)

Determination of the Work Functions of Selected Metals and Inorganic Compounds
Denver Research Institute, Colorado, SC-CR-68-3586 (April
1968), 36 pp.
Chromium, erbium, nickel, titanium, vanadium, erbium
compounds, titanium compounds, hydrides, photoelectric effects

The Electronic Structures of Transition and
Noble Metals
M. H. Cohen and F. M. Mueller
Atomic and Electronic Structure of Metals, Am. Soc. Met.,
Metals Park (1967), pp. 61–81

Charge of Interstitial Ions in Transition Met-
als
E. W. Collings
TID 24210, 1967

Analyse de la structure electronique des sys-
tèmes désordonnés dans l'approximation des
liaisons fortes
Mme. F. Cyrot
Theses Doct. Sci. phys. Paris, 1968, Arch. orig. Centre
Document. C.N.R.S., No. 1893 (Nov. 10, 1967)

Energy Spectra of Secondary Electrons from
Metals
Charles H. Sherman and Nan E. Gordon
(Parke Mathematical Laboratories, Inc., Carlisle, Mass.)
Science Rept. No. 8, Contract AF19(628)–2419 AFCRL-
67–0651 (Nov. 1967), 22 pp.
Data recently obtained at AFCRL as well as data from the
literature have been included

Handbook of Thermionic Properties: Electronic
Work Functions and Richardson Constants of
Elements and Compounds
V. S. Fomenko
Plenum Press, New York (1966), 151 pp.

Temperature Dependence of the Densities of
States in Tantalum, Niobium and Tungsten
H. Merz and K. Ulmer
Phys. Letters, 22:251 (1966)

Determination of the Work Functions of Selected Metals and Inorganic Compounds. Summary Report: Determination of Thermionic

Work Functions of Titanium, Vanadium, Chromium, and Erbium
  A. D. Paddock
    Denver Research Institute, Colo., DRI-2344 (Aug. 1966),
      57 pp.

Struttura elettronica nei metalli di transizione cubici a corpo centrato
  M. Asdente
    Nuovo Cimento, Suppl. 3:506-521 (1965)

Electronic Structure and Superconductivity of Transition Metals and Their Alloys
  B. R. Coles
    Rev. Mod. Phys., 36:139 (1964)

La densité d'états des métaux de transition
  K. Ulmer
    Phys. Verh. D.P.G., Dtsch., 8:276 (1964)

Absorption Contrast Effect in BCC Metals
  S. E. Bronisz and D. L. Douglas
    (University of California, Los Alamos Scientific Laboratory, Los Alamos, New Mexico)
    J. Appl. Phys., 34:713 (1963)

"The Problem of the Transition Metals" Electronic Structure and Alloy Chemistry of the Transition Elements
  W. Hume-Rothery
    Interscience, New York (1963), p. 3-28

Spin-Dependent Scattering and Resistivity of Magnetic Metals and Alloys
  T. van Peski-Tinbergen and A. J. Dekker
    (Instituut voor Kristalfysica der Rijksuniversiteit, Groningen, Nederland)
    Physica, 29:917-937 (1963)

The Electronic Structure and Magnetic Properties of Transition Metals and Alloys
  J. Friedel
    AI-trans. 35, January 1964, J. Phys. Radium, 23:501-510 (1962)

The Effects of Chemical Impurities and Physical Imperfections on the Thermoelectricity of Metals
  W. B. Pearson
    Chap. 10 of Ultra-High-Purity Metals, Seminar of Am. Soc. for Metals (Oct. 21 and 22, 1961), ASM, Metals Park, Ohio (1962)

Relationships Between Electronic Structure and Physical Properties in Transition Metals
  G. Airoldi and M. Asdente
    Energia Nucleare (Milan), 8:342-360 (May 1961)

An Interpretation of Some of the Properties of the Transition Metals and Their Alloys
  D. A. Robins
    J. Less-Common Metals, 1:396-410 (1959)

1.f.2. Group IV

Electronic Density of States and Crystal Structure in IVA Transition Metals
  H. Merz
    Phys. Letters, 33A:53-54 (1970)
    Ti, Zr, and Hf

Lattice Dynamics of Hexagonal-Transition Metal Based on New Electron Force Model
  B. Sharan and R. P. Bajpai
    J. Phys. Soc. Japan, 29:46-49 (1970)
    Y and Ti

Lattice Dynamics of Zirconium and Hafnium
  B. Sharan and R. P. Bajpai
    J. Phys. Soc. Japan, 29:66-68 (1970)

Electron Momentum Density of Titanium
  R. J. Weiss
    Phys. Rev. Letters, 24:883-884 (1970)

Mössbauer-Effect Studies in Hafnium-Metal Single Crystals
  P. Boolchand, B. L. Robinson, and S. Jha
    Phys. Rev., 187(2):475-478 (1969)

Positron Annihilation in Yttrium and Zirconium
  R. P. Gupta and T. L. Loucks
    Phys. Rev., 176:848-850 (1968)

The Band Structure of Hexagonal Close-Packed Metals II. - Scandium, Titanium, Yttrium, and Zirconium
  S. L. Altmann and C. J. Bradley
    Proc. Phys. Soc., 92:764-775 (1967)

Electronic Structure of Zirconium
  T. L. Loucks
    Phys. Rev., 159:544-551 (1967)

Spectra of TiV, TiVI and TiVII in the Region 100-400 Å
  Lars Ake Svensson and Jan Olof Ekberg
    Arkiv Fys., 37(7):65-84 (1967)

Electronic Structure of Zirconium
  T. L. Loucks
    (Ames Lab., Iowa), IS-1482 (Dec. 16, 1966), 34 pp.

X-Ray Spectrographic Investigation of the Electron Structure of the Elements from Zr to Ag
  V. V. Nemoshkalenko and V. P. Krivitskii
    Bull. Acad. Sci. USSR Phys. Ser., 31(6):999-1004 (1968)
    Izv. Akad. Nauk SSSR, Ser. Fiz., 31(6):985-989 (June 1967)
    Eighth Conf. X-ray Spectroscopy, Apatity (June 28-July 3, 1966)

Fermi Surface of Zirconium
  S. L. Altmann and C. J. Bradley
    Phys. Rev., 135:A1253-A1256 (1964)

Field-Microscopic Investigations of Tungsten-Molybdenum Mixed-Crystal Systems
  E. K. Caspary and E. Krautz
    (OSRAM-Studiengesellschaft, Augsburg, Ger.)
    Z. Naturforsch., 19a:591-595 (May 1964)

Vacancies in Quenched Molybdenum Single Crystals
  J. D. Meakin, A. Lawley, and R. C. Koo
    Appl. Phys. Letters, 5:133 (1964)

1.f.3. Group V

X-Ray K-Absorption Edge of Niobium in Niobium Metal and Its Oxides
  V. G. Bhide and M. K. Bahl
    J. Chem. Phys., 52:4093-4096 (1970)

Characteristic Energy Loss Studies of V_2O_5 and Vanadium
L. Fiermans and J. Vennik
Phys. Stat. Sol., 41:621–629 (1970)

Experimental Study of the Fermi Surfaces of Niobium and Tantalum
M. H. Halloran, J. H. Condon, J. E. Graebner, J. E. Kunzler, and F. S. L. Hsu
Phys. Rev., B 1(2):366–372 (1970)

Electronic Structure of Niobium and Tantalum
L. F. Mattheiss
Phys. Rev., B 1(2):373–380 (1970)

The Fermi Surface in Niobium
G. B. Scott and M. Springford
Proc. Roy. Soc. London, A 320:115–130 (1970)

Band Calculations for Vanadium
J. R. Anderson, J. W. McCaffrey, and D. A. Papaconstantopoulos
Solid State Commun., 7:1439–1441 (1969)

Pressure Dependence of the Band Structure of Vanadium
D. A. Papaconstantopoulos, Joseph W. McCaffrey, and J. R. Anderson
Bull. Am. Phys. Soc., 14:360 (1969)

Fermi Surface Calculations for Nb and Ta
L. F. Mattheiss
Bull. Am. Phys. Soc., 13:508 (1968)

Magnetoresistance and Fermi Surface Anisotropy in Vanadium Metal
K. S. Nelson, J. L. Stanford, and F. A. Schmidt
Phys. Letters, 28A:402–403 (1968)

High-Field Galvanomagnetic Properties of Niobium
W. A. Reed and R. R. Soden
Phys. Rev., 173:677–679 (1968)

De Haas-Van Alphen Effect and Fermi Surface in Niobium
G. B. Scott, M. Springford, and J. R. Stockton
Phys. Letters, 27A:655–656 (1968)

Modifications to the Orthogonalized-Plane-Wave Method for Use in Transition Metals: Electronic Band Structure of Niobium
R. A. Deegan and W. D. Twose
Phys. Rev., 164:993–1005 (1967)

X-Ray K Spectra and Energy Band Structure of Metallic Vanadium
S. A. Nemnonov and L. D. Finkelshtein
Fiz. Metallov i Metalloved, 21:211–216 (1966)

Anomalous Densities of States in Normal Tantalum and Niobium
A. F. G. Wyatt
Low Temperature Physics LT9, Part A (J. G. Daunt, D. O. Edwards, F. J. Milford, and M. Yaqub, eds.), Plenum Press, New York (1965), pp. 411–414

The Specific Heats and Resistivities of Molybdenum, Tantalum, and Rhenium
R. E. Taylor and R. A. Finch
J. Less-Common Metals, 6:283–294 (1964)

Electronic and Nuclear Polarization in Vanadium by Slow Neutron Scattering
C. G. Shull and R. P. Ferrier
Phys. Rev. Letters, 10:295–297 (1963)

The Specific Heats and Resistivities of Molybdenum, Tantalum, and Rhenium from Low to Very High Temperatures
R. E. Taylor and R. A. Finch
(Atomic International, Div. of North American Aviation, Inc., Canoga Park, Calif.), Contract AT(11-1)-GEN-8, NAA-SR-6034 (Sept. 15, 1961), 37 pp.

The Physical Properties of Niobium, Tantalum, Molybdenum, and Tungsten
B. B. Argent and G. J. C. Milne
J. Less-Common Metals, 2:154–162 (1960)

Field Dependence of Photoelectric Emission from Tantalum
J. L. Gumnick and D. W. Juenker
J. Appl. Phys., 31:102 (1960)

Metallochemical Properties of Niobium
I. I. Kornilov
NP-tr-704 Dokl. Akad. Nauk SSSR, 135:1399–1401 (1960), 6 pp.

1.f.4. Group VI

De Haas-Shubnikov Effect in Tungsten
H. J. Trodahl and J. A. Woollam
Solid State Commun., 9:291–294 (1971)

Highly Resolved X-Ray Structures and the Electronic Structure of Tungsten
G. Bohm and R. Heyd
Phys. Stat. Sol., 39:K19–K21 (1970)

High-Speed (Subsecond) Measurement of Heat Capacity, Electrical Resistivity, and Thermal Radiation Properties of Molybdenum in the Range 1900 to 2800 K
A. Cezairliyan, M. S. Morse, H. A. Berman, and C. W. Beckett
J. Res. Natl. Bur. Std., 74A:65–92 (1970)

The Radio-Frequency Size Effect and the Fermi Surface of Molybdenum
Joseph R. Cleveland
Thesis (Ames Lab., Iowa), IS-T-356 (May 1970), 127 pp.

Radio-Frequency Size Effect in Molybdenum
J. R. Cleveland and J. L. Stanford
Bull. Am. Phys. Soc., 15:263 (1970)

Evidence for Large Spin-Orbit Coupling in Metallic Molybdenum
J. R. Cleveland and J. L. Stanford
Phys. Rev. Letters, 24:1482–1485 (1970)

Observation of Structures in the Chromium $K\alpha_1$ Line with Spherically Bent Crystal Spectrometer
K. Das Gupta and P. F. Gott
Phys. Letters, 33A:276–277 (1970)

Surface States on Tungsten
E. W. Plummer and J. W. Gadzuk
Phys. Rev. Letters, 25:1493–1495 (1970)

Secondary Electron Emission of Molybdenum Crystals
A. R. Shul'man, V. V. Korablev, and Yu. A. Morozov
Fiz. Tverd. Tela, 12:758-762 (1970)
Soviet Phys. — Solid State, 12(3):586-589 (1970)

Work Function Measurements on Macroscopic Tungsten Specimens
L. van Someren
Surface Sci., 20(2):221-234 (1970)

dc Electronic Size Effects in Tungsten
J. C. Abele and D. E. Soule
Bull. Am. Phys. Soc., 14:307 (1969)

Oscillatory Magnetoresistance and Magnetic Breakdown in Antiferromagnetic Chromium
A. J. Arko, J. A. Marcus, and W. A. Reed
Phys. Rev., 185:901-904 (1969)

Investigation of the Fermi Surface in Molybdenum on Basis of the Radio Frequency Size Effect
V. V. Boiko, V. A. Gasparov, and I. G. Gverdtsiteli
Zh. Éksp. Teor. Fiz., 56(2):489-501 (1969).
Sov. Phys. — JETP, 29(2):267-273 (1969)

New Even Levels and Classified Lines in the First Spectrum of Tungsten (W_1)
C. H. Corliss
J. Res. Natl. Bur. Stds., A 73(3):277-279 (1969)

Thermal Diffuse X-Ray Scattering Measurements of the Fermi Surface of Chromium
James M. Costello and John W. Weymouth
Phys. Rev., 184(3):694-701 (1969)

Effect of Plastic Deformation on the Angular Distribution of Photons during Positron Annihilation in a Molybdenum Crystal
I. Ya. Dekhtyar, V. S. Mikhalenkov, and S. G. Sakharova
Fiz. Tverd. Tela, 11:3266-3269 (1969)
Sov. Phys. — Solid State, 11:2647-2649 (1970)

Photoemission Studies of the Electronic Structure of Transition Metals
D. E. Eastman
J. Appl. Phys., 40:1387 (1969)

Oscillatory Size Effect in Tungsten
R. F. Girvan and P. C. Canepa
Bull. Am. Phys. Soc., 14:29 (1969)

Secondary Electron Emission from the (111) Face of a Tungsten Crystal
M. V. Gomoyunova and B. Z. Aliev
Fiz. Tverd. Tela, 11:3619-3621 (1969)
Soviet Phys. — Solid State, 11:3036-3037 (1970)

Quantum Oscillations in the Ultrasonic Attenuation in Molybdenum
E. H. Gregory and Jean Buttet
Bull. Am. Phys. Soc., 14:401 (1969)

Spectral Density Functions of Field Emission Flicker Noise Caused by Potassium Adsorbed on Tungsten
Ch. Kleint and R. Meclewski
Acta Phys. Pol., 36:97-105 (July 1969)

Connection between the Anomalous Hall Effect and the Structure of the Fermi Surface
E. I. Kondorskii
Zh. Éksper. Teor. Fiz., 55(2):558-566 (1968)
Sov. Phys. — JETP, 28(2):291-295 (1969)

Photoemission Studies of the Electronic Structure of Ru and Mo
K. A. Kress and G. J. Lapeyre
Bull. Am. Phys. Soc., 14:359 (1969)

Photoemission and Reflectance Studies of the Electronic Structure of Molybdenum
K. A. Kress and G. J. Lapeyre
in: Proceedings of the Electronic Density of States Symposium, NBS, Gaithersburg, Md., Nov. 3-6, 1969

de Haas-van Alphen Measurements in Molybdenum
G. Leaver and A. Myers
Phil. Mag., 19:465 (1969)

Role of Work Function in Electron Ejection by Metastable Atoms: Helium and Argon on (111) and (110) Tungsten
D. A. Maclennan and T. A. Delchar
J. Chem. Phys., 50(4):1772-1778 (1969)

Observation of Fine Structures of Chromium $K_{\alpha 1.2}$ Lines with a High Resolution Three Crystal Spectrometer
D. M. Shah and K. Das Gupta
Phys. Letters 29A:570-571 (1969)

Electron Tunneling into Chromium Single Crystals
L. Y. L. Shen
J. Appl. Phys., 40:5171-5173 (1969)

Fermi Surface Curvature in Tungsten
D. E. Soule and J. C. Abele
Bull. Am. Phys. Soc., 14:845 (1969)

Determination of Mean Inner Potential from Kikuchi Patterns of Tungsten Single Crystals
K. H. von Grote
Optik, 28:374-388 (1969)

Effect of Heating on the Photoelectron Yield of Polycrystalline Tungsten in the Vacuum Ultraviolet
B. J. Waclawski and L. R. Hughey
J. Opt. Soc. Am., 59:1494-1495 (Nov. 1969)

Lattice Vibrations in Molybdenum
C. B. Walker and P. A. Egelstaff
Phys. Rev., 177:1111-1122 (1969)

High-Field Galvonomagnetic Effects in Antiferromagnetic Chromium
Aloysius John Arko
Ph.D. thesis, Northwestern Univ., Evanston, Ill. (1968), 157 p.
University Microfilms, Ann Arbor, Mich., Order No. 69-1789

Fermi Surface and Correlation Effects in Optical Properties of Molybdenum
G. A. Bolotin, V. M. Maevskii, and B. A. Charikov
Fiz. Metal. Metalloved., 25:629-636 (1968)

Thermal Diffuse X-Ray Scattering Study of the Fermi Surface in Antiferromagnetic Chromium
James M. Costello and John W. Weymouth
(Nebraska Univ., Lincoln, Dept. of Phys.), Contract AT(11-1)-525, COO-525-28 (Feb. 1968), 11 pp.

Fermi Surface in Molybdenum
A. A. Galkin, S. E. Zhevago, T. B. Butenko, and E. P. Degtyar
Ukr. Fiz. Zh., 13:1106-1108 (1968)

The de Haas-van Alphen Effect and the Fermi Surface of Tungsten
R. F. Girvan, A. V. Gold, and R. A. Phillips
J. Phys. Chem. Solids, 29:1485-1502 (1968)

Investigation of the de Haas-van Alphen Effect in Antiferromagnetic Chromium
John Edwin Graebner
Ph. D. Thesis from Northwestern University, Evanston, Illinois (1968), 134 pp.

De Haas-van Alphen Effect in Antiferromagnetic Chromium
J. E. Graebner and J. A. Marcus
Phys. Rev., 175:659-673 (1968)

Study of Fermi Surfaces of Molybdenum During Cyclotron Resonance Measurements
R. Herrmann
Phys. Stat. Sol., 25:427-435 (1968)

Zur Fermifläche von Wolfram und Molybdän
R. Herrmann
Phys. Stat. Sol., 25:661 (1968)

Photoemission Investigation of the Electronic Structure of Chromium
G. J. Lapeyre and K. A. Kress
Phys. Rev., 166:589-598 (1968)

The First Spectrum of Tungsten (W_1)
D. D. Laun and C. H. Carliss
J. Res. Natl. Bur. Std., Phys. Chem., 72(6):609-755 (1968)

Study of Thermionic Converter Electrode Surfaces Using the Electron Mirror Microscope (Final Report, 15 July 1964 — 14 July 1967)
K. N. Maffit
(Litton Systems, Inc., Applied Science Div., Minneapolis, Minnesota), Contract AF 19(628)-4302, Rept. 3183; AD-668530 (Feb. 1968), 98 pp.
Work function differences on polycrystalline tungsten

Experimental Determination of Fermi Surface Cross-Section for Molybdenum and Tungsten from Size Effect
V. E. Startsev, N. V. Volkenshtein, and G. A. Nikitina
Fiz. Metallov. Metalloved., 26:261-268 (1968)

Work Function of W Single Crystal Planes
J. C. Tracy
Thesis, Cornell University, Ithaca, N. Y. (1968)

Quantum Oscillations in the Attenuation of Transverse Ultrasonic Waves in Field-Cooled Chromium
W. D. Wallace and H. V. Bohm
J. Phys. Chem. Solids, 29:721-734 (1968)

Radio-Frequency Size Effect in Molybdenum
V. V. Boiko, V. A. Gasparov, and I. G. Gverdtsiteli
JETP Letters, USA, 6:212-214 (1967)
Zh. Éksper. Teor. Fiz. Pisma (USSR), 6:737-741 (1967)

The Fermi Surface of Tungsten
R. F. Girvan
Dissertation Abstr., B., (USA), 27:3245 (1967)

Experimental Study of the Surface Structure and Electronic Properties of Single Crystal Molybdenum and Tungsten Ribbons
Donald R. Morgan and William E. Blass
Progress Report COO-1417-3 covering the period Nov. 1, 1966 - Aug. 1, 1967, St. Mary's College, Winona, Minnesota

Decrease of Mössbauer Recoil-Free Fraction in Small Tungsten Particles
R. Roth and E. M. Hoerl
Osterreichische Studiengesellschaft für Atomenergie G.m.b.H., Seibersdorf, Austria, SGAE-M-9/1967 (July 1967), 6 pp.

Thermionic Emission from Planar Monocrystalline Tungsten
S. I. Schreiner
Star, 5:601 (1967)
NASA-CR-80879; UCRL-16690

Electron Emission from W and Mo
D. W. Vance
27th Annual Conference Physical Electronics, American Physical Society, held at Massachusetts Institute of Technology, Cambridge, Mass. (March 20-22, 1967), pp. 90-94

Fermi Surface of Tungsten
Robert F. Girvan
Contract W-7405-eng-82, IS-T-103 (July 1966), 243 pp.

Electron Energy Levels in Metallic Chromium
V. V. Krivitskii and I. Ya. Nikiforov
Bull. Acad. Sci. USSR. Phys. Ser., 31(6):985-988 (1968)
Izv. Akad. Nauk SSSR, Ser. Fiz., 31(6):970-973 (June 1967)
Eighth Conf. X-ray Spectroscopy, Apatity (June 28-July 3, 1966)

Magneto-Acoustic Investigation of the Fermi Surface in Molybdenum
P. A. Bezuglyi and S. I. Zhevago
Zh. Éksper. Teor. Fiz., 49(5):1457-1462 (1965)
Sov. Phys. — JETP, 22(5):1002-1005 (1965)

Magnetoacoustic Effect in Tungsten and Molybdenum near 1 Gcps
C. K. Jones and J. A. Rayne
Low Temperature Physics LT9, Part B (J. G. Daunt, D. O. Edwards, F. J. Milford and M. Yaqub, eds.), Plenum Press, New York (1965), pp. 790-793

Fermi Surfaces of Cr, Mo and W by the APW Method
T. L. Loucks
Contract W-7405-eng-82, IS-1101 (March 8, 1965), 35 pp.

Relativistic-Electronic Structure in Crystals. II. Fermi Surface of Tungsten
T. L. Loucks
Phys. Rev., 143:506-512 (1966)
Phys. Rev. Letters, 14:693 (1965)

Fermi Surface in Tungsten
L. F. Mattheiss
Phys. Rev., 139:A1893 (1965)

Extremal Dimensions of Cyclotron Orbits in Tungsten
    W. M. Walsh, C. C. Grimes, G. Adams, and L. W. Rupp
    Low Temperature Physics LT9, Part B (J. G. Daunt, D. O. Edwards, F. J. Milford, and M. Yaqub, eds.), Plenum Press, New York (1965), pp. 765-769

The de Haas-van Alphen Effect in Chromium
    B. R. Watts
    Low Temperature Physics LT9, Part B, (J. G. Daunt, D. O. Edwards, F. J. Milford, and M. Yaqub, eds.), Plenum Press, New York (1965), pp. 776-778

Fermi Surface in Molybdenum
    W. M. Lomer
    Proc. Phys. Soc. (London), 84:327-330 (1964)

Fermi Surface of Tungsten from Magnetoacoustic Measurements
    John A. Rayne
    (Westinghouse Research Laboratories, Pittsburgh, Pennsylvania), Phys. Rev., 133:A1104-A1108 (1964)

The Fermi Surface of Tungsten and Molybdenum by the de Haas-van Alphen Effect
    Don Merle Sparlin
    Thesis from Northwestern Univ., Evanston, Illinois (1964), 139 pp.

The Specific Heats and Resistivities of Molybdenum, Tantalum, and Rhenium
    R. E. Taylor and R. A. Finch
    J. Less-Common Metals, 6:283-294 (1964)

Photoelectric Yield of W
    J. E. Wheaton
    J. Opt. Soc. Am., 54:1287 (1964)

Resistivity and Electron Microstructure of Quenched Tungsten
    F. W. Kuether
    Bull. Am. Phys. Soc. Ser. II, 8:66 (A) (1963)

The Electronic Structure of Chromium Group Metals from Low Field de Haas-van Alphen Studies
    G. B. Brandt and J. A. Rayne
    Phys. Letters, 3:148-149 (1962)

The Fermi Surface Areas of Chromium, Molybdenum, and Tungsten
    E. Fawcett and D. Griffiths
    Phys. Chem. Solids, 23:1631-1635 (1962)

The Specific Heats and Resistivities of Molybdenum, Tantalum, and Rhenium from Low to Very High Temperatures
    R. E. Taylor and R. A. Finch
    (Atomic International, Div. of North American Aviation, Inc., Canoga Park, Calif.) Contract AT(11-1)-GEN-8, NAA-SR-6034 (Sept. 15, 1961), 37 pp.

The Physical Properties of Niobium, Tantalum, Molybdenum, and Tungsten
    B. B. Argent and G. J. C. Milne
    J. Less-Common Metals, 2:154-162 (1960)

A Set of Effective Coordination Number (12) Radii for the β-Wolfram Structure Elements
    S. Geller
    Acta Cryst., 9:885 (1956)

Kohn Anomalies in Tungsten and Other Cr-Group Metals
    T. M. Rice and B. I. Halperin
    Bell Telephone Laboratories, Murray Hill, N. J., preprint

The Fermi Surface of Tungsten and Molybdenum by the de Haas-van Alphen Effect
    Don M. Sparlin
    Dissertation Abstr., 65-3313, 139 pp.

1.g. Magnetic Properties

1.g.1. General

Superconducting Transition Temperature of Disordered Transition Metal Films
    M. M. Collver
    Ph. D. thesis, Univ. California, Berkeley, UCRL-19186 (April 1970)
    Mo, Nb, Ta, V, Nb-Zr

Bibliography of Magnetic Materials and Tabulation of Magnetic Transition Temperatures
    T. F. Connolly and Emily D. Copenhaver
    ORNL-RMIC-7 (Rev. 2) (July 1970), 116 pp

Magnetostriction of Paramagnetic Transition Metals. I. Group 4 — Ti and Zr; Group 5 — V, Nb, and Ta; Group 6 — Mo and W
    E. Fawcett
    Phys. Rev., B2:1605-1613 (1970)

Superconducting Transition Temperatures of Disordered Nb, W, and Mo Films
    J. E. Crow, M. Strongin, and R. S. Thompson
    Phys. Letters, 30A:161-62 (1969)

Positron Annihilation in Superconductive Metals
    I. Ja. Dekhtjar
    Phys. Letters, 28A:771-772 (1969)

Generalized Susceptibilities and Magnetic Properties of Some bcc Transition Metals
    W. E. Evenson, G. S. Fleming, and S. H. Liu
    Phys. Rev., 178:930-932 (1969)
    Groups V and VI

Magnetic Susceptibility of Transition Metals
    F. M. Galperin
    Phys. Letters, 29A:418-419 (1969)

Über die Temperaturabhängigkeit der paramagnetischen Suszeptibilität von Scandium, Titan, Vanadium, Chrom und Mangan
    R. Kohlhaas and W. D. Weiss
    Z. Angew. Physik, 28:16-20 (1969)

High Pressure Study on Superconductivity of Transition Metals and Alloys
    Ching-wu Chu
    Thesis from University of California, San Diego, California (1968), 163 pp.

Magnetostriction of Paramagnetic Transition Metals and Alloys
    E. Fawcett and G. K. White
    J. Appl. Phys., 39:576 (1968)
    Groups IV, V, VI, and VIII

Magnetic and Inelastic Scattering of Neutrons by Metals
    T. J. Rowland and Paul A. Beck, eds.
    Gordon and Breach Science Publishers, New York (1968), 256 pp.
    Metallurgical Society Conferences

Resonant States of Transition-Metal Elements
    G. Boato
    Nuovo Cimento Suppl., 5(4):1159-1177 (1967)
    52nd National Congress of the Italian Physical Society, Trieste (Oct. 24-28, 1966)
    Review, Friedel-Anderson theory, magnetic and nonmagnetic states

Critical Current and Current Densities of Selected Superconducting Materials
Donald L. Grigsby
Electronic Properties Information Center, Hughes Aircraft Co., Culver City, California
EPIC Interim Rept., 21 (Feb. 1966)

Magnetic Susceptibility and Electronic Specific Heat of Transition Metals and Alloys. IX. Recalculation for Zr, Nb, and Mo Metals
Masao Shimizu, Atsushi Katsuki, and Kazuhiko Ohmori
J. Phys. Soc. Japan, 21:1922 (1966)

Magnetostriction of Nonferromagnetic Metals
N. C. Anderholm and E. R. Peck
J. Appl. Phys., 36:2293 (1965)

La susceptibilité magnetique du titane, du vanadium, du chrome et du manganese entre 100 et 2000°K
R. Kohlhaas and W. D. Weiss
Phys. Verh. D.P.G., Dtsch., 5:127 (1965)

Die magnetische Suszeptibilität von Titan und Vanadin zwischen 100 and 2000°K
Rudolf Kohlhaas and Wolf Dieter Weiss
Z. Naturforsch., 20A:1227–1229 (1965)

Low Temperature Paramagnetic Susceptibility of Transition Metals
N. V. Volkenshtein and E. V. Galoshina
Fiz. Metallov. i Metalloved., 20:368–372 (1965)

Neutron Diffraction Studies of Magnetic Moments in Dilute Transition Metal Alloys
M. F. Collins and G. G. Low
(Solid State Physics Division, U. K. A. E. A. Research Group, Atomic Energy Research Establishment, Harwell)
Z. Physik, 25:596 (1964)

Galvanomagnetic Properties of Single Crystals of Transition Metals
N. E. Alekseevskii, V. S. Egorov, G. E. Karstens, and B. N. Kazak
Sov. Phys. — JETP, 16:519–520 (1963)

Defauts pontuels dan les solides ferromagnetiques et ordre directionnel
L. Neel
(Faculté des Sciences, Grenoble), J. Phys., 24:513 (1963)

Magnetic Properties
H. M. Rosenberg
Low Temperature Solid State Physics, Clarendon Press, Oxford (1963), pp. 285–347

The Susceptibility of Metals — the de Haas — van Alphen Effect
H. M. Rosenberg
Low Temperature Solid State Physics, Clarendon Press, Oxford (1963), pp. 348–358

Spin-Dependent Scattering and Resistivity of Magnetic Metals and Alloys
T. van Peski-Tinbergen and A. J. Dekker
(Instituut voor Kristalfysica der Rijksuniversiteit, Groningen, Nederland), Physica, 29:917–937 (1963)

Intrinsic and External Permeabilities and Susceptibilities of Gyromagnetic Materials
R. A. Waldron
Brit. J. Appl. Phys., 14:700–703 (1963)

Ferromagnetism and Antiferromagnetism in Non-Ferrous Metals and Alloys
J. Crangle
(Univ. of Sheffield, Eng.), Met. Rev., 7:133–174 (1962)

The Electronic Structure and Magnetic Properties of Transition Metals and Alloys
J. Friedel
AI-trans. 35, January 1964, J. Phys. Radium, 23:501–510 (1962)

Ultra-High Purity Metals
Papers presented at a seminar of American Society of Metals (October 21 and 22, 1961), American Society of Metals, Metals Park, Ohio (1962)

Magnetoresistance of Transition Metals in the High-Field Limit
E. Fawcett
Phys. Rev. Letters, 7:370–372 (L) (1961)

The Variation with Temperature of the Magnetic Susceptibility of Some of the Transition Elements
H. Kojima, R. S. Tebble, and D. E. G. Williams
Proc. Roy. Soc. (London), A 260:237–250 (1961)

Propriétes des metaux et alliages aux bases temperatures (Properties of Metals and Alloys at Low Temperatures)
E. Justi
CEA-Tr-A-789
Z. Metallk., 51(1):1–17 (1960)

1.g.2. Group IV

Magnetic Susceptibility and Low-Temperature Specific Heat of High-Purity Titanium
E. W. Collings and J. C. Ho
Phys. Rev. B, 2(2):235–244 (1970)

Calculation of Paramagnetic Susceptibilities for H.C.P. Transition Metals
N. Mori
J. Phys. Soc. Japan, 29(2):366–371 (1970)
Ti, Zr, Hf

Magnetic Susceptibility of Titanium and of Some Ti — O Alloys
R. J. Wasilewski
J. Appl. Phys., 40:2677–2678 (1969)

Hafnium Is Not a Bulk Superconductor
J. E. Cox
Phys. Letters, 28A:326–327 (1968)

Anisotropy of Paramagnetic Susceptibility of Zirconium
N. V. Volkenshtein, E. V. Galoshina, and N. I. Shchogolikhina
Fiz. Metallov i Metalloved., 25:180–183 (1968)

Thermal Dependence of Hall Effect and Paramagnetic Permeability of Zirconium and Rhenium
D. I. Volkov, T. M. Kozlova, V. N. Prudnikov, and E. V. Kozis
Zh. Éksp. Teor. Fiz., 55:2103–2107 (1968)
Sov. Phys. — JETP, 28(6):113–116 (1969)

Superconducting Transition Temperatures of Zirconium and Zirconium Isotopes
Delbert M. Jones
Dissertation Abstr., 26:5506 (1966)

Localized Magnetic Impurity States in Ti, Zr, and Hf
J. A. Cape and R. R. Hake
Phys. Rev., 139:142A (1965)

Preparation and Properties of Thin-Film Hard Superconductors
J. Edgecumbe, L. G. Rosner, and D. E. Anderson
(Department of Electrical Engineering, University of Minnesota, Minneapolis, Minnesota)
J. Appl. Phys., 35:2198 (1964)

Magnetic Susceptibility and Electronic Specific Heat of Transition Metals and Alloys. VII. Zr and Rh Metals
M. Shimizu and A. Katsuki
J. Phys. Soc. Japan, 19:1856 (1964)

Magnetic Susceptibility and Electronic Specific Heat of Transition Metals and Alloys. IV. V and Ti Metals and V-Cr and V-Ti Alloys
M. Shimizu, A. Katsuki, and T. Takahashi
J. Phys. Soc. (Japan), 18:1192-1203 (1963)

Superconductivity of Titanium and Its Alloys
U. Zwicker
Al-Trans.-64; TT-66-11857
Z. Metallkunde, 54:477-483 (1963)

Hall Effect, Resistivity, and Magnetoresistivity of Th, U, Zr, Ti, and Nb
T. G. Berlincourt
Phys. Rev., 114:969 (1959)

Hall Effect in Titanium, Chromium, Vanadium, and Manganese
Simon Foner
Phys. Rev., 107:1513-1516 (1957)

1.g.3. Group V

Magnetization and Susceptibility Measurements of Polycrystalline Niobium
H. Brechna, M. A. Allen, and J. K. Cobb
J. Appl. Phys., 42:103-105 (1971)

Alternating-Current Loss Measurements in Thin-Film Type II Super-conductors
D. W. Deiss, J. R. Gavaler, C. K. Jones, and A. Patterson
J. Appl. Phys., 42:21-26 (1971)
Nb and NbN

Investigation of Time Dependent Flux Transport Voltages in Niobium Foils
C. Heiden and H. P. Friedrich
Solid State Commun., 9:323-326 (1971)

Experimental Tests of Proposed Relations Between the Critical Field and Dislocation Cell Structure of Superconducting Niobium
D. C. Hill and R. M. Rose
Met. Trans., 2:585-589 (1971)

Longitudinal Ultrasonic Attenuation in Tantalum at Frequencies up to 1 GHz
J. M. Perz and W. A. Roger
Can. J. Phys., 49:296-301 (1971)

Ultrasonic Attenuation and Velocity Measurements in the Normal and Superconducting States of Niobium
J. Trivisonno, J. Washick, et al.
NASA-CR-1742 (Jan. 1971)

Nuclear Acoustic Resonance in Niobium Metal
J. Buttet and E. H. Gregory
Phys. Letters, 33A:265-266 (1970)

Microstructural Effects in Critical Fields of Niobium Films
M. J. Connella, R. J. Deck, and R. H. Morriss
J. Appl. Phys., 41:5346-5348 (1970)

Surface Studies on Niobium and Some Implications for Superconductivity
J. M. Dickey, H. H. Farrell, and O. F. Kammerer
Phys. Letters, 32A:483-484 (1970)

Deformation-Induced Anisotropy of the Critical Current in Single Crystal Niobium
J. A. Good and E. J. Kramer
Phil. Mag., 22:329-357 (1970)

Measurement of 50 Hz Dissipation in Superconducting Niobium Strips
R. Grigsby and R. J. Slaughter
J. Phys. D: Appl. Phys., 3:898-905 (1970)

The Second Energy Gap in Superconducting Niobium
J. W. Hafstrom and M.L.A. MacVicar
Massachusetts Institute of Technology, Cambridge, Mass., Rept. TR-3; AD-712072 (Sept. 1970), 29 pp.

Far-Infrared Absorption in Bulk Samples of Superconducting Niobium and Tantalum
D. Hemming, R. Sati, and J. D. Leslie
Can. J. Phys., 48:1254-1258 (1970)

Dislocation Cell Structure and Anomalous Critical Fields for Superconductivity in Severely Deformed Niobium
D. C. Hill and R. M. Rose
Appl. Phys. Letters, 17:66-67 (1970)

Evidence for "Internal Surface" Superconductivity in Severely Deformed Niobium
D. C. Hill, J. G. Kohr, and R. M. Rose
Phys. Letters, A, 31:157-158 (1970)

Electrical Resistivity and Peak Effect in Superconducting Niobium
R. P. Huebener and R. T. Kampwirth
J. Low Temp. Phys., 2:113-120 (1970)

Magneto-Optical Observation of the Magnetic Flux Structure in Superconducting Niobium
R. P. Huebener, V. A. Rowe, and R. T. Kampwirth
J. Appl. Phys., 41:2963-2967 (1970)

Déformation hétérogène de monocristaux de niobium à très basse température
Pierre Kubin and Bernard Jouffrey
Compt. Rend., Ser. B, 270:812-814 (1970)

Temperature Dependence of the Magnetic Field
at a Vortex Center in Type-II Superconducting
Vanadium by NMR
  Albert Kung
  Phys. Rev. Letters, 25:1006-1007 (1970)

Hypersonic Attenuation in Niobium in the Mixed
State
  James N. Lange
  J. Phys. Chem. Solids, 31:1693-1700 (1970)

Nuclear Spin-Lattice Relaxation in the Mixed
State of Superconducting Niobium
  D. E. MacLaughlin and D. Rossier
  Phys. Kondensierten Materie D, 11:43-65 (1970)

Distribution of the Energy Gap in k Space for
Superconducting Nb
  M. L. A. MacVicar
  Phys. Rev., B, 2:97-100 (1970)

Changes of the Magnetic Properties of Superconducting Niobium by Electron Irradiation at
4.5°K
  Constantin Papastaikoudis
  Inst. fuer Festkoerper- und Neutronenphysik, Kernforschungs-
    anlage Juelich, West Germany, JUL-647-FN (March
    1970), 58 pp.

Electronic Tunneling into Niobium Single Crystals,
Final Report, 1 Feb. 1965 - 31 Jan. 1970
  R. M. Rose
  Massachusetts Institute of Technology, Cambridge, Mass.,
    AD-704802 (Jan. 1970), 25 pp.

Thermal Conductivity of Pure Type II Supercon-
ductors
  E. Schmidbauer, H. Wenzl, and E. Umlauf
  Z. Phys., 240:30-41 (1970)
  In German

Flux-Flow Noise Spectrum in Type II Supercon-
ductors
  S. W. Shen
  Appl. Phys. Letters, 17:415-416 (1970)
  Vanadium foil

The Energy Gap of the Lower Band of Super-
conducting Niobium at T = 0°K
  I. M. Tang
  Phys. Letters, 31A:480-481 (1970)

Penetration Depths in Pure Superconducting
Niobium
  I-Ming Tang
  Phys. Letters, A 32:185-186 (1970)

Surface Distortions in Twinned Niobium Crystals
  R. J. Wasilewski
  Met. Trans., 1:1617-1627 (June 1970)

Zur Prüfung der Supraleitungs-Homogenität
grösserer Bleche — Ergebnisse an Niob
  R. Weber
  Z. Angew. Physik, 29:195-197 (1970)

Nonlocal Characteristics of the Bulk Upper
Critical Field of Niobium
  S. J. Williamson and L. E. Valby
  Phys. Rev. Letters, 24:1061-1064 (1970)

Superconductivity of Niobium Films
  Yuji Asada and Hiroshi Nose
  J. Phys. Soc. Japan, 26:347 (1969)

AC Energy Losses Above and Below H_{c1} in Nio-
bium and Niobium-25 at.% Zirconium
  W. T. Beall, Jr., and R. W. Meyerhoff
  J. Appl. Phys., 40:2052-2059 (1969)

A. C. Losses in Niobium Single Crystals in the
Meissner State
  P. R. Brankin and R. G. Rhodes
  Brit. J. Appl. Phys. (J. Phys. D), Ser. 2, Vol. 2:1775-1778
    (1969)

Attenuation of Microwave Phonons in the Mixed
State of Niobium
  R. Buisson and J. D. N. Cheeke
  Phys. Letters, 29A:200-201 (1969)

Effect of Magnetic-Field Cooling through the
Anomaly Temperature in Vanadium Metal
  Toshinobu Chiba and Tadayasu Mitsui
  J. Phys. Soc. Japan, 27:1451-1454 (1969)

Pinning of Fluxoids by Dislocations in Niobium
Single Crystals. II. Effect of Temperature and
Magnetic Field
  Herbert Freyhardt
  Z. Metallkunde, 60(5):409-412 (1969)

Ultrasonic Attenuation in Magnetically Induced
Structures in Superconductors
  M. Gottlieb, M. Garbuny, and C. K. Jones
  IEEE Trans. Son. Ultrason., 16:159-169 (1969)
  Nb and MoRe

An Investigation of Low Energy Structure in Nio-
bium Single Crystals by Tunnelling
  John W. Hafstrom and Robert M. Rose
  Mass. Inst of Tech., Cambridge, Mass., AD-698271; TR-11
    (Nov. 1969), 72 pp.

Evidence for a Second Energy Gap in Super-
conducting Niobium
  J. W. Hafstrom, R. M. Rose, and M. L. A. MacVicar
  Massachusetts Inst. of Tech., Cambridge, Mass., AD-698269
    (Nov. 1969), 8 pp.

Der Einfluss einer Zugspannung auf die Mag-
netisierungskurve von supraleitendem Niob
  B. Hillenbrand
  Z. Angew. Phys., 27:12-18 (1969)

Fast Neutron Elastic and Inelastic Scattering
of Vanadium
  B. Holmqvist, S. G. Johansson, et al.
  Thesis, Aktiebolaget Atomenergi, Sweden, AE-375 (Nov. 1969)

Nernst Effect and Flux Flow in Superconductors.
1. Niobium. 2. Lead Films
  R. P. Huebner and A. Seher
  Phys. Rev., 181:701-716 (1969)

Superconductivity in Niobium and Niobium-
Tantalum Alloys
  Akira Ikushimo and Takao Mizusaki
  J. Phys. Chem. Solids, 30:873-879 (1969)

The Anisotropy of the Critical Current Density
of Superconducting Oxygen-Doped Niobium
(Columbium)
  K. A. Jones and R. M. Rose
  Trans. Met. Soc. AIME, 245:67-73 (1969)

Superconductivity in Niobium Containing Ferro-
magnetic Gadolinium or Paramagnetic Yttrium
Dispersions
  C. C. Koch and G. R. Love
  J. Appl. Phys., 40:3582 (1969)

Nucléation de l'état mixte dans le niobium super-
conducteur
  B. P. Letellier, J. C. Renard, and Ya. A. Rocher
  J. Phys., 39:819-822 (1969)

The Fermi Surface and Delta (K) of Super-
conducting Nb
  Margaret L. A. MacVicar
  Massachusetts Inst. of Tech., Cambridge, Mass., AD-698270
    (Nov. 1969), 14 pp.

Harmonic Generation and Submillimeter Wave
Mixing with the Josephson Effect
  D. G. McDonald, V. E. Kose, K. M. Evenson, J. S. Wells,
    and J. D. Cupp
  Appl. Phys. Letters, 15:121 (1969)

Nucleation of Superconductivity in Tantalum in
a Decreasing Magnetic Field (K_0 = 0.340-0.01)
  J. P. McEvoy, D. P. Jones, and J. G. Park
  Phys. Rev. Letters, 22:229-231 (1969)

Surface Superconductivity in High Purity Nio-
bium
  Jerome E. Ostenson
  Ames Laboratory, Ames, Iowa, Contract W-7405-eng-82,
    IS-T-312 (May 1969), 79 pp.

Critical Phenomena in Sheath Superconductivity
of Nb
  J. E. Ostenson and D. K. Finnemore
  Phys. Rev. Letters, 22:188-190 (1969)

The Resistive Critical Field of Severely Drawn
Niobium
  J. R. Pearson and R. M. Rose
  Appl. Phys. Letters, 15(7):219-220 (Oct. 1, 1969)

Superconductivity of Molybdenum
  Daltro Garcia Pinatti
  Thesis, Rice University, Houston, Texas (1969), 112 pp.

Temperature Dependence of AC Losses in Super-
conducting Niobium
  E. C. Rogers, B. E. Redmonds, and G. D. Chan
  Cryogenics, 9(6):431-437 (1969)

Anomalous Nuclear Spin-Lattice Relaxation in
the Mixed State of Superconducting Niobium
  D. Rossier
  Phys. Rev. Letters, 22(24):1300-1303 (1969)

Nuclear Magnetic Resonance in Single-Crystal
Niobium
  H. E. Schone
  Phys. Rev., 183:410-413 (1969)

"Dip Effects" in the Flux-Flow Resistance in
High-Purity Niobium
  B. Slettenmark and H. U. Astrom
  J. Appl. Phys., 40:3985 (1969)

Helicon Resonances and Hall Angle Measure-
ments in the Mixed State of Niobium
  B. Slettenmark, H. U. Astrom, and P. Weissglas
  Solid State Commun., 7:1337-1340 (1969)

Resistance of Tantalum in the Intermediate
(Superconducting-Normal) State
  Frank T. J. Smith and Harvey C. Gatos
  J. Appl. Phys., 40:2232 (1969)

Superconducting Tunneling Effects in the Mixed
State of Nb
  Nobuo Tsuda
  J. Phys. Soc. Japan, 27:856 (1969)

Creep-like Flux Flow in Superconducting
Niobium
  H. A. Ullmaier
  Solid State Commun., 7:1565-1567 (1969)

Magnetization, Flux Flow Resistance and Hall
Effect in Superconducting Vanadium
  N. Usui, T. Ogasawara, K. Yasukochi, and S. Tomoda
  J. Phys. Soc. Japan, 27:574 (1969)

Orientation and Temperature Dependent Super-
conducting Properties in Niobium and Vanadium
Single Crystals
  I. Williams and A. M. Court
  Solid State Commun., 7:169-171 (1969)

Evidence that the Transition at $H_{c2}(T)$ in Nio-
bium May Assume a Third- or Higher-Order
Character as T Approaches Absolute Zero
  S. J. Williamson
  Bull. Am. Phys. Soc., 14:437 (1969)

Evidence that the Transition at H_{c2} in Niobium
May Not Be Second Order Near Absolute Zero
Temperature
  S. J. Williamson
  Phys. Letters, A, 28(10):665-666 (1969)

Einfluss von Neutronenbestrahlung bei 4.6°K auf
die Supraleitungseigenschaften von Niob
  H. Berndt, N. Kartascheff, and H. Wenzl
  Z. Angew Phys., 24:305-313 (1968)

Superconducting Properties of Sputtered Nio-
bium and Niobium Alloy Thin Films
  J. J. Bessot, J. C. Burlurut, B. P. Letellier, and Y. A.
    Rocher
  Proceedings of the Second Colloquium on Thin Films, Budapest,
    June 26-July 2, 1967 (E. Mahn, ed.), Göttingen, Vandenhoeck
    and Ruprecht (1968), pp. 361-366

Losses in Thin Films of Nb
  B. A. Christensen
  Dissertation, Minnesota University, Minneapolis (1968)

Etude de l'attenuation des ultrasons dans le
niobium
  R. Combescot
  These Doct. 3eme Cycle, Spec. Phys. Atom. Statist., Paris
    (1968), 34 pp.

Domain Boundaries in Vanadium
  F. D'Aragona Secco, R. C. Rau, and R. L. Ladd
  Phys. Stat. Sol., 30:1-3 (1968)

Surface Structure and a.c. Losses in Super-
conductive Niobium
  R. M. Easson and P. Hlawiczka
  British J. Appl. Phys., 1:1477-1485 (1968)

Precision Measurement of Anisotropy in the Upper Critical Field of Superconducting Niobium
D. E. Farrell, B. S. Chandrasekhar, and S. Huang
Phys. Rev., 176:562-566 (1968)

Transport Properties of the Type-II Superconductors Niobium and Vanadium
Anthony Thomas Fiory
Thesis, Rutgers — The State University, New Brunswick, N. J. (1968), 151 pp.

Intrinsic Type-2 Superconductivity in Pure Niobium
R. A. French
Cryogenics, 8:301-308 (1968)

Effect of Temperature and of Substitutional Mo on the Magnetic Behavior of Superconducting Nb
R. A. French, and J. Lowell
Phys. Rev., 173:504-509 (1968)

Defects in Superconducting Nb
B. B. Goodman and G. Kuhn
J. Phys., 29:240-252 (1968)

Hysteresis in the Ultrasonic Attenuation in Pure Nb near H_{c1}
M. Gottlieb, M. Garbuny, and C. K. Jones
Phys. Letters, 28a:148-149 (1968)

Evidence for a Dependence on Crystalline Orientation of the Superconducting Properties of Niobium in the Mixed State
C. E. Gough
Solid State Commun., 6:215-217 (1968)

Anisotropic Surface Impedance in the Mixed State of Superconducting Niobium
W. D. Hibler, III, and B. W. Maxfield
(Cornell University) Abstract submitted to the American Physical Society for the Washington Meeting (April 22-25, 1968)

Nernst Effect in Superconducting Niobium
R. P. Huebner and A. Seher
Solid State Commun., 6:403-406 (1968)

Superconductivity in Niobium and Niobium-Rich Niobium-Tantalum Alloys
A. Ikushima, T. Mizusaki, and T. Odaka
Phys. Letters, 26A:582-583 (1968)

Influence of the Purity of Superconducting Niobium on the Shape of the Magnetization Curve
V. R. Karasik, L. A. Nisel'son, I. V. Petrusevich, A. I. Shal'nikov, and I. Yu. Shebalin
Zh. Éksper. Teor. Fiz., Pis'ma Redakt., SSSR, 8(9):479-481 (1968)
Sov. Phys. — JETP Lett., 8(9):294-295 (1968)

Reinheit von Nb und dessen Supraleitfähigkeit
V. R. Karasik et al.
Zh. Eksp. Teor. Fiz., Pis'ma Redakt., 8:294-295 (1968)

Supraleitung von Vanadium, Niob und Tantal under hohem Druck
D. Kohnlein
Z. Phys., 208:142-158 (1968)

Pulsed NMR Study of the Field Induced Gapless State in Superconducting Niobium
D. E. MacLaughlin and D. Rossier

(Paris Univ., Orsay, France), Proceedings of Eleventh Conf. on Low Temperature Physics, held at Univ. of St. Andrews, Aug. 21-28, 1968, Vol. 2 (J. F. Allen, D. M. Finlayson, and D. M. McCall, ed.), pp. 943-946

Anisotropic Energy-Gap Measurements on Superconducting Niobium Single Crystals by Tunneling
M. L. A. MacVicar and R. M. Rose
J. Appl. Phys., 39:1721 (1968)

Superconductive Transition Temperature of Reactively Sputtered Niobium Films
Takeshi Mitsuoka, Tsutomu Yamashita, Tokuro Nakazawa, and Yutaka Onodera, Yukinori Saito, and Takeshi Anayama
J. Appl. Phys., 39:4788-4791 (1968)

Ultrasonic Attenuation in the Mixed State of Superconducting Vanadium
H. Ozaki, K. Kajimura, and T. Ishiguro
Phys. Letters, 28A:300-301 (1968)

Ultrasonic Attenuation in Superconducting Tantalum
Zoltan W. Sarafi
NASA-TN-D-2806 (Sept. 1968), 23 pp.

Destruction of Superconductivity in Ta
Y. Shibuya et al.
Adv. Cryogenic Engng., 13:181-183 (1968)

Ultrasonic Investigation of the Isolated-Vortex State near H_{c1} Vanadium
A. C. E. Sinclair and J. R. Leibowitz
Phys. Rev., 175:596-598 (1968)

Surface Superconductivity in Tantalum
Frank T. J. Smith and Harry C. Gatos
J. Appl. Phys., 39:3793-3797 (1968)

Hall Effect in Type II Superconductor Vanadium
N. Usui, T. Ogasawara, and K. Yasukochi
Phys. Letters, 27A:139 (1968)

The Ratio of the Upper Two Critical Fields in Pure Nb
G. W. Webb
Solid State Commun., 6:33-36 (1968)

Temperature and Orientation Dependent Surface Critical Current in Superconducting Niobium Single Crystals
I. Williams
Phys. Letters, 28A:409-410 (1968)

Solid State Physics
Ko Yasukochi, Kazuko Sekizawa, Takeshi Ogasawara, Nabumitsu Usui, Motoyoshi Ishizuka, et al.
(Nihon University, Tokyo, Japan), Plasma, Solid State, and Theoretical Physics (1968), pp. 35-46
Magnetic properties of superconducting Nb, Ta, and V

Energy Gap in Superconducting Nb
D. Bonnet
Physics Letters, 25A:452-453 (1967)

Relation Between the Superconducting Properties of Niobium and Its Microhardness
A. Kh. Chizhov
Fiz. Metal. Metalloved., 23:862-865 (1967)

Study of the Superconducting State of Nb by Neutron Diffraction
D. Cribier et al.
Progr. Low Temp. Phys., 5:161-180 (1967)

Flux Penetration and A. C. Losses in Superconducting Niobium
  R. M. Easson and P. Hlawiczka
  Brit. J. Appl. Phys., 18:1237-1249 (1967)

Thermomagnetic Effects in Superconducting Niobium
  A. T. Fiory and B. Serin
  Phys. Rev. Letters, 19:227-229 (1967)

Surface Flux Trapping in Superconducting Niobium
  R. B. Flippen
  Bull. Am. Phys. Soc., 12:38 (1967)

Features Concerning the Lower Critical Field of an Intrinsic Type II Superconductor (Niobium)
  R. A. French
  Cryogenics, 7:52-53 (1967)

Magnetische Flussgradienten in verformten Niob-Einkristallen bei 4.2°K
  Herbert Freyhardt and Peter Haasen
  Z. Metallkunde, 58:856-863 (1967)

The Effect of Structure on the Superconducting Properties of Vanadium and Niobium Foils
  G. J. van Gurp
  Philips Res. Repts., 22:10-35 (1967)

Magnetocaloric Effect and the Upper Critical Field of Superconducting Niobium
  Taiichiro Ohtsuka and Nobuyoshi Takano
  (Inst. of Solid State Phys., Univ. of Tokyo, Tokyo, Japan), Technical Rept. of ISSP Series A No. 263 (July 1967)

Superconducting Transition and Low-Field Magnetoresistance of a Niobium Single Crystal at 4.2°K
  Gastron Perriot
  (Commissariat a l'Energie Atomique, Saclay, France) CEA-R-3149 (Jan. 1967), 60 pp.

Superconducting Transition and Magnetoresistance of Nb Single Crystal at 4.2°K
  G. Perriot
  J. Phys. (France), 28:472-480 (1967)

Zum Einfluss von Versetzungen auf die Supraleitung von Niob im Magnetfeld
  E. Rexer, E. Nembach, and P. Haasen
  Realstruktur und Eigenschaften von Reinstoffen (1967), pp. 885-892

Losses of Superconducting Niobium in Low-Frequency Fields
  Y. A. Rocher and J. Septfonds
  Cryogenics, 7:96-102 (1967)

Superconducting Properties of High-Purity Niobium
  D. K. Finnemore, T. F. Stromberg, and C. A. Swenson
  Phys. Rev., 149:231 (1966)

The Relation Between Hydrogen Content and Paramagnetic Susceptibility in the Vanadium — Hydrogen System
  T. R. P. Gibb and W. A. Morder
  Inorg. Chem., 5:1947-1949 (1966)

Gyromagnetic Effect in Vanadium
  R. Huguenin and D. Baldock
  Phys. Rev. Letters, 16:795 (1966)

Enthalpy of Impure Nb in the Mixed Superconductive State
  S. H. Goedemoed, C. van Kolmeschate, P. H. Kes, and D. de Klerk
  Physica, 32:1183-1188 (1966)

Superconductivity of Nb Under Uniaxial Stress
  K. Luders
  Z. Phys., 193:73-75 (1966)

Investigation of the Interaction of Magnetic Flux Lines with Dislocations and Surfaces in Niobium, Part I
  E. Nembach
  Phys. Stat. Sol., 13:543-558 (1966)
  ORNL-TR-1600

Superconducting Critical Fields of Niobium Thin Films
  Hiroshi Nose and Yuji Asada
  Proc. Memorial Lecture Mtg. on 10th Anniv. of National Research Institute of Metals, Tokyo (1966)

Low-Temperature Thermodynamic Properties of Vanadium. I. Superconducting and Normal States
  Ray Radebaugh and P. H. Keesom
  Phys. Rev., 149:209 (1966)

The Temperature Variation of the Ginzburg-Landau Parameter in Niobium Single Crystals
  Loren C. Skinner
  (Mass. Inst. of Tech., Cambridge, Mass.), TR-1, AD-627910 (Jan. 1966), 50 pp.

Superconductivity of Nb
  R. J. Smith
  NASA-TN-D-3838 (1966)

The Superconducting Properties of High Purity Niobium
  T. F. Stromberg
  Dissertation Abstr., 26, 5510 (1966)

Superconducting Properties of Impure Tantalum
  C. Sulkowski and J. Mazur
  Acta Phys. Polon., 29:107 (1966)

Notes on the Superconductivity of Niobium
  Nob Tsuda, Shigetoshi Koike, and Taira Suzuki
  (Tokyo Univ., Japan, Inst. for Solid State Physics) Tech. Rept. Ser. A, No. 213 (August 1966), NP-16304, 20 pp.

Effect of Heat Treatment on Superconducting Properties of Niobium
  I. Williams and J. A. Catterall
  British J. Appl. Phys., 17:205 (1966)

Effect of Heat Treatment on the Superconducting Properties of Cold-Worked Niobium
  I. Williams and J. A. Catterall
  British J. Appl. Phys., 17:505 (1966)

Ultrasonic Investigation of Flux Penetration into Superconducting Niobium
  M. Gottlieb and C. K. Jones
  Estinghouse Research Labs., Scientific Paper 65-1JO-PHONY-P1 (1965), to be published in Physics Letters

The Influence of Fiber Structure on the Superconducting Behavior of Cold-Rolled Columbium
  D. Kramer and C. G. Rhodes
  Trans. Met. Soc. AIME, 233:192-198 (1965)

The Superconducting Penetration Depth in Niobium
B. W. Maxfield
Thesis, Rutgers University, New Brunswick, New Jersey (1965), III-79 pp.

Reversible Magnetocaloric Effect in Superconducting Niobium
T. Ohtsuka
Phys. Letters, 17:194 (1965)

Calorimetric Study of Vanadium in the Superconducting, Normal, and Mixed States
R. Radebaugh and P. H. Keesom
Rapport PRF 3324, Semiconductor Research Semiannual Report April 1 to September 30, 1965, Purdue University, Lafayette, Indiana (1965), pp. 23-24

(I). Similarity Principle and Isotope Effect in Superconducting Indium; (II). Low Temperature Heat Capacity of Niobium
N. M. Senozan
UCRL-11901 (1965), IV-100 pp.

Evidence for Two Energy Gaps in High-Purity Superconducting Nb, Ta, and V
L. Y. Lung Shen, N. M. Senozan, and N. E. Phillips
Phys. Rev. Letters, 14:1025 (1965)

A Study of Superconducting Domain Structures Using Optical Polarization Techniques
M. W. P. Strandberg
Massachusetts Inst. Technol., Res. Lab. Electron., Quart. Progr. Rept. No. 79 (1965)

The Superconducting Properties of High Purity Niobium
Thorsten Fredrick Stromberg
(Ph. D. Thesis, Iowa State Univ., Ames Lab.), IS-T-22 (July 1965), 107 pp.

Magnetic Susceptibilities of V, Ar, and Nb
H. Suzuki and S. Miyahara
J. Phys. Soc. Japan, 20:2102 (1965)

Orientation Dependence of the Superconducting Behavior of Deformed and Undeformed Niobium Single Crystals
J. A. Catterall, I. Williams, and J. F. Duke
British J. Appl. Phys., 15:1369 (1964)

Preparation and Properties of Thin-Film Hard Superconductors
J. Edgecumbe, L. G. Rosner, and D. E. Anderson
(Department of Electrical Engineering, University of Minnesota, Minneapolis, Minnesota)
J. Appl. Phys., 35:2198 (1964)

The Penetration Depth in Several Hard Superconductors
T. J. Greytak and J. H. Wernick
J. Phys. Chem. Solids, 25:535 (1964)

Superconducting and Normal Specific Heats of a Single Crystal of Nb
H. A. Leupold and H. A. Boorse
Phys. Rev., 134:1322A (1964)

Thermal and Magnetic Properties of Second Kind Superconductors. III. Specific Heat of Nb in a Magnetic Field
T. McConville and B. Serin
Rev. Mod. Phys., 36:112 (1964)

Domain Patterns Due to Interstitial Impurities in Tantalum
L. Stals and J. Van Landuyt
(SCK-CEN, Mol., Belg.), Phys. Stat. Sol., 5:K1-3 (1964)

Some Remarks on the Upper Critical Field of the Superconductivity of Niobium
G. Van den Berg and H. Van Rongen
Physica, 30:1229 (1964)

Domain Patterns in Columbium (Niobium)
J. Van Landuyt and S. Amelinckx
(Solid State Physics Department, S.C.K.-C.E.N., Mol, Belgium), Appl. Phys. Letters, 4:15 (1964)

Study of Superconducting Thin Films
Peter Fowler
(National Research Corp., Cambridge, Mass.) N-65-17508; JPL-950303; NASA-CR-60873 (March 1963), 35 pp.

Ultrasonic Determination of the Superconducting Energy Gap in Tantalum
M. Levy and I. Rudnick
Phys. Rev., 132:1073 (1963)

NMR, Magnetic Susceptibility and Electronic Specific Heat of Nb and Mo Metals and Nb-Tc and Nb-Mo Alloys
D. O. Van Ostenburg, D. J. Lam, M. Shimizu, and A. Katsuki
J. Phys. Soc. (Japan), 18:1744-1754 (1963)

Magnetic Susceptibility and Electronic Specific Heat of Transition Metals and Alloys. IV. V and Ti Metals and V-Cr and V-Ti Alloys
M. Shimizu, A. Katsuki, and T. Takahashi
J. Phys. Soc. (Japan), 18:1192-1203 (1963)

Magnetic and Thermal Behavior in the Superconductor State of a Single Crystal of Niobium
Gerard Kuhn
(Centre de Recherches sur les Tres Basses Températures, C.N.R.S., Université de Grenoble), Compt. Rend., 255:2923-2925 (1962)

Hall Effect, Resistivity, and Magnetoresistivity of Th, U, Zr, Ti, and Nb
T. G. Berlincourt
Phys. Rev., 114:969 (1959)

Hall Effect in Titanium, Chromium, Vanadium, and Manganese
Simon Foner
Phys. Rev., 107:1513-1516 (1957)

Measurement of the Superconducting Energy Gap for Vanadium by Means of Ultrasonic Techniques
John L. Brewster
Dissertation Abstr., 63-6831, 102 pp.

1.g.4. Group VI

Optical Studies of Antiferromagnetism in Chromium and Its Alloys
A. S. Barker, Jr., and J. A. Ditzenberger
Phys. Rev. B, 1:4378-4400 (1970)

Kondo Effect in Mo(Fe) Alloys from Mössbauer-Effect Measurements
M. P. Maley and R. D. Taylor
Phys. Rev. B, 1:4213-4223 (1970)
Very dilute Fe impurities in Mo

Effects of the Néel Transition on the Thermal and Electrical Resistivities of Cr and Cr:Mo Alloys
  G. T. Meaden, K. V. Rao, and K. T. Tee
  Phys. Rev. Letters, 25:359-362 (1970)

Attempt to Find a Magnetic Field Dependence of the Neel Temperature of Chromium
  R. L. Melcher and W. D. Wallace
  Solid State Commun., 8:1535-1538 (1970)
  No variation detected

Kohn Anomalies in Tungsten and Other Cr-Group Metals
  T. M. Rice and B. I. Halperin
  Phys. Rev. B, 1(2):509-516 (1970)

Neel Temperature of the Spin Density Wave in Chromium and Its Alloys
  Akio Shibatani
  J. Phys. Soc. Japan, 29:93-101 (1970)

Lattice Anisotropy in Antiferromagnetic Chromium
  M. O. Steinitz, L. H. Schwartz, J. A. Marcus, E. Fawcett, and W. A. Reed
  J. Appl. Phys., 41:1231 (1970)

Study of Transition Temperatures in Superconductors, Final Report, Mar. 11, 1968 - Mar. 10, 1970
  L. J. Vieland and R. W. Cohen
  RCA Princeton, N. J., NASA-CR-110432 (Mar. 10, 1970), 132 pp.
  β-W, Nb_3Sn

Spin Waves and the Order-Disorder Transition in Chromium
  J. Als-Nielsen and O. W. Dietrich
  Phys. Rev. Letters, 22(7):290-292 (1969)

Magnetic Studies of Annealed and Alloyed Chromium by Neutron Diffraction
  G. E. Bacon and N. Cowlam
  J. Phys. C (Solid State Physics), 2:238-251 (1969)

Critical Magnetic Field Curve of Superconducting Tungsten
  W. C. Black, R. T. Johnson, and J. C. Wheatley
  J. Low Temp. Phys., 1(6):641-667 (1969)

Interaction of Magnon and Alfven Waves in Antiferromagnetic Chromium
  A. Caille and P. R. Wallace
  Solid State Commun., 7:1283-1286 (1969)

The Effect of Grain Size on the Superconducting Transition Temperature of the Transition Metals
  J. J. Hanak, J. I. Gittleman, J. P. Pellicane, and S. Bozowski
  Phys. Letters, 30A:201-202 (1969)
  Mo

Magnetostriction and Anomalous Thermal Expansion of Chromium
  E. W. Lee and M. A. Asgar
  Phys. Rev. Letters, 22:1436-1439 (1969)

Genesis of the Spin-Flip Resistivity Phenomenon in Chromium
  G. T. Meaden and N. H. Sze
  Phys. Letters, 30A:294-295 (1969)

Direct Thermal Evidence for a First-Order Change at the Spin-Flip Transition of Chromium
  G. T. Meaden and N. H. Sze
  Phys. Rev. Letters, 23:1242-1244 (1969)

Antiferromagnetism in Chromium and Its Alloys
  T. M. Rice, A. S. Barker, Jr., B. I. Halperin, and D. B. McWhan
  J. Appl. Phys., 40:1337 (1969)

Spin Density Wave in Chromium and Its Alloys
  A. Shibatani, K. Motizuki, and T. Nagamiya
  Phys. Rev., 177:984-1000 (1969)

Magnetomorphic Size Effect in Tungsten
  D. E. Soule and J. C. Abele
  Phys. Rev. Letters, 23:1287-1291 (1969)

Lattice Relaxation in Cr at Temperatures Above T_N after Rapid Heating from the Intermediate Temperature Phase
  Bengt Stebler and Claes-Goran Andersson
  Phys. Rev. Letters, 22:466-470 (1969)

The Effect of Antiferromagnetic Ordering on the Lattice Dimensions of Chromium
  M. O. Steinitz, L. H. Schwartz, and J. A. Marcus
  Bull. Am. Phys. Soc., 14:348 (1969)

Spin Directions in Pure Chromium
  S. A. Werner, A. Arrott, and M. Atoji
  J. Appl. Phys., 40:1447-1449 (1969)

High-Field Galvanomagnetic Effects in Antiferromagnetic Chromium
  A. J. Arko, J. A. Marcus, and W. A. Reed
  Phys. Rev., 176:671-683 (1968)

Changes in Diffraction Contrast in Antiferromagnetic Chromium at Temperatures around the Neel Point
  C. G. Andersson and B. Stebler
  Arkiv Fys., 37:117-125 (1968)

Antiferromagnetic Energy Gap in Chromium
  A. S. Barker, B. I. Halperin, and T. M. Rice
  Phys. Rev. Letters, 20:384-387 (1968)

Superconductivity in Thin Films of Tungsten
  S. Basavaiah
  Ph. D. thesis, Pennsylvania University, Philadelphia (1968), 189 pp.
  University Microfilms, Ann Arbor, Mich., Order No. 69-5606

Superconductivity in Evaporated Tungsten Films
  S. Basavaiah and S. R. Pollack
  Appl. Phys. Letters, 12:259-260 (1968)

Superconductivity in β-Tungsten Films
  S. Basavaiah and S. R. Pollack
  J. Appl. Phys., 39:5548 (1968)

Critical-Field Measurement of Superconducting Tungsten as Related to the Cerium-Magnesium-Nitrate Temperature Scale
  W. C. Black
  Phys. Rev. Letters, 21:28-31 (1968)

Nesting Fermi Surfaces and Magnetism
  T. L. Loucks
  Bull. Am. Phys. Soc., 13:366 (1968)
  Cr

Microhardness of Chromium near the Neel Point
   O. A. Nabutovskaya
   Fiz. Tverd. Tela, 9(8):2452 (1967)
   Sov. Phys. – Solid State, 9(8):1928 (1968)

Kondo Effect and Superconductivity in Molybdenum
   N. V. Volkenshtein and V. E. Startsev
   Zh. Éksp. Teor. Fiz. Pis'ma Redakt., SSSR, 7(11):426–427 (1968)
   Sov. Phys. – JETP Lett., 7(11):334–335 (1968)

Effects of Pressure and a Magnetic Field on Chromium Studied by Neutron Diffraction
   S. A. Werner, A. Arrott, and M. Atoji
   J. Appl. Phys., 39:671–673 (1968)

Critical Magnetic Fields of Superconducting Tungsten
   William Carter Black, Jr.
   Ph. D. Thesis, University of Illinois, Urbana, Illinois (1967), 171 pp.

Magnetic Properties of Group VIA Transition Metals Containing Dilute Concentrations of Cobalt
   J. G. Booth, K. C. Brog, and W. H. Jones
   Proc. Phys. Soc. London, 92:1083–1089 (1967)

Superconductivity of F. C. C. Tungsten and Molybdenum Films Obtained by Vacuum Sputtering
   K. L. Chopra
   Bull. Am. Phys. Soc., 12:57 (1967)

Type-I Superconductivity in Molybdenum
   R. A. French
   Phys. Stat. Sol., 21:K35 (1967)

Ultrasonic Investigation of Energy Gap Anisotropy in Superconducting Molybdenum
   C. K. Jones and J. A. Rayne
   (Westinghouse Research Labs., Pittsburgh, Penna.) Scientific Paper 67-9J2-PHONY-P1, Proprietory Class 3 (October 31, 1967)
   Phys. Letters, 26A:75 (1967)

Measurements of Magnetization in Superconductive Molybdenum
   Richard G. Mallon and H. E. Rorschach, Jr.
   Phys. Rev., 158:418–423 (1967)

Pressure Dependence of Itinerant Antiferromagnetism in Chromium
   D. B. McWhan and T. M. Rice
   Phys. Rev. Letters, 19:846–849 (1967)

Magnetic Anisotropy in Antiferromagnetic Chromium
   Ramiro Alfonso Montalvo
   Ph. D. Thesis, Northwestern University, Evanston, Illinois (1967), 117 pp.

Neutron-Diffraction Study of Cr and Cr Alloys
   S. A. Werner, A. Arrott, and H. Kendrick
   J. Appl. Phys., 38:1243 (1967)

Temperature and Magnetic-Field Dependence of the Antiferromagnetism in Pure Chromium
   S. A. Werner, A. Arrott, and H. Kendrick
   Phys. Rev., 155:528–539 (1967)

Galvanomagnetic Effects in Antiferromagnetic Chromium
   A. J. Arko and J. A. Marcus
   Bull. Am. Phys. Soc., 11:169 (1966)

Superconductivity of Tungsten
   R. T. Johnson, O. E. Vilches, J. C. Wheatley, and S. Gygax
   Phys. Rev. Letters, 16:101 (1966)

Superconductivity of Tungsten Films at Temperature near 4°K
   O. F. Kammerer and Myron Strongin
   Proc. Intern. Symposium Grundprobleme der Physik dünner Schichten, Göttingen (1965), pp. 511–515 (1966)

Magnetic Form Factor of Chromium
   R. M. Moon and W. C. Koehler
   J. Appl. Phys., 37:1036 (1966)

Ultrasonic Attenuation near the Neel Temperature of Chromium
   E. J. O'Brien and J. Franklin
   J. Appl. Phys., 37:2809–2812 (1966)

The Effect of Field-Cooling on the Magnetic Susceptibility of Chromium
   A. R. Pepper and R. Street
   Proc. Phys. Soc., 87:971 (1966)

High Magnetic Field Susceptibility of Chromium
   Richard Stevenson
   J. Phys., 44:283 (1966)

Etude par diffraction des neutrons de la tr nsformation magnétique à basse température dans le chrome
   R. A. Alikhanov and L. S. Smirnov
   Gosudarstvennyi Komitet po Ispol'zovaniyu Atomnoi Energii SSSR, Institut Teoreticheskoi i Eksperimental'noi Fiziki, Moscow, Report ITEF-393 (1965), 10 pp.

Magnetoresistance in Antiferromagnetic Chromium
   A. J. Arko and J. A. Marcus
   Low Temperature Physics Proc. IX International Conf., Columbus, Ohio, 1964, Plenum Press, New York (1965), pp. 748–751

The Antiferromagnetic Phases of Field-Cooled Chromium
   T. J. Bastow and R. Street
   Proc. Phys. Soc., 86:1143 (1965)

Some Observations on the Higher Temperature Antiferromagnetic Phase of Chromium
   P. J. Brown, C. Wilkinson, J. B. Forsyth, and R. Nathans
   Proc. Phys. Soc., 85:1185 (1965)

An Investigation of the Superconducting Properties of Molybdenum
   I. G. Dyakov and A. D. Shvets
   Zh. Éksperim. i Teor. Fiz., 49(4):1091–1093 (1965)
   Sov. Phys. – JETP, 22(4):759–761 (1966)

Anomalies in the Magnetic Structure of Chromium
   V. S. Golovkin, V. N. Bykov, and V. A. Levdik
   Zh. Éksperim. i Teor. Fiz., 49:1083–1090 (1965)
   Sov. Phys. – JETP, 22(4):754–758 (1966)

1.h. Optical Properties

1.h.1. General

Solar Absorptance and Spectral Reflectance of
Mo, Ta
  E. W. Spisz et al.
  NASA TN D-5353 (1969)

Optical Properties of Some Transition Metals
  D. W. Juenker, L. J. LeBlanc, and C. R. Martin
  J. Opt. Soc. Am., 58:164-171 (1968)

Optical Properties of the Transition Metals in
the Vacuum Ultraviolet Region
  M. M. Kirillova, L. V. Nomerovannaya, G. A. Bolotin,
    V. M. Maevskii, M. M. Noskov, and M. S. Bolotina
  Fiz. Metallov Metalloved., SSSR, 25(3):459-468 (1968)
  Ti, Zr, V, Nb, Co, Rh

Emissionseigenschaften von Ta, Mo, Nb und
NbC
  B. A. Kruschtalev and A. M. Rakov
  Teploobmen. Gidrodin. Teplofiz. Svoistv. (1968), pp. 198-219

Optical Properties of Transition Metals at Infrared Wavelengths
  J. Reichman and C. Feldman
  (Grumman Aircraft Engineering Corp, Bethpage, N. Y.)
  presented at Am. Phys. Soc. Spring Meeting, Berkeley
    (March 1968), RE-338J (August 1968)

Optical Properties and Electronic Structure of
Metals and Alloys
  F. Abeles, ed.
  Proc. Intern. Colloquium, Paris, Sept. 13-16, 1965, North-
    Holland, Amsterdam (1966)

Quantum Absorption of Light in Transition Metals
  F. A. Abramskii
  Izv. Vysshikh Uchebn. Zavedenii, Fiz., 9(4):174-175 (1966)
  Ti, V, Mn, Zr, Nb

Optical Constants of Incandescent Refractory
Metals
  Bentley T. Barnes
  J. Opt. Soc. Am., 56(11):1546-1550 (1966)
  W, Mo, Ta, Nb

Optical Constants of Transition Metals in the
Infrared
  A. P. Lenham and D. M. Treherne
  J. Opt. Soc. Am., 56:1137 (1966)

The Optical Properties of the Transition Metals
  A. P. Lenham and D. M. Treherne
  Proc. Intern. Colloq. Optical Properties and Electronic Struc-
    ture of Metals and Alloys, Paris, 1965, pp. 196-201 (1966).

Emission Parameters of Ta and Mo Single Crys-
tals
  O. D. Protopopov, E. V. Mikheeva, B. N. Sheinberg, and G. N.
    Shuppe
  Fiz. Tverd. Tela, 8(4):1140-1146 (1966)
  Sov. Phys. — Solid State, 8(4):909-914 (1966)

Thermoemissivity Characteristics of Transition
Metals and Their Compounds
  G. V. Samsonov, Yu. B. Paderno, and V. S. Fomenko
  Foreign Technology Div., Wright-Patterson AFB, Ohio,
    FTD-HT-66-178 TT-67-60511, Oct. 4, 1966
  Ukr. Fiz. Zh., 10(6):622-629 (1965)

Thermal Radiative Properties of Selected Ma-
terials
  W. D. Wood, H. W. Deem, and C. F. Lucks
  DMIC Report 177, Vol. 1 of 2; AD-294345 (November 1962)

High Temperature Spectral Emissivity Studies
  Thomas R. Riethof
  (Space Sciences Lab., General Electric Company). R61SD004;
    AD-250274 (January 1961), 34 pp.

Optical Absorption Measurements of the Transi-
tion Metals Ti, V, Cr, Mn, Fe, Co, Ni in the
Region of 3p Electron Transitions
  B. Sonntag, R. Haensel, and C. Kunz
  DESY-69/6

1.h.2. Group IV

Observation of Plasma Resonance at 790 Å
in Reflection from Thin Zirconium Films
  J. E. Rudisill, A. Matsui, and G. L. Weissler
  Optics Communications, 2:39-40 (1970)

Reflectivity Measurements on Zirconium
  L. T. Larson
  Trans. AIME, 245:2047-2050 (1969)

Photoabsorption der Metalle Ti, Bi, S, Cu für
Photonenenergien zwischen 40 eV und 300 eV
  B. Sonntag
  DESY-F41/1 (Feb. 1969)

Reflexion et transmission dans l'infrarouge de
la zircone monoclinique
  Bernard Piriou and Jean Tsakiris
  Compt. Rend., 261:3079 (1965)

L'emissivite optique totale du rhenium, du
rhodium, du palladium et de titane
  W. Landensperger and D. Stark
  Z. Phys., 180:178-183 (1964)

1.h.3. Group V

Harmonic Mixing of Microwave and Far-Infrared
Laser Radiation Using a Josephson Junction
  D. G. McDonald, A. S. Risley, J. D. Cupp, and K. M. Evenson
  Appl. Phys. Letters, 18:162-164 (1971)
  Nb point contact

Temperature Dependence of the X-Point
  L. N. Latyev, V. Ya. Chekhovskii, and E. N. Shestakov
  Phys. Stat. Sol., 38:K149-151 (1970)

Spectral Emissivity of Tantalum in the Red and
in the Green, and the Change in Emissivity Resulting from a Change in Surface Structure Induced by Heat Treatment
  J. M. Martinez and A. H. Madjid
  J. Appl. Phys., 41:5322-5326 (1970)

Propriétés optiques de couches minces de rho-
dium et de vanadium dans l'ultraviolet lointain
  André Seignac and Simone Robin
  Compt. Rend., 271B:919-922 (1970)

Optical Properties of Niobium
  A. J. Golovashkin, I. E. Leksina, G. P. Motulevich, and
    A. A. Shubin
  Zh. Éksp. Teor. Fiz., 56:51-64 (1969)
  Sov. Phys. — JETP, 29(1):27-34 (1969)

tivity; alloys, biological materials, elements, gases, graphites, insulators, liquids, metals, nuclear materials, plasmas, polymers, reference materials, semiconductors, solids

Structure and Properties of Thick Condensates of Nickel, Titanium, Tungsten, Aluminum Oxides, and Zirconium Dioxide in Vacuum
B. A. Movchan and A. V. Demchishin
Fiz. Metal. Metalloved., 28:653-660 (1969)

Mean Square Amplitude of Thermal Vibrations in Cubic Metals at Melting Point
P. K. Sharma and Narain Singh
Chem. Phys. Letters, 4:1-2 (1969)

Thermal Conductivity, Metallic Elements and Alloys
Y. S. Touloukian, R. W. Powell, C. Y. Ho, and P. G. Klemens, eds.
TPRC Data Series, Thermophysical Properties of Matter, Vol. 1, Plenum Press, New York (1969)

The Encyclopedia of the Chemical Elements
Clifford A. Hampel, ed.
Reinhold Book Corp., New York; Amsterdam; London (1968), 849 pp.

Handbook of Differential Thermal Analysis
W. J. Smothers and Yao Chiang
Chemical Publishing Co., New York (1966), 633 pp.
Equipment and methods, qualitative and quantitative

X-Ray Diffraction Study of Deformation by Filing in B. C. C. Refractory Metals
E. N. Aqua and C. N. J. Wagner
Phil. Mag., 9:565 (1964)

Thermally Activated Dislocation Motion in FCC and Refractory BCC Metals
R. H. Chambers
Appl. Phys. Letters, 2:165 (1963)

Properties of Solids and Liquids
E. C. Crittenden
Cryogenic Technology, R. W. Vance, John Wiley and Sons, Inc., New York (1963), pp. 60-94

Selected Values of Thermodynamic Properties of Metals and Alloys
R. Hultgren, R. L. Orr, P. D. Anderson, and K. K. Kelley
Wiley Series on the Science and Technology of Materials (J. H. Holloman, J. E. Burke, B. Chalmers, and J. A. Krumhansl, eds.), Wiley-Interscience, New York (1963), 963 pp.
Continuously updated by loose leaf sheets

Preparation of Wires for Examination by Transmission Electron Microscopy
E. S. Meieran and D. A. Thomas
(Department of Metallurgy, Massachusetts Institute of Technology, Cambridge, Mass.). This work was sponsored by the United States Air Force, Aeronautical Systems Division, Contract No. 33(616)-6838. Trans. AIME, 227:284 (1963)

The Engineering Properties of Tungsten and Tungsten Alloys
F. F. Schmidt and H. R. Ogden
(Defense Metals Information Center, Battelle Memorial Institute, Columbus 1, Ohio), DMIC-Rept. 191 (1963)

Electron Microscopy
Fifth International Congress for Electron Microscopy held in Philadelphia, Pa. (August 29 to Sept. 5, 1962). Volume 1, Non-Biology (Sydney S. Breese, Jr., ed.), Academic Press, New York (1962)

Internal Friction Peaks in Cold-Worked Metals
R. R. Hasiguti, N. Igata, and G. Kamoshita
Acta Met., 10:442-447 (1962)

Current Information on the Refractory Metals
C. E. King
Contrast AF33(657)-11214, AD-436333; ERR-FW-049A (Dec. 1, 1962), 80 pp.

On Some Properties of Electron Beam Melted Metals
E. Rexer
(Electron Beam Technology, Alloyd Electronics Corp., Cambridge, Mass.), Proc. Fourth Symp., p. 245-275 (1962)

Data on Sound Velocities in Pure Metals
Karl Heinz Schramm
Z. Metallk., 53:729-735 (Nov. 1962)

Ultra-High Purity Metals
Papers presented at a seminar of American Society of Metals (October 21 and 22, 1961), American Society of Metals, Metals Park, Ohio (1962)

Fabrication and Properties of Tungsten and Tungsten Alloy Single Crystals
Linde Company, Contract NOw 61-1671, Mar. 31, 1962, 74 pp.
AD 273 951
U.S. Govt. Res. Rept., 37:45 (A) (1962)

Relationships Between Electronic Structure and Physical Properties in Transition Metals
G. Airoldi and M. Asdente
Energia Nucleare (Milan), 8:342-60 (May 1961)

Investigation of the Mechanical Properties of Ultra-Pure Tungsten, Final Report, Jan. 1, 1958 to June 30, 1961
D. P. Ferriss, R. M. Rose, and J. Wulff
Contract DA-19-020-ORD-4543, NP-10583 (June 30, 1961)
AROD-1970: 1, 2, 3, 4, and 5

Substructure and Mechanical Properties of Refractory Metals
B. S. Lement, D. A. Thomas, S. Weissmann, W. S. Owen, and P. B. Hirsch
(Manufacturing Labs., Inc., Cambridge, Mass.) Contract AF33(616)-6838, WADD-TR-61-181 (Apr. 1961), 255 pp.

Zirconium
OTS selective bibliography, supplement to CTR-344, SB-464, Office of Technical Services (May 1961)

The Physical Properties of Niobium, Tantalum, Molybdenum, and Tungsten
B. B. Argent and G. J. C. Milne
J. Less-Common Metals, 2:154-162 (1960)

Thermophysical Properties of Solid Materials, Volume 1 — Elements
Alexander Goldsmith, Thomas E. Waterman, and Harry J. Hirschhorn
(Armour Research Foundation) WADC 58-476, Vol. 1; AD-247193 (August 1960)

Propriétés des métaux et alliages aux basses
températures (Properties of metals and alloys
at low temperatures)
  E. Justi
  CEA-tr-A-789
  Z. Metallk., 51(1):1-17 (1960)

Soviet Research in Field Electron and Ion
Emission, 1955-1959; An Annotated Bibliography
  T. W. Marton and R. Klein
  (National Bureau of Standards, Washington, D. C.), National
    Bureau of Standards Tech. Note 75 (Oct. 1960)

Tantalum and Its Alloys
  Z. Takao and K. Narita
  Metals (Japan), 30:14-20 (1960)

An Interpretation of Some of the Properties of
the Transition Metals and Their Alloys
  D. A. Robins
  J. Less-Common Metals, 1:396-410 (1959)

Compilation of Calculated Data Useful in Pre-
dicting Metallurgical Behavior of the Elements
in Binary Alloy Systems
  E. Teatum, K. Gschneider, Jr., and J. Waber
  (Los Alamos Scientific Laboratory of the University of
    California, Los Alamos, New Mexico), Contract W-7405-ENG-
    36, LA-2345 (August 1959)

Tungsten – Bibliography 1953-1958, Physical
Properties and Phase Diagram
  Sylvania Electric Products, Inc., Chemical and Metallurgical
    Div., Towanda, Pa. (1959), 39 pp.

Molybdenum at a Glance
  C. A. Hampel, ed.
  Rare Metals Handbook, Reinhold Publishing Corp., New York,
    (July 29, 1957), p. 2001

Distribution of Microhardness Values Within a
Single Crystalline Grain of Metal
  A. A. Bockvar and O. S. Zhadaeva
  AERE-Trans-11/5/350, Izvest. Akad. Nauk SSSR, Otdel. Tekh.
    Nauk, 419-424 (1947), 6 pp.

Thermophysical Properties of Solid Materials.
Volume 1 – Elements (Melting Temperature
above 1000°F)
  Alexander Goldsmith, Thomas E. Waterman, and Harry J.
    Hirschhorn
  (Armour Research Foundation), WADC 58-476; AD-247193

A Compilation of Some of the Physical Properties of the Metallic and Semi-Metallic Elements and a Study of Some of Their Interrelationships
  Karl A. Gschneider, Jr.
  (Department of Physics and Materials Research Laboratory,
    University of Illinois, Urbana, Illinois, and University of
    California, Los Alamos Scientific Laboratory, Los Alamos,
    New Mexico), LADC-6220
  To be published in Solid State Phys., Vol. 16

1.i.2. Diffusion

1.i.2.a. General

Anomalous Diffusion in Beta Zirconium, Beta
Titanium and Vanadium
  M. C. Nauk and R. P. Agarwala
  J. Phys. Chem. Solids, 30:2330-2334 (1969)

Thermomigration in Solid Metals
  R. A. Oriani
  J. Phys. Chem. Solids, 30:339-351 (1969)

Diffusion of C in Ta and Zr
  H. Suzuko
  Bull. Tokyo Inst. Tech., 90:105-115 (1969)

Thermotransport of Oxygen and Nitrogen in β-Zir-
conium, β-Titanium, Niobium, and Tantalum
  Daniel Lodewijk Vogel
  Thesis, Technische Hogeschool, Eindhoven, Netherlands, NP-
    18486 (1969)

Hydrogen Effects in Refractory Metals
  W. T. Chandler and R. J. Walter
  Refractory Metal Alloys. Metallurgy and Technology
    (I. Machlin, ed.), Plenum Press, New York (1968), pp. 197-249
  Mo, W, Nb, Ta

Diffusion Rates in Inorganic Nuclear Materials
  A. L. Dragoo
  J. Res. National Bur. Stand., Phys. Chim., U.S.A., 72:157-173
    (1968)
  Be, Mg, Mo, Nb, Ta, Th, Ti, Zr

Diffusion in Metals
  N. L. Peterson
  Solid State Physics – Advances in Research and Applications,
    Vol. 22 (Frederick Seitz, David Turnbull, and Henry
    Ehrenreich, eds.), Academic Press, New York and London
    (1968), pp. 409-512

Reciprocal Diffusion of Tantalum and Tungsten
  I. A. Tregubov, L. N. Kuzina, and O. S. Ivanov
  Dokl. Akad. Nauk SSSR, 180(2):423-424 (1968)

Diffusion of Phosphorus in Niobium and Molyb-
denum
  B. A. Vandyshev and A. S. Panov
  Phys. Metals Metallogr., 26(3):138-143 (1968)

The Solubility of Hydrogen in Transition Metals
and Alloys
  Y. Ebisuzaki and M. O'Keeffe
  Progr. Solid State Chem., 4:187-211 (1967)

Surface Self-Diffusion in FCC and BCC Metals:
a Comparison of Theory and Experiment
  N. A. Gjostein
  Surfaces and Interfaces, I. Chemical and Physical Charac-
    teristics (Burke, Weiss, Reed, eds.), Syracuse University
    Press, Syracuse, New York (1967), p. 271

Diffusion of Alkali Metals in Mo and Nb
  M. G. Karpman
  Metalloved. Term. Obr. Metallov., No. 3, pp. 46-49 (1967)

The Solid State Chemistry of Hydrogen Com-
pounds, Final Technical Progress Report
  M. O'Keeffe, Y. Ebisuzaki, and W. J. Kass
  (Dept. of Chemistry, Arizona State Univ., Tempe, Ariz.),
    Contract Nonr-2794(04), AD-65389 (June 1967), 48 pp.

Mutual Diffusion of Niobium and Metals of the
IVa, Va, and VIa Groups
  D. A. Prokoshkin, E. V. Vasileva, and L. L. Vergasova
  AD-885007; FTD-HT-23-967-68 (Nov. 1968)
  Fiz. Metallov Metalloved., 23(6):1134-1136 (1967)

Hydrogen Permeability of Some Transition Metals and Metals of the First Group of the Periodic System
Yu. I. Zvezdin and Yu. I. Belyakov
UCRL-Trans-10257
Fiz.-Khim. Mekh. Mater., 3:349-351 (1967), 6 pp.

Diffusion of C in Mo, Nb, and Ti
A. I. Nakonetschnikov, L. V. Pavlinov, and V. N. Bykov
Fiz. Metallov Metalloved., 22:234-238 (1966)

Thermal Diffusion of Oxygen in Refractory Metals
G. D. Rieck and D. L. Vogel
(Lab. of Phys. Chem., Tech. Univ., Eindhoven, Netherlands), Plansee Proc. 1964 — Metals for the Space Age, 5th Plansee Seminar (F. Benesovsky, ed.), 1964, Reutte/Tyrol, Austria (1965)

Grain Boundary and Lattice Diffusion of Chromium in Zirconium
R. P. Agarwala, S. P. Murarka, and M. S. Anand
(Atomic Energy Establishment, Trombay, Bombay, India), A.E.E.T./C.D./22 (1964)

Self-Diffusion and the Melting Parameters of Cubic Metals
G. B. Gibbs
(Central Electricity Generating Board, Berkeley Nuclear Laboratories, Berkeley, Glos., England), Acta Met., 12:673-675 (1964)

Diffusion in Body-Centered Cubic Metals Zirconium, Vanadium, Niobium, and Tantalum
T. S. Lundy
ORNL-3617 (June 1964); Ph. D. thesis, University of Tennessee

Diffusion in Body-Centered Cubic Metals
Papers presented at the Intern. Conf. on Diffusion in Body-Centered Cubic Materials, Gatlinburg, Tenn., American Society for Metals, Metals Park, Ohio (Sept. 16-18, 1964)
1. Application of Diffusion Theory to the Body-Centered Cubic Structures, A. D. LeClaire. 2. A Summary of Impurity Diffusion in the Beta Phase of Titanium, D. Graham. 3. A Summary of ORNL Work on Diffusion in Beta Zirconium, Vanadium, Columbium, and Tantalum, T. S. Lundy, J. I. Federer, R. E. Pawel, and F. R. Winslow. 4. Binary Interdiffusion in Body-Centered Cubic Transition Metal Systems, C. S. Hartley, J. E. Steedly, Jr., and L. D. Parsons. 5. The Diffusion of Carbon in Alpha Iron, C. G. Homan. 6. The Electron-Concentration Concept and Diffusion, Niels Engel. 7. The Vibrational Frequency Spectrum of Some Body-Centered Cubic Metals, G. Leibried. 8. The Measurement of Diffusion Coefficients by Internal Friction, R. Gibala and C. A. Wert. 9. Diffusion of Interstitial Impurities in Body-Centered Cubic Metals, D. N. Beshers. 10. Diffusion in Body-Centered Cubic Transition Metals — A Theoretical Critique, D. Lazarus. 11. The Isotope Effect in Sodium Self-Diffusion, L. W. Barr, and J. N. Mundy. 12. Tracer Diffusion in Gamma Uranium, S. J. Rothman and N. L. Peterson. 13. Re-examination of Silver Diffusion in Silver-Rich AgMg, W. C. Hagel and J. H. Westbrook. 14. Simultaneous Diffusion of Silver and Magnesium in Stoichiometric Monocrystalline Beta-AgMg, H. A. Domian and H. I. Aaronson. 15. Diffusion in Magnetic Materials, R. J. Borg. 16. Vanadium Self-Diffusion, R. F. Peart. 17. Tracer Diffusion Studies in Molybdenum, John Askill. 18. Self-Diffusion in Chromium, John Askill.
19. Diffusion of Titanium-44 and Vanadium-48 in Titanium, J. F. Murdock. 20. The Analysis of Diffusion Data, D. Y. F. Lai. 21. Correlation of Diffusion Data as a Periodic Function of Atomic Number, R. H. Moore

Investigation of Diffusion of Sulfur in Some Refractory Metals
M. M. Pavlyuchenko et al.
Dokl. Akad. Nauk BSSR, 8:157-160 (1964)
TT-70-57010 (1970), 10 pp. Israel Program for Scientific Translations, Ltd., Jerusalem
Diffusion into Co, Cr, Mo, and W

The Measurement of Volume Diffusion by the Method of Autoradiography
T. J. Renouf
Phil. Mag., 9:781 (1964)

Diffusion in BCC Metals
R. A. Wolfe and H. W. Paxton
(Carnegie Inst. of Tech., Pittsburgh)
Contract Nonr-760(08), NP-13738 (Mar. 1, 1964), 27 pp.

Diffusion Today
F. R. Winslow and K. H. Osthagen
(Members of AIME, with Oak Ridge National Laboratory, Oak Ridge, Tenn., and Battelle Memorial Institute, Columbus, Ohio, respectively), AEC Contract W-7405-eng-92, J. Metals, 15:855 (1963)

Diffusion Studies of Vacancies and Impurities
David Lazarus
J. Phys. Rad., 23:772 (1962)

On the Theory of Impurity Diffusion in Metals
A. D. LeClaire
(Solid State Physics Division, Atomic Energy Research Establishment, Harwell), Phil. Mag., 7:141 (1962)

Bibliography on Self-Diffusion of Pure Metals in the Solid State, 1950-1960
Compiled by Robert L. Andelin and Lois E. Godfrey
Contract W-7405-ENG. 36, LAMS-2562 (1961)

Preliminary Measurements on the Thermal and Electrical Conductivities of Molybdenum, Niobium, Tantalum, and Tungsten
R. P. Tye
J. Less-Common Metals, 3:13 (1961)

Solid State Diffusion in Metals and Alloys
S. D. Gertsriken and I. Ya. Dekhtyar
AEC-tr-6313, translated from a publication of the State Publishing House for Physical-Mathematical Literature, Moscow (1960), 538 pp.

Thermal Properties of Refractory Materials
G. W. Lehman
(Atomics International Div. of North American Aviation, Inc., Canoga Park, Calif.), Contract AF33(616)-6794, WADD-TR-60-581 (July 1960), 19 pp.

Diffusion in Refractory Metals
N. L. Peterson
(Advanced Metals Research Corp., Somerville, Mass.), Contract AF33(616)-7382, WADD-TR-60-793 (Dec. 1, 1960), 164 pp.
AD-257860 (March 1960)

Diffusion of Silicon in Titanium, Tantalum, Molybdenum, and Iron
G. V. Samsonov and M. S. Koval'chenko
AD 602 404 (1963)
Dopovidi Akad. Nauk Ukr. RSR (Kiev), No. 1, p. 32 (1959)

Methods of Studying Diffusion by Means of Radioactive Isotopes
  J. Cadek and E. Janda
  (Ferrous Metallurgy Research Institute, Prague)
  AERE-TRANS 840
  Hutnicke Listy, 12:1008-1020 (1957)
  Theory and review, 55 references

Mechanism of Diffusion in Hard Metals
  I. L. Mirkin
  AEC-tr-3269
  Tsentral' Nauch.-Issledovatel' Inst. Tekhnol. i Mashino-
    stroen., 79:5-24 (1957), 34 pp.

Diffusion of Boron, Carbon, and Nitrogen into Transition Metals of IV, V, and VI Groups of the Periodic Table
  G. V. Samsonov and V. P. Latisheva
  AEC-tr-2949
  Dokl. Akad. Nauk SSSR, 109:582 (1956)

Investigation of the Diffusion of Boron and Carbon in Certain Metals of Transition Groups
  G. V. Samsonov and V. P. Latisheva
  AEC-tr-3321 (translated by Lydia Venters), Fiz. Metal. i
    Metalloved. Akad. Nauk SSSR, Ural. Filial, 2:309-319 (1956),
    20 pp.

The Study of Volume and Grain Boundary Diffusion in Metals by the Autoradiographic Method
  S. Z. Kokshtein, S. T. Kishkin, L. M. Moroz, and T. I. Gudkov
  AEC-tr-2233 (translated by S. J. Rothman), Dokl. Akad. Nauk
    SSSR, 102:73-75 (1955), 7 pp.

Diffusion of Metals, Exchange Reactions
  Wolfgang Seith and Theodor Heumann
  (University of Münster), AEC-TR-4506, Second Revised and
    Enlarged Edition, Springer-Press, Berlin/Göttingen/Hei-
    delberg (1955)

Physical Property Studies
  D. L. McElroy, T. G. Kollie, and W. Fulkerson
  ORNL-3470, pp. 27-29

Magnetic Resonance Studies of Self-Diffusion in Simple Solids
  C. P. Slichter
  Nuovo Cimento N. 1 del Suppl. al Vol. 9, Serie X, pp. 104-111

1.i.2.b. Group IV

Tracer Diffusion of Silver and Scandium in β-Titanium
  J. Askill
  Phys. Stat. Sol., 43B:K1 (1971)

Hydrogen Solubility in Alpha Titanium
  R. S. Vitt and Kanji Ono
  Met. Trans., 2:608 (1971)

Diffusion of Hydrogen in Films of Titanium Deposited on Copper Supports
  G. Germai and J. M. Peters
  J. Inorg. Nucl. Chem., 32:3411-3414 (1970)
  In French

The Oxidation Mechanism of Zirconium, Final Report, 15 May 1967-14 May 1970
  F. J. Keneshea
  Stanford Research Institute, Menlo Park, Calif., AD-708573;
    AROD-7098:2-MC (June 1970), 5 pp.

Electrotransport of Carbon, Nitrogen, and Oxygen in Zirconium
  F. A. Schmidt, O. N. Carlson, and C. E. Swanson, Jr.
  Met. Trans., 1:1371-1373 (1970)

Parameters of the Diffusion of Oxygen in Beta-Titanium
  L. F. Sokiryanskiy et al.
  NASA's Titanium Alloys for Modern Technology (March 1970),
    pp. 239-249
  N70-24659 11-17

Thermomigration and Electromigration in Zirconium
  D. R. Campbell and H. B. Huntington
  Phys. Rev., 179:601-612 (1969)

The Effect of High Pressure on Self-Diffusion in β-Ti
  Rondo N. Jeffery and David Lazarus
  Bull. Am. Phys. Soc., 14:389 (1969)

Self and Solute Diffusion in Zirconium. I. The Diffusion of ^{60}Co in Beta Zr
  G. V. Kidson and G. J. Young
  Phil. Mag., 20:1047-1055 (1969)

Self and Solute Diffusion in Zirconium. II. The Effect of Cobalt Additions on the Diffusion of ^{95}Zr and ^{60}Co in Beta-Zirconium
  G. V. Kidson and J. S. Kirkalady
  Phil. Mag., 20:1057-1073 (1969)

Heat of Transport in Thermal Diffusion on Hydrogen in Titanium
  M. Kitada and S. Koda
  Scripta Met., 3(8):583-584 (1969)

Diffusion of Hydrogen in Titanium and the Beta-Titanium Alloy VT15
  B. A. Kolachev, O. P. Nazimov, and L. N. Zhuravlev
  Izv. Vysshikh Uchebn. Zaveden., Tsvetnaya Met., 4:104-109
    (July-Aug. 1969)

Reaction Diffusion Between Two Metals Including Titanium and Niobium
  G. D. Rieck, G. F. Bastin, and F. J. J. van Loo
  High Temperature Materials (F. Benesovsky, ed.), 6th Plansee
    Seminar, Reutte/Tirol, June 24-28 (1968), pp. 357-371,
    publ. 1969

Effect of Polymorphic Transformation on the Diffusion of Oxygen in Titanium
  L. F. Sokiryanskii, D. V. Ignatov, and A.Ya. Shinyaev
  Fiz. Metallov Metalloved., SSSR, 28(2):287-291 (1969)

Self-Diffusion in Beta Ti and Beta Hf
  N. E. W. De Reca and C. M. Libanati
  Acta Met., 16:1297-1305 (1968)

Thermo- und Elektrotransport in Beta-Titan und Beta-Zirkonium
  H. Dubler and H. Wever
  Phys. Stat. Sol., 25:109 (1968)

Comments on Self-Diffusion in Alpha Zirconium
  F. Dyment, D. Fainstein, and C. M. Libanati
  Scripta Met., 2:201-203 (1968)

Self-Diffusion of Ti, Zr, and Hf in Their HCP Phases, and Diffusion of ^{95}Nb in HCP Zr
  F. Dyment and C. M. Libanati
  J. Mat. Sci., 3:349-359 (1968)

Mobility of Silver in Titanium
P. P. Kuz'menko and V. I. Lozovoi
Fiz. Metallov Metalloved., 26(2):382-384 (1968)
Phys. Metals Metallogr., 26(2):202-204 (1968)

On Self Diffusion in Alpha Zirconium
M. C. Naik and R. P. Agarwala
Scripta Met., 2:315-317 (1968)

The Diffusion of H_2 in Ti
T. P. Papazoglou and M. T. Hepworth
Trans. Met. Soc. AIME, 242:682-685 (1968)

Diffusion of Molybdenum and Cerium in Beta-Zirconium
A. R. Paul, M. S. Anand, M. C. Naik, and R. P. Agarwala
Kernforschungsanlage Jülich, Intern. Conf. on Vacancies and
    Interstitials in Metals (Sept. 1968), pp. 105-119
N69-13085 03-17

Diffusion of Gold in Titanium
Yu. D. Strongonov, S. E. Salibekov, M. Kh. Levinskaya, and
    S. A. Prokofev
Izv. Akad. Nauk SSSR, Neorg. Mater., 4(12):2068-2073 (1968)
Inorg. Mater., 4(12):1799-1803 (1968)

Thermal Mass Transport and Electromigration
in Beta-Zr
D. R. Campbell
(Rensselaer Polytechnical Institute, 1967), Dissertation Abstr.,
    28B(9):3843-3844 (1967-1968)

Thermodynamic Properties of Hydrogen and
Deuterium in Alpha-Titanium
T. A. Giorgi
Nuovo Cimento, Suppl. 1, Vol. 5, pp. 472-482 (1967)
Intern. Symp. on Sorption-Desorption Phenomena in High
    Vacuum, Rome, Italy (Mar. 14-17) 1967; CONF-670326

Effect of Pressure on the Diffusion of Fe in Ti
and Ti + 10% Fe
R. F. Peart
Phys. Stat. Sol., 20:545 (1967)

On the Anomalous Diffusion Behavior of B. C. C.
Zirconium
G. Tiwari and B. Sharma
Acta Met., 15:155-157 (1967)

Carbon Diffusion in β Zirconium
R. A. Andriyevskiy, V. N. Zagryazkin, and G. Y. Mesh-
    cheryakov
Phys. Met. Metallog., 21:146-149 (1966)

Environmental Effects on the Diffusion of ^{182}Ta
in bcc Titanium
J. Askill
Phys. Stat. Sol., 16:K63 (1966)

Diffusion of Yttrium in Zirconium
G. B. Federov, F. I. Zhomov, and E. A. Smirnov
Met. i Metalloved. Chistykh Metal., 5:22-24 (1966)

Thermotransport of Oxygen in Beta-Zirconium
G. D. Rieck and D. L. Vogel
Acta Met., 14:1703-1714 (1966)

Contribution à l'étude de la diffusion du bore
dans le zirconium
A. Walder
Thèse Doct. 3eme Cycle, Spec. Chim. metallurg., Paris (1966),
    Chatillon-sous-Bogneux, ONERA (1966), 86 pp.

The Volume of the Self-Diffusion Parameters of
Zirconium
G. B. Federov
Met. i Metalloved. Chistykh Metal., Moscow, 4:30-34 (1963)
AERE-Trans-1030 (March 1965)

Diffusion of ^{181}Hf in bcc Hafnium
F. R. Winslow and T. S. Lundy
Trans. Met. Soc. AIME, 233:1790 (1965)

Diffusion of Iron in Zirconium
A. M. Blinkin and V. V. Vorobev
Ukr. Fiz. Zh., 9:91-95 (Jan. 1964)

Diffusion of Boron and Carbon in Refractory
Transition Metals
A. P. Epik
Dopovidi Akad. Nauk Ukr. RSR, 1:67-70 (1964)

Diffusion in Body-Centered Cubic Metals Zir-
conium, Vanadium, Niobium, and Titanium
Ted Sadler Lundy
Contract No. W-7405-eng-26, ORNL-3617 (June 1964)

On "Self-Diffusion in Alpha Titanium"
S. P. Murarka and R. P. Agarwala
Acta Met., 12:1096 (1964)

Diffusion of Titanium-44 and Vanadium-48 in
Titanium
James F. Murdock
ORNL-3616 (1964)

Diffusion of Oxygen in Hafnium
J. P. Pemsler
J. Electrochem. Soc., 111:1185-1186 (Oct. 1964)

On the Reactive Diffusion of Boron and Carbon
in Refractory Transition Metals
G. V. Samsonov
Dopovidi Akad. Nauk Ukr. RSR, 1:68-70 (1964)

The Physical Metallurgy of Zirconium
D. L. Douglas
(General Electric Co., Pleasanton, California), At. Energy
    Rev., 1:71-237 (1963)

Diffusion of ^{95}Zr and ^{95}Cb in bcc zirconium
J. I. Federer and T. S. Lundy
Trans. Met. Soc. AIME, 227:592-597 (June 1963)

Diffusion in Titanium and Titanium-Niobium
Alloys
G. B. Gibbs, D. Graham, and D. H. Tomlin
Phil. Mag., 8:1269 (1963)

Diffusion of Oxygen in Tungsten and Some Other
Transition Metals
A. J. Jacobs
Nature, 200:1310-1311 (1963)

On the Anomalous Self-Diffusion in Body-Cen-
tered Cubic Zirconium
G. V. Kidson
(Research Metallurgy Branch, Atomic Energy of Canada, Ltd.,
    Chalk River, Ontario), Can. J. Phys., 41:1563-1570 (1963)

Self-Diffusion of Alpha-Titanium
C. M. Libanati and F. Dyment
NRC-TT-1213 (1965)
Acta Met., 11:1263-1267 (1963)

Diffusion of Solute Elements in Beta-Titanium
R. F. Peart and D. H. Tomlin
Acta Met., 10:123-134 (1962)

Self-Diffusion in Body-Centered Cubic Zirconium
G. Kidson and J. McGurn
(Research Metallurgy Branch, Atomic Energy of Canada Limited, Chalk River, Ontario), Can. J. Phys., 39:1146 (1961)

Diffusion of Oxygen in Alpha-Titanium
A. V. Revyakin
Izvest. Akad. Nauk SSSR, Otdel. Tekh. Nauk Met. i Toplivo, No. 5, pp. 113-116 (1961)

Thermal Diffusion of Oxygen and Nitrogen in Zirconium
G. D. Rieck and H. A. C. M. Bruning
Nature, 190:1181 (1961)

Metallurgy and Metallography of Pure Metals
V. S. Emel'yanov and A. I. Evstyukhin
Met. i Metalloved. Chistykh Metal. (1960), 395 pp. JPRS-9473

Diffusion in Refractory Metals
N. L. Peterson
(Advanced Metals Research Corp., Sommerville, Mass.) Contract AF33(616)-7382, WADD-TR-60-793 (Dec. 1, 1960), 164 pp
AD-257860 (March 1960)

Diffusion of Aluminum, Tin, Vanadium, Molybdenum, and Manganese in Titanium
D. Goold
(Imperial Chemical Industries, Ltd., Metals Division, Birmingham 6), J. Inst. Metals, 88:444 (1959-1960)

1.i.2.c. Group V

Internal Friction Due to Long-Range Diffusion of Hydrogen and Deuterium in Vanadium
R. Cantelli, F. M. Mazzolai, and M. Nuovo
(Consiglio Nazionale delle Ricerche, Instituto di Acustica "O. M. Corbino," Rome, Italy)
Preprint from Conf. on Ultrasonic Attenuation and Internal Friction in Crystalline Solids, to appear in J. Chem. Phys. Solids, received Feb. 1970

Anelastic Relaxation Effects in Cold-Worked and Hydrogen Doped Vanadium
G. Connelli and F. M. Mazzolai
Consiglio Nazionale delle Ricerche, Istituto di Acustica "O. M. Corbino," Rome, Italy
Preprint from Conf. Ultrasonic Attenuation and Internal Friction in Crystalline Solids, to appear in J. Chem. Phys. Solids, received Feb. 1970

Quasielastic Neutron Scattering by Hydrogen in Niobium
W. Gissler, G. Alefeld, and T. Springer
J. Phys. Chem. Solids, 31:2361-2369 (1970)
Diffusion of hydrogen in niobium

Diffusion of Ti-44 into Niobium Single Crystals
J. Pelleg
National Aeronautical Establishment, Ottawa, LTR-ST-407 (May 1970), 32 pp.

Diffusion of ^{44}Ti into Niobium Single Crystals
J. Pelleg
Phil. Mag., 21(172):735-742 (1970)

Dislocation Diffusion in Niobium (Columbium) Single Crystals
T. C. Reuther and M. R. Achter
Met. Trans., 1:1777-1778 (1970)

The Diffusion Coefficients of Hydrogen and Deuterium in Vanadium, Niobium, and Tantalum by Gorsky-Effect Measurements
G. Schaumann, J. Volkl, and G. Alefeld
Phys. Stat. Sol., 42:401-413 (1970)

Diffusional Relaxation of H and D in V and Nb
J. Volkl, G. Schaumann, and G. Alefeld
J. Phys. Chem. Solids, 31:1805-1809 (1970)

Rev. Phys. Appliquée
Vol. 5, No. 3 (June 1970)
Entire issue on Nb and Nb alloys

Diffusion zwischen Al und Nb
F. Brossa, G. Musso, and H. W. Schleicher
Met. Ital., 61:167-172 (1969)

Hydrogen Penetration in Ta
P. E. Giua and C. D. Messino
Nuovo Cimento, 1:305-307 (1969)

Internal Friction in Vanadium after Heating at Low Air Pressure
M. A. Mondino, D. Vassallo, and M. C. de Achterberg
J. Mater. Sci., 4:1117-1121 (1969)

Solubility and Diffusion of Oxygen in Tantalum
M. Parkman, R. Pape, R. McRac, D. Brayton, and L. Reed
(Varian Associates, San Carlos, Calif.), NASA-CR-1276 (Feb. 1969)

Diffusion of ^{51}Cr into Niobium Single Crystals
J. Pelleg
Phil. Mag., 19:25-32 (1969)

On the Self-Diffusion of Columbium
Joshua Pelleg and G. M. Lindberg
Trans. Met. Soc. AIME, 245:1654 (1969)

Solution of Hydrogen in Niobium
J. A. Pryde and C. G. Titcomb
Trans. Faraday Soc., 65:2758-2765 (1969)

Thermodynamic Data and Kinetics of Evolution for Dilute Solutions of Hydrogen in Tantalum
J. A. Pryde and I. S. T. Tsong
Trans. Faraday Soc., 65:2766-2771 (1969)

The Diffusion of Hydrogen and Deuterium in Niobium and Vanadium
Gunter Schaumann
Inst. fuer Festkoerper- und Neutronenphysik, Kernforschungsanlage, Jülich, W. Germany, JUL-606-FN (July 1969), 90 pp.

Diffusion of Carbon in Vanadium
Son and Pongun
Nippon Kinzoku Gakkaishi, 33:1-3 (1969)

Diffusion of S in V and Ta
B. A. Vandyschev and A. S. Panov
Izv. Akad. Nauk SSSR, Met., 1:244-246 (1969)

Diffusion of H_2 and D_2 in Ta
H. Zuchner and E. Wicke
Z. Phys. Chem. (Frankfurt), 67(1/3):154-158 (1969)

Isotope Effect for Diffusion of Iron in Vanadium
M. G. Coleman, C. A. Wert, and R. F. Peart
Phys. Rev., 175:788-795 (1968)

Kinetics of O_2 Dissolution in Nb and Ta
G. Horz
Z. Metallk., 59:283-288 (1968)

Diffusion of Cr-51 into Poly- and Single Crystals of Niobium
J. Pelleg and G. M. Lindberg
NRC-10636, NRC-LR-511 (Nov. 1968), 52 pp.

Diffusion of Carbon in Tantalum, Niobium, and Vanadium
P. Son, M. Miyake, and T. Sano
Tech. Rep. Osaka Univ. (Japan), 18(823-852):317-324 (1968)

Diffusion of Phosphorus into Niobium and Molybdenum
V. A. Vandshev and A. S. Panov
Fiz. Metallov Metalloved., 26:517-521 (1968)

Dynamics of Hydrogen Impurities in Nb
G. Verlan, R. Rubin, and W. Kley
Proceedings Symp. Neutron Inelastic Scattering, Ispra (1968), pp. 223-231

Grain Boundary Penetration of Niobium by Lithium
William Frederick Brehm, Jr.
Ph. D. thesis, Cornell Univ., Ithaca, New York (1967)
University Microfilms, Ann Arbor, Mich., Order No. 68-4657

Isotope Effect for Diffusion of Iron in Vanadium
M. G. Coleman
Ph. D. thesis, Univ. of Illinois, Chicago (1967)
University Microfilms, Ann Arbor, Mich., Order No. 68-1725

Diffusion of Alkali Metals in Molybdenum and Niobium
M. G. Karpman, G. V. Shcherbedinskii, G. N. Dubinin, and G. P. Benediktova
Metalloved. term. Obr. Metallov. SSSR, No. 3, pp. 46-49 (1967)

Mutual Diffusion of Niobium and Certain Refractory Metals
D. A. Prokoshkin, E. V. Vasil'eva, and L. L. Vergasova
Metalloved. term. Obr. Metallov., SSSR, No. 3, pp. 44-46 (1967)

Electrotransport of Carbon, Nitrogen, and Oxygen in Vanadium
F. A. Schmidt and J. C. Warner
J. Less-Common Metals, 13:493-500 (1967)

Carbon Diffusion in Nb
S. Z. Bokshtein, S. T. Kishkin, L. M. Moroz, and V. S. Chaplygina
Izv. Akad. Nauk SSSR, Metally, 4:139-142 (July-August 1966)

Relaxation Effect due to Diffusion of Interstitial Hydrogen in Tantalum and Niobium
Gaetano Cannelli and Livio Verdini
La ricerca scientifica, 36:98-105 (1966)

Migrazione di imperfezioni Puntiformi e Rilassamento Anelastico Dovuto Alle Dislocazioni nel Tantalio
Fabio Massimo Mazzolai and Mario Nuvo
Consiglio Nazionale Delle Ricerche, Rome (1966)

Diffusion of H in Ta
B. A. Merisoc and B. I. Chotkevitsch and A. I. Karnus
Fiz. Metallov. Metalloved., 22:308-309 (1966)

Diffusion of Carbon in Tantalum
P. Son, S. Ihara, M. Miyake, and T. Sano
Nippon Kinzoku Gakkaishi, 30(12):1137-1140 (1966)

Electrotransport of Oxygen, Nitrogen, and Carbon in Vanadium
Jairus C. Warner
(Ames Lab., Ames, Iowa), IS-T-133 (November 1966), 80 pp.

Tracer Diffusion of ^{113}Sn in Niobium
J. Askill
Phys. Stat. Sol., 9:K167 (1965)

Diffusion of Oxygen in Niobium
A. P. Litman
Phys. Stat. Sol., 11:K47 (1965)

Diffusion of ^{48}V in Vanadium
T. S. Lundy and C. J. McHargue
Trans. AIME, 233:243 (1965)

Diffusion of ^{95}Nb into Tantalum Single Crystals
R. E. Pawel and T. S. Lundy
Acta Met., 13:345 (1965)

The Diffusion of ^{95}Nb and Ta^{182} in Tantalum
R. E. Pawel and T. S. Lundy
Phys. Chem. Solids, 26:937-942 (1965)

Diffusion of ^{48}V and ^{59}Fe in Vanadium
R. F. Peart
J. Phys. Chem. Solids, 26:1153 (1965)

Vanadium Self-Diffusion
R. F. Peart
Proc. Intern. Conf. Diffusion in Body-Centered Cubic Materials, Gatlinburg (1964), pp. 235-245
American Society for Metals, Metals Park, Ohio (1965)

Solubility of Nitrogen in Tantalum
P. Bunn and C. Wert
Trans. Met. Soc. AIME, 230:936 (1964)

Diffusion of Boron and Carbon in Refractory Transition Metals
A. P. Epik
Dopovidi Akad. Nauk Ukr. RSR, 1:67-70 (1964)

Diffusion in Body-Centered Cubic Metals Zirconium, Vanadium, Niobium, and Tantalum
Ted Sadler Lundy
Contract No. W-7405-eng-26, ORNL-3617 (June 1964)

A Submicron Sectioning Technique for Analyzing Diffusion Specimens of Tantalum and Niobium
R. E. Pawel and T. S. Lundy
J. Appl. Phys., 35:435 (1964)

Determination and Application of Thermophysical Properties of Refractory Metals
G. D. Rieck
The Science and Technology of Tungsten, Tantalum, Molybdenum, Niobium, and Their Alloys, Pergamon Press, Oxford (1964), pp. 205-217

On the Reactive Diffusion of Boron and Carbon in Refractory Transition Metals
G. V. Samsonov
Dopovidi Akad. Nauk Ukr. RSR, 1:68-70 (1964)

Activation Energies for Diffusion and Diffusion Coefficients of Oxygen and Nitrogen in Molybdenum
Ma Ying-liang and Son Jiu-yih
UCRL-Trans-10103
Chin Shu Hsuch Pao, 7:68-76 (1964)

Diffusion of Oxygen in Tungsten and Some Other
Transition Metals
  A. J. Jacobs
  Nature, 200:1310-1311 (1963)

Tracer Diffusion in Niobium and Molybdenum
  R. F. Peart, D. Graham, and D. H. Tomlin
  Acta Met., 10:519-523 (1963)

Solid Solubility Limits of Y and Sc in the Ele-
ments W, Ta, Mo, Nb, and Cr
  A. Taylor
  (Westinghouse Electric Corp., Research Laboratories, Beulah
    Road, Churchill Borough, Pittsburgh 35, Pa.), Contract No.
    AF33(616)-8315, ASD-TDR-63-204 (1963)

Investigation of Diffusion of Silicon and Titani-
um in Niobium
  P. M. Arzhanyi, R. M. Volkova, and D. V. Prokoshkin
  FTD-TT-64-710; TT-65-61565; AD-610798
  Tr. Inst. Met. im A. A. Baikova, Akad. Nauk SSSR, No. 11,
    p. 78 (1962)

Diffusion Rates and Solubilities of Interstitials
in Refractory Metals
  W. D. Klopp and Vincent D. Barth
  (Defense Metals Information Center, Battelle Memorial Insti-
    tute, Columbus, Ohio), DMIC-Memo-50 (April 4, 1960),
    10 pp.

Diffusion in Refractory Metals
  N. L. Peterson
  (Advanced Metals Research Corp., Somerville, Mass.) Con-
    tract AF33(616)-7382, WADD-TR-60-793 (Dec. 1, 1960),
    164 pp.
  AD-257860 (March 1960)

Diffusion of Interstitial Solutes in the Group V
Transition Metals
  R. W. Powers and Margaret V. Doyle
  J. Appl. Phys., 30:514-524 (1959)

The Self-Diffusion of Niobium — III. Final
Prog. Rept., July 1, 1958 to January 1, 1959
  R. Resnick, L. S. Castleman, and L. Seigle
  (Sylvania Electric Products, Inc., Research Labs., Bayside,
    N. Y.), SEP-252 (March 30, 1959), 14 pp.

Self Diffusion in Tantalum
  P. L. Gruzin and V. I. Meshkov
  AEC-tr-2926
  Voprosy Fiz. Metal. i Metalloved. Akad. Nauk Ukrain. SSR,
    Sbornik Nauch. Rabot, 570 (1955), 3 pp.

Thermodynamic Properties of Molybdenum and
Columbium: An Annotated Bibliography
  Compiled by Jack B. Goldman
  (Lockheed Missiles and Space Company, a group division of
    Lockheed Aircraft Corporation, Sunnyvale, California),
    6-90-62-122, Special Bibliography SB-62-69

1.i.2.d. Group VI

Investigation of Surface Diffusion of Copper
Atoms on Molybdenum by Secondary Ion-Ion
Emission
  A. D. Abramenkov, V. V. Slezov, L. V. Tanatarov, and Ya. M.
    Fogel
  Sov. Phys. — Solid State, 12:2365-2368 (1971)

The Solubility of Nitrogen in Solid Chromium
  T. Mills
  J. Less-Common Metals, 23:317-324 (1971)

Diffusion von Stickstoff in Molybdän und Wolf-
ram
  H. Jehn and E. Fromm
  J. Less-Common Metals., 21:333-336 (1970)

Diffusion of Cesium in Mono- and Polycrystal-
line Molybdenum
  A. A. Korolev and L. V. Paulinov
  Fiz. Metal. Metalloved., 29:1326-1328 (1970)
  In Russian; English translation in Phys. Metals Metallography

Measurement of the Surface-Diffusion Activation
Energy of Potassium on Tungsten (Field Emission)
  R. Meclewski
  Acta Phys. Pol., 37:41-47 (Jan. 1970)

Surface Diffusion of Au on W in Strong Electric
Fields
  G. G. Vladimirov and I. L. Sokol'skaya
  Fiz. Tverd. Tela, 12(5):1553-1554 (1970)
  Sov. Phys. — Solid State, 12(5):1224-1225 (1970)

Concentration Dependence of the Migration En-
ergy of Ti over W
  G. G. Vladimirov, B. K. Medvedev, and I. L. Sokol'skaya
  Fiz. Tverd Tela, 12(5):1423-1426 (1970)
  Sov. Phys. — Solid State, 12(5):1118-1120 (1970)

Surface Self-Diffusion of Tungsten
  F. P. Bowden and K. E. Singer
  Nature, 222:977-979 (1969)

The Heat of Solution and Diffusivity of Nitrogen
in Molybdenum
  J. H. Evans and B. L. Eyre
  Acta Met., 17:1109-1115 (1969)

Diffusion Saturation of Mo with N
  J. M. Lachtin and others
  Izv. Vyssh. Uchebn., Zaved. Mashinostro, No. 4, pp. 117-121
    (1969)

Tracer Diffusion in Tungsten
  R. E. Pawel and T. S. Lundy
  Acta Met., 17(8):979-988 (1969)
  Nb, Ta, and W isotopes

Diffusion of ^{51}Cr into Polycrystalline Niobium
  Joshua Pelleg
  J. Less-Common Metals, 17:319-324 (1969)

Thermodynamics of the Solubility and Permeation
of Hydrogen in Metals at High Temperature and
Low Pressure
  D. S. Shupe and R. E. Stickney
  J. Chem. Phys., 51(4):1620-1625 (1969)
  Mo and W

Electrical Transport and the Soret Effect in
Molybdenum
  D. N. Vasilkovskii and I. V. Zakurdaev
  Fiz. Tverd. Tela, 10(12):3640-3646 (1968)
  Sov. Phys. — Solid State, 10(12):2886-2890 (1969)

A Technique for Measuring the Diffusivity and
Solubility of Gases in Metals with Application to
Nitrogen in Tungsten
  Richard L. Wagner
  Thesis, Oak Ridge National Lab., Oak Ridge, Tenn. (July 1969)

Electrical and Thermal Mass Transport in a
Molybdenum Crystal and Surface Rearrangement
of a Refractory Metal During Directed Diffusion
I. V. Zakurdaev
Fiz. Tverd. Tela, 11:3463-3473 (1969)
Sov. Phys. – Solid State, 11:2905-2913 (1970)

The Heat of Solution and Diffusivity of Nitrogen
in Molybdenum
J. H. Evans and B. L. Eyre
Kernforschungsanlage Jülich, Intern. Conf. on Vacancies
and Interstitials in Metals, Vol. 2, N69-12994 03-17 (Sept.
1968), pp. 858-869

Permeation of Hydrogen through Tungsten and
Molybdenum
R. Frauenfelder
J. Chem. Phys., 48:3955-3965 (1968)

Permeation, Diffusion, and Solution of Nitrogen
in Tungsten and Molybdenum
R. Frauenfeld
J. Chem. Phys., 48:3966-3971 (1968)

The Diffusion of Nickel during Nickel-Induced
Recrystallization in Doped Tungsten
S. Friedman and J. Brett
Trans. Met. Soc. AIME, 242:2121-2127 (1968)

Discussion of "The Solubility Limit and Dif-
fusivity of Carbon in Molybdenum"
C. P. Kempter
Trans. Met. Soc. AIME, 242:1483 (1968)

Die Diffusion von Silizium in Molybdän
E. Lassner, R. Puschel, and F. Benesovsky
Planseeberichte für Pulvermetallurgie combined with Powder
Metallurgy Bulletin, 16(2):91-103 (1968)

Hydrogen Solubility in Tungsten at High Tem-
peratures and Pressures
A. A. Mazaev, R. G. Avarbe, and Yu. N. Vil'k
Izv. Akad. Nauk SSSR, Metally, No. 6, pp. 223-226 (1968)

Internal Friction Method of Studying the Diffusion Mobility of Carbon in Tungsten and Molybdenum
V. Y. Shchelkonogov, L. N. Aleksandrov, V. A. Piterimov,
and V. S. Mordyuk
Fiz. Metallov Metalloved., 25:80-84 (1968)
Phys. Metals Metallogr., 25:68-72 (1968)

Diffusion of Sulfur in Molybdenum
B. A. Vandyshev and A. S. Panov
Fiz. Metallov Metalloved., 25:321-325 (1968)
Phys. Metals Metallogr., 25:130-133 (1968)

Diffusion of Phosphorus into Niobium and Molyb-
denum
B. A. Vandyshev and A. S. Panov
Fiz. Metal. Metalloved., 26:517-521 (1968)

Tracer Diffusion of W in Mo
J. Askill
Phys. Stat. Sol., 23:K21-23 (1967)

Grain Boundary Diffusion in Tungsten
K. G. Kreider and G. Bruggeman
Trans. Met. Soc. AIME, 239:122-126 (1967)

Effect of Hydrogen on the Diffusion of Carbon
in Chromium
S. V. Zemsky and B. P. Lyakhin
Fiz. Metall. i Metalloved., 23(5):913-916 (1967)

Solubility of Carbon in Solid Chromium
S. V. Zemsky and A. P. Fokin
Zh. Fiz. Khim., 41:93-97 (1967)

Grain Boundary Self-Diffusion in Chromium Near
the Melting Point
John Askill
Appl. Phys. Letters, 9:82 (1966)

Atomic View of Surface Self-Diffusion: Tung-
sten on Tungsten
Gert Ehrlich and F. G. Hudda
J. Chem. Phys., 44:1039 (1966)

Über das Verhalten von Silicium in Molybdän
E. Lassner and R. Puschel
Mikrochim. Acta Suppl. II (1967), p. 219
Kolloquiums über metallkundliche Analyse mit besonderer
Berücksichtigung der Elektronenstrahl-Microanalyse,
Wien (Oct. 25-27, 1966)

Surface Diffusion of Tungsten
G. M. Neumann, W. Hirschwald, I. N. Stranski
Z. Naturforsch., 21A:807-811 (1966)

Selbstdiffusion und Diffusionsmechanismus in
Wolfram
G. M. Neumann and W. Hirschwald
Z. Naturforsch., 21A:812.(1966)

Diffusion of C in W
V. Zh. Shchelkonogov
Uch. Zap. Mosk. Gos. Univ., 50(1):3-9 (1966)

Carbon Diffusion in Chromium
S. V. Zemskii and M. N. Spasskii
Phys. Metals Metallog., 21:129-132 (1966)

Adsorption and Surface Diffusion of Pt on W
Ya. V. Zubenko and I. L. Sokolskaya
Izv. Akad. Nauk SSSR, Ser. Fiz., 30(5) (1966)

The Effect of an Argon Environment on the Dif-
fusion of ^{60}Co in Molybdenum
J. Askill
Phys. Stat. Sol., 9:K113 (1965)

Self-Diffusion in Chromium
J. Askill and D. H. Tomlin
Phil. Mag., 11:467 (1965)

Volume and Boundary Diffusion of Tungsten in
Molybdenum
S. Z. Bokshtein, M. B. Bronfin, and S. T. Kishkin
Diffusion Processes, Structure, and Properties of Metals,
Consultants Bureau, New York (1965), pp. 16-23

Diffusion of Rhenium in Molybdenum
M. B. Bronfin
Diffusion Processes, Structure, and Properties of Metals,
Consultants Bureau, New York (1965), pp. 24-26

Surface Diffusion of Ge and Si on W
J. Nikliborc, T. Radon, and J. Zebrowski
Acta Phys. Polon., 26:1023-1025 (1965)

Self-Diffusion in Single-Crystal Tungsten and Diffusion of Rhenium Tracer in Single-Crystal Tungsten
  Robert L. Andelin
  Dissertation Abstr., 64-1439, 155 pp.

Diffusion of Boron and Carbon in Refractory Transition Metals
  A. P. Epik
  Dopovidi Akad. Nauk Ukr. RSR, 1:67-70 (1964)

Field-Aided Electron Beam Zone Refining of Tungsten
  D. R. Hay and E. Scala
  (Department of Engineering Physics and Materials Science, Cornell University, Ithaca, N. Y.), First World Conf. Electron and Ion Beam Technology (May 4-7, 1964), Toronto, Canada

Self-Diffusion in Molybdenum
  L. V. Pavlinov and V. N. Bykov
  Fiz. Metal. Metalloved. SSSR, 18:459-461 (1964)

Determination and Application of Thermophysical Properties of Refractory Metals
  G. D. Rieck
  The Science and Technology of Tungsten, Tantalum, Molybdenum, Niobium, and Their Alloys, Pergamon Press, Oxford (1964), pp. 205-217

On the Reactive Diffusion of Boron and Carbon in Refractory Transition Metals
  G. V. Samsonov
  Dopovidi Akad. Nauk Ukr. RSR, 1:68-70 (1964)

The Recovery of the Electrical Resistivity of Cold-Worked Tungsten
  H. Schultz
  (OSRAM — Studiengesellschaft, Augsburg, Germany), Acta Met., 12:649-664 (1964)

Self-Diffusion in Molybdenum
  J. Askill and D. H. Tomlin
  (Physics Department, The University, Reading, Berks.), Phil. Mag., 8:997 (1963)

Study of Diffusion of Carbon and Molybdenum in Chromium
  P. L. Gruzin, S. V. Zemskii, and I. B. Rodina
  AERE-Trans-1032
  Met. Metalloved. Chistykh Metal., Sb. Nauchn. Rabot, 4:243 (1963)

Diffusion of Oxygen in Tungsten and Some Other Transition Metals
  A. J. Jacobs
  Nature, 200:1310-1311 (1963)

Thermal Diffusivity in Tungsten at Temperatures Between 1600 and 2960°C
  O. A. Kraev and A. A. Stel'makh
  (Inst. of Thermophysics, Siberian Branch, Academy of Sciences, USSR), High Temp., 1:5-8 (July-Aug. 1963)

Diffusion of Chromium in Nickel
  S. P. Murarka, M. S. Anand, and R. P. Agarwala
  (Chemistry Div., Atomic Energy Establishment, Trombay, Bombay, India), A.E.E.T./C.D./17 (1963)

Solid Solubility Limits of Y and Sc in the Elements W, Ta, Mo, Nb, and Cr
  A Taylor
  (Westinghouse Electric Corp., Research Laboratories, Beulah Road, Churchill Borough, Pittsburgh 35, Pa.), Contract No. AF33(616)-8315, ASD-TDR-63-204 (1963)

Investigation of Autodiffusion in Tungsten
  W. Danneberg
  SLAC-Trans-61
  Metall, 15:977-981 (1961)

Diffusion in Refractory Metals
  N. L. Peterson
  Contract AF33(616)-7382, WADD-TR-60-793 (Dec. 1, 1960), 164 pp
  AD-257860 (March 1960)

Die Geschwindigkeit der Selbstdiffusion in reinem Chrom
  Harold W. Paxton and Ewald G. Gondolf
  Arch. Eisenhüttenw., 30:55-60 (1959)

Thermodynamic Properties of Molybdenum and Columbium: An Annotated Bibliography
  compiled by Jack B. Goldman
  (Lockheed Missiles and Space Company, a group division of Lockheed Aircraft Corporation, Sunnyvale, California), 6-90-62-122, Special Bibliography SB-62-69

1.i.3. Elastic Properties

1.i.3.a. General

Elastic Moduli of Transition Metals
  Francois Ducastelle
  ONERA-T.P. 848 (1970), 20 pp.
  Model accounts for the variation of the atomic volume and elastic moduli along a transition series

Relation of the C' Elastic Modulus to Stability of B.C.C. Transition Metals
  E. S. Fisher and D. Dever
  Acta Met., 18:265-269 (1970)

Evaluation of Third-Order Elastic Constants for Cubic Crystals
  M. W. Guinan and A. D. Ritchie
  J. Appl. Phys., 41(5):2256-2258 (1970)

Tables internationales de constantes sélectionées 16. Métaux données thermiques et mécaniques Mecaniques
  Simonne Allard
  Pergamon Press, New York (1969)

Elasticité des métaux paramagnétiques
  O. Fischer, M. Peter, and S. Steinemann
  Helv. Phys. Acta, 42:459-484 (1969)

Bibliography of Second and Third Order Elastic Constants
  Herbert Pomerance
  Contract No. W-7405-eng-26, ORNL-RMIC-9, December 1968

Elastic Properties of W, Mo, Nb
V. A. Dreschpak
Dopov. Akad. Nauk Ukr. SSSR A, 29:917-921 (1967)

Third-Order Elastic Moduli of Polycrystalline
Metals from Ultrasonic Velocity Measurements
R. T. Smith, R. Stern, and R. W. B. Stephens
J. Acoust. Soc. Am., 40:1002-08 (1966)
W, Mo

Comparison of Measured and Predicted Bulk
Moduli of Tantalum and Tungsten at High Tem-
peratures
N. Soga
J. Appl. Phys., 37:3416 (1966)

Dynamic Young's Modulus Measurements Above
1000°C on Some Pure Polycrystalline Metals and
Commercial Graphites
P. E. Armstrong and H. L. Brown
Trans. AIME, 230:962 (1964)

Dynamic Method of Measuring Young's Modulus
of Elasticity
K. D. Baveja
J. Sci. Instr., 41:662 (1964)

Variation of the Elastic Moduli at the Super-
conducting Transition
G. A. Alers and D. L. Waldorf
IBM J. Res. Develop, 6:89-93 (1962)

Effect of Plastic Deformation and Annealing
Temperature on Superconductive Properties
J. H. Hauser and E. Buehler
Phys. Rev., 125:142-148 (1962)

Current Information on the Refractory Metals
C. E. King
Contract AF33(657)-11214, AD-436333; ERR-FW-049A (Dec.
1, 1962), 80 pp.

Data on Sound Velocities in Pure Metals
Karl Heinz Schramm
Z. Metallk., 53:729-735 (Nov. 1962)

A Deformation (Strain) Apparatus for the Measurement of the Electrical Resistance of Metals
at Lowest Temperatures, and Investigations into
the Plastic Behavior of Copper Monocrystals
Wolfgang Schule, Otto Buck, and Eberhard Koster
SCL-T-434 (1962)
Z. Metallk., 53:172-177 (1962)

The Effect of Temperature and Alloying Conditions on the Deformation of Metal Crystals
R. W. K. Honeycombe
(University of Sheffield, England), Progr. in Materials Sci.,
9:93-130 (1961)

The Mechanism of Plastic Deformation Followed
by Electron Microscopy
Aurel Berghezan and Angeline Fourdeux
Compt. Rend., 248:1333 (1959)

Accurate Measurement of Dynamic Elastic
Constants for Very Small Specimens
G. Bradfield
Proc. Intern. Congr. Acoustics (L. Cremer, ed.), Elsevier
Publishing Co., Amsterdam, Netherlands

A Compilation of Some of the Physical Prop-
erties of the Metallic and Semi-Metallic Ele-
ments and a Study of Some of Their Interrela-
tionships
Karl A. Gschneider, Jr.
(Department of Physics and Materials Research Laboratory,
University of Illinois, Urbana, Illinois, and University of
California, Los Alamos Scientific Laboratory, Los Alamos,
New Mexico), LADC-6220
To be published in Solid State Phys., Vol. 16

1.1.3.b. Group IV

Effects of Changes in Volume and c/a Ratio on
the Pressure Derivatives of the Elastic Moduli
of H.C.P. Ti and Zr
E. S. Fisher and M. H. Manghnani
J. Phys. Chem. Solids, 32:657-667 (1971)

Lattice Dynamics of Zirconium
H. F. Bezdek, R. E. Schmunk, and L. Finegold
Phys. Stat. Sol., 42:275-280 (1970)

Hydrostatic Pressure Derivatives of the Single-
Crystal Elastic Moduli of Zirconium
E. S. Fisher, M. H. Manghnani, and T. J. Sokolowski
J. Appl. Phys., 41:2991-2998 (1970)

Compilation of Elastic Wave Modes in Hexagonal
Metals
E. G. Henneke II and R. E. Green, Jr.
J. Appl. Phys., 40:3626 (1969)

Anomalous Elastic Behavior of Titanium at Low
Temperatures
G. A. Alers and J. A. Karbon
J. Appl. Phys., 39:4348 (1968)

Effect of Hydrogen on the Elasticity Modulus of
Titanium and a β-Titanium Alloy VT15
B. A. Kolachev, K. M. Konstantinov, and A. A. Bukhanova
Fiz. Met. Metalloved., 26:572-574 (1968)

The Physical Metallurgy of Zirconium
D. L. Douglass
(General Electric Co., Pleasanton, California), At. Energy
Rev., 1:71-237 (1963)

Relationship of Elastic Shear Moduli to the Phase
Transformations in Zirconium, Titanium, and
Hafnium
E. S. Fisher and C. J. Renken
(Argonne National Laboratory, Argonne, Illinois), Bull. Am.
Phys. Soc. Ser. II, 8:65 (A) (1963)

Preferred Orientation and Anisotropy in Titanium
W. T. Roberts
(Research Department, I.C.I. Metals Division, Witton,
Birmingham, Great Britain), J. Less-Common Metals,
4:345 (1962)

Precipitation of Zirconium Hydride in Alpha
Zirconium Crystals
D. G. Westlake and E. S. Fisher
(Argonne National Laboratory, Argonne, Illinois), Trans.
AIME, 224:254 (1962)

Adiabatic Elastic Moduli of Single Crystal Alpha
Zirconium
E. S. Fisher and C. J. Renken
J. Nucl. Mater., 4:311 (1961)

Elastic Constants of Nb by Resonance at Sonic
Frequencies
R. J. Wasilewski
J. Phys. Chem. Solids, 26:1643 (1965)

Low-Temperature Deformation of Body-Centered
Cubic Metals. I. Yield and Flow Stress Mea-
surements
J. W. Christian and B. C. Masters
Proc. Roy. Soc., A281:223-239 (1964)

Elasticity of Tantalum Single Crystals
C. S. Hartley
J. Less-Common Metals, 6:245-248 (1964)
ASD-TR-63-870

Effects of Magnetic Field on the Ultrasonic
Attenuation in Superconducting Niobium
Akira Ikushima, Taira Suzuki, Nobuo Tanaka, and Sasao
Nakajima
J. Phys. Soc. Japan, 19:2235 (1964)

Substructure and Mechanical Properties of
Refractory Metal Wires
Roy Kaplow, John F. Peck, Frank T. J. Smith, and David A.
Thomas
(Massachusetts Inst. of Tech., Cambridge), Contract AF33(657)-
11238, ML-TDR-64-56 (Feb. 1964), 52 pp.

The Effect of Chromium and Iron on the Elastic
Properties of Niobium
I. M. Nedyukha and V. G. Chernii
Fiz. Metal. i Metalloved. SSSR, 18:599-604 (1964)

Ultrasonic Measurements of Elastic Moduli of
Polycrystalline Tantalum and Niobium
L. Palmieri
Appl. Materials Res., 3:139 (1964)

A New Method for Preparation of Strain-Free
Single Crystal Tensile Samples of High-Melting
Metals and Some Results on the Plastic Defor-
mation of Niobium Single Crystals
E. Votava
(Union Carbide European Research Associates, Brussels)
Phys. Stat. Sol., 5:421-434 (1964)

Niobium Hardness at 20-2120°C
V. A. Borisenko
"Sbornik, Voprosy Vysokotemperaturnoi Prochnosti v
Mashinostroenii," Kiev, Academy of Sciences (1963),
pp. 31-35

Elastic Constants of Tantalum, Tungsten, and
Molybdenum
F. H. Featherston and J. R. Neighbours
Phys. Rev., 130:1324-1333 (1963)

Work-Hardening Mechanisms in Body-Centered
Cubic Metals
Donald P. Gregory
(Pratt and Whitney Aircraft Div., United Aircraft Corp.,
Connecticut Aircraft Nuclear Engine Lab., Middletown),
Contract AF33(616)-7855, AD-411124, ASD-TDR-62-354,
Pt. II (Mar. 1963), 48 pp.

The Influence of Lattice Defects on the Electrical Resistance and the Hall-Effect in Tungsten and Tantalum
E. Krautz and H. Schultz
Z. Angew. Phys., 15:1-4 (1963)

The Engineering Properties of Tantalum and
Tantalum Alloys
F. F. Schmidt and H. R. Ogden
(Battelle Memorial Institute, Defense Metals Information
Center, Columbus, Ohio), DMIC-Rept. 189 (September
13, 1963)
Contract AF 33(616)-7747, 112 pp.

Heat Treatment of Niobium
G. V. Zakharova and L. P. Zhorova
JPRS-22404
Tsvetnyye Metally (Nonferrous Metals), No. 5, Publishing
House of Metallurgical Literature, Moscow (1963), pp. 53-
58

Effect of Oxygen on the Lattice Constant, Hard-
ness and Ductility of Vanadium
S. A. Bradford and O. N. Carlson
Trans. Quart. Am. Soc. Metals, 55:169 (1962)

Calculated Lattice Specific Heats for Seven
BCC Elements (Mo, V)
C. B. Clark
Phys. Rev., 125:1898-1902 (1962)

The Effects of Carbon on the Hardness, Micro-
structure, and Cold Working Properties of High-
Purity Niobium
F. R. Cortes and A. L. Feild, Jr.
J. Less-Common Metals, 4:169-180 (Apr. 1962)

Deformation of Tantalum Single Crystals
D. P. Ferriss, R. M. Rose, and J. Wulff
Trans. Met. Soc. AIME, 224:975 (1962)

Study of Effects of Impurities on the Mechanical
Properties of Niobium. Quarterly Prog. Rept.
No. 2, July 1, 1962 through September 30, 1962
Torleif Lindtveit, Paul Storvik, and Ingard Kvernes
(Sentralinstitutt for Industriell Forskning, Blindern, Norway),
NP-12633 (1962), 9 pp.

Comparison of Tensile Properties of Zone-
Refined Niobium Rod and Sheet
J. D. W. Rawson and J. McG. Regart
J. Inst. Metals, 90:448 (1962)

Low Temperature Tensile Properties of Zone-
Refined Niobium (Columbium)
J. F. Wellings and R. Maddin
(University of Pennsylvania, School of Metallurgical En-
gineering, Philadelphia), NP-11795 (1962), 19 pp.

Hardness Anisotropy of Columbium
D. L. Douglass
Trans. Am. Soc. Metals, 54:322-330 (1961)

Young's Modulus of Columbium at Elevated Temperatures
David P. Laverty and Edward B. Evans
"Columbium Metallurgy," New York, Interscience Publishers
(1961), pp. 299-307

A Metal for Service at Both Cryogenic and Ultra-
high Temperatures
H. R. Ogden and I. Perlmutter
Metal Progr., 80:97-99 (Nov. 1961)

Effect of Grain Size, Strain Rate, and Temperature on the Yield Strength of Columbium
E. S. Tankins and R. Maddin
"Columbium Metallurgy," New York, Interscience Publishers
(1961), pp. 343-363

Elastic Moduli of Vanadium
G. A. Alers
Phys. Rev., 119:1532-1533 (1960)

Effects of Interstitial Impurity Levels on Mechanical Properties of Columbium at Low Temperatures
M. D. Carver, J. T. Dunham, and H. Kato
(Bureau of Mines, Albany Metallurgy Research Center, Ore.), BM-RI-5872 (Feb. 1960), 17 pp.

Mechanical and Physical Properties of the Refractory Metals, Tungsten, Tantalum, and Molybdenum above 4000°F
L. F. Glasier, Jr., R. D. Allen, and I. L. Saldinger
(Aerojet-General Corp., Azusa, Cal.), Rept. M-1826 (April 1959), 89 pp.

1.i.3.d. Group VI

A Measurement of the X-Ray Debye Temperature of Tungsten in the Range $4.2°K \leq T \leq 297°K$
Farrell M. Kilbane
Thesis, Ohio State University, Columbus, Ohio (1969), 109 pp.

Anelastic Effect in Single Crystals of Mo
M. J. Murray
Phil. Mag., 20:561-568 (1969)

Elastic Constants of Molybdenum-Rich Rhenium Alloys in the Temperature Range −190°C to +100°C
D. L. Davidson and F. R. Brotzen
(Rice University, Houston, Texas), preprint, RMIC received July 1968

Temperature Dependence of the Internal Friction of Mo Wires
A. A. Belyakov, V. P. Elyutin, and E. I. Mozzhukhin
Fiz. Metal. Metalloved., 23:306-311 (1967)

Temperature Dependence of the Elastic Constants of Molybdenum
J. M. Dickinson and P. E. Armstrong
J. Appl. Phys., 38:602 (1967)

Single-Crystal Elastic Properties of Tungsten from 24° to 1800°C
Robert Lawrie and A. M. Gonas
J. Appl. Phys., 38:4505 (1967)

Elastic Moduli and Ultrasound Velocites of Tungsten as a Function of Temperature
Ronald G. Peterson
(Argonne National Laboratory, Argonne, Ill.), Contract W-31-109-eng-38 (Dec. 1967) ANL-7272, 10 pp.

Internal Friction of Tungsten Single Crystals
R. H. Schnitzel
Trans. AIME, 233:186 (1965)

Transverse Magnetoresistance of High Purity Chromium Foils
T. J. Bastow and R. Street
Phil. Mag., 10:269 (1964)

Rolled Molybdenum Single Crystals: Deformation Textures, Recrystallation, and Transition Temperature
R. Blickensderfer, R. Siemens, G. Asai, and H. Kato
Bureau of Mines report of investigations 6539 (1964)

Investigation of Room-Temperature Slip in Zone-Melted Tungsten Single Crystals
R. G. Garlick and H. B. Probst
Lewis Research Center, National Aeronautics and Space Administration, Cleveland, Ohio, Trans. Met. Soc. AIME, 230:1120 (1964)

Ultrasonic Attenuation in Tungsten and Molybdenum up to 1 Gc/sec
C. K. Jones and J. A. Rayne
(Westinghouse Research Lab., Pittsburgh, Pa.), Scientific Paper 64-IJO-107 (November 2, 1964)

Étude de la restauration d'un pic de frottement interne du tungstène écroui. Détermination de la migration des défauts ponctuels aux stades III et IV, par des mesures en pendule de torsion
Bernard Martinet
Helv. Phys., Acta, 37:673 (1964)

Frottement interne du tungstène polycristallin ou monocristallin soumis à un mode de vibration de flexion
Bernard Secretan
Helv. Phys. Acta, 37:699 (1964)

Tensile Properties of Pyrolytic Tungsten from 1370 to 2980°C in Vacuum
J. L. Taylor and D. H. Boone
J. Less-Common. Metals, 6:157-164 (1964)

Anomalies in the Elastic Constants and Thermal Expansion of Chromium Single Crystals
D. I. Bolef and J. de Klerk
Phys. Rev., 129:1063-1067 (1963)

Obtaining Molybdenum Single Crystals and Their Properties
V. S. Emel'yanov, A. I. Evstyukhin, G. A. Leont'ev, and A. N. Semenikhin
JPRS-22697, pp. 39-45
Met. i Metalloved. Chistykh Metal., Sb. Nauchn. Rabot, No. 4 (May 31, 1963)
OTS-64-21263, pp. 39-45

Elastic Constants of Tantalum, Tungsten, and Molybdenum
F. H. Featherston and J. R. Neighbours
Phys. Rev., 130:1324-1333 (1963)

Work-Hardening Mechanisms in Body-Centered Cubic Metals
Donald P. Gregory
(Pratt and Whitney Aircraft Div., United Aircraft Corp., Connecticut Aircraft Nuclear Engine Lab., Middletown), Contract AF33(616)-7855, AD-411124, ASD-TDR-62-354, Pt. II (Mar. 1963), 48 pp.

Effect of Purity on the Tensile Properties of Tungsten Single Crystals from −196°C to 29°C
R. C. Koo
Acta Met., 11:1083 (1963)

The Influence of Lattice Defects on the Electrical Resistance and the Hall-Effect in Tungsten and Tantalum
E. Krautz and H. Schultz
Z. Angew. Phys., 15:1-4 (1963)

Elastic Constants of Single-Crystal Mo and W Between 77° and 500°K
D. I. Bolef and J. de Klerk
J. Appl. Phys., 33:2311-2314 (1962)

Dislocation Relaxation Spectra in Plastically Deformed Refractory BCC Metals
R. H. Chambers and J. Schultz
Acta Met., 10:466-483 (1962)

Calculated Lattice Specific Heats for Seven BCC Elements (Mo, V)
C. B. Clark
Phys. Rev., 125:1898-1902 (1962)

Transitions in Chromium
Jean-Francois Marin
U. S. Atomic Energy Commission Contract No. AT(04-3)221 and Office of Naval Research, Contract No. Nonr-220(30), AD-14740 (Jan. 1962)

Internal Friction and Young's Modulus in Irradiated Tungsten
D. R. Muss and J. R. Townsend
J. Appl. Phys., 33:1804-1807 (1962)

The Recrystallization and Ductile-Brittle Transition Behavior of Tungsten. Effect of Impurities on Polycrystals Prepared from Single Crystals
B. C. Allen, D. J. Maykuth, and R. I. Jaffee
J. Inst. Metals, 90:120-128 (1961)

Mechanical and Physical Properties of the Refractory Metals, Tungsten, Tantalum, and Molybdenum Above 4000°F
L. F. Glasier, Jr., R. D. Allen, and I. L. Saldinger
(Aerojet-General Corp., Azusa, Cal.), Rept. M-1826 (April 1959), 89 pp.

The Purification of Tungsten by Electron-Bombardment Floating-Zone Melting
Walter R. Witzke
(Lewis Research Center, National Aeronautics and Space Administration, Cleveland, Ohio), Transactions of the Vacuum Metallurgy Conf. (1959), p. 140

Influence of Recrystallation on Damping and Shear Modulus of Tungsten Wires
W. Schilling and R. Haspel
AERE-Trans-11/3/5/1150 (translated by R. C. Murray)
Z. Metallk., 48:32-34 (1957), 8 pp.

1.i.4. Melting Points, Phase Transformations, Specific Heat, etc.

1.i.4.a. General

Physical Metallurgy of Refractory Metals and Alloys
E. M. Savitskii and G. S. Burkhanov
Consultants Bureau, New York (1970), 287 pp.

Melting Points and Phase Transformations of the Elements, Draft Bibliography
Research Materials Information Center
(Solid State Division, Oak Ridge National Lab., Oak Ridge, Tenn.), Feb. 1969

Specific Heat, Metallic Elements and Alloys, TPRC Data Series on Thermophysical Properties of Matter, Vol. 4
Y. S. Touloukian and C. Y. Ho
Plenum Press, New York (1969), 830 pp.

Measuring Specific Heat of Nb and W
C. Affortit and R. Lallement
Rev. Hautes Temp. Refract., 5:19-26 (1968)

Thermodynamic Properties of Interstitial Elements in the Refractory Metals Semiannual Report Dec. 1, 1967-May 31, 1968
P. P. Bansal, N. C. Birla, A. C. Huang, and L. Seigle
(State Univ. of New York, Stony Brook, College of Engineering), Contract NGL-33-015-035, NASA-CR-97615, SAR-4 (May 1968), 17 pp.

The Chemistry of Titanium and Vanadium, An Introduction to the Chemistry of the Early Transition Elements
R. J. H. Clark
Elsevier, New York (1968), 327 pp.

Technology and Uses of Beryllium, Cesium, Germanium, Hafnium, Niobium (Columbium), Rare Earths (Including Yttrium and Scandium), Tantalum, Titanium, and Zirconium
K. A. Gschneidner, Jr.
(Ames. Lab., Ames, Iowa), Contract W-7405-eng-82, IS-1757 (Jan. 1968), 158 pp.

Low Temperature Melting of Elements under High Pressure and Its Progression in the Periodic Table
Naoto Kawai and Yukio Inokuti
Japan J. Appl. Phys., 7(9):989-1004 (1968)

Physical Properties of Very Pure Refractory Metals
J. P. Langeron
Annales Des Mines (Fr.) (April, 1968), pp. 76-90

Electronic Heat Capacity of Transition Metals
V. S. Neshpor and G. V. Samsonov
Fiz. Metallov Metalloved., 25(6):1132-1134 (1968)

A Bibliography of Refractory Metals, AGARD Biblio. No. 5
Robert Syre, compiler
K. J. Spencer, ed. (1968)

Unusual Oxidation States of Transitional Elements
George W. Watt
(Univ. of Texas, Austin, Dept. of Chemistry), Contract AT-(40-1)-1639, ORO-1639-17 (June 1, 1968), 15 pp.

Thermodynamic Properties of Interstitial Elements in the Refractory Metals
P. P. Bansa and others
NASA CR 88526 (1967)

High-Temperature Materials
Eduard Nikitovich, Osip Solomonovish Gurvich, and Lyudmila Fedorovna
Izdatel'stvo Metallurgiya, Moscow (1967), 214 pp.
Properties of W, Mo, Re, Ta, Nb (in Russian)

Low Temperature Specific Heat of Transition Metals and Alloys
F. Heiniger, E. Bucher, and J. Muller
Phys. Kondens. Materie, 5:243-284 (1966)

Behavior of the Physical Properties of Transition Metals
G. V. Samsonov and V. N. Paderno
Izv. Vyssh. Ucheb. Zaved., Fiz. SSSR, 2:64-67 (1966)

i. Physical Properties

The Energy and Entropy of Vacancy Formation in Body-Centered Cubic Refractory Metals
M. Hoch
GE-TM-65-12-2; TID-22851 (November 1965), 8 pp.

Die Atomwärme von Titan, Vanadin und Chrom im Bereich hoher Temperaturen
Rudolf Kohlhaas, Martin Braun, and Otmar Vollmer
Z. Naturforsch., 20A:1077-1079 (1965)

Handbook on the Properties of Niobium, Molybdenum, Tantalum, Tungsten, and Some of Their Alloys
R. Syre
Advisory Group for Aeronautical Res. and Develop., Paris, France, AGARDograph-94 (May 1965), 302 pp.

Lattice Dynamics of Transition Metals
A. D. B. Woods
Inelastic Scattering of Neutrons, Vol. 1, Intern. Atomic Energy Agency, Vienna (1965), pp. 87-94

Self-Diffusion and the Melting Parameters of Cubic Metals
G. B. Gibbs
(Central Electricity Generating Board, Berkeley Nuclear Laboratories, Berkeley, Glos., England), Acta Met., 12:673-675 (1964)

Specific Heat of Type II Superconductors in a Magnetic Field
T. McConville and B. Serin
Phys. Rev. Letters, 13:365 (1964)

A Critical Review of Refractories
Edmund K. Storms
(Los Alamos Scientific Lab., New Mexico), Contract W-7405-eng-36, LA-2942 (March 1964), 252 pp.

Large Crystals Grown by Recrystallization
K. T. Aust
The Art and Science of Growing Crystals, John Wiley and Sons, Inc., New York (1963), pp. 452-478

Work-Function Measurements and Lattice Defects
J. A. Dillon, Jr., and R. M. Oman
(Department of Physics, Brown University, Providence, Rhode Island), reprinted from 1963 Transactions, The Tenth National Vacuum Symposium, American Vacuum Society

A New Electrical Resistivity Technique for the Study of Phase Transformations
P. H. Dixon and F. H. Fern
(Metallurgy Division, U.K.A.E.A. Research Group, Atomic Energy Research Establishment, Harwell, Berks.), AERE-R 4419 (August 1963)

The Specific Heat of Metals Between 1200° and 2400°K
G. C. Lowenthal
Australian J. Phys., 16:47-67 (1963)

Current Information on the Refractory Metals
C. E. King
Contract AF33(657)-11214, AD-436333; ERR-FW-049A (Dec. 1, 1962), 80 pp.

Ultra-High-Purity Metals
Papers presented at a seminar of American Society of Metals (October 21 and 22, 1961), American Society of Metals, Metals Park, Ohio (1962)

Selected Applications of High Temperature X-Ray Studies in the Metallurgical Field
H. J. Goldschmidt
(The B.S.A. Group Research Centre, Birmingham, England), Advances in X-Ray Analysis, Vol. 5, Plenum Press, New York (1962) (Review), pp. 191-212

Vacuum-Arc Evaporation of Refractory Metals
M. S. P. Lucas, H. A. Owen, Jr., W. C. Stewart, and C. R. Vail
(Superconducting Circuits Laboratory, Department of Electrical Engineering, Duke University, Durham, North Carolina), Rev. Sci. Instr., 32:203-204 (1961)

Thermodynamic Properties of 65 Elements — Their Oxides, Halides, Carbides, and Nitrides
Charles E. Wicks and F. E. Block
(Bureau of Mines, Albany Metallurgy Research Center, Ore.), NP-13622, Bureau of Mines Bulletin, 605 (Dec. 1961)

Differences in Lattice Specific Heats in the Normal and Superconducting Phases
H. A. Boorse, A. T. Hirshfeld, and H. Leupold
Phys. Rev. Letters, 5:246-248 (L) (1960)

Beobachtung und Registrierung veränderlicher Vorgänge im Emissions-Elektronenmikroskop
H. Duker
(Max-Planck-Institut für Metallforschung, Abt. Sondermetalle, Stuttgart, und Institut für experimentelle und angewandte Physik der Universität Tübingen), Radex Rundschau, No. 6 (1960)

Beobachtung von Umwandlungs- und Oxydations-Vorgängen im Elektronen-Emissions-Mikroskop
H. Duker
Proc. Eur. Reg. Conf. on Electron Microscopy, Delft, Vol. 1:456 (1960)

A Compendium of the Properties of Materials at Low Temperature (Phase I)
Victor J. Johnson
WADD Tech. Rept. 60-56, Part II, Materials Central Contract AF33(616)-58-4, Project 7360 (1960)

Debye Characteristic Temperatures — Table and Bibliography
M. W. Holm
(Phillips Petroleum Co., Atomic Energy Division, Idaho Operations Office, U. S. Atomic Energy Commission), Contract No. AT(10-1)-205, IDO-16399 (August 1957)

A Compilation of Some of the Physical Properties of the Metallic and Semi-Metallic Elements and a Study of Some of Their Interrelationships
Karl A. Gschneider, Jr.
(Department of Physics and Materials Research Laboratory, University of Illinois, Urbana, Illinois, and University of California, Los Alamos Scientific Laboratory, Los Alamos, New Mexico), LADC-6220
To be published in Solid State Phys., Vol. 16

1.i.4.b. Group IV

Thermal Conductivity of Metals at High Temperature by the Jain and Krishnan Method: V. Zirconium
S. C. Jain, V. Sinha, and B. K. Reddy
J. Phys. D: Appl. Phys., 3:1359-1362 (1970)

Melting Point Measurements in the Tri-Arc
Furnace
   T. B. Reed
   in Lincoln Lab., Mass. Inst. of Tech. Solid State Research
      Report (1970:3); ESD-TR-70-234 (August 1970), pp. 15-17
   Zr, Ti

Materials (Including Zirconium — Crystal Lattice
Dynamics)
   Nuclear Technology Branch Yearly Progress Report for
      Period Ending June 30, 1968 (R. L. Heath, ed.), Idaho
      Nuclear Corp., Idaho Falls, Contract AT(10-1)-1230,
      IN-1218 (Dec. 1968), pp. 141-179

The Less Common Refractory Metals (Rhenium,
Hafnium, Technetium, Noble Metals)
   J. Maltz
   National Aeronautcis and Space Administration, Lewis Res.
      Center, Cleveland, Ohio, NASA-TM-X-61073 (1967), 34 pp
   Conf. presented at AIME Symp. Met. and Technol. of Re-
      fractory Metal Alloys, Washington, D. C. (April 25-26, 1968)

The Solid-Phase Transformation in Hafnium
   Glen N. Bates and George Barnes
   Appl. Phys. Letters, 11:75 (1967)

Evaluation of Specific Heats of Titanium, Zir-
conium, and Hafnium
   R. P. Gupta and B. Dayal
   Phys. Stat. Sol., 13:257 (1966)

Rapid Phase Transformations in Titanium In-
duced by Pulse Heating
   Robert Parker
   Trans. AIME, 233:1545 (1965)

The Transformation Temperature of Hafnium
   P. A. Romans, O. G. Passche, and H. Kato
   J. Less-Common Metals, 8:213 (1965)

Titanium and Its Alloys
   JPRS-31091; TT-65-31589
   Metallovednie i Termicheskaya Obrabotka Metallov (USSR),
      5:15-21; 33-35; 45-50; 53-63 (1965)

Preparation and Properties of Thin-Film Hard
Superconductors
   J. Edgecumbe, L. G. Rosner, and D. E. Anderson
   (Department of Electrical Engineering, University of Minnesota,
      Minneapolis, Minnesota), J. Appl. Phys., 35:2198 (1964)

Single-Crystal Elastic Moduli and the HCP →
BCC Transformation in Ti, Zr, and Hf
   E. S. Fisher and C. J. Renken
   (Argonne National Laboratory, Argonne, Illinois), Phys. Rev.,
      135:A482 (1964)

Growth of Whiskers Due to Solid-to-Solid Phase
Transformation in Zirconium
   M. M. Nieto and A. M. Russell
   J. Appl. Phys., 35:461 (1964)

The $\beta \to \alpha$ Transformation in Titanium
   M. J. Bibby and J. Gordon Parr
   (Department of Mining and Metallurgy, University of Alberta,
      Edmonton, Alberta, Canada), J. Inst. Metals, 92:341
      (1963-1964)

The Observation of Dislocations Observed in
Zirconium during $\alpha \to \beta$ Phase Transformation
   Jacqueline Devaud, Jean Pollard, and Pierre Lehr
   (C. E. C. M. Laboratoire de Vitry du C. N. R. S.), Compt.
      Rend., 256:4426 (1963)

The Physical Metallurgy of Zirconium
   D. L. Douglass
   (General Electric Co., Pleasanton, California), At. Energy
      Rev., 1:71-237 (1963)

Relationship of Elastic Shear Moduli to the
Phase Transformations in Zirconium, Titanium,
and Hafnium
   E. S. Fisher and C. J. Renken
   (Argonne National Laboratory, Argonne, Illinois), Bull. Am.
      Phys. Soc. Ser. II, 8:65 (A) (1963)

Solid-State Transitions in Titanium and Zir-
conium at High Pressures
   A. Jayaraman, K. Klement, Jr., and G. C. Kennedy
   (Institute of Geophysics and Planetary Physics, University of
      California, Los Angeles, California), Phys. Rev., 131:644
      (1963)

Low Temperature Specific Heats of Titanium,
Zirconium, and Hafnium
   G. D. Kneip, Jr., J. O. Betterton, Jr., and J. O. Scarbrough
   Phys. Rev., 130:1687-1692 (1963)

Magnetic Susceptibility and Electronic Specific
Heat of Transition Metals and Alloys. IV, V,
and Ti Metals and V-Cr and V-Ti Alloys
   M. Shimizu, A. Katsuki, and T. Takahashi
   J. Phys. Soc. (Japan), 18:1192-1203 (1963)

Metallurgy and Metallography of Pure Metals
   V. S. Emel'yanov and A. I. Evstyukhin
   JPRS-9473, Met. i Metalloved. Chistykh Metal. (1960), 395 pp.

Effect of Some Rare Earth Elements on the
HCP — BCC Transformation of Zirconium
   L. Ianniello and A. Burr
   J. Appl. Phys., 33:2689-2690 (1962)

Exposure Diagrams for X-Ray Film Work on
Titanium and Zirconium
   K. Sagel, AEC-tr-3128, translated by K. S. Bevis from Metall.
   AEC-tr-3128, translated from K. S. Bevis from Metall., 11,
      769 (1957), 2 pp.

Specific Heat of Zirconium by a Pulse Heating
Method
   A. H. Klein and G. C. Danielson
   (Ames Lab., Ames, Iowa), NP-14281, pp. 47-54

1.i.4.c. Group V

The X-Ray Debye Temperature of V, Ni, Cu, Nb
and Ag from the Measured Integrated Intensities
at 300 and 4 K
   M. V. Linkoaho
   (Lab. of Physics, Technical University of Helsinki, Otaniemi,
      Finland), Research Report 7/1970

Low Temperature Specific Heat of Annealed
High-Purity Niobium in Magnetic Fields
   J. Ferreira da Silva, E. A. Burgemeister, and Z. Dokoupil
   Physica, 41:409-439 (1969)

Phase Transition in Vanadium
   V. A. Finkel, V. I. Galamazda, and G. P. Koutun
   Zh. Éksp. Teor. Fiz., 57(4):1065-1068 (1969)
   Sov. Phys. — JETP, 30(4):581-583 (1970)

Solubility of Nitrogen in Vanadium
Frank M. Monroe and James R. Cost
Trans. Met. Soc. AIME, 245:1079-1082 (1969)

X-Ray Debye Temperatures for Al, Nb, and Pb
Siv Grimvall and G. Grimvall
Acta Cryst., 24A, Pt. 6:612-613 (1968)

Niobium Physico-Chemical Properties of Its
Compounds and Alloys
O. Kubaschewski, ed.
Atomic Energy Review, Special Issue No. 2, International
Atomic Energy Agency, Vienna (1968)

Niobium: Thermochemical Properties
V. I. Lavrentev and Ya. I. Gerassimov
At. Energy Rev., Special Issue No. 2, pp. 7-44 (July 1968)

Determination of the Debye Characteristic
Temperature of Vanadium from the Bloch-Grü-
eisen Relation
D. G. Westlake and L. C. R. Alfred
J. Phys. Chem. Solids, 29:1931-1934 (1968)

Low Temperature Specific Heats of Annealed
High Purity Niobium in Magnetic Fields
J. Ferreira da Silva, E. A. Burgemeister, and Z. Dokoupil
Phys. Letters, 25A:354 (1967)

Low Temperature Specific Heats of Unannealed
Niobium Wires in Magnetic Fields
J. Ferreira de Silva, N. W. J. van Duykeren, and
Z. Dokoupil
Physica, 32:1253 (1966)

Density of Niobium
John T. Harding
J. Appl. Phys., 37:928 (1966)

Specific Heat of Tantalum in the Temperature
Range 1200-2900°K
Ya. A. Kraftmakher
J. Appl. Mech. Tech. Phys., 28:276-281 (1966)

Low-Temperature Thermodynamic Properties of
Vanadium. II. Mixed State
Ray Radebough and P. H. Keesom
Phys. Rev., 149:217 (1966)

Low-Temperature Heat Capacity of Niobium
N. M. Senozan
(Univ. California, Berkeley), Dissertation Abst. 26, No. 7,
p. 3623 (1966)

Thermodynamic Properties of Niobium from
0°K
V. A. Kirillin, A. E. Sheindlin, V. Ya. Chekovskoi, and I. A.
Zhukova
Advances in Thermophysical Properties at Extreme Tem-
peratures and Pressures, The American Society of
Mechanical Engineers, New York (1965), pp. 152-155

The Thermodynamic Properties of Niobium in the
Temperature Range from 0°K to the Melting
Point 2740°K
V. A. Kirillin, A. E. Sheindlin, V. Ya. Chekhovskoi, and
I. A. Zhukova
Templofiz. Vyssokikh Temp. (USSR), 3(6):860-865 (1965)

I. Low Temperature Heat Capacities of Vanadium, Niobium, and Tantalum. II. Low Temperature Heat Capacity of Ferromagnetic Chromic
Tribromide

Shen Yun Lung
(Lawrence Radiation Laboratory, Univ. of California, Berkeley, Cal.), UCRL-161117 (1965), 91 pp.

Preparation and Properties of Thin-Film Hard
Superconductors
J. Edgecumbe, L. G. Rosner, and D. E. Anderson
(Department of Electrical Engineering, University of Min-
nesota, Minneapolis, Minnesota), J. Appl. Phys., 35:2198
(1964)

On the Specific Heat of Niobium in the Mixed
State at Temperatures Far Below the Transi-
tion Temperature
J. Ferreira da Silva, J. Scheffer, N. W. J. van Duykeren,
and Z. Dokoupil
Phys. Letters, 12:166 (1964)

Mixed-State Specific Heat of Vanadium at Very
Low Temperatures
P. H. Keesom and Ray Radebaugh
Phys. Rev. Letters, 13:685 (1964)

Superconducting and Normal Specific Heats of
a Single Crystal of Niobium
H. A. Leupold and H. A. Boorse
Phys. Rev., 134:A1322 (1964)

Thermal and Magnetic Properties of Second Kind
Superconductors. III. Specific Heat of Nb in a
Magnetic Field
T. McConville and B. Serin
Rev. Mod. Phys., 36:112 (1964)

Determination and Application of Thermophysi-
cal Properties of Refractory Metals
G. D. Rieck
The Science and Technology of Tungsten, Tantalum, Molyb-
denum, Niobium, and Their Alloys, Pergamon Press,
Oxford (1964), pp. 205-217

The Specific Heats and Resistivities of Molyb-
denum, Tantalum, and Rhenium
R. E. Taylor and R. A. Finch
J. Less-Common Metals, 6:283-294 (1964)

Phase Transformations in Niobium Involving
Interstitials
L. I. van Torne and G. Thomas
(Inorganic Materials Research Division, Lawrence Radiation
Laboratory and Department of Mineral Technology,
University of California, Berkeley, California), Acta Met.,
12:601 (1964)

Calculated Vibrational Spectra and Specific Heats
of Lithium and Vanadium
J. B. Henricks and H. N. Riser
Phys. Rev., 130:1377-1380 (1963)

Relation Between Specific Heat and Emissivity
of Tantalum at Elevated Temperatures
Michael Hock and H. V. L. Narasimhamurty
(Univ. of Cincinnati, Cincinnati, Ohio), ASD-TDR-63-371
(1963), 31 pp.
AD-414194

The Engineering Properties of Tantalum and
Tantalum Alloys
F. F. Schmidt and H. R. Ogden
(Battelle Memorial Institute, Defense Metals Information
Center, Columbus, Ohio), DMIC-Rept. 189 (September 13,
1963)
Contract AF33(616)-7747, 112 pp.

Magnetic Susceptibility and Electronic Specific
Heat of Transition Metals and Alloys. IV, V,
and Ti Metals and V-Cr and V-Ti Alloys
   M. Shimizu, A. Katsuki, and T. Takahashi
   J. Phys. Soc. (Japan), 18:1192-1203 (1963)

NMR, Magnetic Susceptibility and Electronic
Specific Heat of Nb and Mo Metals and Nb-Tc
and Nb-Mo Alloys
   D. O. Van Ostenburg, D. J. Lam, M. Shimizu, and A. Katsuki
   J. Phys. Soc. (Japan), 18:1744-1754 (1963)

Calculated Lattice Specific Heats for Seven BCC
Elements (Mo, V)
   C. B. Clark
   Phys. Rev., 125:1898-1902 (1962)

Specific Heat of Tantalum and Tin near the
Superconducting Critical Temperature
   John F. Cochran
   Ann. Phys. (New York), 19:186-218 (1962)

Heats of Formation of Metal Halides: Penta-
bromides of Niobium and Tantalum
   P. Gross, C. Hayman, D. L. Levi, and G. L. Wilson
   (Fulmer Research Institute Ltd., Stoke Poges, Bucks.),
      Faraday Soc. Trans., 58:890 (1962)

Superconducting and Normal Specific Heats of
Niobium
   A. T. Hirshfeld, H. A. Leupold, and H. A. Boorse
   Phys. Rev., 127:1501-1507 (1962)

Magnetic and Thermal Behavior in the Super-
conductor State of a Single Crystal of Niobium
   Gerard Kuhn
   (Centre de Recherches sur les Tres Basses Températures,
      C.N.R.S., Université de Grenoble), Compt. Rend., 255:
      2923-2925 (1962)

Thermodynamic Properties of Seven Metals at
Zero Pressure (Nb, Ta)
   J. W. Carter
   (Los Alamos Scientific Laboratory, New Mexico), Contract
      7405-eng-36, LAMS-2640 (September 1961), 62 pp.

Development of Ultrapure Refractory Materials
   Peter T. B. Shaffer
   Contract NOrd-17175, Progress rept. No. 24, Feb. 1, 1961-
      April 30, 1961 (Apr. 28, 1961), 12 pp.

The Specific Heats and Resistivities of Molyb-
denum, Tantalum, and Rhenium from Low to
Very High Temperatures
   R. E. Taylor and R. A. Finch
   (Atomic International, Div. of North American Aviation, Inc.,
      Canoga Park, Calif.), Contract AT(11-1)-GEN-8, NAA-SR-
      6034 (Sept. 15, 1961), 37 pp.

Mechanical, Oxidation, and Thermal Property
Data for Seven Refractory Metals and Their
Alloys, Final Report
   T. E. Tietz and J. W. Wilson
   (Lockheed Aircraft Corp., Missiles and Space Div., Sunny-
      vale, Calif.), Contract NOas-60-6119-c, NP-11108 (Sept.
      15, 1961), 315 pp.

The Physical Properties of Niobium, Tantalum,
Molybdenum, and Tungsten
   B. B. Argent and G. J. C. Milne
   J. Less-Common Metals, 2:154-162 (1960)

Metallochemical Properties of Niobium
   I. I. Kornilov
   NP-tr-704, Dokl. Akad. Nauk SSSR, 135:1399-1401 (1960), 6 pp.

Thermal Properties of Refractory Materials
   G. W. Lehman
   (Atomics International Div. of North American Aviation, Inc.,
      Canoga Park, Calif.), Contract AF33(616)-6794, WADD-
      TR-60-581 (July 1960), 19 pp.

Thermal Properties of Graphite, Molybdenum,
and Tantalum to Their Destruction Temperatures
   N. S. Rasor and J. D. McClelland
   J. Phys. Chem. Solids, 15:17-26 (1960)

Vapor Pressure of Niobium
   R. Speiser, P. Blackburn, and H. L. Johnston
   J. Electrochem. Soc., 106:52-53 (1959)

Measurements of Thermal Properties (Nb), July
1, 1957 to March 31, 1958
   I. B. Fieldhouse, J. C. Hedge, and J. I. Lang
   (Illinois Institute of Technology, Chicago, Armour Research
      Foundation), WADC-TR-58-274, May 1, 1958, Contract
      AF33(616)-3701, 87 pp.; AD-206 892

Thermodynamic Properties of Molybdenum and
Columbium: An Annotated Bibliography
   Compiled by Jack B. Goldman
   (Lockheed Missiles and Space Company, a group division of
      Lockheed Aircraft Corporation, Sunnyvale, California),
      6-90-62-122, Special Bibliography SB-62-69

1.i.4.d. Group VI

Lattice Vibrational and Thermal Properties of
Chromium
   J. L. Feldman
   Bull. Am. Phys. Soc., 15:277 (1970)

Thermophysical Properties of Tungsten
   B. N. Ivanov
   Teplofiz. Vysok. Temp., 7(5)898-905 (1969)
   High Temperature, 7(5):834-840 (1969)

Phonon Dispersion in Transition Metals
   P. K. Sharma, Satya Pal, and R. P. Gupta
   Rev. Roumaine Phys., 14(3):247-260 (1969)
   Mo, W, and Fe

Selected Values of Chemical Thermodynamic
Properties — Tables for Elements 35 Through 53
in the Standard Order of Arrangement
   D. D. Wagman, W. H. Evans, V. B. Parker, I. Halow, S. M.
      Bailey, and R. H. Schumm
   NBS Tech. Note 270-4 (May 1969)
   Cr, Mo, W

Experimental Studies of Mo Enthalpy at Maximum
Temperatures
   V. Ya. Chekhovskoi and V. A. Petrov
   Teplofiz. Vys. Temp., 6(4):752-753 (July-Aug. 1968)
   High Temperature, 6(4):717-718 (1968)

Enthalpy of W
   R. A. Hein and P. N. Flagella
   GEMP, 578 (1968)

Concerning the Combined Solid-State Solubility of Tungsten and Boron in Molybdenum
V. I. Kharitonov, M. S. Makunin, and F. I. Shamrai
Tsvetn. Metal., No. 9, pp. 84-86 (1968)

Enthalpy and Heat Capacity of Molybdenum at Extremely High Temperatures
V. A. Kirillin, A. E. Sheindlin, V. Ya. Chekhovskoi, and V. A. Petrov
(Inst. High Temperatures Academy of Sciences USSR, Moscow), Proceedings of the Fourth Symposium on Thermophysical Properties held at College Park, Maryland, April 1-4, 1968, ASME, New York (1968), pp. 54-57

Physical Characterization of Molybdenum Single Crystals for Irradiation Experiments
H. E. Kissinger, J. L. Brimhall, and B. Mastel
Mat. Res. Bull., 2:437-448 (1967)

Tungsten and Its Compounds
G. D. Rieck
Pergamon Press, Oxford, England (1967)

Magnetic Structures of Field-Cooled and Stress-Cooled Chromium
T. J. Bastow and R. Street
Phys. Rev., 141:510 (1966)

Electrochemical Properties of Mo and W
E. V. Speranskaya, V. É. Mertsalova, and I. I. Kulev
Usp. Khim., 35(12):2129-2150 (1966)

Some Properties of Floating Zone Melted Mo, W, and Ta
T. Takaai
Nippon Kinzoku Gakkaishi, 20(11):1022-1026 (1966)

Measurements of the Thermal Variation of the X-Ray Debye Temperature of Pure Nickel and Chromium
Ronald H. Wilson, Earl F. Skelton, and J. Lawrence Katz
Acta Cryst., 21:635 (1966)

Volume Measurements on Chromium to Pressure of 30 Kilobars
W. E. Evenson and H. T. Hall
(Brigham Young Univ., Provo, Utah), AROD 2723:7, AD 626 427, Contract DA-ARO(D)-31-124-G633 (Sept. 7, 1965)

Vapor Pressure and Heat of Sublimation of Tungsten
R. Szwarc, E. R. Plante, and J. J. Diamond
J. Res. Natl. Bur. Stds-69A, 17 (1965)

Molybdenum and Tungsten — Selective Bibliography of Government Research Reports and Translations
Clearinghouse for Federal Scientific and Technical Information, Springfield, Va., SB-415, Suppl. 1 (1965), 48 pp.

Properties of Some Refractory Metals. I. Enthalpy and Heat Capacity
J. B. Conway
(Nuclear Materials and Propulsion Operation, General Electric Company, Cincinnati, Ohio), Contract AT(40-1)-2847, GE-TM-64-2-12; TID-24314 (Pt. 1) (Feb. 25, 1964), 30 pp.

Effect of the Degree of Purity of Some Refractory Metals on Their Melting Points
V. M. Pan
Sb. Nauch. Tr. Inst. Metallofiz., Akad. Nauk Ukr. SSR, 20:130-132 (1964)

Determination and Application of Thermophysical Properties of Refractory Metals
G. D. Rieck
(Technical Univ., Eindhoven, Netherlands), AGARD Conf. on Refractory Metals, Oslo, Norway, June 1963, Pergamon Press, New York (1964), p. 205

The Specific Heats and Resistivities of Molybdenum, Tantalum, and Rhenium
R. E. Taylor and R. A. Finch
J. Less-Common Metals, 6:283-294 (1964)

Vapor Pressure of Molybdenum
P. A. Vozzella, A. D. Miller, and M. A. DeCrescente
(Pratt and Whitney Aircraft, Middletown, Conn.), J. Chem. Phys., 41:589-590 (1964)

NMR, Magnetic Susceptibility and Electronic Specific Heat of Nb and Mo Metals and Nb-Tc and Nb-Mo Alloys
D. O. Van Ostenburg, D. J. Lam, M. Shimizu, and A. Katsuki
J. Phys. Soc. (Japan), 18:1744-1754 (1963)

Heat Capacity and Enthalpy of Molybdenum in the Temperature Interval 0 to 2500°
V. Ya. Chekhovskoi
Inzh.-Fiz. Zh., 5(6):43-47 (1962)

Experimental Study of the Enthalpy of Tungsten within the Temperature Range of 350-2000°C
V. Ya. Chekhovskoi, B. Ya. Shumyatskii, and K. A. Yakimovich
ANL-Trans-451
Inzh.-Fiz. Zh., 5(10):13-18 (1962)

Calculated Lattice Specific Heats for Seven BCC Elements (Mo, V)
C. B. Clark
Phys. Rev., 125:1898-1902 (1962)

Yielding and Plastic Flow in Single Crystals of Tungsten
D. P. Ferriss, R. M. Rose, and J. Wulff
Trans. Met. Soc., AIME, 224:981 (1962)

Thermodynamic Properties of Tungsten in the Temperature Range 0-2400°C
V. A. Kirillin, A. E. Sheindlin, and V. Ya. Chekhovskoi
Teploenerg., 9(3):63-66 (1962)
ANL-Trans-452

Experimental Determination of the Enthalpy and Heat Capacity of Molybdenum Up To 2337°C
V. A. Kirillin, A. E. Sheindlin, and V. Ya. Chekhovskoi
Intern. J. Heat Mass Transfer, 5:1-9 (1962)

Transitions in Chromium
Jean-Francois Marin
U. S. Atomic Energy Commission Contract No. AT(04-3)221 and Office of Naval Research, Contract No. Nonr-220(30), AD-14740 (Jan. 1962)

The Recrystallization and Ductile-Brittle Transition Behavior of Tungsten. Effect of Impurities on Polycrystals Prepared from Single Crystals
B. C. Allen, D. J. Maykuth, and R. I. Jaffee
J. Inst. Metals, 90:120-128 (1961)

Low Temperature Specific Heat of Molybdenum
C. A. Bryant and P. H. Keeson
J. Chem. Phys., 35:1149-1150 (1961)

Experimental Determination of the Enthalpy of
Molybdenum in the 700-2337°C Range
  V. A. Kirillin, A. E. Sheindlin, and V. Ya. Chekhovskoi
  Dokl. Akad. Nauk SSSR, 139:645-647 (1961)

Development of Ultrapure Refractory Materials
  Peter T. B. Shaffer
  Contract NOrd-17175, Progress rept., No. 24, Feb. 1, 1961-
    April 30, 1961 (Apr. 28, 1961), 12 pp.

The Specific Heats and Resistivities of Molyb-
denum, Tantalum, and Rhenium from Low to
Very High Temperatures
  R. E. Taylor and R. A. Finch
  (Atomic International, Div. of North American Aviation, Inc.,
    Canoga Park, Calif.), Contract AT(11-1)-GEN-8, NAA-SR-
    6034 (Sept. 15, 1961), 37 pp.

Mechanical Oxidation, and Thermal Property
Data for Seven Refractory Metals and Their
Alloys, Final Report
  T. E. Tietz and J. W. Wilson
  (Lockheed Aircraft Corp. Missiles and Space Div., Sunny-
    vale, Calif.), Contract NOas-60-6119-c, NP-11108 (Sept.
    15, 1961), 315 pp.

The Physical Properties of Niobium, Tantalum,
Molybdenum, and Tungsten
  B. B. Argent and G. J. C. Milne
  J. Less-Common Metals, 2:154-162 (1960)

Thermal Properties of Refractory Materials
  G. W. Lehman
  (Atomics International Div. of North American Aviation, Inc.,
    Canoga Park, Calif.), Contract AF33(616)-6794, WADD-
    TR-60-581 (July 1960), 19 pp.

Thermal Properties of Graphite, Molybdenum,
and Tantalum to Their Destruction Temperatures
  N. S. Rasor and J. D. McClelland
  J. Phys. Chem. Solids, 15:17-26 (1960)

Measurements of the Thermal Properties of
Metals at Elevated Temperatures
  R. L. Rudkin, W. J. Parker, and R. W. Westover
  USNRDL-TR-419 (May 1960)

Thermal Properties of Molybdenum Between
2700°R and 4000°R
  R. L. Rudkin
  (U. S. Naval Radiological Defense Laboratory, San Francisco,
    Calif.), USNRDL-TR-433 (June 20, 1960), 23 pp.

Deformation Behavior of Zone-Melted Tungsten
Single Crystals
  H. W. Schadler
  Trans. Met. Soc. AIME, 218:649 (1960)

Thermodynamic Properties of Molybdenum and
Columbium: An Annotated Bibliography
  Compiled by Jack B. Goldman
  (Lockheed Missiles and Space Company, a group division of
    Lockheed Aircraft Corporation, Sunnyvale, California),
    6-90-62-122, Special Bibliography SB-62-69

Impurity Atom-Dislocation Interactions and Sub-
sequent Effects on Mechanical Properties of
Refractory Metals
  M. A. Adams and H. Nesor
  (Materials Research Corp., Orangeburg, New York), Contract
    No. AF33(616)-7596, ASD-TDR-62-11

1.i.5. Thermal Conductivity

1.i.5.a. General

Thermal Conductivity, Metallic Elements and
Alloys
  TPRC Data Series on Thermophysical Properties of Matter,
    Vol. 1, Plenum Press, New York (1970)

The Thermal Conductivity of the Transition
Metals
  A. F. Kiseleva, V. A. Klimenko, and Ya. S. Malakhov
  Izv. V. U. Z. Fizika, No. 5, pp. 41-47 (1969)

Thermal Conductivity of Selected Materials,
Part 2
  C. Y. Ho, R. W. Powell, and P. E. Liley
  (National Standard Reference Data System in Cooperation
    with Purdue Univ., Thermophysical Properties Research
    Center, West Lafayette, Ind.), NSRDS-NBS-16-Pt. 2 (Feb.
    1968)
  Ti, Zr, Ta, Nb, Cr, Mo

The State of Knowledge Regarding the Thermal
Conductivity of the Metallic Elements
  R. W. Powell and C. Y. Ho
  Thermal Conductivity — Proceedings of the Seventh Con-
    ference (D. R. Flynn and B. A. Peavy, eds.), NBS Spec.
    Publ. 302 (1968), pp. 1-32

Thermal Conductivity of Selected Materials
  R. W. Powell, C. Y. Ho, and P. E. Liley
  NSRDS-NBS 8 (Nov. 25, 1966)

Heat Conductivity of Pure Metals Below 1°K
  K. Mendelssohn, J. K. N. Sharma, and I. Yoshida
  Bull. Inst. Intern. Froid, Annexe 2, 49-56 (1965)
  Ta, Ti, W

Properties of Some Refractory Metals. II.
Thermal Conductivity
  J. B. Conway
  (General Electric Co., Nuclear Materials and Propulsion
    Operation, Cincinnati, Ohio), GE-TM-64-4-3; TID-24314
    Pt. 2 (1964), 26 pp.
  W, Re, Ta, Mo, Nb, Cr

Platinum as a Thermal Conductivity Standard
  Glen A. Slack
  J. Appl. Phys., 35:339 (1964)

Prediction of Thermal Conductivity of Metallic
Elements and Their Dilute Alloys at Cryogenic
Temperatures
  Ared Cezairliyan
  Contract No. AF33(616)-7617, ASD-TDR-63-291 (Sept. 1963)

Measurement of Thermal Conductivity of Elec-
trical Conductors at High Temperatures
  M. Cutler and G. T. Cheney
  J. Appl. Phys., 34:1714-1718 (1963)

Heat Conductivity of Pure Metals Below 1°K
  G. Davey and K. Mendelssohn
  Phys. Letters, 7:183 (1963)

Reduction of the Lattice Thermal Conductivity
of Superconductors Due to Point Defects
  P. G. Klemens and L. Tewordt
  CONF-301-1, Intern. Conf. on Science of Superconductivity,
    Hamilton, N. Y. (Aug. 1963), 11 pp.

The Unmatched Guard Method of Measuring Thermal Conductivity at High Temperatures
  M. J. Laubitz
    (Division of Applied Physics, National Research Council, Ottawa, Canada), Can. J. Phys., 41:1663 (1963)

Theory of the Thermal Conductivity of Metals, Alloys, and Semiconductors
  John R. Madigan
    Contract No. AF33(616)-7374, ASD-TDR-62-74, Pt. II (Jan. 1963)

The Study of Crystal Imperfections in Thermal Conductivity Measurements
  K. Mendelssohn
    Proceedings of the International Conference on Crystals of the International Conference on Crystal Lattice Defects, 1962, Conference
    J. Phys. Soc. Japan, 18, Suppl. II (1963)

The Effect of Heat Loss on the Flash Method of Determining Thermal Diffusivity
  A. R. Mendelssohn
    Appl. Phys. Letters, 2:19 (1963)

Foundations of the Callaway Theory of Thermal Conductivity
  R. E. Nettleton
    Phys. Rev., 132:2032-2038 (1963)

The Influence of Defects on the Lattice Thermal Conductivity at Low Temperatures
  H. Bross
    NP-tr-963
    (translated by L. F. Secretan) Phys. Stat. Sol., 2:481-516 (1962), 72 pp.

Transient Thermal Diffusivity of Refractory Solids
  J. A. Cape, G. W. Lehman, and M. M. Nakata
    Bull. Am. Phys. Soc. Ser. II, 7:8 (A) (1962)

Anomalous Resistive Transitions and New Phenomena in Hard Superconductors
  M. A. R. LeBlanc
    IBM J. Res. Develop., 6:122-125 (1962).

Thermal Diffusivity and Heat Capacity Measurements at Low Temperatures by the Flash Method
  O. Makarounis and R. J. Jenkins
    USNRDL-TR-599 (1962)

Ultra-High-Purity Metals
    Papers presented at a seminar of American Society of Metals (October 21 and 22, 1961), American Society of Metals, Metals Park, Ohio (1962)

Spectral Emissivity, Total Emissivity, and Thermal Conductivity of Molybdenum, Tantalum, and Tungsten above 2300°K
  Robert D. Allen, Louis F. Glasier, Jr., and Paul L. Jordan
    J. Appl. Phys., 31:1382 (1960)

Propriétés des métaux et alliages aux basses températures (Properties of Metals and Alloys at Low Temperatures)
  E. Justi
    CEA-tr-A-789
    Z. Metallk., 51(1):1-17 (1960), 62 pp.
    Includes original, 17 pp.

Determination of the Coefficient of Thermal Conductivity for Metals in the High Temperature Range
  V. V. Lebedev
    Phys. Metals Metallog., 10:31-34 (1960)

Thermal Properties of Refractory Materials
  G. W. Lehman
    (Atomics International Div. of North American Aviation, Inc., Canoga Park, Calif.), Contract AF33(616)-6794, WADD-TR-60-581 (July 1960), 19 pp.

Some Remarks on Lattice Thermal Conductivity
  D. K. C. MacDonald
    Am. J. Phys., 28:551-556 (1960)

Apparatus for Measuring Thermal Conductivity of Ceramic and Metallic Materials to 1200°C
  W. H. Sutton
    J. Am. Ceram. Soc., 43:81 (1960)

Electrical and Thermal Resistivity of Dislocations
  Peter Carruthers
    Phys. Rev. Letters, 2:336 (1959)

Electrical and Thermal Resistivity of the Transition Elements at Low Temperatures
  G. K. White and S. B. Woods
    Phil. Trans. Roy. Soc., 251:35 (1959)

Measurement of Metal Heat Conductivity in a Wide Range of Temperature during a Single Experiment
  O. A. Kraev
    SCL-T-220
    (translated by Marcel I. Weinrich) Teploenergetika, 4(12):69-72 (1957), 10 pp.

Apparatus for Measuring the Thermal Conductivity of Metals in Vacuum at High Temperatures
  Marvin Moss
    Rev. Sci. Instr., 26:276-280 (1955)

Physical Property Studies
  D. L. McElroy, T. G. Kollie, and W. Fulkerson
    (Oak Ridge National Lab., Tenn.), ORNL-3470, pp. 27-29

1.1.5.b. Group V

Anomalous Thermal Conductivity in Superconducting Niobium
  J. R. Carlson and C. B. Satterthwaite
    Phys. Rev. Letters, 24:461-464 (1970)

Damping of Dislocations in Niobium by Phonon Viscosity
  W. P. Mason and D. E. MacDonald
    (George Washington Univ., Washington, D. C.), AD-707 786 (June 1970), 30 pp.

Thermal Conductivity of Superconducting Vanadium in the Vortex State
  N. H. Zebouni, A. Waleh, and S. M. Wasim
    Bull. Am. Phys. Soc., 15:360 (1970)

Thermal Conductivity of Superconducting Niobium
  S. M. Wasim and N. H. Zebouni
    Phys. Rev., 187:539 (1969)

Thermal Diffusivity of Vanadium at High Temperatures
  V. E. Zinov'ev, R. P. Krentsis, and P. V. Gel'd
  Fiz. Tverd. Tela, 11:3045-3048 (1969)
  Sov. Phys. — Solid State, 11:2475 (1970)

Thermal Conductivity of Superconducting Niobium
  Jon Ross Carlson
  Ph. D. Thesis from Illinois University, Urbana, Illinois (1968), 84 pp.

Thermal Conductivity and Total Emittance of Ta, W, Re
  M. Hoch
  Vortrag: Deutsche keram. Ges., Baden-Baden (1968)

Anisotropy in the Mixed State Thermal Conductivity of Niobium
  J. Lowell and J. B. Sousa
  Phys. Letters, 26A:480-481 (Apr. 8, 1968)

Lattice Component of the Heat Conductivity of Technically Pure Niobium
  B. A. Merisov and V. I. Khotkevich
  Fiz. Metallov. Metalloved., 26:369-371 (1968)

Temperature Effect on Thermal Conductivity of Ta and Nb
  K. G. Akhmemzyanov, N. Z. Pozdnyak, and A. F. Dobrovolskii
  Teplofiz. Vysok. Temp., 5(1):179-181 (1967)
  High Temperature, 5(1):156-158 (1967)

Wärmeleitfähigkeit und kalorische Effekte in supraleitendem Nb
  R. Umlauf
  Z. Phys., 205:S, 415-430, 431-438 (1967)

Thermal Conductivity and Emissivity of Nb at Temperatures above 1000°C
  V. Yu. Voskresenskii, V. E. Peletskii, and D. L. Timrot
  Teplofiz. Vysok. Temp., 4(1):46-49 (1966)
  High Temperature, 4(1):39-42 (1966)

Investigation of Temperature Dependence of Thermal and Electric Conductivity of Tantalum and Niobium
  N. Z. Pozdnyak and K. G. Akhmetzyanov
  (All-Union Correspondence Polytechnic Inst., USSR)
  Vysokotemperaturnye Neorganicheskie Soedineniya (G. V. Samsonov, ed.), Izdatel'stvo Naukova Dumka, Kiev (1965), pp. 48-51

Thermal Diffusivity of Tantalum, Molybdenum, and Niobium at Temperatures above 1800K
  O. A. Kraev and A. A. Stelmakh
  Teplofiz. Vysokikh Temp., 2:302 (1964)

Determination and Application of Thermophysical Properties of Refractory Metals
  G. D. Rieck
  The Science and Technology of Tungsten, Tantalum, Molybdenum, Niobium, and Their Alloys, Pergamon Press, Oxford (1964), pp. 205-217

Thermal Conductivity of Niobium from 1.6°K to 15°K
  T. Amundsen and T. Olsen
  Physica Norvegica, 1:167 (1963)

Thermal Conductivity of Tantalum and Niobium below 1°K
  A. Connolly and K. Mendelssohn
  Proc. Roy. Soc. (London), 266:429-439 (Mar. 27, 1962)

Spectral Emissivity, Total Emissivity, and Thermal Conductivity of Molybdenum, Tantalum, and Tungsten above 2300°K
  R. D. Allen, L. F. Glasier, and P. L. Jordon
  J. Appl. Phys., 31:1382-1387 (1960)

Mechanical and Physical Properties of the Refractory Metals, Tungsten, Tantalum, and Molybdenum above 4000°F
  L. F. Glasier, Jr., R. D. Allen, and I. L. Saldinger
  (Aerojet-General Corp., Azusa, Cal.), Rept. M-1826 (April 1959), 89 pp.

Measurements of Thermal Properties (Nb), July 1, 1957 to March 31, 1958
  I. B. Fieldhouse, J. C. Hedge, and J. I. Lang
  (Illinois Institute of Technology, Chicago, Armour Research Foundation), WADC-TR-58-274, May 1, 1958, Contract AF33(616)-3701, 87 pp.; AD-206 892

Low Temperature Resistivity of Transition Elements: Vanadium, Niobium, and Hafnium
  G. K. White and S. B. Woods
  (Contribution from the Division of Pure Physics, National Research Council, Ottawa, Canada), Can. J. Phys., 35:892-900 (1957)

1.i.5.c. Group VI

Thermal Diffusivity and Thermal Conductivity of Chromium at High Temperatures
  V. E. Zino'ev, R. P. Krentsis, and P. V. Gel'd
  Fiz. Tverd. Tela, 11(7):2012-2014 (1969)
  Sov. Phys. — Solid State, 11(7):1623-1625 (1970)

Thermal Conductivities of Mo and W
  D. A. Howl
  J. Nucl. Mat., 33:138-148 (1969)

The Thermal Conductivity and Total Emittance of Tungsten at 1800 to 2800K, Technical Report 1 Sept. 1967-30 June 1968
  C. K. Jun, Salman Ebrahim, and M. Hoch
  Contract F33615-67-C-1445, AFML-TR-69-275; AD-698357 (Oct. 1969), 52 pp.

Tungsten Thermal Conductivity as Measured by Modified Method of Exponential Temperature Distribution
  V. Ya. Chekhovskoi and V. A. Vertogradskii
  (Akademiya Nauk SSSR, Moscow, Institut Vysokikh Temperatur), CONF-691002-, pp. 300-306

Thermal Conductivity of Molybdenum Thick Films on Beryllium Oxide
S. S. Cole
Am. Ceram. Soc. Bull., 47:806-809 (1968)

The Thermal Conductivity of Chromium above and below the Néel Temperature. An Analysis
J. F. Goff
NBS Spec. Publ., 302:311-321 (1968)

Thermal Properties of Molybdenum Single Crystals of High Temperatures
B. N. Khusainova and L. P. Filippov
Teplofiz. Vys. Temp., 6(5):929-930 (1968)
High Temperature, 6(5):891-892 (1968)

Further Comments on Transport Properties of Chromium
J. P. Moore, R. K. Williams, and D. L. McElroy
Thermal Conductivity (C. Y. Ho and R. E. Taylor, eds.), Plenum Press, New York (1969), pp. 303-314
Proceedings of the 8th Conference, held at Purdue University, October 7-10 (1968)

Measurement and Analysis of the Thermal Conductivity of Tungsten and Molybdenum at 100-400°K
N. G. Backlund
J. Phys. Chem. Solids, 28:2219-2223 (1967)

Anisotropy in the Thermal Conductivity and Thermoelectric Force of Regular Metals (Tungsten) in a Transverse Magnetic Field at 20°K
E. Gruneisen and H. Adenstedt
Contract NASw-1497, NASA-TT-F-11079
Ann. Physik, Leipzig, Ser. 5, 29:597-604 (1937), July 1967

Thermotransport in W
G. M. Neumann
Z. Naturforsch., 22A:393-395 (March 1967)

Measurement and Analysis of the Thermal Conductivity of Tungsten and Molybdenum at 100-400°K
N. G. Backlund
(Dept. of Solid State Physics, KTH, Stockholm 70, Sweden), J. Phys. Chem. Solids (1965)

Measurements of the Thermal Conductivity and Electrical Resistivity of Molybdenum
A. D. Feith
(Nuclear Materials and Propulsion Operation, General Electric Company, Cincinnati, Ohio), GE-TM 65-10-1 (1965)

The Thermal Conductivity of Molybdenum at Elevated Temperatures. Influence of Grain Size
C. K. Jun and M. Hoch
Advances in Thermophysical Properties at Extreme Temperatures and Pressures, American Society of Mechanical Engineers, New York (1965), pp. 296-300

Thermal Diffusivity of Tantalum, Molybdenum, and Niobium at Temperatures above 1800K
O. A. Kraev and A. A. Stelmakh
Teplofiz. Vysokikh Temp., 2:302 (1964)

Determination and Application of Thermophysical Properties of Refractory Metals
G. D. Rieck
The Science and Technology of Tungsten, Tantalum, Molybdenum, Niobium, and Their Alloys, Pergamon Press, Oxford (1964), pp. 205-217

Spectral Emissivity, Total Emissivity, and Thermal Conductivity of Molybdenum, Tantalum, and Tungsten above 2300°K
R. D. Allen, L. F. Glasier, and P. L. Jordon
J. Appl. Phys., 31:1382-1387 (1960)

Thermal Properties of Molybdenum between 2700°R and 4000°R
R. L. Rudkin
(U. S. Naval Radiological Defense Laboratory, San Francisco, Calif.), USNRDL-TR-433 (June 20, 1960), 23 pp.

Mechanical and Physical Properties of the Refractory Metals, Tungsten, Tantalum, and Molybdenum above 4000°F
L. F. Glasier, Jr., R. D. Allen, and I. L. Saldinger
(Aerojet-General Corp., Azusa, Cal.), Rept. M-1826 (April 1959), 89 pp.

1.i.6. Thermal Expansion

1.i.6.a. General

Thermal Expansion of Niobium, Molybdenum and Their Alloy at Low Temperatures
V. P. Lebedev, A. A. Mamalui, V. A. Pervakov, et al.
Ukr. Fiz. Zh., 14(5):746-750 (1969)

The Temperature Dependence of the Thermal Expansion Coefficient of Some Polycrystals
A. P. Nasekovskii
Izv. VUZ Fiz. (USSR), No. 1, pp. 65-68 (1969)
Ti, V, Mo, W

Gitterstruktur und thermische Ausdehnung der Übergangsmetalle Sc, Ti, V, Cr, Mn, Fe, Co, und Ni im Temperaturbereich von −180°C bis 1500°C
Norbert Schmitz-Pranghe
Thesis, Institute of Theoretical Physics, University of Cologne (1967)

Thermal Expansion Characteristics of Several Refractory Metals to 2500°C
J. B. Conway and A. C. Losekamp
Trans. Met. Soc., Am. Inst. Min. Met. Eng., 236(5):702-709 (1966)
W, Re, Ta, Mo, and Nb

Thermal Expansion of Solids at Low Temperatures
G. K. White
Low Temperature Physics Conference, London (1962), pp. 394-396 (1963)

Selected Applications of High Temperature X-Ray Studies in the Metallurgical Field
H. J. Goldschmidt
(The B.S.A. Group Research Centre, Birmingham, England), Advances in X-Ray Analysis, Vol. 5, Plenum Press, New York, 1962 (Review), pp. 191-212

The Measurement of Thermal Expansion of Metals at Low Temperatures
K. Andres
(Institut für kalorische Apparate und Kältetechnik der Eidgenössischen Technischen Hochschule, Zürich), Cryogenics, 2(2):1 (Dec. 1961)

Thermal Expansion of Technical Solids at Low Temperatures, a Compilation from the Literature
Robert J. Corruccini and John J. Gniewek
(National Bureau of Standards, Washington, D. C.) NBS Monograph 29 (May 19, 1961)

Application of a Variable Transformer to the Study of Low Temperature Thermal Expansion
R. H. Carr and C. A. Swenson
(Institute for Atomic Research and Department of Physics, Iowa State University, Ames, Iowa, USA), Cryogenics, 4:76 (1946)

A Compilation of Some of the Physical Properties of the Metallic and Semi-Metallic Elements and a Study of Some of Their Interrelationships
Karl A. Gschneider, Jr.
(Department of Physics and Materials Research Laboratory, University of Illinois, Urbana, Illinois, and University of California, Los Alamos, Scientific Laboratory, Los Alamos, New Mexico), LADC-6220
To be published in Solid State Physics., Vol. 16

1.i.6.b. Group IV

The Anisotropic Thermal Expansion of the Crystal Lattice of Hafnium
M. P. Krug and B. E. Davis
J. Less-Common Metals, 22:363-366 (1970)

The Anisotropy of the Thermal Expansion of α-Titanium
Ram Rao Pawar and V. T. Deshpande
Acta Cryst., 24A:316-317 (1968)

Determination of the Principal Coefficients of Thermal Expansion of α-Zr
J. C. Couterne and G. Cizeron
WAPD-TRANS-45
J. Nucl. Mater. (Amsterdam), 20:75-82 (1966)

Lattice Parameters, Thermal Expansions, and Grüneisen Coefficients of Zirconium, 4.2 to 130°K
J. Goldak, L. T. Lloyd, and C. S. Barrett
Phys. Rev., 144:478-484 (1966)

Dilatometric Investigation of Titanium
R. S. Mints, A. E. Shelest, and Yu. S. Malkov
Titanium and Its Alloys, Israel Program for Scientific Translations, Ltd., Jerusalem (1966), pp. 99-103

Thermal Expansion of Iodide Titanium
H. E. McCoy, Jr.
Trans. ASM, 57:743 (1964)

The Physical Metallurgy of Zirconium
D. L. Douglass
(General Electric Co., Pleasanton, California), At. Energy Rev., 1:71-237 (1963)

Thermal Expansion of Alpha-Zirconium Single Crystals
L. T. Lloyd
(Argonne National Laboratory, 9700 South Cass Avenue, Argonne, Illinois),ANL-6591 (Jan. 1963)

Preferred Orientation and Anisotropy in Titanium
W. T. Roberts

(Research Department, I.C.I. Metals Division, Witton, Birmingham, Great Britain), J. Less-Common Metals, 4:345 (1962)

1.i.6.c. Group V

The Debye Temperature and the Coefficient of Thermal Expansion for V and Nb by X-Ray Diffraction
M. Linkoaho and E. Rantavuori
Phys. Stat. Sol., 37:K89-K91 (1970)

Thermal Expansion of Vanadium and Vanadium Hydride at Low Temperatures
D. G. Westlake and S. T. Ockers
J. Less-Common Metals, 22:225-230 (1970)

Anomalies in the Physical Properties of Vanadium. The Role of Hydrogen
D. G. Westlake
Phil. Mag., 16:905 (1967)
Lattice parameters, electrical resistance, thermal expansion, magnetic susceptibility, elastic constants, and thermoelectric power

Mechanical and Physical Properties of the Refractory Metals, Tungsten, Tantalum, and Molybdenum above 4000°F
L. F. Glasier, Jr., R. D. Allen, and I. L. Saldinger
(Aerojet-General Corp., Azusa, Cal.), Rept. M-1826 (April 1959), 89 pp.

Measurements of Thermal Properties (Nb), July 1, 1957 to March 31, 1958
I. B. Fieldhouse, J. C. Hedge, and J. I. Lang
(Illinois Institute of Technology, Chicago, Armour Research Foundation), WADC-TR-58-274, May 1, 1958, Contract AF33(616)-3701, 87 pp.; AD-206 892

1.i.6.d. Group VI

Anomalous Thermal Expansion of Chromium near T_n
B. Stebler, C.-G. Andersson, and O. Kristensson
Phys. Scripta, 1(5-6):281-285 (1970)

Measurement of Thermal Expansion Coefficient of Tungsten at Elevated Temperatures
R. H. Knibbs
J. Sci. Instr., 2:515-517 (1969)

Thermal Expansion of Chromium Single Crystals at the Neel Temperature
T. Matsumoto and T. Mitsui
J. Phys. Soc. Japan, 27:786 (1969)

Lattice Anisotropy in Antiferromagnetic Chromium
M. O. Steinitz, L. H. Schwartz, J. A. Marcus, E. Fawcett, and W. A. Reed
Phys. Letters, 23(17):979-982 (1969)

X-Ray Determination of Lattice Parameters and Thermal Expansion Coefficients of Aluminum, Silver, and Molybdenum at Cryogenic Temperatures
Claude L. Woodard
Thesis, University of Missouri, Rolla, Mo., 1969, 142 pp.
Available from University Microfilms, Inc., Ann Arbor, Mich., Order No. 70-11, 768

Lattice Constants, Thermal Expansion Coefficients and Densities of Molybdenum and the Solubility of Sulphur, Selenium, and Tellurium in It at 1100°C
M. E. Straumanis and R. P. Shodhan
J. Metallk., 59:492-495 (1968)

Lattice Parameter and Thermal Expansion Coefficient of Mo between 15° and 65°C
M. E. Straumanis and R. P. Shodhan
Trans. Met. Soc. AIME, 242:1185-1186 (1968)

Lattice Expansion of Molybdenum
Ram Rao Pawar
Current Sci., India 36(16):428 (1967)

Present Possibilities of the Activation Analysis of Traces of Metalloids in High-Temperature Metals
Ch. Engelmann, J. Gosset, and Mlle M. Loeillet
Mem. Sci. Rev. Met., 63:1081-1085 (1966)

Thermal Expansion and Atomic Vibration Amplitudes for TiC, TiN, ZrC, ZrN, and Pure Tungsten

C. R. Houska
J. Phys. Chem. Solids, 25:359 (1964)

The Thermal Expansion of Molybdenum, Tungsten, Rhenium, Platinum, and Cadmium at Low Temperatures
K. Andres
(Institut für kalorische Apparate und Kältetechnik, Swiss Federal Institute of Technology, Zürich, Switzerland),
Phys. Letters, 7:315 (1963)

Anomalies in the Elastic Constants and Thermal Expansion of Chromium Single Crystals
D. I. Bolef and J. de Klerk
Phys. Rev., 129:1063-1067 (1963)

Observations on Prismatic and Grown-In Dislocations in Zone-Melted Molybdenum
A. Lawley and H. L. Gaigher
Phil. Mag., 8:1713 (1963)

The Anomalous Thermal Expansion of Chromium
G. K. White
Australian J. Phys., 14:359-367 (1961)

2. Transition Metal Borides

2.a. General

[11]B spin-lattice relaxation times in transition metal diborides
  R. B. Creel and R. G. Barnes
  Phys. Stat. Sol., 41:K27-K29 (1970)

Low-temperature specific-heat measurements of semi-borides of the first series of transition elements
  R. Kuentzler
  Phys. Stat. Sol., 41:291-296 (1970)
  Nineteen borides T_2B of transition elements between 1.2 and 4.2°K

K-spectra of boron in transition-metal diborides and LaB_6, BaB_6, and AsB
  I. I. Lyankhovskaya, T. M. Zimkina, and V. A. Fomichev
  Fiz. Tverd. Tela, 12(1):174-180 (1970)
  Sov. Phys. — Solid State, 12(1):138-143 (1970)

Heat resistance of diborides of tantalum and chromium and their alloys
  B. G. Arabei, K. I. Frolova, and R. S. Tikhonova
  Zashchita Metallov, 5:417-421 (1969)
  Protection of Metals, 5:366-369 (1969)

Electronic structure and magnetism in transition-metal diborides
  J. Castaing, R. Caudron, G. Toupance, and P. Costa
  (ONERA, 92 Chatillon, France), Third International Conf. on Solid Compounds of Transition Elements, Oslo (June 16-20, 1969)

Electronic structure of transition-metal diborides
  J. Castaing, R. Caudron, G. Toupance, and P. Costa
  Solid State Commun., 7:1453-1456 (1969)

Nuclear-magnetic resonance in transition-metal borides
  R. B. Creel
  (Ames Lab., Iowa), IS-T-323 (Nov. 1969), 194 pp.

A method for the production of spherical particles (case I)
  Karl Knotik, Peter Koss, and Heinz Markl
  (Oesterreichische Studiengesellschaft fuer Atomenergie GmbH), British Patent 1,157,012 (July 2, 1969)
  Priority date July 8, 1965, Austria
  Groups IVA, VA, or VIA carbides, nitrides, oxides, sulfides, borides, or silicides

Preparation of carbon-free borides of chromium, molybdenum, zirconium by thermal method
  G. A. Meerson and A. E. Gorbunov
  Izv. Akad. Nauk SSSR, Neorg. Mater., 5(12):2075-2082 (Dec. 1969)
  Inorg. Mater., 5(12):1767-1772 (1969)

Metal-rich metal — metalloid phases
  H. H. Stadelmaier
  Developments in the Structural Chemistry of Alloy Phases, Plenum Press, New York (1969), pp. 141-180
  A review of borides, carbides, nitrides, and oxides of transition elements

Study of the projection of borides and nitrides by the plasma gun
  G. Bentz, G. Provost, and C. Urban
  Bull. Soc. Fr. Ceram., 78:5-17 (1968)
  TiB_2, TaB_2, NbB_2, ZrB_2, TiN, ZrN, TaN

Preparation of crystalline refractory carbides and borides by zone melting
  C. A. Brookes and M. E. Packer
  4th Symposium on Special Ceramics, Stoke-on-Trent, England (P. Popper, ed.), British Ceramic Research Association, Stoke-on-Trent, England (1968), pp. 15-28
  CONF-670731

Band structure and the titanium $L_{II,III}$ x-ray emission and absorption spectra from pure metal, oxides, nitride, carbide, and boride
  D. W. Fischer and W. L. Baun
  J. Appl. Phys., 39:4757-4776 (1968)

Superconductivity of ceramic compounds
  K. M. Ralls, E. R. Parker, and V. F. Zackay
  Ceram. Microstruct. Anal. Significance Product. Proc. 3rd Internation. Mater. Symp., Berkeley, Calif., 1966, John Wiley and Sons, New York (1968), pp. 489-508

Anisotropy in Single-Crystal Refractory Compounds, Vol. 2
F. W. Vahldiek and S. A. Mersol, eds.
Plenum Press, New York (1968)
Proceedings International Symposium on Anisotropy in Single-Crystal Refractory Compounds, June 13-15, 1967, Dayton, Ohio, sponsored by Ceramics and Graphite Branch of Air Force Materials Lab., U. S. Air Force

Study of the characteristic x-ray absorption of borides of the iron group metals
S. M. Karal'nik, E. T. Kachkovskaya, and A. B. Lyashchenko
Issledovanie elektronnykh svoistv metallov i splavov, Izdat. Naukova Dumka, Kiev (1967), pp. 92-96
Ti, V, Cr, Fe, Co

Temperature dependence of thermoelectromotive force and specific electrical resistance of titanium, vanadium, chromium, and their borides, carbides, and nitrides
S. N. L'vov and V. F. Nemchenko
Vysokotemperaturnye Neorganicheskie Soedineniia, transl. by Helen J. Dahlby, Los Alamos Scientific Lab., Los Alamos, New Mexico, LA-TR-67-17 (May 1967), pp. 100-107

Chemistry of Boron and Its Compounds
Earl L. Muetterties, ed.
John Wiley and Sons, New York (1967)

Handbook of Lattice Spacings and Structures of Metals and Alloys, Vol. 2
W. B. Pearson, ed.
Pergamon Press, New York (1967), 1446 pp.; Vol. 1 (1958), 1044 pp.
Includes hydrides, borides, carbides and nitrides

Superconductivity of the borides of transition and rare-earth metals
O. I. Shulishova and I. A. Shcherbak
Izv. Akad. Nauk SSSR, Neorg. Mater., 3(8):1495-1497 (1967)
Inorg. Mater., 3(8):1304-1306 (1967)
ZrB, HfB, and TaBi; critical temps., 3.4, 3.1, and 4.0°K

Handbook of Binary Metallic Systems, Structure and Properties, Vol. II: Physicochemical Properties of the Elements
A. E. Vol
TT 66-51150 (1967), 870 pp.; Vol. I, TT-66-51149 (1966), 635 pp.
Includes hydrides, borides, carbides, nitrides, phase diagrams, and tables of properties

Electrolytic growth and preparation of transition-metal-compound single crystals
Aaron Wold
(Brown University, Providence, R. I.), AD-660615 (1967), 41 pp.
CaB_6, SrB_6, BaB_6, YB_6, MnB, TiB_2, and TaB_2

Electronic structure of some transition-metal borides
M. C. Cadeville and E. Daniel
J. Phys., 27:449 (1966)

An investigation of the physical properties of several transition-metal diborides and of binary diboride solid solutions
G. M. Farrior
(University of Tennessee), Dissertation Abstr., 26(8):4555 (1966)

Metal borides
N. N. Greenwood, R. V. Parish, and P. Thornton
Quart. Rev. (London), 20(3):441-464 (1966)
Review of preparation, structure, bonding and chemical properties; 106 refs.

Thermodynamic properties of transition-metal diborides
L. Kaufman
Met. Soc. AIME Spec. Rep. Ser., 10:S. 193-213 (1966)

Transportreaktionen von Siliciden und Boriden der Übergangsmetalle
J. Nickl, M. Duck, and J. Pieritz
Angew. Chem., 78:484-551 (1966)
TiB_2, ZrB_2, VB_2, CrB_2, and CrB

Borides, Silicides and Phosphides, a Critical Review of Their Preparation, Properties, and Crystal Chemistry
Bertil Aronsson, Torsten Lundstrom, and Stig Rundqvist
John Wiley and Sons, Inc., New York (1965)

Propriétés magnétiques des phosphures de fer et de borures des métaux de la première série de transition
M.-C. Cadeville
Ph. D. thesis, Univ. Strasbourg (1965), 105 pp.
Cr, Mn, Fe, Co, Ni, V

Methods of analysis of metal refractories — borides, carbides, and nitrides of hafnium, niobium, tantalum, titanium, and zirconium
R. E. Dutton, G. J. McKinley, D. McLean, and H. F. Wendt
(Union Carbide Research Inst., Tarrytown, N. Y.), Rept. No. C-29; NP-14952 (Feb. 1965), 25 pp.

Thermal conductivity of IV-VI group transition-metal borides
S. N. L'vov, V. F. Nemchenko, and P. I. Mal'kov
Vysokotemperaturnye Neorganicheskie Soedineniya (G. V. Samsonov, ed.), Izdat. Nauk. Dumka, Kiev (1965), pp. 451-455

Ternary interstitial compounds of boron and nitrogen with transition and post-transition metals, final report
H. H. Stadelmaier
(North Carolina State University at Raleigh), Grant No. DA-AROD-31-124-G94; Dept. of Army Project No. 59901004; U. S. Army Research Office-Durham Project No. 3138-MC (Dec. 15, 1965)

Borides — their chemistry and applications
R. Thompson
Roy. Inst. Chem. (London) Lecture Ser. No. 5 (1965), 40 pp.
Review including classification, structure, preparation, properties, fabrication, 306 refs.

Diffusion of boron and carbon in refractory transition metals
A. P. Epik
Dopovidi Akad. Nauk Ukr. RSR, No. 1, pp. 67-70 (1964)
Ti, Zr, Nb, Ta, Mo, and W

Refractory Ceramics for Aerospace
J. R. Hague, J. F. Lynch, A. Rudnick, F. C. Holden, and W. H. Duckworth, comps. and eds.
American Ceramic Society, Inc., 4055 N. High Street, Columbus, Ohio 43214 (1964)

Hartstoffe
R. Kieffer and F. Benesovsky
Springer-Verlag, Vienna (1963)
Preparation, crystal growth, physical properties, phase
diagrams, crystal structures

High-temperature thermodynamic properties
of some refractory borides
M. Mezaki, E. W. Tilleux, D. W. Barnes, and J. L.
Margrave
Thermodynamic Nuclear Material Proc. Symp., Vienna (1962),
pp. 775–788
Published 1963

Some (mechanical) kinetic and thermodynamic
properties of the refractory metal borides and
nitrides — an annotated bibliography
M. T. Milliken and J. E. Senkin
(Univ. California, Lawrence Radiation Lab), UCRL–7559
(Sept. 1963), 345 pp.
About 1240 refs.

X-ray spectra and the interatomic bond in solid
metallike compounds
V. S. Neshpor
AEC-tr-5873 (1963)
Vysokotemperaturnye Metallokericheskie Materialy, Akad.
Nauk URSR, Kiev (1962), pp. 46–74

Crystallochemical peculiarities of borides,
nitrides, silicides and phosphides of transition
metals
G. V. Samsonov
Poroshkovaya Met., Akad. Nauk Ukr. SSR, No. 2, pp. 65–79
(Mar.–Apr. 1963)

Plenum Press Handbooks of High-Temperature
Materials, No. 1: Materials Index
Peter T. B. Shaffer
Plenum Press, New York (1963)

The thermal properties of twenty-six solid materials to 5000°F or their destruction temperatures
Southern Research Institute, Birmingham, Alabama, Contract
No. AF33(616)-7319, Rept. No. ASD-TDR-62-765; AD-
298061 (Jan. 1963)

Literature survey on synthesis, properties, and
applications of selected boride compounds
B. R. Emrich
Aeronautical Systems Division, Dir. Materials and Processes,
Applications Lab., Wright-Patterson AFB, Ohio, ASD-TDR-
62-873 (Dec. 1962), 122 pp.
TiB_2, ZrB_2, HfB_2, VB_2, NbB_2, TaB_2, CrB_2, and ThB_4

Metal borides, carbides, and nitrides, and com-
plementary refractories — an annotated bibliography
J. B. Goldman
Lockheed Missiles Space Co., AD-296391 (Oct. 1962), 86 pp.

Selected bibliography concerning the properties
and uses of borides; elastic constants of titanium carbide and titanium diboride. Carbides,
nitrides, and silicides
E. E. Childs, J. J. Gilman, and B. W. Roberts
J. Appl. Phys., 32:1405 (1961)

Binary and ternary phase diagrams of columbium, molybdenum, tantalum, and tungsten
J. J. English

(Defense Metals Information Center, Battelle Memorial
Institute, Columbus, Ohio), OTS PB-171421; DMIC Rept.
152 (April 28, 1961)

Thermodynamic properties of 65 elements —
their oxides, halides, carbides, and nitrides
C. E. Wicks and F. E. Block
(Bureau of Mines, Albany Metallurgy Research Center,
Oregon), Bulletin 605; NP-13622 (Dec. 1961), 150 pp.

Research investigation to determine the optimum conditions for growing single crystals
of selected borides, silicides and carbides
A. D. Kiffer
(Linde Co., Union Carbide Corp.), WADD TR 60-52 (April
1960)

Electrical properties of borides, carbide, and
nitrides; the IVa to VIa metals
S. N. L'vov, V. F. Nemchenko, and G. V. Samsonov
Dokl. Akad. Nauk SSSR 135:577–580 (1960)

The oxidation of Cr, Hf, Mo, Nb, Ta, Ti, W, V,
Zr and some of their alloys; properties of
some borides, carbides, nitrides, and oxides
of the same metals
E. A. Cernak, comp.
(Pratt and Whitney Aircraft Div., United Aircraft Corp.,
Middletown, Conn.), CNLM-1802-5 (Dec. 15, 1959), 99 pp.
922 refs.; 1951 to Sept. 1959

Infrared spectra of inorganic solids — II.
Oxides, nitrides, carbides, and borides
E. G. Brame, Jr., J. L. Margrave, and V. W. Meloche
J. Inorg. Nucl. Chem., 5:48–52 (1957)
Li_3N, Cu_3N, BN, AlN, CrN, Mg_3N_2, Zn_3N_2, B_2O_3, Al_2O_3, Ga_2O_3,
Cr_2O_3, B_4C, SiC, Mo_2B, ZrB_2, and TiB_2

Some physicochemical properties of compounds
of the transitional refractory metals with boron,
carbon, and nitrogen, and the characteristics
of their binary alloys
G. V. Samsonov
AEC-tr-3016
Izv. Sektora Fiz. Khim. Anal. Inst. Obshchei Neorg. Khim.,
Akad. Nauk SSSR, 27:97–125 (1956)

Investigation of the diffusion of boron and
carbon in certain metals of transition groups
G. V. Samsonov and V. P. Latisheva
AEC-tr-3321 (translated by Lydia Venters) Fiz. Metal. i
Metalloved. Akad. Nauk SSSR, Ural. Filial, 2:309–319
(1956)

The preparation of pure high-melting carbides,
nitrides, and borides by the deposition method
and the description of their properties
K. Moers
AEC-tr-3239
Z. Anorg. u Allgem. Chem., 198:243–261 (1931)

2.b. Group IV

Thermal diffusivity of TiB_2, ZrB_2, and HfB_2
Thomas M. Branscomb
Thesis, Iowa State University of Science and Technology, Ames,
Iowa, PB-192253; IS-T-368 (Feb. 1970), 92 pp.

Growth of titanium diboride single crystals in
molten aluminum
I. Higashi and T. Atoda
J. Cryst. Growth, 7:251–253 (1970)

Phase diagram of the zirconium – boron system
K. I. Portnoy, V. M. Romashov, and L. N. Burobina
Poroshkovaya Met., 7:68-71 (1970)

Fine particulates to ultrafine-grain ceramics
T. Vasilos and W. Rhodes
Ultrafine Grain Ceramics (J. J. Burke, N. L. Reed, and
V. Weiss, eds.), Syracuse Univ. Press (1970), pp. 137-172

Electrodeposition of ZrB_2
K. E. Anthony and B. J. Welch
Austral. J. Chem., 22:1593-1597 (1969)

Oriented growth and defect structures of single
crystal transition metal diborides
J. S. Haggerty, J. F. Weickus, and D. W. Lee
Pure Appl. Chem., Suppl., pp. 547-555 (1969)
Proceedings 3rd International Symposium on High-Temperature
Technology, Pacific Grove, Calif., CONF-670935
ZrB_2 and HfB_2

Croissance des grains de diborure de zirconium
M. Kinoshita, S. Kose, and Y. Hamano
J. Ceram. Soc. Japan, 77(8):255-263 (1969)

Grain growth of ZrB_2
M. Kinoshita et al.
Yogyo Kyokai-Shi, 77:255-263 (1969)

Magnetoresistance and Hall effect in diborides
of metals in the IVa subgroup
B. I. Malinskii, V. F. Nemchenko, and S. N. L'vov
Izv. VUZ Fiz. (USSR), No. 12, pp. 69-74 (1969)

Elastic properties of polycrystalline TiB_2, ZrB_2,
and HfB_2 from room temperature to 1300°K
D. E. Wiley, W. R. Manning, and O. Hunter, Jr.
J. Less-Common Metals, 18:149-157 (1969)

Boron isotope effect in superconducting zir-
conium dodecaboride
C. W. Chu and H. H. Hill
Science, 159:1227-1228 (1968)

Preparation and characterization of high-quality
single-crystal refractory metal borides and car-
bides
J. S. Haggerty, D. W. Lee, and J. F. Wenckus
Arthur D. Little, Inc., Cambridge, Mass., AFML-TR-68-228
(August 1968), 83 pp.
HfB_2, ZrB_2, HfC, TaC, and ZrC

Growth and characterization of single-crystal
ZrB_2
J. S. Haggerty, J. L. O'Brien, and J. F. Wenckus
J. Crystal Growth, 3(4):291-294 (1968)

Structure and orientation of the second phase
in ZrB_2 crystals
W. J. Leombruno, J. S. Haggerty, and J. L. O'Brien
Mat. Res. Bull., 3:361-368 (1968)

The possible application of the single-zone rep-
resentation to the diborides of group IV transition metals
S. N. L'vov and V. F. Nemchenko
Izv. VUZ Fiz. (USSR), 1:59-62 (1968)
Electrical conductivity, Hall coefficient, Ettingshausen-Nernst
coefficient, thermal emf and thermal conductivity of the
diborides of Ti, Zr, Hf

Defect structure of single-crystal titanium
diboride
S. A. Mersol, C. T. Lynch, and F. W. Vahldiek
Anisotropy in Single-Crystal Refractory Compounds, Vol. 2

(F. W. Vahldiek and S. A. Mersol, eds.), Plenum Press,
New York (1968), pp. 41-94
Proceedings of an International Symposium held June 13-15,
1967, in Dayton, Ohio

Electrolytic deposition of ZrB_2
P. A. Polishchuk
Fiz. Khim. Elektrokhim. Raspl. Solei Shlakov, pp. 327-333
(1968)

High-pressure electrical resistance and com-
pressive strength of TiB_2
F. W. Vahldiek
Anisotropy in Single-Crystal Refractory Compounds, Vol. 2
(F. W. Vahldiek and S. A. Mersol. eds.), Plenum Press,
New York (1968), pp. 339
Proceedings of an International Symposium held June 13-15,
1967, in Dayton, Ohio

The Ti – Fe – B system
T. F. Fedorov and Yu. B. Kuz'ma
Izv. Akad. Nauk SSSR, Neorg. Mater., 3(8):1498-1499 (1967)
Inorg. Mater., 3(8):1307-1308 (1967)

A possible new group of semiconducting com-
pounds
W. P. R. George, C. H. L. Goodman, H. F. Sterling, and
R. W. Warren
Phys. Stat. Sol., 21:205 (1967)
TiB_2 and ZrB_2

Synthesis of TiB_2 by carbon reduction
M. Kinoshita and Y. Hamano
Bull. Govt. Ind. Res. Inst. Osaka, 18(1):72-77 (1967)

Heat capacity of TiB_2 at low temperatures
I. E. Paukov, L. M. Khriplovich, and V. S. Filatkina
Zh. Fiz. Khim., 41(7):1621-1624 (1967)

Préparation de borures au moyen de trifluorure
de bore
Philippe Pichat
Compt. Rend. C 265:385-387 (1967)
Al, Mg, Ca, Ti, and Zr

Electrical resistivity, elastic modulus, and
Debye temperature of titanium diboride
F. W. Vahldiek
J. Less-Common Metals, 12:202-209 (1967)

Étude de la boruration superficielle du zirconi-
um
G. Vuillard and A. Walder
Rev. Int. Hautes Temper. Refract., 4:29-37 (1967)

Physical properties of titanium diboride single-
crystal material, 1957-1966
Technical Information Libraries, Bell Telphone Labs., Bell
Telephone Laboratories No. 107 (March 1967)

A new intermediate compound in the titanium –
boron system, Ti_3B_4
R. G. Fenish
Trans. AIME, 236:804 (1966)

Lattice constants of titanium diboride
I. Higashi and T. Atoda
Sci. Papers Inst. Phys. Chem. Res., Japan, 60(1):32-34 (1966)

The heat of formation of titanium diboride
E. J. Huber, Jr.
J. Chem. Eng. Data, 11:430 (1966)

Synthesis of titanium diboride by carbon re-
duction
M. Kinoshita and Y. Hamano
Yogyo Kyokai Shi, 74(7):244-248 (1966)

Preparation and properties of thin films of transitional-metal borides
K. G. Knauff
Proc. International Symposium Grundprobleme der Physik dünner Schichten, Göttingen, 1965, pp. 207-211 (1966)
Ti, Zr, and La

The microstructure of single-crystal titanium diboride
C. T. Lynch, S. A. Mersol, and F. W. Vahldiek
J. Less-Common Metals, 10:206-219 (1966)

Preparation of titanium diboride crystals from the gas phase
P. Peshev
Bulg. Akad. Nauk, Izv. Inst. Obshta Neorg. Khim., 4:53-67 (1966)

Galvanomagnetic effects in single-crystal ZrB_2
J. Piper
J. Phys. Chem. Solids, 27:1907-1915 (1966)
Magnetoresistance and Hall effect

Vapor-deposited borides of group IVA metals
J. J. Gebhardt and R. F. Cree
J. Am. Ceram. Soc., 48:262-267 (1965)

The thermal conductivities of some electrically conducting compounds
R. W. Powell and R. P. Tye
Proceedings of Symposium on Special Ceramics, Stoke-on-Trent, 1964, Academic Press, London; New York (1965), pp. 243-257
CeS, ZrB_2, TiB_2, TiC, TiB_2-TiC 50/50

Fundamental study of the sintering kinetics of refractory compound phases at high pressure and high temperature, final report
E. V. Clougherty and Larry Kaufman
Contract Nonr-4262(00), AD-637247 (Oct. 1964), 21 pp.
TiB_2

Diffusion of boron and carbon in refractory transition metals
A. P. Epik
Dopovidi Akad. Nauk Ukr. RSR, 1:67-70 (1964)

The heats of formation of zirconium diboride and dioxide
E. J. Huber, Jr., E. L. Head, and C. E. Holley, Jr.
J. Phys. Chem., 68:3040 (1964)

On the reactive diffusion of boron and carbon in refractory transition metals
G. V. Samsonov
Dopovidi Akad. Nauk Ukr. RSR, 1:68-70 (1964)

Mass-spectrometric study of zirconium diboride
O. C. Trulson and H. W. Goldstein
(Union Carbide Research Institute), Contract DA-30-069-ORD-2787, Tech. Rept. C-25 (Oct. 1964)

Thermodynamic properties of inorganic substances VII. The high-temperature heat content of zirconium diboride
R. H. Valentine, T. F. Jambois, and J. L. Margrave
J. Chem. Eng. Data, 9(2):182 (1964)

Preparation of zirconium boride
V. F. Funke and S. I. Yudkovskii
Soviet Powder Met. Metal Ceram., 4:293-296 (1963)
From B_4C, ZrO_2

Titanium borides
R. Kieffer and F. Benesovsky
Hartstoffe (Hard Materials), Springer, Vienna (1963), pp. 176-187
Review, 131 refs.

Atomic thermal vibrations and bonding effects in graphite, diamond and titanium diboride
Mark G. Miksic
Thesis, Polytechnic Institute of Brooklyn, 1962
Available from University Microfilms, Inc., Ann Arbor, Mich., Order No. POLYBRK 62-5662

Investigation of single-crystal and polycrystalline titanium diboride: Metallographic procedures and findings
C. T. Lynch, F. W. Vahldiek, S. A. Mersol, and C. R. Underwood
(Metals and Ceramics Lab., Directorate of Materials and Processes, Aeronautical Systems Div., Wright-Patterson Air Force Base, Ohio), ASD TR 61-350 (Oct. 1961), 52 pp.

The heat of formation of titanium diboride: Experimental and analytical resolution of literature conflict
W. S. Williams
J. Phys. Chem., 65:2213 (1961)

The electron-beam melting of beryllium, boron, boron carbide, tantalum carbide, titanium carbide, tungsten, and zirconium diboride, quarterly progress report No. 8 for Feb. 1-April 30, 1960
R. L. Martin, S. R. Seagle, and O. Bertea
(Reactive Metals, Inc., Niles, Ohio), NP-8816 (May 1960), 21 pp.

The production and certain properties of a boride of hafnium
J. B. Paderno, T. E. Serebriakova, and G. V. Samsonov
AEC-tr-4030
Translated by N. H. Krikorian (Los Alamos Scientific Lab.) from Tsvetnye Metally, 32(11):48-51 (1959), 5 pp.

2.c. Group V

Evidence for localized character of electronic structure in vanadium borides from [51]V nuclear-magnetic resonance
R. G. Barnes, R. B. Creel, and D. R. Torgeson
Solid State Commun., 8:1411-1414 (1970)
V_3B_2, VB, V_5B_6, V_3B_4, V_2B_3, and VB_2

Porosity dependence of the modulus of elasticity of TaB_2
B. Claussen
Mat. Sci. Eng., 4:245:246 (1969)

Boron-rhenium-vanadium system
Yu. B. Kuz'ma and D. A. Kovalyk
Izv. Akad. Nauk SSSR, Neorg. Mater., 5(10):1687-1690 (1969)
Inorg. Mater., 5(10):1428-1431 (1969)
Binary systems V-B and Re-B

Phase and structure relations in the vanadium-boron system
K. E. Spear and P. W. Gilles
High Temp. Sci., 1:86-97 (1969)

Thermionic properties of TaB_2
  V. Kh. Burkhanova, N. A. Gorbatyi, V. M. Sultanov, and
    F. A. Fekhretdinov
  Zh. Tekh. Fiz., 38:1356-1361 (1968)
  Sov. Phys. – Tech. Phys., 13:1107-1111 (1969)

Niobium – physicochemical properties of its compounds and alloys. III. Crystal structures and densities
  Hans Nowotny and Karl J. Seifert
  At. Energy Rev., Spec. Issue, No. 2, 71-172 (1968)

On the borothermic preparation of some vanadium, niobium, and tantalum borides
  P. Peshev, L. Leyarovska, and G. Bliznakov
  J. Less-Common Metals, 15:259-267 (1968)

The chemical and thermodynamic properties of V borides and related systems. I. The V-B system. II. The V-B-N system. III. Chemical-transport studies involving V
  K. E. Spear
  Thesis Univ. of Kansas, 1967; Dissertation Abstr., 28B(8):3245 (1967-1968)

Chemistry of Niobium and Tantalum
  F. Fairbrother
  Topics in Inorganic and General Chemistry, Monograph 10, American Elsevier Publishing Co., New York (1967)

Vacuum thermionic work functions and thermal stability of TaB_2, ZrC, Mo_2C, $MoSi_2$, $TaSi_2$, and WSi_2
  R. G. Wilson and W. E. McKee
  J. Appl. Phys., 38(4):1716 (1967)

Niobium alloys and compounds
  Donald L. Grigsby
  EPIC Data Sheet DS-148 (Jan. 1966)
  Electronic Properties Information Center, Hughes Aircraft Co., Culver City, Calif.

Superconductive anomaly in specific heats of some niobium and vanadium compounds
  K. Ukei and E. Kanda
  Ann. Acad. Sci. Fennicae A VI (Finland), No. 210, pp. 104-107 (1966)
  Proceedings Low-Temperature Calorimetry Conference, Otaniemi, 1966
  NbB_2, NbS_2

Diffusion of boron and carbon in refractory transition metals
  A. P. Epik
  Dopovidi Akad. Nauk Ukr. RSR, 1:67-70 (1964)

On the reactive diffusion of boron and carbon in refractory transition metals
  G. V. Samsonov
  Dopovidi Akad. Nauk Ukr. RSR, 1:68-70 (1964)

$NbB_{1.963}$: The heat capacity and thermodynamic properties from 5 to 350°K
  E. F. Westrum, Jr., and G. A. Clay
  J. Phys. Chem., 67:2385 (1963)

2.d. Group VI

Chaleur spécifuque à basse température et susceptibilité magnétique des borures mixtes de molybdène et de chrome
  J. Castaing, R. Caplain, and P. Costa
  Solid State Commun., 9:297-300 (1971)

Chromium-like antiferromagnetic behavior of CrB_2
  R. G. Barnes and R. B. Creel
  Phys. Letters, 29A(4):203-204 (1969)

Superconducting isotope effect in molybdenum boride and tungsten boride
  J. J. Engelhardt
  Phys. Rev., 179:452-458 (1969)

The crystal structure of CrB_4
  Sven Andersson and Torsten Lundstrom
  Acta Chem. Scand., 22:3103-3110 (1968)

The structure of Ru_2B_3 and $WB_{2.0}$ as determined by single-crystal diffractometry, and some notes on the W-B system
  Torsten Lundstrom
  Arkiv Kemi, 30(11):115-127 (1968)

Superconductivity and the d-shell
  J. J. Engelhardt, G. W. Webb, and B. T. Matthias
  Science, 155(3759):191-193 (1967)
  Isotope effects on the superconducting transition temperature of Mo_2B and W_2B

Polymorphic transition of W_2B_5
  Yu. B. Kuzma, T. I. Serebryakova, and A. M. Plakhina
  Zh. Neorg. Khim., 12(2):559-560 (1967)

Bor-reiche Wolframboride
  H. Nowotny, H. Haschke, and F. Benesovsky
  Monatsh. Chem., 98(3):547 (1967)
  WB_4 and MoB_4

Phase diagram of the system tungsten-boron
  K. I. Portnoi, V. M. Romashov, Yu. V. Levinskii, and I. V. Romanovich
  Soviet Powder Met. Metal Ceram., No. 5, pp. 398-402 (1967)

Tungsten and Its Compounds
  G. D. Riech
  Pergamon Press, Oxford, London, New York (1967)

Propriétés magnétiques des diborures de manganèse et de chrome: MnB_2 et CrB_2
  M. C. Cadeville
  J. Phys. Chem. Solids, 27:667-670 (1966)

Composition and crystallographic data for the highest boride of tungsten
  P. A. Romans and M. P. Krug
  Acta Cryst., 20:313 (1966)
  WB_4

Physical properties of the boride phases of Cr
  T. I. Serebryakova and B. A. Kovenskaya
  Izv. Akad. Nauk SSSR, Neorg. Mater., 2(12):2134-2138 (1966)
  Inorg. Mater., 2(12):1846-1859

Tungsten diboride: Preparation and structure
  H. P. Woods, F. E. Wawner, Jr., and Barbara G. Fox
  Science, 152:75 (1966)

Diffusion of boron and carbon in refractory
transition metals
A. P. Epik
Dopovidi Akad. Nauk Ukr. RSR, 1:67–70 (1964)

On the reactive diffusion of boron and carbon
in refractory transition metals
G. V. Samsonov
Dopovidi Akad. Nauk Ukr. RSR, 1:68–70 (1964)

Crystal structure of Cr_3B_4
M. Elfstrom
Uppsala University, Sweden, AD–262205 (June 1961), 5 pp.

3. Transition Metal Carbides

3.a. General

Growth of transition metal carbide single crystals by recrystallization. I. Preparation of fully-dense transition metal carbide
L. R. Fleischer and J. M. Tobin
J. Cryst. Growth, 8:235-242 (1971)

Growth of transition metal carbide single crystals by recrystallization. II. Transition metal carbide recrystallization and crystal growth
L. R. Fleischer and J. M. Tobin
J. Cryst. Growth, 8:243-246 (1971)

Electronic structure of cubic refractory carbides
Lars Ramqvist
J. Appl. Phys., 42:2113-2120 (1971)
ESCA, x-ray and absorption spectra, 29 refs.

X-ray emission spectra of VC_x, Nb_x, TaC_x, and ZrC
Lars Ramqvist, Borje Ekstig, Elisabeth Kallne, Erik Noreland, and Rolf Manne
J. Phys. Chem. Solids, 32:149-157 (1971)

Transition Metal Carbides and Nitrides
Louis E. Toth
Academic Press, New York and London, 1971, 279 pp.
Preparation, characterization, crystal chemistry, phase relationships, thermodynamics, mechanical properties, electrical and magnetic properties, superconducting properties, band structure and bonding, author and subject index

Strengthening of metals and alloys by refractory fibers grown from the melt
H. Bibring, G. Seigel, and M. Rabinovitch
Office National d'Etudes et de Recherches Aérospatiales, 92 Chatillon, France, Rept. T. P. No. 853 (1970)
Présentée à la Deuxième Conférence Internationale sur la Résistance des Métaux et Alliages, Asilomar, Calif., 30 Aout-4 Sept. 1970
TaC, NbC, TiC; in English

Electronic structure of face-centered cubic titanium and vanadium carbide alloys
R. Caudron, J. Castaing, and P. Costa
Solid State Commun., 8:621-625 (1970)

Chaleur spécifique électronique et susceptibilité magnétique des carbures de transition hexagonaux de formule M_2C
R. Caudron, F. Ducastelle, and P. Costa
J. Phys. Chem. Solids, 31:291-297 (1970)
V_2C, $(Mo-V)_2C$, Mo_2C, Nb_2C

The magnetic susceptibility of solid solutions of zirconium and niobium monocarbides
L. B. Dubrovskaya, A. S. Borukhovich, P. V. Gel'd, I. I. Matveenko, L. V. Kudryasheva, and S. S. Ordan'yan
Fiz. Tverd. Tela, 11(10):3021-3022 (1969)
Sov. Phys. — Solid State, 11(10):2451-2453 (1970)

Molecular-orbital interpretation of the soft x-ray $L_{II,III}$ emission and absorption spectra from some titanium and vanadium compounds
D. W. Fischer
J. Appl. Phys., 41(9):3561-3569 (1970)
Oxides, nitrides, carbides, and borides

Über die Beständigkeit von Übergangsmetallcarbiden gegen Stickstoff bis zu 300 at
R. Kieffer, H. Nowotny, P. Ettmayer, und M. Freudhofmeier
Monatsh. Chem., 101:65-82 (1970)
TiC, ZrC, HfC, VC_{1-x}, NbC, TaC, Cr_3C_2, Mo_2C, WC

Mass-spectrometric determination of the dissociation energy of vanadium and chromium dicarbide and vanadium tetracarbide
Fred J. Kohl and C. A. Stearns
(NASA, Lewis Research Center), NASA-TN-D-5719 (March 1970)

On the crystal chemistry of the close-packed transition-metal carbides. II. A proposal for the notation of the different crystal structures
E. Parthe and K. Yvon
Acta Cryst., 26B:153-156 (1970)

Thermal conductivity of transition-metal carbides
L. G. Radosevich and Wendell S. Williams
J. Am. Ceram. Soc., 53:30-33 (1970)

3. Transition Metal Carbides

Phonon spectra in TaC and HfC
H. G. Smith and W. Glaser
Phys. Rev. Letters, 25:1611-1613 (1970)

Méthode de préparation de lames minces d'alliages composites à fibres de carbure
Jean-Pierre Trottier and Bernard Genty
Office National d'Etudes et de Recherches Aerospatiales, 92 Chatillon, France, Rept. T.P. No. 825 (1970)
Septième Congrès International de Microscopie Electronique, Grenoble, 1970

Self-diffusion of carbon in group IV and V transition-metal carbides
R. A. Andrievskii, V. V. Klimenko, and Yu. F. Khromov
Fiz. Metallov. Metalloved., 28:298-303 (1969)

Microhardness of carbides of certain transition metals
L. N. Bazhenova and A. A. Ivan'ko
Izv. Akad. Nauk SSSR, Neorg. Mater., 5(12):2071-2074 (Dec. 1969)
Inorg. Mater., 5(12):1763-1767 (1969)
TiC, ZrC, HfC, VC, NbC, TaC, Cr_3C_2, Mo_2C, and WC

Comparison of the chemical and self-diffusivities in the transition-metal monocarbides
W. F. Brizes and E. I. Salkovitz
Scripta Met., 3:659-662 (1969)

Carbide single crystals produced by solid-state grain growth
L. R. Fleischer and J. M. Tobin
Pure Appl. Chem., Suppl., pp. 665-672 (1969)
Proceedings 3rd International Symposium on High Temperature Technology, Pacific Grove, Calif., CONF-670935
ZrC, VC, HfC, and NbC

Temperature coefficient of the electric resistance of carbides of titanium, zirconium, and niobium
V. G. Grebenkina and A. Ya. Kuchma
Izv. Akad. Nauk SSSR, Neorg. Mater., 5(4):717-720 (1969)
Inorg. Mater., 5(4):609-611 (1969)

Debye temperatures of some metal monocarbides
C. P. Kempter
Phys. Stat. Sol., 36:K137-K139 (1969)
ZrC, HfC, NbC, TaC

Preparation and properties of interstitial compounds
R. Kieffer
J. Inst. Metals, 97:164-172 (1969)

A method for the production of spherical particles (case 1)
Karl Knotik, Peter Koss, and Heinz Markl
(Oesterreichische Studiengesellschaft fuer Atomenergie GmbH), British Patent 1,157,012 (July 2, 1969)
Priority date July 8, 1965, Austria
In groups IVA, VA, or VIA of the periodic table or metals such as U, Th, or Pu

Estimate of the surface energies of some metal carbides
D. A. Mortimer and M. Nicholas
Atomic Energy Research Establishment, Harwell, England, AERE-M-2247 (Aug. 1969), 10 pp.

Zur elektronischen Struktur der Übergangsmetallcarbide und -nitride
A. Neckel
Vortrag: Physiker-Tagung, Salzburg (1969)

Thermal conductivity of transition-metal carbides
L. G. Radosevich and Wendell S. Williams
(Illinois Univ., Urbana), COO-1198-628, Contract AT(11-1)-1198 (April 1969), 21 pp.

Phonon scattering by conduction electrons and by lattice vacancies in carbides of the transition metals
L. G. Radosevich and W. S. Williams
Phys. Rev., 181:1110-1117 (1969)
TiC_x, ZrC_x, NbC_x

Preparation, properties and electronic structure of refractory carbides and related compounds
L. Ramqvist
Jernkont. Ann., 153, 21 pp, 53 refs. (1969)
Review; re-evaluation of literature data; binding energy; heat of formation

X-ray emission spectra of VC_x, NbC_x, TaC_x, and ZrC
Lars Ramqvist, Boerje Ekstig, Elisabeth Kaellne, Erik Noreland, and Rolf Manne
(Uppsala Univ., Sweden, Inst. of Physics) UUIP-620 (Feb. 1969), 23 pp.

Charge transfer in transition-metal carbides and related compounds studied by ESCA
Lars Ramqvist, Kjell Hamrin, Gunilla Johansson, Anders Fahlman, and Carl Nordling
J. Phys. Chem. Solids, 30:1835-1847 (1969)
Ti and V

Physical properties of the monocarbides of the transition metals in their homogeneity regions. I. TiC_{1-x} and NbC_{1-x}
G. V. Samsonov and G. Sh. Upadkhaya
Poroshkov. Met., 9(5):69-74 (1969)
TiC, NbC

Temperature dependence of the electrical resistivity and thermal emf of niobium and titanium carbide alloys on their homogeneity range
G. V. Samsonov, G. Sh. Upadkhaya, and V. F. Nemchenko
Teplofiz. Vysokikh Temp. (USSR), pp. 449-452
High Temp., 7(3):408-410 (May 1969)

Metal-rich metal-metalloid phases
H. H. Stadelmaier
Developments in the Structural Chemistry of Alloy Phases, Plenum Press, New York (1969), pp. 141-180

Automatic measurements of secondary electron characteristics of TaC, TiC, and ZrC
S. Thomas and E. B. Pattinson
Brit. J. Appl. Phys., 2:1539 (1969)

Estimating the thermodynamic characteristics of refractory carbides from their infrared absorption spectra
V. A. Tskhai and P. V. Gel'd
Zh. Neorg. Khim., 14(10):2666-2761 (Oct. 1969)
Groups IV and V

On the diffusion mechanism in transition-metal monocarbides
V. N. Zagriazkin
LA-tr-70-20, 9 pp.
Translated by Helen J. Dahlby (Los Alamos Scientific Lab., New Mexico) from Fiz. Metal. Metalloved., 28:292-297 (1969)

X-ray $K\alpha$ emission band of carbon in the mono-
carbides of transition metals belonging to the
fourth and fifth groups
E. A. Zhurakovskii
Dokl. Akad. Nauk SSSR, 184(6):1317-1320 (1969)
Sov. Phys. — Dokl., 14(2):168-171 (1969)

Mechanical properties of transition-metal mono-
carbides
W. F. Brizes
NASA-CR-95887 (1968)

Preparation of crystalline refractory carbides
and borides by zone melting
C. A. Brookes and M. E. Packer
In Special Ceramics, Vol. 4 (P. Popper, ed.), Stoke-on Trent,
Eng., British Ceramic Research Association (1968),
pp. 15-28
(CONF-670731)
Groups IV, V, and VI

Directional variations of elastic properties of
some transition-metal monocarbides cubes
H. L. Brown, P. E. Armstrong, J. A. O'Rourke, and C. P.
Kempter
Planseeber. Pulvermet., 16:114-118 (1968)

Low-temperature electrical properties of some
transition metals and transition-metal carbides
F. W. Clinard, Jr., and C. P. Kempter
J. Less-Common Metals, 15:59 (1968)

Structure électroniques des carbures et des
nitrures de métaux de transition cubiques faces
centrées de type NaCl
Paul Costa
(Office National d'Etudes et de Recherches Aérospatiales,
29, Avenue de la Division Leclerc, 92 Chatillon, France),
ONERA-TP-540 (to be publ. in Annales de Physique)
Received RMIC May 1968

Electronic structure of transition carbides and
nitrides
P. Costa
(ONERA. Chatillon-sous-Bagneaux, France), Anisotropy
in Single-Crystal Refractory Compounds, Vol. 1 (Fred W.
Vahldiek and Stanley A. Mersol, eds.), Plenum Press,
New York (1968), pp. 151-159
From Symp. on Anisotropy in Single-Crystal Refractory
Compounds, Dayton, Ohio, CONF-670637-(Vol. 1)

Cohesion in cubic refractory monocarbides,
mononitrides and monoxides
S. P. Denker
J. Less-Common Metals, 14:1 (1968)

Metal carbides
William A. Frad
Advances in Organic Chemistry and Radiochemistry, Vol. II,
Academic Press, New York (1968), pp. 153-247

Verfahren zur Herstellung von harten Überzügen
Hans Gass und Hans Erich Hintermann
(Lab. Suisse de Recherches Horlogeres, Neuenburg, Ger-
many), German Patent 452,205 (May 31, 1968)

Anomalous superconducting properties of car-
bides and nitrides of Group IVa and Va elements
A. L. Giorgi, E. G. Szklarz, and T. C. Wallace
Proc. Brit. Ceram. Soc., No. 10, pp. 183-193 (March 1968)

Crystal structures of refractory carbides and
nitrides, A. Bowman in Fundamentals of Refractory Compounds
H. H. Hausner and M. G. Bowman, eds.
Plenum Press, New York (1968)

Non-stoichiometry and bonding in refractory
monocarbides
Michael Hoch
Anisotropy in Single-Crystal Refractory Compounds, Vol. 1
(Fred W. Vahldiek and Stanley A. Mersol, eds.), Plenum
Press, New York (1968), pp. 163-175
From Symp. on Anisotropy in Single-Crystal Refractory Com-
pounds, Dayton, Ohio, CONF-670637-(Vol. 1)

The microstructure and mechanical behavior of
carbides
Graham E. Hollox
(Research Inst. for Advanced Studies, Div. of Martin Marietta
Corp., Baltimore, Md.), Contract NASw-1290, DA-31-124-
ARO-D-467, RIAS TR 68 10c (May 1968), 59 pp.
Materials Sci. Eng., 3:121-137 (1968)

Discussion of "Thermal properties of tantalum
monocarbide and tungsten monocarbide"
C. P. Kempter and H. L. Brown
Trans. AIME, 242:1751-1752 (1968)

Carbides: Properties, Production, and Appli-
cations
T. Ya. Kosolapova
Metallurgiya, Moscow (1968)
Plenum Press, New York (1971)

Vapor-phase deposition of refractory metals
and their carbides
R. Lorel and B. Pinteau
Galvano, 37:337-341 (1968)

Obtention de carbures et de carbonitrures sim-
ples et mixtes de quelque métaux de transition
B. Mas
Diplomarbeit, Univ. Lyon (1968)

Field ion microscopy of refractory metal car-
bides ($TaC_{0.97}$, $TaC_{0.72}$, WC, W_2C)
J. D. Meakin
Phil. Mag., 17:865 (1968)

Application of the electron microprobe to the
study of refractory metal carbides, oxides, and
silicides
S. H. Moll and G. W. Bruno
Anisotropy in Single-Crystal Refractory Compounds, Vol. 1
(Fred W. Vahldiek and Stanley A. Mersol, eds.), Plenum
Press, New York (1968), pp. 23-36

Thermal conductivity of cubic monocarbides
and mononitrides of transition metals
V. S. Neshpor
Izv. Akad. Nauk SSSR, Neorg. Mater., 4(12):2200-2202 (1968)
Inorg. Mater., 4(12):1915-1917 (1968)

Ordnungsstrukturen bei Übergangsmetall-Car-
biden und -Nitriden
H. Nowotny and F. Benesovsky
Planseeberichte für Pulvermetallurgie, 16(3):204-214 (Oct.
1968)

Preparation of pure and mixed metallic carbides
R. A. Paris and E. Clar
Chim. Ind., 99:255-259 (1968)

A study of superconductivity in interstitial com-
pounds
N. Pessall, R. E. Gold, and H. A. Johansen
J. Phys. Chem. Solids, 29:19-38 (1968)

Emissivity and electrical resistivity of some
refractory carbides
V. A. Petrov, V. Ya. Chekhovskoi, A. E. Sheindlin
Proc. Fourth Symp. Thermophysical Properties, College Park,
Md., Apr. 1-4, 1968, ASME, New York (1968), pp. 270-277

Floating zone technique for growth of carbide single crystals
  W. Precht and G. E. Hollox
  High-Frequency Heating Rev., 2(6):7 (1968)

Superconductivity of ceramic compounds
  K. M. Ralls, E. R. Parker, and V. F. Zackay
  Ceram. Microstruct. Anal. Significance Product. Proc. 3rd Internation. Mater. Symp., Berkeley, Calif., 1966, John Wiley and Sons, New York (1968), pp. 489-508

Magnetic susceptibility of ZrC, HfC, and NbC
  G. V. Samsonov and A. Ya. Kuzma
  Izv. Akad. Nauk SSSR, Neorg. Mater., 4(8):1361-1363 (1968)
  Inorg. Mater., 4(8):1195-1197 (1968)

Behavior of zirconium and niobium carbides in a plasma
  S. Ya. Scharivker and M. I. Olievskii
  Poroshkovaya Met., 8(12):74-81 (1968)

Debye-temperatur von NbC, TaC, und TiC
  J. Sedivy, H. Sichova, and H. Vaneckova
  Freiberger Forschungsh. Met., B, 129:95-100 (1968)

Superconductivity in relation to the chemical bond in the carbides and nitrides of transition metals and their solid solutions
  O. I. Shulishova
  Chemical Bonds in Semiconductors and Thermodynamics, Nauka i Tekhnika, Minsk (1966), pp. 299-316
  Publ. Sept. 1967
  English translation, LA-TR-68-31, 1968

Superconducting properties of transition-metal carbides and nitrides
  Louis E. Toth
  Final Report 1 Oct. 65-1 Dec. 67 (Univ. of Minnesota, Minneapolis School of Mineral and Metallurgical Engineering), AFOSR-68-0265; AD-671944 (Feb. 1968)

Study of volatilization and chemical reactions in the plasma torch. Application to mixtures of refractory oxides and carbon
  H. Triche, C. Butti, and J. Besombes-Vailhe
  Method. Phys. Anal., 4:379-387 (1968)
  Refractory carbides synthesis (TiC, VC, TaC, SiC), in French

Anisotropy in Single-Crystal Refractory Compounds, Volume 2
  Fred W. Vahldiek and Stanley A. Mersol, ed.
  Plenum Press, New York (1968)
  Proceedings of International Symp. on Anisotropy in Single-Crystal Refractory Compounds, held June 13-15, 1967, in Dayton, Ohio; sponsored by Ceramics and Graphite Branch of Air Force Materials Laboratory

Diffusion and chemical kinetics related to problems in high-temperature chemistry of the refractory compounds
  T. C. Wallace
  Fundamentals of Refractory Compounds (H. H. Hausner and M. G. Bowman, eds.), Plenum Press, New York (1968), pp. 133-154

Neutron-diffraction study of the ordering of nonstoichiometric cubic carbides of fifth-group transition metals
  V. G. Zubkov, L. B. Dubrovskaya, P. V. Geld, and others
  Dokl. Akad. Nauk SSSR, 184(4):874-876 (1968)

Superconductivity of solid solutions of carbides and nitrides of transition metals
  N. E. Alekseevskii, G. V. Samsonov, and O. I. Shulishova
  Izv. Akad. Nauk SSSR, Neorg. Mater., 3(1):61-66 (1967)
  Inorg. Mater., 3(1):49-53 (1967)

Diffusion of carbon in carbides of the transition metals. IV. Group VI of the periodic system
  R. A. Andrievskii, V. S. Eremeev, V. N. Zagryazkin, and A. S. Panov
  Izv. Akad. Nauk SSSR, Neorg. Mater., 3(12):2158-2164 (1967)
  Inorg. Mater., 3(12):1884-1889 (1967)

Melting points in the zirconium carbide – hafnium carbide, tantalum carbide – zirconium carbide, and tantalum carbide – hafnium carbide systems
  R. A. Andrievskii, N. S. Strel'nikova, N. I. Poltoratskii, E. D. Khakhardin, and V. S. Smirnov
  Poroshkovaya Met., 1:85-88 (1967)

Phase formation accompanying the interaction of transition metals with carbon in fine layers
  A. A. Babad-Zakhryapin, L. M. Gert, and V. D. Rogozin
  (Inst. of Energetic Physics, Obninsk, USSR), Thermodynamics of Nuclear Materials, 1967, Vienna, International Atomic Energy Agency (1968), pp. 427-434
  From Symp. on Thermodynamics of Nuclear Materials with Emphasis on Solution Systems, Vienna, Austria
  STI/PUB-162; CONF-670915

Some physical properties (electrical resistance and Hall constant in relation to the electronic structures) of alloys in the zirconium carbide-niobium carbide and tantalum carbide-hafnium carbide systems
  I. G. Barantseva, V. N. Paderno, and Yu. B. Paderno
  Poroshkovaya Met., 2:70-73 (1967)

Study of growth parameters for refractory carbide single crystals
  R. W. Bartlett and F. A. Halden
  (Stanford Research Institute, Menlo Park, Calif.), Final Report, March 1, 1964 to June 30, 1967, Contract NASr-49(19), SRI Project FMU-4892 (June 30, 1967)

Low-temperature specific heat of carbides V_2C, Nb_2C, Mo_2C, and VMoC
  R. Caudron, P. Costa, and B. Saulgeot
  (Office National d'Etudes et de Recherches Aérospatiales, Chatillon-sous-Bagneux, France), ONERA-TP-480/1967 (1967), 10 pp.

Carbide synthesis by metal explosions in acetylene
  Eileen Cook and Bernard Siegel
  (Aerospace Corp., El Segundo, Calif.), Rept. TR-0158(3210)-3; SAMSO-TR-67-68, Contract F04695-67-C-0158 (Sept. 1967), 25 pp.
  Also AD-662 580

Lattice parameters, solution range, and production of solid solutions of some complex carbides of transition metals
  E. N. Denbnovetskaya
  Poroshkovaya Met., 3:32-37 (1967)

The chemical diffusion of carbon in the group VI-B metal carbides
  R. J. Fries, J. E. Cummings, C. G. Hoffman, and S. A. Dailey
  (Los Alamos Sci. Lab., N. M.), Contract W-7405-eng-36
  Presented at 6th International Plansee Seminar on High-Temperature Materials, Fundamental and Developments, Reutte, Tirol, Austria, CONF-680603-1; LA-3795 (Aug. 1967), 32 pp.

Interstitial Alloys
H. J. Goldschmidt
Plenum Press, New York, 632 pp.; Butterworths, London (1967)

Electrical properties of solid solutions of carbides of group IV and V transition metals
O. A. Golikova, E. O. Dzhafarov, A. I. Avgustinik, L. V. Kudryashova, and S. S. Ordanyan
Fiz. Tverd. Tela, 9(5):1557 (1967)
Sov. Phys. – Solid State, 9(5):1228-1230 (1967)

Investigation of the carbon K and metal emission bands and bonding for stoichiometric and non-stoichiometric carbides
J. E. Holliday
J. Appl. Phys., 38:4720 (1967)
The technique used and the results of this study will also be of importance to the electronmicroprobe user interested in the detection and quantitative analysis of carbon

Factors affecting the stability of metallic phases
W. Hume-Rothery
Phase Stability in Metals and Alloys (P. S. Rudman, J. Stringer, and R. I. Jaffee, eds.), McGraw-Hill, N. Y. (1967), pp. 3-23

Enthalpies and specific heats of Nb and Zr carbides at 500-2400°K
P. B. Kantor and E. N. Fomichev
Teplofiz. Vysok. Temp., 5(1):48-51 (1967)
High Temperature, 5(1):41-44 (1967)

Methods of producing pure carbides
T. Ya. Kosolapova
Paper presented at the Second Conference on Powder Metallurgy, Cracow (1967)

The thermodynamic stability of metallic phases
O. Kubaschewski
Phase Stability in Metals and Alloys (P. S. Rudman, J. Stringer, and R. I. Jaffee, eds.), McGraw-Hill, N. Y. (1967), pp. 63-83

Formation of stable electronic configurations of some physical properties of transition-metal carbides and nitrides in the range of their homogeneity
A. Ya. Kuchma and G. U. Samsonov
Izv. Akad. Nauk SSSR, Neorg. Mater., 2:1970-1974 (1966)
Inorg. Mater., 2(11):1705-1708 (1966)
NASA-TT-F-10847 (April 1967), 7 pp.

Thermo-emf of niobium and hafnium monocarbides as a function of carbon content in the range of homogeneity
A. Ya. Kuchma
Izv. Akad. Nauk SSSR, Neorg. Mater., 3(5):884-886 (1967)
Inorg. Mater., 3(5):790-792 (1967)

Magnetization of superconducting carbides
A. V. Narlikar and D. Dew-Hughes
J. Mater. Sci., 2:496 (1967)

La chimie structurale des composés des métaux de transition avec les éléments C, Si, Ge, Sn
H. Nowotny, W. Jeistchko, H. Goretzki, A. Wittmann, and H. Vollenkle
Propriétés thermodynam. phys. struct. Dérives semi-metal. Coll. internation. Orsay (1965), pp. 67-84
Paris, C.N.R.S., publ. 1967, pp. 181-191

A Handbook of Lattice Spacings and Structures of Metals and Alloys, Vol. 2
W. B. Pearson
Pergamon Press, Oxford, London, Edinburgh, New York, Toronto, Sydney, Paris, Braunschweig (1967), 1446 pp.

Total hemispherical emissive power monochromatic ($\lambda = 0.65\,\mu$) emissive power, and specific electrical resistivity of zirconium and niobium carbide in the temperature range 1200-3500°K
V. A. Petrov, V. Ya. Chekhovskoi, A. E. Sheindlin, V. A. Nikolaeva, and L. P. Fomina
Teplofiz. Vys. Temper., SSSR, 5(6):995-1000 (1967)
High Temperature, 5(6):889-893 (1967)

Classification of carbides
G. V. Samsonov
RTS-4233, Dec. 1967, 21 pp.
Ukr. Khim. Zh. (Kiev), 31:1005-1015 (1965)

The Refractory Carbides
E. K. Storms
(Los Alamos Sci. Lab., Los Alamos, N. M.), Vol. 2 of Refractory Materials (J. L. Margrave, ed.), Academic Press, N. Y. (1967), 285 pp.

Specific heat and character of the chemical bond in Ti and V carbides, nitrides and oxides
V. V. Tarasov, A. F. Demidenko, and A. K. Maltsev
Izv. Akad. Nauk SSSR, Neorg. Mater., 3(6):957-962 (1967)
Inorg. Mater., 3(6):857-861 (1967)

Preparation and characterization of high-quality single-crystal refractory metal carbides
J. M. Tobin and L. R. Fleischer
Second semi-annual progress report (1 April through 30 Sept. 1967) (Westinghouse Astronuclear Laboratory), Contract AF33(615)-3982, WANL-PR(GG)-017 (Nov. 1967)
See also AFML-TR-67-137, Part II (May 1968), 54 pp.

Carbide single crystals by solid-state grain growth
J. M. Tobin and L. R. Fleischer
Paper presented at High-Temperature Technology, Assilomar (1967)

Handbook of Binary Metallic Systems, Structure and Properties, Vol. II: Physicochemical Properties of the Elements
A. E. Vol
TT 66-51150 (1967), 870 pp.; Vol. I, TT 66-51149 (1966), 635 pp.
Includes hydrides, borides, carbides and nitrides, phase diagrams, and tables of properties

Production of refractory compounds
G. F. Wakefield
Vortrag: Chemical-Vapor Deposition of Refractory Metals, Alloys and Compounds, Gatlinburg (1967)

Carbide systems (HfC, TaC, NbC, ZrC, UC)
H. A. Wilhelm and J. P. Tosadale
(Ames. Lab., Iowa State Univ.) Annual Summary Research Report Ceramic and Mechanical Engineering, Chemical Engineering, Chemistry, Mathematics and Computer Science, Metallurgy, Physics, and Reactor Divisions, IS-1600 (July 1967), p. M-29

Superconductivity of the transition-metal carbides
R. H. Willens, E. Buehler, and B. T. Matthias
Phys. Rev., 159:327 (1967)

Vacuum thermionic work functions and thermal stability of TaB_2, ZrC, MoC, $MoSi_2$, $TaSi_2$, and WSi_2
R. G. Wilson and W. E. McKee
J. Appl. Phys., 38:1716 (1967)

Groups IVB and VB metal carbide-carbon eutectic temperatures
L. M. Adelsberg, L. H. Cadoff, and J. M. Tobin
J. Amer. Ceram. Soc., 49:573 (1966)

Structure of alloys in the system Zr – C – Ta
A. I. Avgustinik and S. S. Ordanyan
J. Appl. Chem. USSR, 39:289-293 (1966)

Study of growth parameters for refractory carbide single crystals
R. W. Bartlett
(Stanford Research Institute, Menlo Park, Calif.) Qtrly. Status Report No. 9, March 1-June 1, 1966, Contract NASr-49(19), NASA-CR-76365

The importance of powder metallurgical techniques and solid-state reactions in the treatment of metallurgical problems
F. Benesovsky
(Metallwerk Plansee A.G., Reutte/Tirol, Austria), Modern Developments in Powder Metallurgy, Vol. 3: Development and Future Prospects, Plenum Press (1966), p. 175

Gleitlinien in Carbidkristallen
F. Benesovsky and A. Ihrenberger
Z. Prak. Metall., 3:126 (1966)

Elastic properties of some polycrystalline transition-metal monocarbides
H. L. Brown, P. E. Armstrong, and C. P. Kempter
J. Chem. Phys., 45:547-549 (1966)
$HfC_{0.967}$, $NbC_{0.964}$, $TaC_{0.094}$, and $WC_{1.007}$

The use of an auxiliary metal bath for the production of high-purity carbide powders of the IVa-VIa group of elements
R. Kieffer and H. Rassaerts
Int. J. Powder Met., 2:15 (1966)

Engineering properties of selected ceramic materials
J. F. Lynch, C. G. Ruderer, and W. H. Duckworth
(Battelle Memorial Institute, Columbus, Ohio), publ. by American Ceramic Society, Inc., 4055 N. High Street, Columbus, Ohio (1966)

Untersuchungen über die Existenzbedingungen und die Eigenschaften kubischer Hartstoffphasen
A. Merz
Abh. dtsch. Akad. Wissensch. Berlin, Kl. Math. Phys. Tech., 1:237-244 (1966)
Carbides and nitrides of refractory metals

Effect of the chemical composition of the monocarbides of group IV and V transition metals in the homogeneous range on the temperature dependence of the resistivity and thermo-emf
V. S. Neshpor, S. V. Airapetyants, S. S. Ordan'yan, and A. I. Avgustinik
Izv. Akad. Nauk SSSR, Neorg. Mater., 2(5):855-863 (1966)
Inorg. Mater., 2(5):728-734 (1966)

Structure of alloys in the system Zr – C – Nb
S. S. Ordanyan, A. I. Avgustinik, and V. Sh. Vigdergauz
J. Appl. Chem. USSR, 39:284-288 (1966)

Thermodynamic data for Mo_2C and TaC
L. B. Prankratz, W. W. Weller, and E. G. King
Bur. Mines RI 6861 (1966)

Thermoemissivity characteristics of transient metals and their compounds
G. V. Samsonov, Yu. B. Paderno, and V. S. Fomenko
Ukr. Fiz. Zh., USSR, 10:622-629 (1965)
FTD-HT-66-178 (October 1966), 14 pp.

Preparation and investigation of some physical properties of HfC – MoC and TaC – MoC solid solutions
O. I. Shulishova and I. A. Shcherbak
Izv. Akad. Nauk SSSR, Neorg. Mater., 2:2145-2150 (1966)
Inorg. Mater., 2(12):1855-1860 (1966)

Superconducting critical temperatures of non-stoichiometric transition-metal carbides and nitrides
L. E. Toth, C. P. Wang, and C. M. Yen
Acta Met., 14:1403 (1966)

Preparation and characterization of high-quality single-crystal refractory metal borides and carbides
J. F. Wenckus
Arthur D. Little, Inc., Cambridge, Mass., Semi-Annual Progress Report, Contract AF33(615)-5130 (Nov. 1966)

Cubic carbides
W. S. Williams
Science, 152:34 (1966)

Thermodynamic stability of monocarbides of transition metals from subgroups IV to VI
R. G. Avarbe
Soviet Powder Met. Metal Ceram., 2:122-128 (1965)

Methods of analysis of metal refractories – borides, carbides and nitrides of hafnium, niobium, tantalum, titanium, and zirconium
R. E. Dutton, G. J. McKinley, D. McLean, and H. F. Wendt
(Union Carbide Inst., Tarrytown, N. Y.) Tech. Report No. C-29, Contract DA-30-069-ORD-2787, NP-14952 (Feb. 1965), 25 pp.

Herstellung der harten und hochschmelzenden Übergangsmetallcarbide aus dem Hilfsmetallbad
R. Kieffer, H. Rassaerts, und O. Schob
Monatsh. Chem., 96:686 (1965)

Measurement of the coefficients of thermal expansion of ZrC, HfC, NbC, and TaC at high temperatures
Yu. B. Paderno, E. M. Dudnik, T. V. Andreeva, I. G. Barantseva, and V. L. Yupko
Vysokotemperaturnye Neorg. Soedineniya, Akad. Nauk Ukr. SSR, Kiev (1965), pp. 293-296
AEC-tr-6867, 6 pp.

Diffusion of carbon in niobium and titanium carbide
A. V. Shcherbedinskaya and A. N. Minkevich
Izv. Vyssh. Ucheb. Zaved. Tsvetn. Met., 8:123 (1965), LA-tr-66-21

Towards a perfect crucible
H. F. Sterling
Standard Telecommunication Labs., Ltd., Harlow, Essex, England, Discovery (April 1965)

A differential thermal analysis apparatus for high temperatures. High-temperature phase reactions in refractory carbide systems
Planseeber. Pulvermet., 13:105 (1965)

Fundamental study of the sintering kinetics of refractory compound phases at high pressure and high temperature
E. V. Clougherty and Larry Kaufman
Final Report, Contract Nonr-4262(00), AD-637247 (Oct. 1964), 21 pp.

Chaleur spécifique électronique de carbures de métaux de transition
R. Conte
These Doct. Univ. Paris, 1964
Gap. Impr. Louis-Jean, 1966, 59 pp.

Diffusion of boron and carbon in refractory transition metals
A. P. Epik
Dopovidi Akad. Nauk Ukr. RSR, 1:67-70 (1964)
Ti, Zr, Nb, Ta, Mo, and W

Thermal expansion of certain group IV and group V carbides at high temperatures
C. R. Houska
J. Am. Ceram. Soc., 47:310 (1964)

Electrical properties of some transition-metal carbides and nitrides
John Piper
(Union Carbide Corp., Research Inst., Tarrytown, N. Y.) Tech. Report No. C-21, Contract DA-30-069-ORD-2787; NP-13944 (April 1964)

Plenum Press Handbooks of High-Temperature Materials, No. 1, Materials Index
P. T. B. Shaffer
Plenum Press, New York (1964), 740 pp.

A Critical Review of Refractories
Edmund K. Storms
(Los Alamos Scientific Lab., New Mexico), Contract W-7405-eng-36, LA-2942 (March 1964), 252 pp.

Process for the manufacture of very pure crystalline carbides, nitrides, or borides
Wacker-Chemie GmbH
British Patent 968,590 (Sept. 1964)

Transient thermal-diffusivity technique for refractory solids
J. A. Cape, G. W. Lehman, and M. M. Nakata
J. Appl. Phys., 34:3550 (1963)

Carbides of elements in the IVA (Ti, Zr, Hf), VA (V, Nb, Ta), and VIA (Cr, Mo, W) families, in carbides
William A. Frad
(Ames Laboratory), IS-711 (August 1963)
Extensive review, preparation and crystal structures, 1893 to 1961

Chemical synthesis via the high-intensity arc process
J. O. Gibson and R. Weidman
Chem. Eng. Prog., 59:53 (Sept. 1963)

Selected Values of Thermodynamic Properties of Metals and Alloys
Ralph Hultgren, R. L. Orr, P. D. Anderson, and K. K. Kelley
John Wiley and Sons, Inc. (1963)
Supplements issued in loose-leaf form at frequent intervals

Thermionic properties of some refractory metal carbides
J. H. Ingold
J. Appl. Phys., 34:2033 (1963)

Hartstoffe
R. Kieffer and F. Benesovsky
(Univ. of Vienna, Reutte/Tirol, Austria and Metallwerk Plansee A. G., Reutte/Tirol, Austria), Springer-Verlag, Vienna (1963)

Some physical, mechanical, and thermodynamic properties of transition-metal refractory carbides
J. E. Senkin and M. T. Milliken
(Lawrence Radiation Lab., Univ. of California, Livermore), Contract W-7405-eng-26, UCRL-7284 (March 1963)

Evidence for high-temperature forms of zirconium and tantalum monocarbides
P. T. B. Shaffer
J. Am. Ceram. Soc., 46:177 (1963)

Thermodynamic properties of 65 elements — their oxides, halides, carbides, and nitrides
C. E. Wicks and F. E. Block
(Bureau of Mines, Albany, Oregon), Bulletin 605, Dec. 1961

Research investigation to determine the optimum conditions for growing single crystals of selected borides, silicides, and carbides
A. D. Kiffer
(Linde Co., Newark, N. J.), Contract AF33(616)-6326, March 1959-Feb. 1960, WADD-TR-60-52

Graphite, carbide, nitride, and sulfide refractories
L. M. Litz
(National Carbon Co., Cleveland, Ohio), Proc. Intern. Symp. High-Temperature Technology, Asilomar Conf. Grounds, Calif., Oct. 6-9, 1959, McGraw-Hill Book Co. (1960), p. 90

Electrical properties of borides, carbides, and nitrides; the IVa to VIa metals
S. N. L'vov, V. F. Nemchenko, and G. V. Samsonov
Dokl. Akad. Nauk SSSR, 135:577-580 (1960)

The electron-beam melting of beryllium, boron, boron carbide, tantalum carbide, titanium carbide, tungsten, and zirconium diboride
R. L. Martin, S. R. Seagle, and O. Bertea
(Reactive Metals, Inc., Niles, Ohio) Qtrly. Prog. Report No. 8, Feb. 1-April 30, 1960, Contract AF33(616)-5603, NP-8816

Thermoelectric power of TiC and VC
S. Noguchi and T. Sato
J. Phys. Soc. Japan, 15:2359 (1960)

An evaluation of data on nuclear carbides
F. A. Rough and W. Chubb
(Battelle Memorial Institute), BMI-1441 (May 31, 1960)

The oxidation of Cr, Hf, Mo, Nb, Ta, Ti, W, V, Zr, and some of their alloys; properties of some borides, carbides, nitrides, and oxides of the same metals, a bibliography
E. A. Cernak
(Pratt and Whitney Aircraft Div., United Aircraft Corp., Middletown, Conn.), CNLM-1802-5 (Dec. 15, 1959)

The preparation of uranium monocarbide and its behavior compared with other high-melting carbides
R. Kieffer, F. Benesovsky, et al.
(United Kingdom Atomic Energy Authority), Trans. of Plansee-berichte für Pulvermetallurgie, 5:33 (1957)
IGRL-T/C-52; AD-147979 (Nov. 1957)

Intermediate stages of the reactions in the formation of carbides of titanium, zirconium, vanadium, niobium, and tantalum
G. V. Samsonov
Ukr. Khim. Zh., 23:287 (1956) for Westinghouse Electric Corp., AEC-tr-3387

Investigation of the diffusion of boron carbon in certain metals of transition groups
G. V. Samsonov and V. P. Latisheva
AEC-tr-3321 (translated by Lydia Venters), Fiz. Metal. i Metalloved. Akad. Nauk SSSR, Ural. Filial, 2:309-319 (1956), 20 pp.

Development of a heat-resisting and refractory material with low specific weight
J. Bingel
Arch. Metallkunde, 1:309 (1947)

Thermodynamic properties of nonstoichiometric vanadium and titanium carbides
V. I. Alekseev, Ye. V. Fiveiskii, and L. A. Shvartsman
LA-4212-TR, pp. 435-447 of STI/PUB-162, 9 pp.

Chemical reactions in the electric arc: Reactive metal carbides
E. D. Calvert, M. M. Kirk, and R. A. Beall
Bureau of Mines, BM-RI-5951

Temperature dependence of thermoelectromotive force and specific electrical resistance of titanium, vanadium, chromium, and their borides, carbides, and nitrides
S. N. L'vov and V. F. Nemchenko
Vysokotemperaturnye Neorganicheskie Soedineniia, pp. 100-107, LA-TR-67-17

3.b. Group IV

Magnetoresistance and Hall effect in carbides of metals in sub-group IVA
B. I. Malinskii, V. F. Nemchenko, and S. N. L'vov
Izv. VUZ Fiz. (USSR), 6:125-129 (1970)
$TiC_{0.98}$, $ZrC_{1.0}$, and $HfC_{0.9}$

Formation of titanium and zirconium carbides by reduction of the chlorides with hydrogen on a graphite substrate
R. S. Ambartsumyan and B. N. Babich
Izv. Akad. Nauk SSSR, Neorg. Mater., 5(2):301-304 (1969)
Inorg. Mater., 5(2):250-253 (1969)

Electrical properties of carbides of group IV transition metals
O. A. Golikova, E. O. Dzhafarov, A. I. Avgustinik, and G. M. Klimashin
Fiz. Tekh. Poluprovod., 3:506-510 (1969)
Sov. Phys. — Semicond., 3:429 (1969)
Electrical conductivity, thermoelectric power, and Hall co-efficient of zirconium, titanium and hafnium carbides as a function of their carbon content

Thermoelectric power and electrical conductivity of titanium and zirconium monocarbides in the temperature range 20-2000°C
O. A. Golikova, E. O. Dzhafarov, A. I. Avgustinik, and G. M. Klimashin
Fiz. Tverd. Tela, 10(1):168-170 (1968)
Sov. Phys. — Solid State, 10(1):124-126 (1968)

Preparation and characterization of high-quality single-crystal refractory metal borides and carbides
J. S. Haggerty, D. W. Lee, and J. F. Wenckus
Arthur D. Little, Inc., Cambridge, Mass.
Final Technical Report, Contract AF33(615)-5130, AFML-TR-68-228 (August 1968), 83 pp.

Determination of free carbon in the carbides of titanium, zirconium, and hafnium
K. D. Modylevskaya and G. Kh. Kotlyar
Ind. Lab., 34:1123 (1968)
Ti, Zr, Hf

Thermodynamic properties of zirconium and hafnium carbides over the range of 298 to 2500°K
A. S. Bolgar, E. A. Guseva, and V. V. Fesenko
Porosh. Met., Akad. Nauk Ukr. SSR, 1:40-43 (1967)

Preparation and characterization of high-quality single-crystal refractory metal borides and carbides
J. S. Haggerty, D. W. Lee, and J. F. Wenckus
Arthur D. Little, Inc., Cambridge, Mass.
Semi-annual Progress Report, Contract AF33(615)-5130 (May 1967)
No crystals at date

Cubic carbides, nitrides, and oxides of the first transition-metal series
F. Petru and V. Brozek
Pokroky Praskove Met. VUPM, No. 4, pp. 3-27 (1967)

Mass-spectrometric investigation of the composition of the vapor above the systems $TiC-C$ and $ZrC-C$
T. S. Starostina, L. N. Sidorov, P. A. Akishim, and N. M. Karasev
Izv. Akad. Nauk SSSR Neorg. Mater., 3(4):727-728 (1967)
Inorg. Mater., 3(4):647-648 (1967)

Low-temperature elastic properties of ZrC and TiC
Roger Chang and L. J. Graham
J. Appl. Phys., 37:3778 (1966)

Thermoelectric emission from HfC and ZrC
N. A. Gorbatyi, G. V. L'vov, and V. A. Perederii
Izv. Akad. Nauk SSSR, Ser. Fiz., 30:1942-1949 (1966)

Thermal expansion and atomic vibration amplitudes for TiC, TiN, ZrC, ZrN, and pure tungsten
C. R. Houska
J. Phys. Chem. Solids, 25:359-366 (1964)

Properties of non-stoichiometric metallic carbides
J. T. Norton and R. K. Lewis
(Advanced Metals Research Corp., Somerville, Mass.)
Final Report Contract NASw-663, NASA-CR-58046 (April 1964), 42 pp.
Group IV carbides, lattice parameter, specific gravity, thermal expansion, and thermoelectric power

3.b.1. Titanium Carbide

Investigation of the charge distribution in titanium carbide using electromigration
D. L. Kohlstedt and W. S. Williams
Phys. Rev., B3:293-305 (1971)

Chemical Vapor Deposition in the Titanium-Carbon Systems
J. J. Nickl and M. Reichle
J. Less-Common Metals, 24:63-72 (1971)
Single crystals and polycrystalline layers

Mechanism of collective recrystallization of titanium carbide
S. A. Bozhko and G. V. Samsonov
Poroshkovaya Met., 10(2):46-52 (1970)

Etude des bandes de précipitation dans le carbure de titane sous stoechiométrique
J.-L. Chermant, P. Delavignette, and A. Deschanvres
J. Less-Common Metals, 21:89 (1970)

Electronic band structure and the K and L x-ray spectra from TiO, TiN, and TiC
D. W. Fischer
J. Appl. Phys., 41:3922-3926 (1970)

Superconductivity of titanium carbide
G. M. Klimashin, V. S. Nashpor, V. P. Nikitin, V. I. Novikov, and S. S. Shalyt
Zh. Eksper. Teor. Fiz., Pisma, 12(3):147-149 (1970)
JETP Letters, 12(3):102-104 (1970)

Concentration-dependent chemical diffusion in TiC
D. L. Kohlstedt and Wendell S. Williams
Bull. Am. Phys. Soc., 15:390 (1970)

Chemical diffusion in titanium carbide crystals
D. L. Kohlstedt, W. S. Williams, and J. B. Woodhouse
J. Appl. Phys., 41(11):4476-4484 (1970)

Mass-spectrometric determination of the dissociation energies of titanium dicarbide and titanium tetracarbide
Carl A. Steams and Fred J. Kohl
(NASA, Lewis Research Center), NASA-TN-D-5653 (March 1970)

Single-crystal growth of titanium carbide by chemical-vapor deposition
T. Takahashi, K. Sugiyama, and H. Itoh
J. Electrochem. Soc., 117:541-545 (1970)

Investigation of ionicity in titanium carbide using electromigration
Wendell S. Williams and D. L. Kohlstedt
Bull. Am. Phys. Soc., 15:390 (1970)

Reactions de formation et propriétés de TiC
J.-L. Chermant
Rev. Int. Hautes Temp., 6:299-312 (1969).

Contribution à l'étude des propriétés physico-chimiques du carbure de titane
J.-L. Chermant
Ph. D. thesis, Caen, CNRS No. 2944 (Jan. 1969), 100 pp.

Effect of high-temperature heating on the thermoelectric characteristics of graphite and titanium carbide
P. S. Kislyi, L. S. Golubyak, L. V. D'yakonova, and O. V. Zaverukha
Teplofiz. Vysok. Temp., 7(5):1023-1025 (1969)
High Temperature, 7(5):955-957 (1969)

Study of electromigration in TiC using electron probe
D. L. Kohlstedt and W. S. Williams
Bull. Am. Phys. Soc., 14:389 (1969)

Study of the electronic structure and interatomic bonds in some compounds and binary alloys by the method of x-ray spectroscopy
S. A. Nemnonov, A. Z. Menshikov, K. M. Kolobova, E. Z. Kurmayev, and V. A. Trapeznikov
Trans. Met. Soc. AIME, 245:1191-1198 (1969)

X-ray study of inner-level shifts and band structure of TiC and related compounds
Lars Ramqvist, Borje Ekstig, Elisabeth Kallne, Erik Noreland, and Rolf Manne
J. Phys. Chem. Solids, 30:1849-1860 (1969)

Diffusion of ^{44}Ti in TiC_x
S. Sarian
J. Appl. Phys., 40:3515 (1969)

Application of statistical model of plastic flow to transition-metal carbides
Wendell S. Williams
Bull. Am. Phys. Soc., 14:440 (1969)

Thermodynamic properties of non-stoichiometric vanadium and titanium carbides
B. I. Alekseev, A. S. Panov, E. V. Fiveiskii, and L. A. Schvarzman
Thermodynamics of Nuclear Materials (IAEA, Vienna, Conf., 1967, published 1968), pp. 435-447

Band structure and the titanium $L_{II,III}$ x-ray emission and absorption spectra from pure metal, oxides, nitride, carbide, and boride
David W. Fischer and William L. Baun
J. Appl. Phys., 39:4757:4776 (1968)

Determination of N in TiC
B. J. Hambridge and A. Parker
AERE AM 108 (1968)

Thermodynamics and kinetics of alloy preparation in the C — TiC system by precipitation from the gaseous phase
V. S. Kilin, A. I. Evstyukhin, and V. S. Dergunova
Zh. Fiz. Khim., 42:785-792 (1968)

Comparison of methods for the determination of nitrogen in titanium carbide
A. Parker, C. Healy, and E. H. Henderson
(Atomic Energy Establishment, Harwell, England), AERE-R-5937 (Oct. 1968), 9 pp.

Formation of silicon and titanium carbides by chemical-vapor deposition
M. L. Pearce and R. W. Marek
J. Amer. Ceram. Soc., 51:84 (1968)

Discussion of "Formation of silicon and titanium carbides by chemical-vapor deposition"
M. L. Pearce and R. W. Marek
J. Am. Ceram. Soc., 51:355 (1968)

Low-temperature thermal conductivity of titanium carbide
L. G. Radosevich and W. S. Williams
Bull. Am. Phys. Soc., 13:510 (1968)

Properties of alloys of NbC and TiC in the homogeneity region
G. V. Samsonov and G. S. Upadkhaya
Poroshkovaya Met., 8(9):70-74 (1968)

Diffusion of carbon in TiC
S. Sarian
J. Appl. Phys., 39:3305 (1968)

Anomalous diffusion of C in $TiC_{0.67}$
S. Sarian
J. Appl. Phys., 39:5036 (1968)

Enthalpy and specific heat of TiC
A. G. Turchanin and V. V. Fesenko
Zh. Fiz. Khim., 42:1026-1028 (1968)

K_α band of carbon x-ray emission in titanium carbides, in diamond and in graphite
E. A. Zhurakovskii
Dokl. Akad. Nauk SSSR, 180(5):1088-1091 (1968)
Sov. Phys. — Dokl., 13(6):578-580 (1968)

Study of the electron and atomic structure of titanium carbide and titanium oxide
M. P. Arbuzov, E. T. Kachkovskaya, and B. V. Khaenko
Chemical Bonds in Semiconductors and Thermodynamics, 1966, Nauka i Tekhnika, Minsk (Sept. 1967), pp. 64-71

Study of x-ray K spectra of titanium in its nitride and carbide
V. I. Chirkov, S. M. Blokhin, and E. E. Vainshtein
Fiz. Tverd. Tela, 9(4):1116-1121 (1967)
Sov. Phys. — Solid State, 9(4):873-877 (1967)

Hall effect in titanium monocarbide at high temperatures
O. A. Golikova, F. L. Feigel'man, A. I. Avgustinik, and G. M. Klimashin
Fiz. Tekh. Poluprovod., 1(2):236-238 (1967)
Sov. Phys. — Semicond., 1(2):187-189 (1967)

Band structure and bonding in TiC
R. G. Lye
Atomic and Electronic Structure of Metals, American Soc. Metals, Metals Park (1967)

Electronic energy band structure of titanium carbide
R. G. Lye
Propriétés Thermodynamiques, Physiques, et Structurales des Dérivés Semi-Metalliques, Colloques Internationaux du Centre National de la Récherche Scientifique No. 157, Orsay, 28 Sept. 1-1 Oct. 1965 (Publ. 1967), p. 208

Titanium carbide formation
S. Ozaki and Y. Iida
Powder Met., 3:43-49 (1967)

Sphäroidisierung von TiC und Chromkarbid im Plasma
V. M. Sleptsov
Izv. Akad. Nauk SSSR, Met., 6:213-215 (1967)

Field-ion microscopy of titanium carbide
D. A. Smith, B. Ralph, and W. S. Williams
Phil. Mag., 16:415 (1967)

Bildung von TiC
T. Takahashi
Kinzoku Hyomen Gijutsu, 18:264-267 (1967)

The microstructure of single-crystal titanium carbide
F. W. Vahldiek
J. Less-Common Metals, 12:429-440 (1967)

The nature of precipitates in boron-doped TiC
John D. Venables
Phil. Mag., 16:873 (1967)

Stacking faults in titanium carbides
J. Venables
Phys. Stat. Sol., 15:413-416 (1966)

The effect of carbon content on certain electrophysical properties of TiC
A. I. Avgustinik, O. A. Golikova, G. M. Klimashin, L. V. Kozlovskii, and V. S. Neshpor
Issled. v Obl. Khim. Silikatov i Okislov, 26:241-244 (1965)

Electrical properties of titanium carbide
O. A. Golikova, A. I. Avgustinik, G. M. Klimashin, and L. V. Kozlovskii
Fiz. Tverd. Tela, 7(9):2860-2862 (1965)
Sov. Phys. — Solid State, 7(9):2317-2318 (1966)

The thermoelectric power of titanium carbide
R. G. Lye
J. Phys. Chem. Solids, 26:407-413 (1965)

[Title Not Given]
E. Rudy, D. P. Harmon, and C. E. Bruckl
Wright-Patterson Air Force Base Tech. Rept. No. AFML-TR-65-2, Part 1, Vol. II (1965)
Phase diagram

Diffusion of boron and carbon in refractory transition metals
A. P. Epik
Dopovidi Akad. Nauk Ukr. RSR, 1:67-70 (1964)

Electronic band structure of TiC, TiN, and TiO
V. Ern and A. C. Switendick
(Massachusetts Inst. of Tech., Cambridge, Mass.), Contracts Nonr-1841(10); AF33(616)-8353; Grant NSF-GP-3241, TR-192; AD-608826 (Oct. 1964), 28 pp.

Physical properties of titanium carbide in the region of homogeneity
S. N. L'vov, V. F. Nemchenko, T. Ya. Kosolapova, and G. V. Samsonov
Dokl. Akad. Nauk SSSR, 157(2):408-411 (1964)
AEC-TR-6507 (1964)
Dokl. — Phys. Chem., 157(2):717-720 (1964)

On the reactive diffusion of boron and carbon in refractory transition metals
G. V. Samsonov
Dopovidi Akad. Nauk Ukr, RSR, 1:68-70 (1964)

A Critical Review of Refractories
E. K. Storms
Los Alamos Scientific Laboratory Rept. LA-2942 (1964)
Phase diagram

Elastic deformation, plastic flow and disloca-
tion in single crystals of titanium carbide
W. S. Williams and R. D. School
J. Appl. Phys., 33:955 (1962)

Electrical conductivity and thermoelectric
effect in single-crystal TiC
L. E. Hollander, Jr.
J. Appl. Phys., 32:996-997 (1961)

3.b.2. Zirconium Carbide

Neutron-diffraction study of zirconium carbo-
hydride
V. N. Bykov, V. S. Golovkin, V. P. Kalinin, V. A. Levdik, and
V. I. Shcherbak
Kristallografiya, 14(5):913-914 (1969)
Sov. Phys. — Cryst., 14(5):785-786 (1970)

X-ray diffraction pycnometric determinations
of the density of zirconium-base interstitial
phases
A. S. Shevchenko, R. A. Andrievskii, V. P. Kalinin, and R. A.
Lyutikov
Poroshkovaya Met., 10(1):89-91 (1970)
Tabulation of carbides, nitrides

Solubility of oxygen in zirconium carbide
Yu. G. Zainulin, S. I. Alyamovskii, G. P. Shveikin, and P. V.
Geld
Izv. Akad. Nauk SSSR, Neorg. Mater., 6(1):118-119 (1970)
Inorg. Mater., 6(1):96-97 (1970)

Magnetic susceptibility and energy band struc-
ture of zirconium monocarbide
A. S. Borukhovich, L. B. Dubrovskaya, I. I. Matveenko, and
P. V. Gel'd
Phys. Stat. Sol., 36:97 (1969)

Microstructure of ZrC
R. D. Carnahan, K. R. Janowski, and R. C. Rosse
Metallography, 2:65-77 (1969)

The determination of oxygen in zirconium
carbide by a spectral method involving use of
the direct-current arc
V. M. Plotnitskii
Zavodsk. Lab. 35:169-172 (1969)
Indust. Lab., 35:202-204 (1969)

Thermal activation in Zr — C alloys
P. D. Gupta and V. S. Arunachalam
Trans. Indian Inst. Metals, 21(4):29-33 (1968)

Thermionic emission of pyrolytic ZrC and W
coatings
T. L. Matskevich, T. V. Krachino, Yu. N. Vil'k, and V. S.
Davydov
Zh. Tekh. Fiz. (USSR), 38(8):1379-1384 (1968)

Beitrag zum System Zr — C
H. Nickel, O. Inanc, and K. Lucke
Z. Metallkunde, 59:935-940 (1968)

Electrical resistivity of ZrC
D. L. Paulson and G. Asai
Bur. Mines, RI:1361 (1968)

A floating zone technique for the growth of car-
bide single crystals
Walter Precht and Graham E. Hollox
(Research Institute for Advanced Studies, Martin Marietta
Corp., Baltimore, Md.), Contract DA-31-124-ARO-D-467
and NASw-1290, RIAS TR-68-9c (May 1968), 19 pp.

Production of zirconium carbides on a pilot
plant scale
R. G. Shumilova and T. Ya. Kosolapova
Poroshkovaya Met., 8(4):86-69 (1968)

Electrolytic purification of ZrC
A. B. Suchkov, G. A. Meerson, and J. G. Olesov
Izv. Akad. Nauk SSSR, Met., 3:106-108 (1968)

Study of alloys belonging to the zirconium —
carbon system
Yu. G. Godin, A. I. Evstyukhin, V. S. Emel'yanov, A. A. Rusa-
kov, and I. I. Suchkov
High-Purity Metals and Alloys. Fabrication, Properties, and
Testing (V. S. Emel'yanov and A. I. Evstyukhin, eds.),
Consultants Bureau, New York (1967), pp. 17-22

Preparation of ZrC from the gas phase
V. S. Kilin and others
Met. Metalloved. Chist. Met., No. 6, 19-29 (1967)

Investigation of the mechanical properties of
transition-metal carbides and borides
Final Report, Arthur D. Little, Inc., Cambridge, Mass.,
Contract AT(30-1)-3411, ALI-3411-3 (Oct. 14, 1967), 64 pp.

Diffusion of carbon through zirconium mono-
carbide
S. Sarian and J. M. Criscione
J. Appl. Phys., 38:1794 (1967)

Preparation and characterization of high-quality
single-crystal refractory metal carbides
J. M. Tobin and L. R. Fleischer
(Westinghouse Astronuclear Laboratory), Second Semi-Annual
Progress Rept. (1 April through 30 Sept. 1967), Contract
AF33(615)-3982, WANL-PR(GG)-017 (Nov. 1967)

Some electrical and thermophysical properties
of zirconium monocarbide as functions of carbon
content over the homogeneous range
A. I. Avgustinik, O. A. Golikova, G. M. Klimashin, V. S.
Neshpor, S. S. Ordan'yan, and V. A. Snet'kova
Izv. Akad. Nauk SSSR, Neorg. Mater., 2(8):1439-1443 (1966)
Inorg. Mater., 2(8):1230-1233 (1966)

Elastic properties of zirconium carbide
H. L. Brown and C. P. Kempter
Phys. Stat. Sol., 18:K21 (1966)

High-temperature thermophysical properties
of zirconium carbide
L. N. Grossman
J. Am. Ceram Soc., 48:236 (1965)

The system zirconium — carbon
R. V. Sara
J. Am. Ceram. Soc., 48:243 (1965)

Phase diagram of the Zr — ZrC system
Yu. N. Vil'k, S. S. Ordan'yan, R. G. Avarbe, A. I. Avgustinik,
T. P. Ryzhkova, and Yu. A. Omel'chenko
Zh. Prikl. Khim., 38:1500-1506 (1965)

Diffusion of boron and carbon in refractory
transition metals
A. P. Epik
Dopovidi Akad. Nauk Ukr. RSR, 1:67-70 (1964)

The properties of pyrolytic ZrC
M. P. Lepie
Trans. Brit. Ceram. Soc., 63:431 (1964)

On the reactive diffusion of boron and carbon in refractory transition metals
G. V. Samsonov
Dopovidi Akad. Nauk Ukr. RSR, 1:68-70 (1964)

An x-ray investigation of polymorphism in ZrC
D. K. Smith and C. F. Cline
J. Am. Ceram. Soc., 46:566 (1963)

Development of ultrapure refractory materials
Peter T. B. Shaffer
Contract NOrd-17175, Progress rept. No. 24, Feb. 1, 1961-
April 30, 1961 (Apr. 28, 1961), 12 pp.

Thermal properties of refractory materials
G. W. Lehman
(Atomics International Div. of North American Aviation, Inc.,
Canoga Park, Calif.), Contract AF33(616)-6794, WADD-TR-
60-581 (July 1960), 19 pp.

3.b.3. Hafnium Carbide

Structure and properties of carbide crystals grown from solution
D. J. Rowcliffe
Am. Ceram. Soc. Bull., 48(4):404 (1969)

Chemical-vapor deposition of hafnium carbide
W. R. Wilcox, J. R. Teviotdale, and R. A. Corley
(Aerospace Corp., El Segundo California), Rept. TR-0158
(3250-10)-3; SAMSO-TR-67-77, Contract F04695-67-C-
0158 (Oct. 1967), 21 pp.
Also AD-661-960
Trans. AIME, 242:588 (1968)

The hafnium-carbon phase diagram
Donald K. Deardorff, R. P. Adams, and Mark I. Copeland
(Bureau of Mines, Albany, Oregon), BM-RI-6983 (July 1967),
20 pp.

Electrical resistivity of fused hafnium carbide alloys at elevated temperatures
D. K. Deardorff and M. I. Copeland
(Bureau of Mines), USBM-RC-1113 (1964), 20 pp.

On the composition and structure of hafnium a carbide
V. I. Zhelankin and V. S. Kutsev
Zh. Strukt. Khim., 4:865-867 (Nov.-Dec. 1963)

[Title Not Given]
E. Rudy
Aerojet-General Corp. Tech. Rept. No. AFML-TR-65-2,
Part 1, Vol. IV
Phase diagram

3.c. Group V

The real structure of the higher carbide of vanadium
M. P. Arbuzov, V. G. Fak, and B. V. Khaenko
Kristallografiya, 15:196-199 (1970)
Sov. Phys. — Cryst., 15:164-166 (1970)

VC, NbC, and TaC with varying carbon content studied by ESCA
Lars Ramqvist, Kjell Hamrin, Gunilla Johansson, Ulrik Gelius,
and Carl Nordling
J. Phys. Chem. Solids, 31:2669-2672 (1970)

On the crystal chemistry of the close-packed transition-metal carbides. I. The crystal structure of the ζ-V, Nb and Ta carbides
K. Yvon and E. Parthe
Acta Cryst., 26B:149 (1970)

Magnetic susceptibility of niobium and tantalum monocarbides at low temperatures
A. S. Borukhovich, L. B. Dubrovich, L. B. Dubrovskaya,
I. I. Matveenko, and P. V. Gel'd
Fiz. Tverd. Tela, 11:830-832 (1969)
Sov. Phys. — Solid State, 11:681 (1969)

Obtaining simple and mixed metallic carbides. II. (Production, structure, and thermodynamic properties of) carbides of vanadium, tantalum, and niobium
Réné A. Paris and Edith Clar
Chim. Ind. (Paris), 102:57-62 (1969)
In French

Ordering in the structure of niobium and tantalum monocarbides
V. G. Zubkov, L. B. Dubrovskaya, P. V. Gel'd, V. A. Tskhai,
and Yu. A. Dorofeev
Fiz. Met. Metalloved., 27:352 (1969)

Diffusion of C in the carbides of Ta
W. F. Brizes
J. Nucl. Mater., 26:227-231 (1968)

The thermodynamic functions of NbC and TaC
I. A. Nikolskaya et al.
Zur. Fiz. Chim., 42:637-640 (1968)

Revision of the vanadium-carbon and niobium-carbon systems
E. Rudy, S. Windisch, and C. E. Brukl
Planseeberichte für Pulvermetallurgie, 16:3 (1968)

Orthorhombic β-phases of niobium and vanadium carbides
S. I. Alyamovskii, G. P. Shveikin, P. V. Gel'd, and N. M.
Volkova
Zh. Neorg. Khim. (Moscow), 12:579-582 (1967)
LA-TR-67-73

Effect of temperature and composition on certain thermodynamic characteristics of vanadium carbides
V. S. Chernyaev, E. N. Shchetnikov, G. P. Shveikin, R. P.
Krentsis, and P. V. Gel'd
Izv. Akad. Nauk SSSR, Neorg. Mater., 3(5):789-796 (1967)
Inorg. Mater., 3(5):705-711 (1967)

Chemistry of Niobium and Tantalum
F. Fairbrother
Topics in Inorganic and General Chemistry. Monograph 10,
American Elsevier Publishing Co., Inc., New York (1967),
250 pp.

Lower-temperature modifications of Nb_2C and V_2C
Erwin Rudy and Charles E. Brukl
J. Am. Ceram. Soc., 50:265 (1967)

Superconductive anomaly in specific heats of some niobium and vanadium compounds
K. Ukei and E. Kanda
Ann. Acad. Sci. Fennicae A VI 210, 104-7 (1966)
Proc. Low Temp. Calorimetry Conf., Otaniemi, 1966

The crystal structures of V_2C and Ta_2C
A. L. Bowman, T. C. Wallace, J. L. Yarnell, R. G. Wenzel, and E. K. Storms
Acta. Cryst., 19:6 (1965)

The free energies of formation of the vanadium, niobium, and tantalum carbides
W. L. Worrell and John Chipman
J. Phys. Chem., 68:860-866 (1964)

Investigation of tantalum carbide, niobium carbide, and vanadium carbide for superconductivity
A. L. Giorgi, E. G. Szklarz, E. K. Storms, and A. L. Bowman
Phys. Rev., 129:1524-1525 (1963)

High-temperature heat content of niobium carbide and of tantalum carbide
L. S. Levinson
(Univ. of Calif.), J. Chem. Phys., 39:1550 (1963)
Letter of Jan. 11, 1965 from Levinson to T. A. Chang, Aerojet-General Corp. gives corrected equation (−2.407 should be −24.070)

Standard heat of formation of higher niobium and tantalum carbides
A. N. Kornilov, V. Ya. Leonidov, and S. M. Skuratov
Vestn. Mosk. Univ., Ser. II, Khim., 6:48-50 (1962)

Thermal decomposition of niobium and tantalum monocarbides
C. P. Kempter and M. R. Nadler
J. Chem. Phys., 32:1477 (1960)

The solubility of carbon and structure of carbide phases in tantalum and columbium
M. L. Pochon, C. R. McKinsey, R. A. Perkins, and W. D. Forgeng
(Electro Metallurgical Co., Niagara Falls, N. Y.), Reactive Metals. Proc. 3rd Ann. Conf., Buffalo, N. Y., May 27-29, 1958 (W. R. Clough, ed.), Met. Soc. Conf., Vol. 2, Interscience Publ. (1959)

3.c.1. Vanadium Carbide

Dislocation etch pits in vanadium carbide monocrystals
R. K. Govila
Phil. Mag., 22(176):431-436 (1970)

Low temperature specific heat of vanadium carbide
D. H. Lowndes, Jr., L. Finegold, and R. G. Lye
Phil. Mag., 21:245-255 (1970)

Vanadium L(II, III) x-ray emission and absorption spectra from metal oxides, nitride, carbide and boride, summary tech. rept., May 1968-April 1969
David W. Fisher
(Air Force Systems Command, Materials Laboratory, Wright-Patterson AFB, Ohio), AFML-TR-69-143, AD-697027 (Sept. 1969), 57 pp.

Vanadium $L_{II,III}$ x-ray emission and absorption spectra from metal, oxides, nitride, carbide, and boride
D. W. Fischer
J. Appl. Phys., 40:4151-4163 (1969)

Specific heat of vanadium carbide, 1-20K
D. H. Lowndes, Jr., L. Finegold, D. W. Bloom, and R. G. Lye
Program and Abstracts of the 3rd Materials Research Symposium, Electronic Density of States, Gaithersburg, Md., Nov. 3-6, 1969, pp. 173-175, National Bureau of Standards, 1969

Low-temperature specific heat of vanadium carbide
D. H. Lowndes, Leonard Finegold, and R. G. Lye
(Research Institute for Advanced Studies, Martin Marietta Corp., Baltimore, Maryland), NASA Contract NASw-1290, Ninth Tech. Rept. to NASA, RIAS Tech. Rept. 69-15c (October 1969)

The thermodynamics of VC
V. I. Malkin and V. V. Pokidyschev
Izv. Akad. Nauk SSSR, Met., No. 2, pp. 183-187 (1969)

Radiation damage of ordered V_6C_5 by electron-microscope beam bombardment
John D. Venables and R. G. Lye
Phil. Mag., 19:565-582 (1969)

Thermodynamic properties of non-stoichiometric vanadium and titanium carbides
B. I. Alekseev, A. S. Panov, E. V. Fiveiskii, and L. A. Shvarzman
Thermodynamics of Nuclear Materials, 1967 (IAEA, Vienna, 1968), pp. 435-447

The thermodynamic activity of C in VC
V. I. Alekseev et al.
Zh. Fiz. Khim., 42:615-619 (1968)

A new phase in the $V - C$ system
S. I. Alyamovskii, P. V. Gel'd, G. P. Shveikin, and E. N. Shchetnikov
Zh. Neorg. Khim., 13:895-896 (1968)

Thermal properties of vanadium carbides
V. S. Chernyaev et al.
Tr. Ural. Politechn. Inst., No. 167, pp. 151-152 (1968)

Properties of boron-doped VC
G. E. Hollox and J. D. Venables
AD-682 605 (1968)

The microstructure and mechanical properties of pure and boron-doped $VC_{0.85}$
G. E. Hollox and J. D. Venables
Trans. Japan Inst. Metals, 9 Suppl., 295-300 (1968)

Formation of vanadium carbide during diffusion transformation
E. L. Kolosova, M. I. Goldshtein, and G. D. Susloparov
Fiz. Metal. Metalloved., 25:681-688 (1968)

Vanadium monocarbide with non-stoichiometric vacancies
D. Kordes
Phys. Stat. Sol., 26:K103-105 (1968)

Non-stoichiometry of vanadium carbide
F. Petru, V. Dufek, V. Brozek, and L. Mrnak
Chem. Prum., 18:177-180 (1968)

A floating zone technique for the growth of carbide single crystals
W. Precht and G. E. Hollox
J. Crystal Growth, 3:818-823 (1968)

Structure of the ordered compound V_6C_5
  J. D. Venables, D. Kahn, and R. G. Lye
  RIAS Technical Report 68-3c, Feb. 1968, NASA Contract
    NASw-1290
  Phil. Mag., 18:177-192 (1968)

X-ray spectrometric study of vanadium compounds with metallic properties
  E. Z. Kurmaev, S. A. Nemnonov, A. Z. Men'shikov, and
    G. P. Shveikin
  Izv. Akad. Nauk SSSR, Ser. Fiz., 31:996-1001 (1967)

Bonding structure and mechanical behavior of vanadium carbide single crystals
  R. G. Lye, G. E. Hollox, and J. D. Venables
  (RIAS-Martin Marietta) Report 67-5, Contracts NASw-1290
    and DA-31-124-ARO-D-467, AD-655 456 (June 1967)

Existence of a superstructure in the vanadium carbide VC_{1-x}
  C. H. DeNovion, R. Lorenzeili, and P. Costa
  Compt. Rend., Ser. A and B, 263, 775-778 (1966)
  La-tr-67-19, 5 pp.

Thermodynamic properties of vanadium and its compounds
  A. D. Mah
  (Bureau of Mines), BM-RI 6727 (1966)

Beitrag zu den Systemen Vanadin — Kohlenstoff und Vanadin — Chrom — Kohlenstoff
  H. Rassaerts, F. Benesovsky, und H. Nowotny
  Planseeber. Pulvermet., 14:178 (1966)

Thermal expansion of some vanadium carbides
  E. K. Storms and C. P. Kempter
  J. Chem. Phys., 42:2043-2045 (1965)

Heat of formation lower carbides of vanadium
  N. M. Volkova and P. V. Gel'd
  Izv. Vyssh. Ucheb. Zaved. Tsvetn. Met., 8:77 (1965) for Los
    Alamos Sci. Lab., Univ. of Calif., LA-tr-66-20

Phase transformation of the higher carbide of vanadium
  N. M. Volkova, P. V. Gel'd, and S. I. Alyamovskii
  Zh. Neorgan. Khim., 10:1758 (1965)

[Title Not Given]
  E. K. Storms
  Los Alamos Scientific Laboratory Rept. LA-2942 (1964)
  Phase diagram

The vanadium — vanadium carbide system
  E. K. Storms and R. J. McNeal
  J. Phys. Chem., 66:1401-1408 (1962)
  Phase diagram

The cubic phases of vanadium carbides
  S. I. Alyamovskii, P. V. Gel'd, and I. I. Matveenko
  Zh. Strukt. Khim., 2:445 (1961)

Heat capacity at low temperatures and entropies of vanadium carbide and vanadium nitride
  C. H. Shomate and K. K. Kelley
  J. Am. Chem. Soc., 71:314-315 (1949)
  Specific heat

3.c.2. Niobium Carbide

Energy bands of niobium carbide
  James B. Conklin, Jr., and Russell W. Simpson
  Bull. Am. Phys. Soc., 15:310 (1970)

Etching of dislocations in niobium carbide
  V. N. Turchin and G. A. Rymashevskii
  Kristallografiya, 15:193-194 (1970)
  Sov. Phys. — Cryst., 15:160-161 (1970)

Elastic properties and thermal expansion of niobium monocarbide to high temperatures
  R. F. Brenton, C. R. Saunders, and C. P. Kempter
  J. Less-Common Metals, 19:273-278 (1969)

X-ray diffraction study of the charge distribution in niobium monocarbide
  M. Merisalo, O. Inkinem, M. Jarvinen, and K. Kurki-Suonio
  J. Phys. C, Proc. Phys. Soc., 2:1984 (1969)

Automatic measurement of secondary electron emission characteristics of TaC, TiC, and ZrC
  S. Thomas and E. B. Pattinson
  British J. Appl. Phys., 2:1539-1547 (1969)

Electrical resistance and Hall effect of niobium monocarbide
  L. B. Dubrovskaya and I. I. Matveenko
  Akad. Nauk SSSR, Ural. Fil. (Trudy Inst. Khim.), 18:124-125
    (1968)

The thermodynamic properties of NbC
  V. V. Fesenko, A. G. Turchanin, and E. A. Guseva
  Zh. Fiz. Khim., 42:2332-2334 (1968)

The emission properties of Ta, Mo, Nb, and NbC
  B. A. Kruschtalev and A. M. Rakov
  Teploobmen. Gidrodin. Teplofiz. Svoistv., 198-219 (1968)

Niobium: physicochemical properties of its compounds and alloys. III. Crystal structures and densities
  Hans Nowotny and Karl J. Seifert
  At. Energy Rev., Spec. Issue No. 2, pp. 71-172 (1968)

Activities of Nb and C in NbC
  G. L. de Poorter
  LA 3744 (1968)

Properties of NbC and TiC in the region of homogeneity
  G. V. Samsonov and G. S. Upadkhaya
  Poroshkovaya Met., 8(9):70-74 (1968)

Low-temperature heat capacities of superconducting niobium and tantalum carbides
  L. E. Toth, M. Ishikawa, and Y. A. Chang
  Acta Met., 16:1183-1187 (Sept. 1968)

Ordering of vacancies in the (niobium) monocarbide $NbC_{0.75}$
  V. G. Zubkov, L. B. Dubrovskaya, P. V. Gel'd, and V. A.
    Tskhai
  Akad. Nauk SSSR, Ural. Fil. (Trudy Inst. Khim.), 18:133-134
    (1968)

Mechanism of conductivity in NbC with defects
  A. E. Avgustinik, O. A. Golikova, V. S. Neshpor, and S. S.
    Ordanyan
  Izv. Akad. Nauk SSSR, Neorg. Mater., 3(2):286-290 (1967)
  Inorg. Mat., 3(2):256-259 (1967)

Electrical conductivity and thermo-emf of Nb carbide between 20 and 2000°C
  O. A. Golikova, N. N. Matveeva, A. I. Avgustinik, and S. S.
    Ordanyan
  Teplofiz. Vysok. Temp., 5(6):1001-1004 (1967)
  High Temperature, 5(6):894-896 (1967)

Thermal expansion of some niobium carbides
C. P. Kempter and E. K. Storms
J. Less-Common Metals, 13:443-447 (1967)

Electrical resistivity of hyperstoichiometric columbium carbide materials at elevated temperatures
D. L. Paulson and G. Asai
(Bureau of Mines, Metallurgy Research Ctr., Albany, Oregon), Contract AT(11-1)-599, Topical Report 1, Nov. 1966-1 Nov. 1967, USBM-RC-1327 (Jan. 1968), 31 pp.

Enthalpy and heat capacity of NbC
A. G. Turchanin, S. S. Ordonyan, and V. V. Fesenko
Porosh. Met., 7:23-27 (1967)

Preparation of nitride and carbonitride of Nb
G. A. Meerson, E. M. Rakitskaya, V. N. Bulgakov, and S. A. Ladygo
Izv. Akad. Nauk SSSR, Neorg. Mater., 2(8):1429-1433 (1966)
Inorg. Mater., 2(8):1220-1224 (1966)

Heat capacities of $NbC_{0.702}$, $NbC_{0.825}$, $NbC_{0.980}$ and Nb_2C below 320°K
T. A. Sandenaw and E. K. Storms
(Los Alasmos Lab.,), LA-DC-6565; STAR 4, 1990 (A) (June 8, 1966)

[Title Not Given]
E. Rudy and D. P. Harmon
Aerojet-General Corp. Tech. Rept. AFML-TR-65-2, Part I, Vol. V (Dec. 1965)
Phase diagram

Diffusion of boron and carbon in refractory transition metals
A. P. Epik
Dopovidi Akad. Nauk Ukr. RSR, 1:67-70 (1964)

On the reactive diffusion of boron and carbon in refractory transition metals
G. V. Samsonov
Dopovidi Akad. Nauk Ukr. RSR, 1:68-70 (1964)

Structure des carbures de niobium
N. Terao
Japan. J. Appl. Phys., 3:104 (1964)

The effects of carbon on the hardness, microstructure, and cold working properties of high-purity niobium
F. R. Cortes and A. L. Feild, Jr.
J. Less-Common Metals, 4:169-180 (Apr. 1962)

The niobium-carbon system
H. Kimura and Y. Sasaki
Trans. Natl. Res. Inst. Metals (Tokyo), 3:31 (1961)

Development of ultrapure refractory materials
Peter T. B. Shaffer
Contract NOrd-17175, Progress rept., No. 24, Feb. 1, 1961-April 30, 1961 (Apr. 28, 1961), 12 pp.

Columbium-carbon system
R. P. Elliott
American Soc. Metals, Preprint No. 179 (1960)

Lattice dimensions of NbC as a function of stoichiometry
C. P. Kempter, E. K. Storms, and R. J. Fries
J. Chem. Phys., 33:1873-1874 (1960)

Thermodynamic data for columbium (niobium) carbide
L. B. Pankratz, W. W. Weller, and K. K. Kelley
(Bureau of Mines), BM-RI 6446

3.c.3. Tantalum Carbide

Field ion microscopy of tantalum-carbon alloys
P. Rao
UCRL-19630; CONF-701003-2 (May 1970), 5 pp.
Presented at 28th Meeting on Electron Microscopy Society of America, Houston, Texas, Oct. 5-9, 1970

Structure and properties of tantalum carbide crystals
D. J. Rowcliffe and W. J. Warren
J. Mat. Sci., 5:345-350 (1970)

Use of hollow-cathode dc plasma discharge float zoning for the growth of materials with high melting points: the growth of single crystals of Ta_2C
R. N. Storey and R. A. Laudise
J. Crystal Growth, 6:261-265 (1970)

Etude par diffraction électronique de la structure des carbures de tantale
Nobuzo Terao
Japan. J. Appl. Phys., 9:1263 (1970)

Interstitial ordering and mechanical properties of tantalum – carbon alloys
Prakash Rao
Ph.D. Thesis, Lawrence Radiation Lab., University of California, Berkeley, UCRL-19113 (Dec. 1969), 120 pp.

Structure and properties of carbide crystals grown from solution
D. J. Rowcliffe
Am. Ceram. Soc. Bull., 48(4):404 (1969)

Heat content and heat capacity of tantalum carbide in the homogeneity region
A. S. Bolgar, E. A. Guseva, V. A. Gorbatyuk, and V. V. Fesenko
Porosh. Met., Akad. Nauk Ukr. SSR, 4:60-62 (1968)

Partial dislocations in a nonstoichiometric tantalum carbide
J. L. Martin and B. Jouffrey
J. Phys. (France), 29(10):911-916 (1968)

A study of non-stoichometry in metal carbides using field ion microscopy
J. D. Meakin and D. Raghavan
(Franklin Institute Research Labs., Philadelphia, Penn.), 2nd Progress report, Contract AT(30-1)-3716, NYO-3716-2, April 1968

Grain growth during hot-pressing of tantalum carbide
E. Roeder and J. Hornstra
J. Am. Ceram. Soc., 51:224 (1968)

Hall coefficient of tantalum carbide as a function of carbon content and temperature
G. Santoro and R. T. Dolloff
J. Appl. Phys., 39:2293 (1968)

Low-temperature heat capacities of superconducting niobium and tantalum carbides
L. E. Toth, M. Ishikawa, and Y. A. Chang
Acta Met., 16:1183-1187 (Sept. 1968)

3. Transition Metal Carbides

Elastic constants of tantalum monocarbide, $TaC_{0.90}$
R. W. Bartlett and C. W. Smith
J. Appl. Phys., 38:5428 (1967)

Effect of stoichiometry on the thermal expansion of TaC_x
R. J. Fries and L. A. Wahman
J. Am. Ceram. Soc., 50:475-477 (1967)

Standard heat of formation of tantalum carbides from TaC phase
A. N. Kornilov, I. D. Zaikin, S. M. Skuratov, L. B. Dubrovskaya, and G. P. Shveikin
Zh. Fiz. Khim., 41:346-350 (1967)
Russian J. Phys. Chem., 41:172-174 (1967)

Stacking faults in a non-stoichiometric face-centered cubic TaC
J. L. Martin, B. Jouffrey, and P. Costa
Phys. Stat. Sol., 22:349 (1967)

Un matériel réfractaire résistant du choc thermique: le carbure de tantale imprégné
Andre Hivert
(Office National d'Etudes et de Récherches Aérospatiales, Chatillon, France) Presented at a colloquium on materials for use in space, C.N.E.S., Paris, 7-11 Feb. 1966; T. P. No. 323 (1966)

Magnetic properties of cubic tantalum carbide
L. B. Dubrovskaya and I. I. Matveenko
Phys. Metals Metallography, 19:42-46 (1965)

[Title Not Given]
E. Rudy and D. P. Harmon
Aerojet General Corp. Tech. Rept. AFML-TR-65-2, Part I, Vol. V (Dec. 1965)
Phase diagram

Ordered domains and the c/a ratio of $Ta_{64}C$
R. Villagrana and G. Thomas
Appl. Phys. Letters, 6:61 (1965)

Diffusion of boron and carbon in refractory transition metals
A. P. Epik
Dopovidi Akad. Nauk Ukr. RSR, 1:67-70 (1964)

Superconducting thin films of niobium, tantalum, tantalum nitride, tantalum carbide, and niobium nitride
D. Gerstenberg and P. M. Hall
J. Electrochem. Soc., 111:936 (1964)

On the reactive diffusion of boron and carbon in refractory transition metals
G. V. Samsonov
Dopovidi Akad. Nauk Ukr. RSR, 1:68-70 (1964)

Variation of some properties of tantalum carbide with carbon content
Gilbert Santoro
Trans. AIME, 227:1361 (1963)

Development of ultrapure refractory materials
Peter T. B. Shaffer
Contract NOrd-17175, Progress rept., No. 24, Feb. 1, 1961-April 30, 1961 (Apr. 28, 1961), 12 pp.

The specific heats at low temperatures of tantalum oxide and tantalum carbide
K. K. Kelley
J. Am. Chem. Soc., 62:818-819 (1940)

3.d. Group VI

The chemical diffusion of carbon in the group VI-B metal carbides
R. J. Fries, J. E. Cummings, C. G. Hoffman, and S. A. Daily
Contract W-7405-ENG-36, CONF-680603-1; LA-3795-MS (April 1968)
Also in 6th Plansee Seminar Preprints, Vol. 2, 1968 (Plansee Society for Powder Metallurgy, Reutte, Austria)

Rhombic modifications of W_2C and Mo_2C compounds
V. S. Telegus, E. I. Gladyshevskii, and P. I. Kripyakevich
Kristallografiya, 12(5):936-938 (1967)
Sov. Phys. — Crystallogr., 12(5):813-815 (1968)

The high-temperature terminal solubility of carbon in molybdenum, tungsten, and rhenium
E. Fromm and U. Roy
Phys. Stat. Sol., 9:K83 (1965)

Molybdenum and tungsten (Price list — selective bibliography of government research reports and translations)
Clearinghouse for Federal Scientific and Technical Information, Springfield, Va., SB-415, Suppl. 1 (1965), 48 pp.

3.d.1. Chromium Carbide

Study of conditions for the preparation of the carbides $(Cr, Fe)_7C_3$ and $(Cr, Fe)_{23}C_6$
M. Kh. Freid and V. A. Suprunov
Izv. VUZ Khim. Tekhnol., 12(8):1016-1020 (1969)

Preparation of Cr_7C_3 and $Cr_{23}C_6$
M. Kh. Freid and V. A. Suprunov
Izv. VUZ Khim. Tekhnol., 12:224-227 (1969)

Heats of formation of chromium carbides
Alla D. Mah
BM-RI17217 (Jan. 1969)

Crystal structure refinement of Cr_3C_2
S. Rundqvist and G. Runnsjo
Acta Chem. Scand., 23:1191-1199 (1969)

Solubility of carbon in solid chromium
S. V. Zemsky and A. P. Fokin
Zh. Fiz. Khim., 41:93-97 (1967)

Study of the evaporation and thermodynamic properties of chromium carbides
A. S. Bolgar, V. V. Fesenko, and S. P. Gordienko
Soviet Power Met. Metal Ceram., No. 2, pp. 159-165 (1966)

Die Kristallstruktur von $Cr_3(C, N)_2$ und Cr_2VC_2
P. Ettmayer, G. Vinek, and H. Rassaerts
Monatsh. Chem., 97:1258 (1966)

Cr borocarbide (preparation and properties)
L. Ya. Markovskii, N. V. Vekshina, and Yu. D. Kondrashev
J. Appl. Chem. USSR, 39:923-926 (1966)

Free energy of formation of chromium carbide, Cr_3C_2
Molly Gleiser
J. Phys. Chem., 69:1771-1772 (1965)

3.d.2. Molybdenum Carbide

Low-temperature modification of molybdenum carbides
V. N. Eremenko, T. Ya. Velikanova, V. E. Listovnichii, and S. A. Komarova
Izv. Akad. Nauk SSSR, Neorg. Mater., 6:11-14 (1970)
At 1170 ± 13°C

The formation and stability of group IVA carbides and nitrides in molybdenum
N. E. Ryan and J. W. Martin
J. Less-Common Metals, 17:363-376 (1969)

Diffusion of carbon in molybdenum carbide
V. S. Eremeev and A. S. Panov
Izv. Akad. Nauk SSSR, Neorg. Mater., 4(9):1507-1512 (1968)

Superconduction in Mo_2C
N. Morton
Cryogenics, 8:30-31 (1968)

Heat capacity at low temperatures, absolute entropy and enthalpy of molybdenum carbide
I. E. Paukov, P. G. Strelkov, and V. S. Filatkina
Zh. Fiz. Khim., 42:2962-2963 (1968)

Low-temperature heat capacities of superconducting molybdenum carbides
L. E. Toth and J. Zbasnik
Acta Met., 16(9):1177-1182 (Sept. 1968)

Phase relationships and defect structure in dimolybdenum carbide
F. W. Vahldiek and S. A. Mersol
Anisotropy in Single-Crystal Refractory Compounds, Vol. 1 (Fred W. Vahldiek and Stanley A. Mersol, eds.), Plenum Press, New York (1968), pp. 199-248

Activity of carbon in molybdenum carbide
V. I. Alekseev, Yu. N. Surovoi, and L. A. Shvarzman
Porosh. Met., 7:64-70 (1967)

On the production of Mo_2C
A. Domsa, L. Szabo, S. Coman, and V. Coman
Bull. Politechn. Cluj, 10:87-95 (1967)

Vaporization of Mo_2C
R. J. Fries
J. Chem. Phys., 46:4463 (1967)

The constitution of binary Mo − C alloys
E. Rudy, S. Windisch, A. J. Stosick, and J. R. Hoffman
Trans. Met. Soc. AIME, 239:1247 (1967)

High superconducting transition temperatues in the molybdenum carbide family of compounds
L. E. Toth
J. Less-Common Metals, 13:129-131 (1967)

Investigation of the deposition of molybdenum carbide films from molybdenum hexacarbonyl using an electron beam
B. A. Vishnyakov and K. A. Osipov
Fiz. Tverd. Tela, 8(12):3706-3707 (1966)
Sov. Phys. − Solid State, 8(12):2976-2977 (1967)

Obtaining thin films of molybdenum carbide from molybdenum hexacarbonyl under the action of an electron beam
B. A. Vishnyakov and K. A. Osipov
Fiz. Tverd. Tela, 9(5):1545-1546 (1967)
Sov. Phys. − Solid State, 9(5):1216-1217 (1967)

Effect of paramagnetic impurities on the superconducting behavior of cubic molybdenum carbide
R. H. Willens and E. Buehler
J. Appl. Phys., 38:405 (1967)

Some peculiarities of carbon diffusion in molybdenum carbide
A. V. Shovensin, G. V. Shcherbedinskii, and A. N. Minkevich
Poroshkovaya Met., 11:46-51 (1966)

Dislocation etch pits and phase relationships in dimolybdenum carbide single crystals
F. W. Vahldiek, S. A. Mersol, and C. T. Lynch
Japan. J. Appl. Phys., 5:663 (1966)

Microhardness anisotropy, slip, and twinning in Mo_2C single crystals
F. W. Vahldiek, S. A. Mersol, and C. T. Lynch
Trans. AIME, 236:1490 (1966)

Superconducting critical temperatures of the carbides and nitrides with the NaCl structure: Superconductivity of MoC
L. E. Toth, J. Johnston, and E. R. Parker
J. Phys. Chem. Solids, 26:517 (1965)

Diffusion of boron and carbon in refractory transition metals
A. P. Epik
Dopovidi Akad. Nauk Ukr. RSR, 1:67-70 (1964)

On the reactive diffusion of boron and carbon in refractory transition metals
G. V. Samsonov
Dopovidi Akad. Nauk Ukr. RSR, 1:68-70 (1964)

The structure of dimolybdenum carbide by neutron-diffraction technique
E. Parthe and V. Sadagopan
Acta. Cryst., 16:202 (1963)

Structure of molybdenum monocarbide
A. E. Koval'skii and S. V. Semenovskaya
Kristallografiya, 4(6):923 (1959)
Sov. Phys. − Cryst., 4(6):878-879 (1960)
French translation: CNRS, Org. 58-B.04.783, G/R-3062

Impurity atom-dislocation interactions and subsequent effects on mechanical properties of refractory metals
M. A. Adams and H. Nesor
(Materials Research Corporation, Orangeburg, New York), Contract No. AF33(616)-7596, ASD-TDR-62-11

3.d.3. Tungsten Carbide

Studies on the formation process of tungsten carbide powder from tungsten powder
Akio Hara and Masaya Miyake
Planseeberichte für Pulvermetallurgie, 18:91-110 (1970)

Slip system of tungsten carbide crystals at room temperature
S. B. Luyckx
Acta Met., 18:233-236 (1970)

New method for preparing unusually oriented tungsten tips and their growth mechanism
Fumio Okuyama and Tadatosi Hibi
Japan. J. Appl. Phys., 9:15 (1970)
W and WC surface film

Thermodynamic analysis of the process of production of refractory tungsten-based carbonyl materials
V. G. Syrkin and A. A. Uzl'skii
Zh. Fiz. Khim., 43:2766-2770 (1970)
In Russian. Reaction schemes for production of tungsten oxides and carbides

Atomic and molecular diffraction of helium and deuterium from a tungsten carbide surface characterized by low-energy electron diffraction
W. H. Weinberg and R. P. Merrill
Phys. Rev. Letters, 25:1198-1201 (1970)

Résultats expérimentaux sur la cinétique de croissance de cristaux de carbure de tungstène en présence de cobalt
Michel Coster and Alfred Deschanvres
Compt. Rend., C269:221-223 (1969)

Formation of carbides on tungsten wires
W. J. Croft and K. J. Nygaard
J. Sci. Instr., 2:1012 (1969)

Preparation of tungsten carbide by electro-deposition
John M. Gomes and M. M. Wong
(Bureau of Mines, Washington, D. C.), BM-RI-7247 (1969)

Analysis of Co in WC by atomic absorption spectrometry
S. L. Levine
Atomic Absorption Newsletter, 8(3):58-59 (1969)

Carbon self-diffusion in polycrystalline tungsten carbide
Charles Paul Buhsmer, Jr.
Ph. D. Thesis from Alfred University, New York College of Ceramics (1968), 167 pp.
Univ. of Microfilms, Ann Arbor, Michigan, Order No. 68-16616

Thermal expansion of tungsten monocarbide
V. T. Deshpande, R. R. Pawar, and S. V. Suryanarayana
Current Science, 5(19):543-545 (1968)

Face-centered cubic WC
V. N. Filimonenko and L. C. Pivovarov
Metalloved. Term. Obr. Met., 9:58 (1968)

Carburizing tungsten in the field ion microscope
R. D. French and M. H. Richman
Phil. Mag., 18(153):471-481 (Sept. 1968)

Growth of tungsten carbide monocrystals
Alvin P. Gerk
(Illinois University, Urbana), Contract AT(11-1)-1198, COO-1198-501 (Feb. 1968), 51 pp.
Thesis

Growth of tungsten carbide monocrystals
A. P. Gerk and J. J. Gilman
J. Appl. Phys., 39:4497 (1968)

Neutron-diffraction study of W_2C rhombic modification
Yu. Z. Nozik, Yu. V. Lipin, and B. V. Kuvaldin
Latv. PSR Zinat. Akad. Vestis Fiz. Techn. Ser. (USSR), No. 6, pp. 30-33 (1968)

Field ion microscopy of carbides of tungsten formed externally and in situ
M. H. Richman and W. D. Sproul
(Brown Univ., Providence, R. I.), AD-678755; ARPA-E61; AISI-20012, Contract ARPA SD-86 (Oct. 1968), 21 pp.
Metallography, 2:149-159 (1969)

Electron-beam gun for evaporation
T. Szucs and G. Vago
Finommechanika, 7:293-298 (1968)

Crystal structure of W_2C
K. Yvon, H. Nowotny, and F. Benesovsky
Monatsh. Chem., 99:726-729 (1968)

Field ion microscope investigation of the diffusion of carbon in tungsten
Robert Dexter French
(Brown Univ.), Thesis (1967), 96 pp.

Question of the cubic tungsten carbides
E. Krainer and J. Robitsch
Translated by C. P. Kempter (Los Alamos Scientific Lab., New Mexico) from Planseeber. Pulvermet., 15:179 (1967)
LA-tr-68-27, 3 pp.

Metallographic investigations on WC single crystals
M. Malli and E. Hillnhagen
Prakt. Metallog., 4:24-27 (1967)

Revealing hard-metal (tungsten carbide alloys and c.) structures by interference films deposited from the vapour phase
Walter Peter, Erich Kohlhaas, and Otto Jung
Praktische Metallographie, 4:284-290 (1967)

Tungsten carbide-cobalt alloys
L. K. Pivovarov, E. A. Shchetilina, A. V. Varaksina, and O. I. Serebrova
Izv. Akad. Nauk SSSR, Metal., 2:177-182 (1967)

Tungsten and Its Compounds
G. D. Rieck
(Tech. Univ. of Eindhoven, Netherlands), Pergamon Press (1967)

Production and properties of WC single crystals
O. Ruediger
In: Second Conference on Powder Metallurgy, Krakow, Zakladzie Graficznym Politechniki Slaskiej Gliwicach, Poland (1967), pp. 196-215 (I. Czesc, S. Stolarz, W. Missol, W. Zolkowski, J. Kurzeja, and J. Cyunczyk, eds.). In German. CONF-670988, Vol. 1

Phasengleichgewichte im Bereich der kubischen Karbidphase im System Wolfram-Kohlenstoff
E. Rudy and J. R. Hoffman
Planseeber. Pulvermet., 15:174 (1967)

Evidence for zeta Fe_2N-type sublattice order in W_2C at intermediate temperatures
Erwin Rudy and S. Windisch
J. Am. Ceram. Soc., 50:272 (1967)

Production of titanium carbide-tungsten carbide and niobium carbide-tungsten carbide solid solutions and study of their superconductivity
O. I. Shulishova and I. A. Shcherbak
Poroshkovaya Met., 6:16-20 (1967)

[Title Not Given]
E. Rudy, S. Windisch, and J. R. Hoffman
Aerojet General Corp. Tech. Rept. AMFL-TR-65-2, Part I, Vol. VI (1966)
Phase diagram

Determination of the glide planes in tungsten monocarbide
J. Corteville, J. C. Monier, and L. Pons
Compt. Rend., 260:2773 (1965)

Hardness anisotropy and slip in WC crystals
D. N. French and A. A. Thomas
Trans. AIME, 233-950 (1965)

Phase equilibrium in the system tungsten-carbon
R. V. Sara
J. Am. Ceram. Soc., 48:251 (1965)

Determination of the slip systems in single crystals of tungsten monocarbide
T. Takahashi and E. J. Freise
Phil. Mag., 12:1 (1965)

Diffusion of boron and carbon in refractory transition metals
A. P. Epik
Dopovidi Akad. Nauk Ukr. RSR, 1:67-70 (1964)

On the reactive diffusion of boron and carbon in refractory transition metals
G. V. Samsonov
Dopovidi Akad. Nauk Ukr. RSR, 1:68-70 (1964)

Demonstration of the plasticity of tungsten monocarbide
J. Corteville and L. Pons
Compt. Rend., 257:1915-1918 (1963)

Free energy of formation of tungsten carbide, WC
Molly Gleiser and John Chipman
Trans. AIME, 224:1278 (1962)

Neutron and x-ray diffraction studies on the structure of WC and a comparison of it with earlier electron diffraction data
E. Parthe and V. Sadagopan
(Univ. of Pennsylvania) Status Rept. No. 1, Feb. 1-Sept. 30, 1961, Contract AF49(638)-1027, AFOSR-1721; AD-269689

4. Binary Transition Metal Nitrides

4.a. General

Transition Metal Carbides and Nitrides
Louis E. Toth
Academic Press, New York and London, 1971, 279 pp.
Preparation, characterization, crystal chemistry, phase relationships, thermodynamics, mechanical properties, electrical and magnetic properties, superconducting properties, band structure and bonding, author and subject index

Molecular-orbital interpretation of the soft x-ray $L_{II,III}$ emission and absorption spectra from some titanium and vanadium compounds
David W. Fischer
J. Appl. Phys., 41(9):3561-3569 (1970)
Oxides, nitrides, carbides, and borides

Chemical determination of nitrogen in the nitrides of boron, titanium, zirconium, chromium, molybdenum, niobium, tantalum and vanadium
C. Healy and A. Parker
Atomic Energy Research Establishment, Harwell, England, AERE-R-6491 (Sept. 1970), 14 pp.

Superconducting properties of niobium-titanium-nitride thin films
J. R. Gavaler, D. W. Deis, J. K. Hulm, and C. K. Jones
Appl. Phys. Letters, 15:329-331 (1969)

Preparation and superconducting properties of thin films of transition-metal interstitial compounds
J. R. Gavaler, J. K. Hulm, M. A. Janocko, and C. K. Jones
J. Vacuum Sci. Tech., 6:177-179 (1969)

A method for the production of spherical particles (case 1)
Karl Knotik, Peter Koss, and Heinz Markl
Österreichische Studiengesellschaft für Atomenergie GmbH
British Patent 1,157,012 (July 2, 1969), Priority date July 8, 1965, Austria
Groups IVa, Va, or VIa carbides, nitrides, oxides, sulfides, borides, or silicides

Phase studies of the products from the nitridation of titanium, zirconium, hafnium, and chromium
M. D. Lyutaya and O. P. Kulik
Zavod. Lab., 35(1):25-27 (1969)

Zur elektronischen Struktur der Übergangsmetall-carbide und -nitride
A. Neckel
Paper presented at Physiker-Tagung, Salzburg (1969)

Study of the electronic structure and inter-atomic bonds in some compounds and binary alloys by the method of x-ray spectroscopy
S. A. Nemnonov, A. Z. Menshikov, K. M. Kolobova, E. Z. Kurmayev, and V. A. Trapeznikov
Trans. AIME, 245:1191-1198 (1969)

Determination of nitrogen in high-temperature resistant metal nitrides and carbonitrides
J. Rottmann and H. Nickel
Z. Anal. Chem., 247:208-220 (1969)

Nitrogen determination in high-melting nitrides
H. Schneider and W. Hein
In: High Temperature Materials (F. Benesovsky, ed.), Metallwerk Plansee AG, Reutte/Tirol, Austria (1969), pp. 1028-1030
In German

Metal-rich metal-metalloid phases
H. H. Stadelmaier
Developments in the Structural Chemistry of Alloy Phases, Plenum Press, New York (1969), pp. 141-180
Review, nitrides, transition elements

Study of the projection of borides and nitrides by the plasma gun
G. Bentz, G. Provost, and C. Urban
Bull. Soc. Fr. Ceram., No. 78, pp. 5-17 (1968)
TiB_2, TaB_2, NbB_2, ZrB_2, TiN, ZrN, TaN

Superconductive material of NbN and ZrN
Ahmed El Bindari
(Avco Corp.), U. S. Patent 3,392,126 (July 9, 1968)

Crystal structures of refractory carbides and nitrides
A. Bowman
Fundamentals of Refractory Compounds (H. H. Hausner and M. G. Bowman, eds.), Plenum Press, New York (1968), pp. 49-66

Electronic structure of transition carbides, nitrides, and borides
  P. Costa
  Anisotropy in Single-Crystal Refractory Compounds, Vol. 1 (Fred W. Vahldiek and Stanley A. Mersol, eds.), Plenum Press, New York (1968)
  Proc. Intern. Symp. on Anisotropy in Single-Crystal Refractory Compounds, Dayton, Ohio (June 13–15, 1967), p. 151

Structure électronique des carbures et des nitrures de métaux de transition cubiques faces centrées de type NaCl
  Paul Costa
  (Office National d'Etudes et de Recherches Aérospatiales, Chatillon, France), TP-540
  Also will be published in Ann. Phys. (Received at RMIC May 1968)
  Includes synthesis

Band structure and the titanium $L_{II,III}$ x-ray emission and absorption spectra from pure metal, oxides, nitride, carbide, and boride
  D. W. Fischer and W. L. Baun
  J. Appl. Phys., 39:4757–4776 (1968)
  57 refs.

Thermodynamic properties of transition-metal nitrides
  K. A. Gingerich
  J. Chem. Phys., 49:14–18 (1968)

Anomalous superconducting properties of carbides and nitrides of group IVa and Va elements
  A. L. Giorgi, E. G. Szklarz, and T. C. Wallace
  Proc. Brit. Ceram. Soc., 10:183–193 (March 1968)

Preparation und Supraleitungseigenschaften von Niobnitrid sowie Niobnitrid mit Titan-, Zirkon- und Tantalzusatz
  G. Horn and E. Sauer
  Z. Physik, 210:70–79 (1968)

Crystal chemistry of the chalcogenides and pnictides of the transition elements
  F. Hulliger
  Struct. Bonding (Berlin), 4:83–229 (1968)
  532 refs.

Thermal conductivity of cubic monocarbides and mononitrides of transition metals
  V. S. Neshpor
  Izv. Akad. Nauk SSSR, Neorg. Mater., 4(12):2200–2202 (1968)
  Inorg. Mater., 4(12):1915–1917 (1968)

Ordnungsstrukturen bei Übergangsmetall-Karbiden und -Nitriden
  H. Nowotny and F. Benesovsky
  Planseeber. Pulvermet., 16(3):204–214 (1968)

A study of superconductivity in interstitial compounds
  N. Pessall, R. E. Gold, and H. A. Johansen
  J. Phys. Chem. Solids, 29:19–38 (1968)
  IV and V nitrides, except Ta; IV and V carbides; synthesis

Superconductivity of ceramic compounds
  K. M. Ralls, E. R. Parker, and V. F. Zackay
  Ceram. Microstruct. Anal. Significance Product., Proc. 3rd Intern. Mater. Symp., Berkeley, Calif., 1966, John Wiley and Sons, Inc., New York (1968), pp. 489–508

Superconductivity and the nature of the bond in carbides and nitrides of transition metals and their solution with NaCl-type structure
  O. I. Shulishova
  LA-TR-68-31 (1968)
  Khim. Svyaz Poluprov. Termodin., Inst. Fiz. Tverd. Tela Poluprov., Akad. Nauk Beloruss., SSSR (1966), pp. 299–316

Superconducting properties of transition-metal carbides and nitrides, final report Oct. 1, 1965– Dec. 1, 1967
  Louis E. Toth
  (School of Mineral and Metallurgical Engineering, University of Minnesota, Minneapolis), AFOSR-68-0265; AD-671-944 (Feb. 15, 1968)

Chemistry and Physics of the Nitrides
  Naukova Dumka, Kiev (1968)
  Electronic structure and classification of nitrides, G. V. Samsonov, pp. 9–20
  Production and physicochemical properties of nitrides, M. D. Lyutaya, G. V. Samsonov, and O. P. Kulik, pp. 21–46
  Diffusion and reaction of nitrogen with the transition metals of group IV-VI of the periodic system, R. A. Andrievskii and I. I. Spivak, pp. 53–61
  Study of the azotization of Zr, Nb, and Ta, K. I. Portnoi, S. E. Salibekov, Yu. V. Levinskii, Yu. D. Strogonov, V. D. Khvostikov, and M. Kh. Levinskaya, pp. 69–75
  Investigation of the conditions of formation and certain properties of the nitrides of rare metals, M. D. Lyutaya and I. G. Chernysh, pp. 76–83
  Enthalpy of formation and specific heat of titanium nitrides, S. M. Ariya, M. P. Morozova, and M. M. Khernburg, pp. 130–133
  Superconductivity of the nitrides of zirconium and niobium in the homogeneity range, O. I. Shulishova, pp. 157–161
  Thermoemission properties of the nitrides of the transition metals, G. V. Samsonov, B. S. Fomenko, and T. S. Verkhoglyadova, pp. 162–167

Temperature dependence of thermoelectromotive force and specific electrical resistance of titanium, vanadium, chromium, and their borides, carbides, and nitrides
  S. N. L'vov and V. F. Nemchenko
  LA-TR-67-17 translated from Vysokotemperaturnye Neorganicheskie Soedineniia, pp. 100–107

Determination of nitrogen in transition-metal nitrides by tube-furnace oxidation and gas-chromatographic measurement
  R. A. Meyer, E. P. Parry, and J. H. Davis
  Anal. Chem., 39:1321 (1967)

Handbook of Lattice Spacings and Structures of Metals and Alloys, Vol. 2
  W. B. Pearson
  Pergamon Press, New York (1967), 1446 pp.; Vol. 1 (1958), 1044 pp.
  Includes hydrides, borides, carbides and nitrides

Cubic carbides, nitrides, and oxides of the first transition-metal series
  F. Petru and V. Brozek
  Pokroky Praskove Met. VUPM, 4:3–27 (1967)

Specific heat and character of the chemical bond in Ti and V carbides, nitrides, and oxides
  V. V. Tarasov, A. F. Demidenko, and A. K. Maltsev
  Izv. Akad. Nauk SSSR, Neorg. Mater., 3(6):957–962 (1967)
  Inorg. Mater., 3(6):857–861 (1967)

Handbook of Binary Metallic Systems, Structure
and Properties, Vol. II: Physicochemical Prop-
erties of the Elements
  A. E. Vol
  TT 66-51150 (1967), 870 pp.; Vol. I, TT 66-51149 (1966),
    635 pp.
  Includes hydrides, borides, carbides and nitrides, phase dia-
    grams, and tables of properties

Second International Conference on Solid Compounds of Transition Elements
  Enschede, Netherlands (June 12-16, 1967), 140 pp.

Anomalous superconducting properties of re-
fractory carbides and nitrides of group IVa and
Va elements
  A. L. Giorgi, E. G. Szklarz, and T. C. Wallace
  (Los Alamos Sci. Lab., New Mexico), LA-DC-8022; CONF-
    661207-1 (1966)
  Presented at the British Ceram. Soc. Meeting, London, 19 pp.

Methods of analysis of metal refractories —
borides, carbides and nitrides of hafnium, niobium, tantalum, titanium, and zirconium
  R. E. Dutton, G. J. McKinley, D. McLean, and H. F. Wendt
  (Union Carbide Res. Inst., Tarrytown, N. Y.), Tech. Rept.
    No. C-29; NP-14952 (Feb. 1965), 25 pp.

Refractory Ceramics for Aerospace, a Materials
Selection Handbook
  J. R. Hague, J. F. Lynch, A. Rudnick, F. C. Holden, and
    W. H. Duckworth, eds.
  The American Ceramic Society, Columbus, Ohio (1964)

Electrical properties of some transition-metal
carbides and nitrides
  J. Piper
  Nucl. Metal., 10:29-43 (1964)

A critical review of refractories
  Edmund K. Storms
  (Los Alamos Sci. Lab., New Mexico), LA-2942 (1964)

Binary nitrogen compounds of the elements: a
literature survey, Appendix II
  Rosemary G. Ehl, J. P. Piper, Phyllis R. Clopper, and J. L.
    Margrave
  (Rice University, and University of Wisconsin), TID-19898
    (Oct. 1963), 85 pp.
  Supplement to WACD Tech. Note 59-115, prepared at Univ. of
    Wisconsin in June 1959 by Rosemary Ehl, R. J. Sime, and
    J. L. Margrave
  Thermodynamic, structural, preparative or electrical and
    magnetic data

Hartstoffe
  R. Kieffer and F. Benesovsky
  Springer-Verlag, Vienna (1963)
  Preparation, crystal growth, physical properties, phase
    diagrams, crystal structures

Some kinetic and thermodynamic properties of
the refractory metal borides and nitrides, an
annotated bibliography
  M. Temple Milliken and Jean E. Senkin, comp.
  LRL, Univ. of Calif., Livermore, Calif., UCRL-7559 (Sept.
    1963)

X-ray spectra and the interatomic bond in solid
metallike compounds
  V. S. Neshpor
  AEC-tr-5873 (1963)
  Vysokotemperaturnye Metallokeramicheskia Materialy, Akad.
    Nauk URSR, Kiev (1962), pp. 46-74 .

Crystallochemical peculiarities of borides,
nitrides, silicides, and phosphides of transition metals
  G. V. Samsonov
  Poroshkovaya Met., Akad. Nauk Ukr. SSR, No. 2, pp. 65-79
    (Mar.-Apr. 1963)

Plenum Press Handbook of High-Temperature
Materials. No. 1. Materials Index
  Peter T. B. Shaffer
  Plenum Press, New York (1963)

The thermal properties of twenty-six solid materials to 5000°F or their destruction temperatures
  Southern Research Institute (Birmingham, Ala.), ASD-TDR-
    62-765; AD 298061 (Jan. 1963)

Binary and ternary phase diagrams of colum-
bium, molybdenum, tantalum and tungsten
  J. J. English
  (Defense Metals Information Center, Battelle Memorial Inst.,
    Columbus, Ohio), DMIC Rept., 152 (April 28, 1961); OTS PB
    171421

Thermodynamic properties of 65 elements —
their oxides, halides, carbides, and nitrides
  Charles E. Wicks and F. E. Block
  (Bureau of Mines, Albany Metallurgy Res. Center, Oregon),
    NP-13622 (Dec. 1961), 150 pp.

Graphite, carbide, nitride, and sulfide refrac-
tories
  Lawrence M. Litz
  Proc. Intern. Symp. High Temp. Technology, Asilomar Conf.
    Grounds, Calif., Oct. 6-9, 1959,
  McGraw-Hill Book Co., Inc., New York (1960), pp. 90-112

Electrical properties of borides, carbides, and
nitrides; the IVa to VIa metals
  S. N. L'vov, V. F. Nemchenko, and G. V. Samsonov
  Dokl. Akad. Nauk SSSR, 135:577-580 (1960)

Acid stability and methods of analysis of titani-
um, zirconium, niobium, and tantalum nitrides
  O. I. Popova and G. T. Kabannik
  Zh. Neorg. Khim., 5:930-934 (1960)

The oxidation of Cr, Hf, Mo, Nb, Ta, Ti, W, V,
Zr, and some of their alloys; properties of
some borides, carbides, nitrides, and oxides of
the same metals, a bibliography
  E. A. Cernak, comp.
  (Pratt and Whitney Aircraft Div., United Aircraft Corp.,
    Middletown, Conn.), CNLM-1802-5 (Dec. 15, 1959), 99 pp.
  922 refs.; 1951 to Sept. 1959

Infrared spectra of inorganic solids — II.
Oxides, nitrides, carbides, and borides
  E. G. Brame, Jr., J. L. Margrave, and V. W. Meloche
  J. Inorg. Nucl. Chem., 5:48-52 (1957)
  Li_3N, Cu_3N, BN, AlN, CrN, Mg_3N, Zn_3N, B_2O_3, Al_2O_3, Ga_2O_3,
    Cr_2O_3, B_4C, SiC, Mo_2B, ZrB_2, and TiB_2

Some physicochemical properties of compounds
of the transitional refractory metals with boron,
carbon, and nitrogen, and the characteristics of
their binary alloys
  G. V. Samsonov
  AEC-tr-3016, translated from Izv. Sektora Fiz. Khim. Anal.
    Inst. Obshchei Neorg. Khim., Akad. Nauk SSSR, 27:97-125
    (1956), 39 pp.

Diffusion of boron, carbon, and nitrogen into transition metals of IV, V, and VI groups of the periodic table
G. V. Samsonov and V. P. Latisheba
AEC-tr-2949, Dokl. Akad. Nauk SSSR, 109:582 (1956)

The preparation of pure high-melting carbides, nitrides, and borides by the deposition method and the description of their properties
K. Moers
AEC-tr-3239
Z. Anorg. Allgem. Chem., 198:243-261 (1931), 22 pp.

4.b. Group IV

Vapour-phase crystallization and some physical properties of titanium nitride
W. Synielnikowa, T. Niemyski, J. Panczyk, and E. Kierzek-Pecold
J. Less-Common Metals, 23:1-6 (1971)

Rotational analysis of the red electronic emission system of titanium nitride
T. M. Dunn, L. K. Hanson, and K. A. Rubinson
Can. J. Phys., 48:1657-1663 (1970)

Electronic band structure and the K and L x-ray spectra from TiO, TiN, and TiC
D. W. Fischer
J. Appl. Phys., 41:3922-3926 (1970)

Enthalpy of nitrogen solid solution formation in α-zirconium and zirconium nitride in the homogeneity region
E. I. Galbraikh, O. P. Kulik, A. A. Kuznetsov, M. D. Lyutaya, and M. P. Morozova
Poroshkovaya Met. (USSR), 9:62-66 (1970)

Phase equilibria investigation of binary, ternary, and higher order systems. Part X. The crystal structures of Hf_3N_2 and Hf_4N_3
Erwin Rudy
(Aerojet-General Corp., Sacramento, Calif.), AD-708176 (March 1970), 22 pp.

X-ray diffraction pycnometric determinations of the density of zirconium-base interstitial phases
A. S. Shevchenko, R. A. Andrievskii, V. P. Kalinin, and R. A. Lyutikov
Poroshkovaya Met., 10(1):89-91 (1970)
Table of nitrides, carbides, and mixed hydrides

Structure and electrical properties of sputtered films of hafnium and hafnium compounds
F. T. J. Smith
J. Appl. Phys., 41(10):4227-4231 (1970)

Thermal conductivity of titanium nitride in the homogeneous range
M. I. Aivazov, A. Kh. Muranevich, and I. A. Domashnev
Teplofizika Vysokikh Temperatur., 7(5):893-897 (Sept.-Oct. 1969)
High Temperature, 7(5):830-834 (1969)

Etude et structure d'une nouvelle phase du sous-nitrure de titane Ti_2N
G. Lobier and J.-P. Marcon
Compt. Rend., 268:Ser.C, 1132:1135 (1969)

Solubility of oxygen in titanium nitride
Yu. A. Nechaev and V. M. Kamyshov
Izv. Akad. Nauk SSSR, Metally No. 6, pp. 50-53 (Nov.-Dec. 1969)

Oxidation kinetics of single-crystal titanium nitride by optical measurements
M. L. Pearce and C. Basch
J. Am. Ceram. Soc., 52:496-498 (1969)

Diffusion of N in ZrN
I. I. Spivak
Izv. Akad. Nauk SSSR, Neorg. Mater., 5(6):1138-1139 (1969)
Inorg. Mater., 5(6):967-968 (1969)

The dissociation energy of TiN
C. A. Stearns and F. J. Kohl
NASA TN D 5027 (1969)

Enthalpy of formation and specific heat of titanium nitrides
S. M. Ariya, M. P. Morozova, and M. M. Kernburg
Chemistry and Physics of the Nitrides, Naukova Dumka, Kiev (1968), pp. 130-133

X-ray spectroscopic study of titanium monoxide in the homogeneous region and of titanium nitride
I. A. Brytov, M. A. Rumsh, and A. S. Parobets
Fiz. Tverd. Tela, 10(3):794-800 (1968)
Sov. Phys. − Solid State, 10(3):621-626 (1968)

Formation of titanium nitride by means of a nitrogen plasma jet
Osamu Matsumoto
J. Electrochem. Soc. Japan, 36:153-159 (1968)

Formation of titanium nitride by means of a nitrogen plasma jet
Osamu Matsumoto
J. Electrochem. Soc. Japan, 36(4):213-218 (1968)

Bonding, imperfect structure, and properties of the refractory nitrides of titanium, zirconium, and hafnium
M. E. Straumanis
Anisotropy in Single-Crystal Refractory Compounds, Vol. 1 (Fred W. Vahldiek and Stanley A. Mersol, eds.), Plenum Press, New York (1968), p. 121
Proc. Intern. Symp. Anisotropy in Single-Crystal Refractory Compounds, Dayton, Ohio, June 13-15, 1967

Study of the electron and atomic structure of titanium carbide and titanium oxide
M. P. Arbuzov, E. T. Kachkovskaya, and B. V. Khaenko
Khimicheskaya Svyaz' v Poluprovodnikakh i Termodinamika (Chemical Bonds in Semiconductors and Thermodynamics), 1966, Nauka i Tekhnika, Minsk (Sept. 1967) pp. 64-71
TiC, TiN, TiO

Study of x-ray K spectra of titanium in its nitride and carbide
V. I. Chirkov, S. M. Blokhin, and É. E. Vainshtein
Fiz. Tverd. Tela, 9(4):1116-1121 (1967)
Sov. Phys. − Solid State, 9(4):873-877 (1967)

Bonding, lattice parameter, density and defect structure of TiN containing an excess of N
M. E. Straumanis, C. A. Faunce, and W. J. James
Acta Met., 15(1):65-71 (1967)

Nitriding of titanium and its alloys
S. I. Muroi and M. Someno
Nippon Kinzoku Gakkai Shi, 30(1):26-31 (1966)

Electronic band structure of TiC, TiN, and TiO
V. Ern and A. C. Switendick
(MIT, Cambridge, Mass.), TR-192; AD-608826 (Oct. 1964), 28 pp.

Thermal expansion and atomic vibration amplitudes for TiC, TiN, ZrC, ZrN, and pure tungsten
C. R. Houska
Phys. Chem. Solids, 25:359-366 (1964)

Static atomic displacements resulting from vacancies in defect structures $TiN_x \square_{1-x}$
C. R. Houska
(Union Carbide Res. Inst.), Tech. Rept. No. C-22 (April 1964)

A blue zirconium nitride
R. Juza, A. Gabel, H. Rabenau, and W. Klose
Z. Anorg. Allgem. Chem., 329:136-145 (1964)
LA-TR-64-23 (1964)

Preparation of titanium nitride
E. K. Kleespies and T. A. Henrie
(Bureau of Mines, Nevada), BM-RI-6447 (1964), 12 pp.
Carbon reduction of rutile and ilmenite in ammonia

Thermal conductivity of titanium carbide, zirconium carbide, and titanium nitride at high temperatures
R. E. Taylor and J. Morreale
J. Am. Ceram. Soc., 47(2):69 (Feb. 1964)

Metallurgy and Mellography of Pure Metals
V. S. Emel'yanov and A. I. Evstyukhin
JPRS-9473, Met. i Metalloved. Chistykh Metal. (1960), 395 pp.

Single crystals of TiN by van Arkel method
K. Wilke
Monatsber. Deut. Akad. Wiss., Berlin, 2:425-426 (1960)

Preliminary report on the thermodynamic properties of selected light-element compounds (supplement to NBS reports 6297 and 6484)
National Bureau of Standards 6645 (Jan. 1, 1960)
Includes Ti nitrides and carbides

Dependence of heats and free energies of formation of zirconium nitrides on the composition and structure
E. I. Smagina, V. S. Kutsev, and B. F. Ormont
Dokl. Akad. Nauk SSSR, 115(2):354-357 (1957)
Dokl. – Phys. Chem., 115(2):487-490 (1957)

Single crystals of ZrN by van Arkel method
I. E. Campbell, C. F. Powell, D. H. Nowicki, and B. W. Gonser
Trans. Electrochem. Soc., 96:318 (1949)

Crystal growth of TiN by van Arkel method
F. H. Pollard and P. Woodward
J. Chem. Soc. London (1948), p. 1708

4.c. Group V

Alternating-current loss measurements in thin-film type II superconductors
D. W. Deis, J. R. Gavaler, C. K. Jones, and A. Patterson
J. Appl. Phys., 42:21-26 (1971)
Nb and NbN

Very high critical current and field characteristics of niobium nitride thin films
J. R. Gavaler, M. A. Janocko, A. Patterson, and C. K. Jones
J. Appl. Phys., 42:54-57 (1971)

Characteristics of NbN Dayem bridges
M. A. Janocko, J. R. Gavaler, C. K. Jones, and R. D. Blaugher
J. Appl. Phys., 42:182-185 (1971)
Josephson junction

New phases of niobium nitride
N. Terao
J. Less-Common Metals, 23:159-169 (1971)
Nb_4N_5 and Nb_5N_6

Eigenschaftsänderungen von Vanadium durch gelösten Stickstoff oder Sauerstoff
Gerhard Horz
Z. Metallkunde, 61:371-378 (1970)

Superconductivity in the niobium nitride system
Robert B. Laibowitz and V. Sadagopan
Bull. Am. Phys. Soc., 15:321 (1970)

Neutron diffraction study of vanadium nitride
E. Z. Vintaikin, V. B. Dmitriev, I. A. Tomilin, et al.
Dokl. Akad. Nauk SSSR, 193(5):1022-1024 (1970)

Nuclear-magnetic resonance study of the niobium-nitrogen system NbN_x
Robert A. Bennett, H. O. Hopper, and U. Roy
J. Appl. Phys., 40:2441-2444 (1969)

Superconducting properties of niobium titanium nitride thin films
J. R. Gavaler, D. W. Deis, J. K. Hulm, and C. K. Jones
Appl. Phys. Letters, 15:329-331 (1969)
Prepared with various Nb/Ti ratios by a high-purity reactive sputtering process

A new etchant for thin films of tantalum and tantalum compounds
J. Grossman and D. S. Herman
J. Electrochem. Soc., 116:674 (1969)
TaN and Ta_2O_3

Changes in concentration in binary metal-gas solutions during simultaneous evaporation of gas and metal. I. Theoretical considerations. II. Discussion on the changes of concentration in the systems $V - O$, $Nb - O$, $Ta - O$, $V - N$, $Nb - N$, and $Ta - N$
G. Horz
Z. Metallkunde, 60:115-126 (1969)

Heat capacity of tantalum nitride between 15 and 300°K
V. V. Nogteva, I. E. Paukov, P. G. Strelkov, and V. S. Filatkina
Zh. Fiz. Khim., 43:1108 (1969)

The upper critical field H_{c2} of NbN film prepared by reactive sputtering
Yukinori Saito, Takeshi Anayama, Kazuhiko Yasohama, Ko Yasukouchi, and Yutaka Onodera
Appl. Phys. Letters, 14:285 (1969)

Investigation of ordering of interstitial impurities in tantalum using an electron diffraction microscope
M. P. Usikov and A. G. Khachaturyan
Kristallografiya, 13(6):1045-1055 (1968)
Sov. Phys. — Cryst., 13(6):910-918 (1969)
Ta—O and Ta—N systems

The high field properties of pure niobium nitride thin films
D. W. Deis, J. R. Gavaler, J. K. Hulm, and C. K. Jones
AED-Conf. 68-283-012, 12 pp.

Dislocation-interstitial interaction in the niobium nitrogen system
Charles Curtis Dollins
Ph. D. thesis, Illinois Univ., Urbana, Ill. (1968), 89 pp.
University Microfilms, Ann Arbor, Mich., Order No. 69-1333

Superconducting properties of NbN thin films
J. Gavaler
(Westinghouse Res. Center, Pittsburgh), p. 22 in QTSR-34, Francis Bitter National Magnet Laboratory, MIT (Dec. 1968)

Sur deux nouveaux nitrures de tantale
Jean-Claude Gilles
Compt. Rend., Ser. C, 266:546-547 (1968)
Ta_4N_5 and Ta_5N_6

Energy-gap measurement of niobium nitride
K. Komenou, T. Yamashita, and Y. Onodera
Phys. Letters, 28A:335-336 (1968)

Thin films of niobium nitride and tantalum nitride deposited by reactive evaporation
J. R. Rairden
Electrochem. Tech., 6:269-272 (1968)

Properties of some high field superconductors in fields up to 240 kG
E. Saur and H. Wizgall
Cryogen. Engng. Present Status, Future Develop., Proc. 1st Intern. Conf., Tokyo, Kyoto, Japan, 1967, S. L. Heywood-Temple Industr. Publ. (1968), pp. 156-158
Nb_3Sn, V_3Si, V_3Ga, NbN

Solid State Physics
Ko Yasukochi, Kazuko Sekizawa, Takeshi Ogasawara, Nobumitsu Usui, Motoyoshi Ishizuka, and others
Nihon University (Tokyo, Japan) Plasma, Solid State, and Theoretical Physics (1968), pp. 35-46
NbN made by reactive sputtering had upper critical field of 200 kG

Constitution of a portion of the niobium (columbium)-nitrogen system
R. W. Guard, J. W. Savage, and D. G. Swarthout
Trans. AIME, 239:643 (1967)

Präparation und Supraleitungseigenschaften von reinem sowie zirkon- und titanhältigem Niobnitrid
K. Hechler and E. Saur
Z. Physik, 205:392-399 (1967)

Critical data of niobium nitride in transverse magnetic fields
K. Hechler, E. Sauer, and H. Wizgall
Z. Physik, 205:400-408 (1967)

Study of electron state in vanadium nitride by intensity measurements of x-ray diffraction
Sukeaki Hosoya, Tomoe Yamagishi, and Masayasu Tokonami
(Inst. for Solid State Phys., Univ. of Tokyo, Japan), ISSP-A-275 (Sept. 1967)

The solid solutibility of nitrogen in Nb and Nb-rich Nb — Hf, Nb — Mo, and Nb — W alloys.
Part I: The binary system Nb — N
A. Taylor and N. J. Doyle
J. Less-Common Metals, 13:399-412 (1967)

Physical properties of niobium and tantalum nitrides
T. S. Verkhoglyadova, S. N. L'vov, V. F. Nemchenko, and G. V. Samsonov
Izv. VUZ Fiz. (USSR), No. 8, pp. 31-34 (1967)
Lattice dimensions, thermal and electrical conductivity, Hall coefficient, thermal emf and microhardness

Über das Tantalnitrid Ta_3N_5 und das Tantaloxidnitrid TaON
G. Brauer, J. Weidlein, and J. Strahle
Z. Anorg. Allgem. Chem., 348:298 (1966)

Untersuchungen im System Niob-Stickstoff. III. Kinetik der Entgasung von Niob-Stickstoff-Mischkristallen im Hochvakuum
E. Gebhardt, W. Durrschnabel, and G. Horz
J. Nucl. Mater., 18:149-160 (1966)

Anomalies in the electrical resistivity of vanadium nitride
L. Glasser and J. Hoy
J. Phys. Chem., 70:281 (1966)

Deposition of tantalum, tantalum oxide, and tantalum nitride with controlled electrical characteristics
E. Krikorian and R. J. Sneed
J. Appl. Phys., 37:3674-3681 (1966)

Tantalum and tantalum compounds in thin-film microcircuitry
David A. McLean
J. Electrochem. Chem., 34:1 (1966)
Film deposition

Preparation of nitride and carbonitride of Nb
G. A. Meerson, E. M. Rakitskaya, V. N. Bulgakov, and S. A. Ladygo
Izv. Akad. Nauk SSSR, Neorg. Mater., 2(8):1429-1433 (1966)
Inorg. Mater., 2(8):1220-1224 (1966)

Herstellung und Charakterisierung von ultrafeinen Karbiden, Nitriden, und Metallen
E. Neuenschwander
J. Less-Common Metals, 11:365-375 (1966)
TaC, NbC, TiC, TaN, and W

Synthese und Eigenschaften des roten Tantalnitrids Ta_3N_5
G. Brauer and J. R. Weidlein
Angew. Chem., 77:218-219 (1965)

Some superconducting properties of several carbides and nitrides of the transition metals
L. E. Toth, V. F. Zackay, M. Wells, J. Olson, and E. R. Parker
Acta Met., 13:379 (1965)
Cubic solid solutions of TaN and TaC

Deposition of tantalum films with an openended vacuum system
J. W. Balde, S. S. Charschan, and J. J. Dineen
Bell System Tech. J., 43:127 (1964)

Solubility of nitrogen in tantalum
P. Bunn and C. Wert
Trans. Met. Soc. AIME, 230:936 (1964)

Superconducting thin films of niobium, tantalum, tantalum nitride, tantalum carbide, and niobium nitride
D. Gerstenberg and P. M. Hall
J. Electrochem. Soc., 111:936 (1964)

Tantalum-film technology
D. A. McLean, N. Schwartz, and E. D. Tidd
Proc. IEEE, 52:1450-1462 (1964)

Gleichgewichsuntersuchungen im System Tantal-Stickstoff
Eckehard Fromm
Thesis, Technische Hochschule Stuttgart (1961)

Heats of formation of niobium dioxide, niobium subnitride and tantalum subnitride
Alla D. Mah
J. Am. Chem. Soc., 80:3872-3874 (1958)

On some properties of vanadium nitride crystals
V. A. Epel'baum and B. F. Ormont
Zh. Fiz. Khim., 21:10 (1947)
AEC-tr-5805

Superconductive compounds with extremely high transition temperatures: columbium hydride and columbium nitride
G. Aschermann, E. Friederich, E. Justi, and J. Kramer
TT-66-12282
Physikalische Zeitschrift (Germany), 42:349-360 (1941)

4.d. Group VI

The solubility of nitrogen in solid chromium
T. Mills
J. Less-Common Metals, 23:317-324 (1971)

The preparation of beta-molybdenum nitride
Ronald Karam and Roland Ward
Inorganic Chemistry, 9:1385 (1970)

Pressure-temperature relations in the chromium-nitrogen system
T. Mills
J. Less-Common Metals, 22:373-381 (1970)

Phase diagram of the chromium-Cr_2N system
V. M. Svechnikov, G. F. Kobzenko, V. G. Ivanchenko, and E. K. Martinchuk
Dop. Akad. Nauk Ukr. SSR, A9:833-837 (1970)
In Ukrainian

Contribution à l'étude de CrN. Susceptibilité magnétique et effet de la pression
Daniel Bloch, Paul Mollard, and Jean Voiron
Compt. Rend., Ser. B, 269:553-555 (1969)

Reaction of Cr with N
L. Cadiou and J. Paidassi
Compt. Rend., 286C:743 (1969)

Contribution à l'étude de CrN à basses températures par diffraction de rayons X et de neutrons
Mahmoud Nasr Eddine, Francoise Sayetat, and E. F. Bertaut
Compt. Rend., Ser. B, 269:574-577 (1969)

High-temperature solubility of nitrogen in tungsten
E. Fromm and H. Jehn
Translated by T. Watt for Sandia Labs., Albuquerque, N. M., from J. Less-Common Metals, 17:124-126 (1969)
SC-T-4025

The effect of composition and purity on the Neel temperature of CrN
T. Mills
(Aeronautical Res. Labs., Mellbourne, Australia), ARL/MET-74 (March 1969), 21 pp.

Structure of cubic tungsten nitride
F. Günther and H. G. Schneider
Kristallografiya, 11(4):683-685 (1966)
Sov. Phys. — Cryst., 11(4):585-587 (1967)

Formation of tungsten nitride by means of nitrogen plasma jet
O. Matsumoto
Denki Kagaku, 35(7):488-492 (1967)

Tungsten and Its Compounds
G. D. Rieck
Pergamon Press, New York (1967)

The preparation of the nitrides of chromium
T. Mills and T. G. Hill
(Aeronautical Res. Labs., Melbourne, Australia), ARL/MET-44 (Sept. 1966), 10 pp.

Antiferromagnetic structure of CrN
L. M. Corliss, N. Elliott, and J. M. Hastings
Phys. Rev., 117:929 (1960)

5. Binary Transition Metal Oxides

5.a. General

Oxides of Rare Earth, Titanium, Zirconium, Hafnium, Niobium, and Tantalum
  Allen M. Alper, ed.
  Refractory Materials, Vol. 5, Part 2
  Academic Press, New York (1970)

Structural, electrical, and magnetic properties of vacancy stabilized cubic "TiO" and "VO"
  M. D. Banus and T. B. Reed
  The Chemistry of Extended Defects in Non-Metallic Solids, North-Holland Publishing Co., Amsterdam (1970), pp. 388-522

Theory of Mott transition
  M. Cyrot
  Solid State Commun., 8:1255-1256 (1970)

Some reactions of tungsten (VI) oxide and molybdenum (VI) oxide with liquid sulfur
  John W. Goodrum
  (Air Force Cambridge Research Labs., L. G. Hanscom Field, Bedford, Mass.), AFCRL-70-0417 (July 23, 1970), 31 pp.
  In situ generation of species for crystal growth in solution

Transition-metal oxides, amorphous semiconductors, semiconducting glasses, Ovshinsky effect, and other switching (memory) materials (a literature review)
  John T. Milek
  (Hughes Aircraft Co., Culver City, Calif.), Interim Report No. 72 (Sept. 1970)
  131 refs.

Preparation and some characteristics of self-supporting 300 to 2500-Å oxide films
  A. Aladjem and D. G. Brandon
  J. Vacuum Sci. Tech., 6:635-637 (1969)
  International Conference on Thin Films, Boston, Mass., CONF-690435
  Ta, Nb, W, Ti, and Zr oxide films

Arc techniques in the preparation of inorganic materials
  R. E. Loehman, C. N. R. Rao, J. M. Honig, and C. E. Smith
  J. Sci. Ind. Res. (India), 28:13-16 (1969)
  TiO_x, NbO_x, VO_x, Ti_3O_5, NbO, Ti_2O_3 single crystals

Mixed oxides prepared with an induction plasma torch. Part 2. Chromia/titania
  T. I. Barry, R. K. Bayliss, and L. A. Lay
  J. Mater. Sci., 3:239-243 (1968)

Growing single crystals of refractory oxides
  V. K. Yanovskii
  Zh. Vsesoyuz. Khim. Obshchest. D. I. Mendeleeva, SSSR, 13:134-142 (1968)

Electrolytic separation of transition-metal oxide crystals
  W. Kunnmann, A. Ferretti, R. J. Arnott, and D. B. Rogers
  U. S. Patent 3,382,161 (May 7, 1968)

Crystallization of titanium, zirconium, and hafnium oxides and some titanate and zirconate compounds under hydrothermal conditions
  V. A. Kuznetzov
  J. Crystal Growth, 3:405-410 (1968)

Aspects on the problems of synthesis and structure of some oxide or oxide-like compounds formed by the transition elements in the groups IV, V, and VI of the periodic table
  Sten Andersson
  Arkiv for Kemi, 26:521 (1967)
  Review: 103 refs.

A study of the anodic oxides of titanium, niobium, and tantalum: growth mechanism, interfacial phenomena, contribution to the knowledge of their structure
  Francois Kover
  Ph. D. thesis, University Paris, France (June 1967), 100 pp.

Cubic carbides, nitrides, and oxides of the first transition-metal series
  F. Petru and V. Brozek
  Pokroky Praskove Met. VUPM, No. 4, pp. 3-27 (1967)

Chemical structure, atomic constants, homogeneity, and electron structure

Mass transport in oxides
J. B. Wachtman, Jr., and A. D. Franklin
(National Bureau of Standards, Wash., D. C.), NBS-SP-296 (August 1968), 213 pp.
Presented at the Symposium held at Gaithersburg, Md., Oct. 22-25, 1967

Growth and some mechanical properties of filamentary single crystals (whiskers) of NiO, WO_3, $W_{20}O_{58}$, $W_{18}O_{49}$, and WO_2
I. Ahmad and G. P. Capsimalis
Crystal Growth (H. Steffen Peiser, ed.), Pergamon Press, Oxford; London, New York (1967), p. 325
Proceedings of an International Conference on Crystal Growth, Boston (June 20-24, 1966)

Crystal growth by chemical transport reactions. IV. New results on the growth of binary, terminary and mixed-crystal chalcogenides
R. Nitsche
Crystal Growth (H. Steffen Peiser, ed.), Pergamon Press, Oxvord, London, New York (1967), pp. 215-220
Proceedings of an International Conference on Crystal Growth, Boston (June 20-24, 1966)

Crystal growth techniques
E. A. D. White
GEC Journal, 31:43-53 (1964)

The structure of anodic films — I. An electron diffraction examination of the products of anodic oxidation on tantalum, niobium, and zirconium
P. H. G. Draper and J. Harvey
Acta Met., 11:873 (1963)

The growth of oxide single crystals from the fluxed melt
E. A. D. White
Technique of Inorganic Chemistry, Vol. IV (Hans B. Jonassen and Arnold Weissberger, eds.), Interscience Publishers, New York (1965), p. 31

Current Information on the Refractory Metals
C. E. King
Contract AF33(657)-11214, AD-436333; ERR-FW-049A (Dec. 1, 1962), 80 pp.

Halides, oxides, and sulfides of the transition metal
F. J. Morin
J. Appl. Phys., Suppl. to Vol. 32, pp. 2195-2197 (1961)

Beobachtung von Umwandlungs- und Oxydationsvorgängen im Elektronen-Emissions-Mikroskop
H. Duker
Proc. Eur. Reg. Conf. on Electron Microscopy, Delft, 1:456 (1960)

5.b. Material Preparation and Crystal Growth

5.b.1. Group IV

5.b.1.a. Titanium

Growth of the intermediate oxides of titanium from borate fluxes under controlled oxygen fugacities
Robert F. Bartholomew and William B. White
J. Crystal Growth, 6:249-252 (1970)

Mechanism of heterogeneous deposition of thin-film rutile
R. N. Ghoshtagore
J. Electrochem. Soc., 117(4):529-534 (1970)

Growth characteristics of rutile film by chemical vapor deposition
R. N. Ghoshtagore and A. J. Noreika
J. Electrochem. Soc., 117:1310-1314 (1970)

Preparation and characterization of submicron hafnium oxide
K. S. Mazdiyasni, L. M. Brown
J. Am. Ceram. Soc., 53:43-51 (1970)

Plasma-grown rutile single crystals and their distinctive properties
J. D. Chase and L. J. Van Ruyven
J. Crystal Growth, 5:294-298 (1969)

Reduction of TiO_2 powders and rutile single crystals
D. M. J. Compton and T. E. Firle
J. Am. Ceram. Soc., 52:515-516 (1969)

Formation and nature of radical species in the oxidation of precipitated titanium dioxide
R. D. Iyengar and R. Kellermann
Z. Phys. Chem., 64:345-349 (1969)

Stabilization of amorphous films of titanium oxide
A. I. Korobov, E. V. Semenova, N. V. Troitskaya, and B. D. Galkin
Izv. Akad. Nauk SSSR, Neorg. Mater., 5(7):1206-1209 (1969)
Inorg. Mater., 5(7):1027-1029 (1969)

Evaporation of TiO_2
G. A. Semenov
Izv. Akad. Nauk SSSR, Neorg. Mater., 5(1):67-73 (1969)
Inorg. Mater., 5(1):54-59 (1969)

Crystal growth from amorphous phase in thin films
M. Shiojiri, H. Morikawa, and E. Suito
Japan. J. Appl. Phys., 8:1077-1081 (1969)

Influence de la pression d'oxygène sur la température de solidification de certains oxydes des éléments de transition
Jean-Pierre Coutures and Marc Foex
Compt. Rend., 267:1577-1580 (1968)

Chemical-vapor deposition of thin-film dielectrics
D. R. Harbison and H. L. Taylor
(Texas Univ., Electronics Research Center), Grants AF-AFOSR-776-67, NGR-44-013-043, AFOSR-68-1796; AD-674 020 (July 1968), 7 pp.

A study of the phase diagrams and reaction of titanium and zirconium with oxygen
I. I. Kornilov, V. V. Glazova, and E. M. Kenina
Diagrammy Sostoyaniya Metallicheskikh Sistem (E. M. Savitskii, ed.), Moscow, Izdatel'stvo Nauka (1968), pp. 145-153

A study of the phase diagrams and reaction of titanium and zirconium with oxygen
M. S. Model
Diagrammy Sostoyaniya Metallicheskikh Sistem (E. M. Savitskii, ed.), Moscow, Izdatel'stvo Nauka (1968), pp. 164-166
Izv. Akad. Nauk SSSR, Metally, No. 6, pp. 143-157 (1968)

Tri-arc furnace for Czochralski growth with a cold crucible
T. B. Reed and E. R. Pollard
J. Crystal Growth, 2:243-247 (1968)

Time-dependent electrical conduction in rutile single crystals
L. J. van Ruyven and J. D. Chase
Appl. Phys. Letters, 12:214 (1968)

Vapor deposition of TiO_2
M. Yokozawa, H. Iwasa, and I. Teramoto
Japan J. Appl. Phys., 7:96-97 (1968)

High-temperature formation of anatase
W. C. Beard and W. R. Foster
J. Am. Ceram. Soc., 50:493 (1967)

Brookite fibers and their preparation
Kenneth L. Berry
(E. I. duPont de Nemours and Co.), U. S. Patent 3,338,677 (Aug. 29, 1967)

Preparation of single-crystal oxides MO_2
A. Harari, J. Thery, and M. R. Collongues
Rev. Hautes Temp. Refract., 4:207-209 (1967)
Growth of TiO_2, ZrO_2, HfO_2, CeO_2, ThO_2, and $ThTi_2O_6$ crystals

Hydrothermal crystal growth of the structure oxides TiO_2, GeO_2, and SnO_2
M. L. Harvill
Dissertation Abstracts 27, 3246B (1966-1967)
University Microfilms, Ann Arbor, Michigan

The growth of crystals by chemical transport of material. IV. The transport system Ti_2O_3/HCl
Z. Hauptman, D. Schmidt, and S. K. Banerjee
Collection Czech. Chem. Commun., 32:2421 (1967)

The growth of rutile (TiO_2) single crystals by chemical transport with $TeCl_4$
T. Niemyski and W. Piekarczyk
J. Crystal Growth, 1:177-182 (1967)

Growth of single Ti_2O_3 crystals from the melt
T. B. Reed, R. E. Fahey, and J. M. Honig
(MIT) TN-1967-38; ESD-TR-67-343, August 11, 1967, Contract AF19(628)-5167

Crystallization of amorphous films prepared by vacuum-evaporation
M. Shiojiri
Mem. Fac. Industr. Arts Kyoto Tech. Univ. (Sci. Technol.) (Japan), 16:1-18 (1967)

Growth of rutile (TiO_2) single crystals from Li_2O-MoO_3 flux
Tokuko Sugai, Shuzo Hasegawa, and Gisaku Ohara
Japan. J. Appl. Phys., 6:901-902 (1967)

Plasma-beam purifies and grows single crystals of refractory compounds
Technical release from Materials Research Corporation, Orangeburg, N. Y. (Sept. 15, 1967)

Arc-transfer method of crystal growth
J. R. Drabble and A. W. Palmer
J. Appl. Phys., 37:1778-1780 (1966)

Procèsses de formation du cristal de TiO_2 par hydrolyse de $TiCl_4$. Propr. physiques du cristal
T. Iida, K. Yamaoka, S. Noziri, and H. Nozaki
J.Chem.Soc.Japan,Industr.Chem.Sect., 69(11):2087-2095(1966)

Resistance heated crystal puller for operation at 2000°C
T. B. Reed and R. E. Fahey
Rev. Sci. Instr., 37:59-61 (1966)

Application of the isothermal flux evaporation method
Rustum Roy
Materials Res. Bull., 1:299-302 (1966)

Crystallization of amorphous titanium dioxide films prepared by vacuum evaporation
M. Shiojiri
J. Phys. Soc. Japan, 21:335 (1966)

Observations on the crystallization process of amorphous thin films (TiO_2)
M. Shiojiri, H. Morikawa, and E. Suito
Sixth International Congress for Electron Microscopy, Kyoto (1966), p. 467

Vapor growth of SiO_2 and TiO_2 films
H. Teshima, Y. Tarui, Y. Yoshida, and H. Katayama
Bull. Electrotech. Lab. (Japan), 30:45-58 (1966)

New oxy-hydrogen burner for flame fusion
J. A. Adamski
J. Appl. Phys., 36:1784 (1965)

Dislocations and stacking faults in rutile crystals grown by flame-fusion methods
D. J. Barber and E. N. Farabaugh
(National Bureau of Standards, Washington, D. C.), J. Appl. Phys. preprint (1965)

Titanium-dioxide dielectric films prepared by vapor reaction
A. E. Feuersanger
Proc. IEEE, 52:1463-1465 (1965)

Refining and growth of rutile single crystals by r.f. zone melting
J. Holt
J. Appl. Phys., 16:639 (1965)

Preparation of single crystals of refractory compounds from binary or multi-component systems and effect of temperature conditions on their form and faceting
V. N. Izvekov, L. A. Sysoev, Ya. A. Obukhovskii, and B. I. Birman
Rost Kristallov. Vol. 6B (N. N. Sheftal' and A. V. Shubnikov, eds.) (1965), pp. 116-121
Growth of Crystals, Vol. 6B, Consultants Bureau, New York (1968), pp. 106-111

Hydrothermal synthesis of rutile
V. A. Kuznetsov and V. V. Panteleev
Kristallografiya, 10(3):445 (1965)
Sov. Phys. — Cryst., 10(3):369 (1965)

Growth of controlled-composition nonstoichiometric rutile from borate fluxes
J. S. Berkes, W. B. White, and R. Roy
Ceram. Bull., 43:255 (1964)

The growth and zone refining of single-crystal rutile
P. F. Chester, J. Holt, B. J. Maddock, and A. B. Willott
General Electricity Research Lab. Rept. RD/L/R 1267 (Dec. 1964)

The crystal growth by the flame fusion method.
Preparation of Al_2O_3, TiO_2, NiO, CoO, MnO,
$BaTiO_3$, and $SrTiO_3$ single crystals
  Yoshihide Nakazumi
  Butsuri, 19:284 (1964)
  Translated from Nakazumi Crystals Corp., 11-2 Masumi-cho,
    Ikeda-shi, Osaka-fu, Japan

Studies on the formation and properties of high-
temperature dielectrics
  A. Von Hippel and others
  ML-TDR-64-219; AD-603899 (July 1964)

Crystal-growth techniques
  E. A. D. White
  G. E. C. Journal, 31:43 (1964)

Growth of crystals by flame fusion
  A. Linz, E. F. Farrell, M. J. Redman, A. Berkebile, and
    A. Vetrovs
  MIT Tech. Rept., 185 (Dec. 1963), Contracts AF33(616)-8353
    and AF19(638)-395

Zone-melting of oxides in a carbon-arc image
furnace
  P. Kooy and H. J. M. Couwenbery
  Engineers Digest, 23:87-89 (May 1962)

Growth of rare-earth-doped rutile single crys-
tals. Part I: Preparation of rutile single
crystals containing rare earths
  K. Shiroki and K. Hayashi
  TT-66-13886
  Kogyo Kagaku Zasshi (Japan), 65:1733-1735 (1962)

Progress Report No. 30, MIT
  A. R. Von Hippel and others
  AD-273 448 (January 1962)

Studies on the brookite-rutile transformation
  C. N. R. Rao, S. R. Yoganarasimhan, and P. A. Faeth
  Trans. Faraday Soc., 57:504 (1961)

Diffusion of oxygen in alpha-titanium
  A. V. Revyakin
  Izvest. Akad. Nauk SSSR, Otdel. Tekh. Nauk Met. i Toplivo,
    No. 5, pp. 113-116 (1961)

Crystal growth of TiO_2 and V_2O_3 by flame
fusion
  A. R. Von Hippel and others
  AD-241 907 (July 1960), pp. 29, 33-36
  Contracts AF33(616)-5920 and Nonr-1841(10)
  Progress Report No. 27,

Crystallization of anhydrous titania dissolved
in cryolite in the form of rutile
  P. Mergault and G. Branche
  Compt. Rend., 238:914 (1954)

[Title Not Given]
  W. Hartwig and Ch. Herfort
  Wiss. Zschr. d. Humboldt-Univ. zu Berlin II, 67 (1952/1953)

[Title Not Given]
  C. H. Moore, L. Merker, and L. E. Lynd
  Amer. Mineralogist, 35:127 (1949)

Notes on the properties of synthetic rutile
single crystals
  S. Zerfoss, R. G. Stokes, and C. H. Moore, Jr.
  J. Chem. Phys., 16:1166 (1948)

5.b.1.b. Zirconium

Stabilization of cubic ZrO_2 by MnO and partial
substitution of Ti, Nb, or Ta for Zr

  Gerhard Bayer
  J. Am. Ceram. Soc., 53:294 (1970)

Production of cubic zirconium dioxide under hy-
drothermal conditions
  V. G. Chukhlantsev and Yu. M. Galkin
  Russ. J. Inorg. Chem., 14:161-163 (1969)

Crystallization of ZrO_2 and HfO_2 under hydro-
thermal conditions
  V. A. Kuznetsov and O. V. Sidorenko
  Kristallografiya, 13(4):748-749 (1968)
  Sov. Phys. − Cryst., 13(4):651-652 (1969)

A study of the phase diagrams and reaction of
titanium and zirconium with oxygen
  I. I. Kornilov, V. V. Glazova, and E. M. Kenina
  Diagrammy Sostoyaniya Metallicheskikh Sistem (E. M. Savit-
    skii, ed.), Moscow, Izdatel'stvo Nauka (1968), pp. 145-153

A study of the phase diagrams and reaction of
titanium and zirconium with oxygen
  M. S. Model
  Diagrammy Sostoyaniya Metallicheskikh Sistem (E. M.
    Savitskii, ed.), Moscow, Izdatel'stvo Nauka (1968), pp. 164-166
  Izv. Akad. Nauk SSSR, Metally, No. 6, pp. 143-157 (1968)

Determination of impurities in zirconium di-
oxide
  G. A. Pevtsov and others
  FTD-HT-23-471-69; AD-702280
  Metody Analiza Khim. Reaktivov i Preparatov (Moscow),
    No. 15, pp. 58-60 (1968)

Localized cooling in flux crystal growth
  A. B. Chase and Judith A. Osmer
  J. Am. Ceram. Soc., 50:325 (1967)

Preparation of single-crystal oxides MO_2
  A. Harari, J. Thery, and M. R. Collongues
  Rev. Hautes Temp. Réfractaires, 4:207-209 (1967)
  Growth of TiO_2, ZrO_2, HfO_2, CeO_2, ThO_2, and $ThTi_2O_6$ crystals

The preparation and structure of zirconia spheres
  A. Madeyski and W. W. Smeltzer
  Mat. Res. Bull., 2:427-435 (1967)

Translucent ZrO_2 prepared at high pressures
  F. W. Vahldiek
  J. Less-Common Metals, 13:530 (1967)

High-pressure hot pressing of oxides
  F. W. Vahldiek and C. T. Lynch
  Sintering and Related Phenomena. Proceedings of the Internatl.
    Conf., Univ. Notre Dame (G. C. Kuczynski, N. A. Hooton,
    and C. F. Gibbon, eds.), Gordon and Breach, Science Pub-
    lishers, New York (1967), pp. 637-664
  CONF-650644

Plasma-beam purifies and grows single crystals
of refractory compounds
  Technical release of Sept. 15, 1967, from Materials Research
    Corp., Orangeburg, New York 10962

Growth of single crystals of ZrO_2 and HfO_2 from
PbF_2
  A. B. Chase and J. A. Osmer
  (Aerospace Corp., El Segundo, Calif.), TR-669(9230-02)-5;
    SSD-TR-66-66 (April 1966)
  Contract AF04(695)-669
  Am. Mineralogist, 51:1808 (1966)

Untersuchungen über Kristalle der Zusammen-
setzung $ZrO_2 \cdot 2MoO_3$
  W. Kleber and J. Doerschel
  Z. anorg. allgem. Chem., 347:289-293 (1966)

Growth and properties of zirconia and titania whiskers from fused salt baths
R. C. Johnson and J. K. Alley
BM-RI-6667 (1965)

Crystal modifications produced by quenching zirconia-containing systems and their crystal growth
A. Krauth and H. Meyer
Ber. Deut. Keram. Ges., 42:61-72 (1965)

The crystal structure of baddeleyite (monoclinic ZrO_2) and its relation to the polymorphism of ZrO_2
D. K. Smith and H. W. Newkirk
Acta Cryst., 18:983 (1965)

A high-intensity carbon-arc-image furnace and its application to single-crystal growth of refractory oxides
G. J. Goldsmith, M. Hopkins, and M. Kestigian
J. Electrochem. Soc., 111:260 (1964)

Growth and evaluation of boron suboxide and zirconium dioxide single crystals
F. A. Halden and R. Sedlacek
ARL 63-2, Jan. 1963, Contract AF33(616)-7967, Project 7022, Task 7022-02

Zone-melting of oxides in a carbon-arc image furnace
P. Kooy and H. J. M. Couwenberg
Engrs. Digest, 23:8709 (May 1962)

Growth of refractory crystals using the induction plasma torch
Thomas B. Reed
(Lincoln Laboratory, Massachusetts Institute of Technology, Lexington 73, Massachusetts), J. Appl. Phys., 32:2534 (1962)

Preparation of small samples of ductile titanium and zirconium from the isotopic oxides by iodide refining
N. D. Veigel and J. M. Blocher, Jr.
(Battelle Memorial Institute, Columbus, Ohio), J. Electrochem. Soc., 109:647 (1962)

Metallurgy and Metallography of Pure Metals
V. S. Emel'yanov and A. I. Evstyukhin
JPRS-9473, Met. i Metalloved. Chistykh Metal. (1960), 395 pp.

5.b.1.c. Hafnium

Separation of zirconium and hafnium, and the preparation of hafnium dioxide by the method of selective extraction of the thiocyanate from ethyl acetate
I. V. Vinarov, E. S. Gertsenshtein, G. I. Beek, and I. E. Kovaleva
LIB-Trans-269 (translated by Gavriloff)
Ukr. Khim. Zh., 34:153-155 (1968)

Preparation of single-crystal oxides MO_2
A. Harari, J. Thery, and M. R. Collongues
Rev. Hautes Temp. Réfractaires, 4:207-209 (1967)
Growth of TiO_2, ZrO_2, HfO_2, CeO_2, ThO_2, and $ThTi_2O_6$ crystals

High-pressure hot pressing of oxides
F. W. Vahldiek and C. T. Lynch
Sintering and Related Phenomena. Proceedings of the International Conf., Univ. Notre Dame (G. C. Kuczynski, N. A. Hooton and C. F. Gibbon, eds.), Gordon and Breach, Science Publishers, New York (1967), pp. 637-664
CONF-650644

Growth of single crystals of ZrO_2 and HfO_2 from PbF_2
A. B. Chase and J. A. Osmer
(Aerospace Corp.), TR-669 (9230-02)-5; SSD-TR-66-66 (April 1966)
Am. Mineralogist, 51:1806 (1966)

Oxide crystal growth by flux evaporation
W. H. Grodkiewicz and D. J. Nitti
J. Am. Ceram. Soc., 49:576 (1966)

Préparation de monocristaux d'oxyde d'hafnium par la méthode des sels fondus
Vutien Loc, Anne-Marie Anthony and Roger Bouaziz
Compt. Rend., 262:1715-1717 (1966)

A high-intensity carbon-arc-image furnace and its application to single-crystal growth of refractory oxides
G. J. Goldsmith, M. Hopkins, and M. Kestigian
J. Electrochem. Soc., 111:260 (1964)

Electron-beam purification of hafnium
W. E. Anable and R. A. Beall
(Bureau of Mines, Albany Metallurgy Research Center, Albany, Oregon), USBM-U-929 (June 1962)

5.b.2. Group V

5.b.2.a. Vanadium

Growth of vanadium pentoxide single crystals and their dislocation structure
A. A. Abdullaev, L. M. Belyaev, I. V. Binarov, G. F. Dobrzhanskii, and R. G. Yankelevich
Kristallografiya, 14:1095-1097 (1969)
Sov. Phys. – Cryst., 14:957 (1970)

Crystal growth of vanadium dioxide
N. Kimizuka, M. Saeki, and M. Nakahira
Mat. Res. Bull., 5:403-408 (1970)

Growth of V_5O_9 single crystals
K. Nagasawa, Y. Bando, and T. Takada
Japan. J. Appl. Phys., 9:407 (1970)

Hydrothermal synthesis of $V_3O_7 \cdot H_2O$
F. Theobald and R. Cabala
Compt. Rend., C, Sci. Chim., 270:2138-2141 (1970)

Growth of VO_2 single crystals by chemical transport reaction
Yoshichika Bando, Koichi Nagasawa Yasutoshi Kato, and Toshio Takada
Japan. J. Appl. Phys., 8:633-634 (1969)

Reactively sputtered thin films in the vanadium-oxygen system using triode sputtering
D. H. Hensler, A. R. Ross, and E. N. Fuls
J. Electrochem. Soc., 116:887-889 (1969)

Growth of V_3O_5 and V_6O_{11} single crystals
Koichi Nagasawa, Yoshichika Bando, and Toshio Takada
Japan. J. Appl. Phys., 8:1267 (1969)

Purification of V_2O_5
F. A. Schmidt, M. S. Davis, and O. N. Carlson
Annual Summary Research Report Ceramic and Mechanical

Engineering, Chemical Engineering, Chemistry, Mathematics and Computer Science, Metallurgy, Physics, and Reactor Divisions, July 1, 1968-June 30, 1969, Ames Laboratory, Ames, Iowa, Contract W-7405-eng-82, IS-2100 (July 1969)
A process for removing silicon from V_2O_5 by dissolution in aqueous HF has been developed and the purified oxide used to prepare vanadium metal

Single-crystal growth of VO_2 by isothermal flux-evaporation
S. Aramaki and R. Roy
J. Mater. Sci., 3:643-645 (1968)

Examples of flux growth of crystals
S. Ashida
Kobutsugaku Zasshi, 8:407 (1968)
Growth of VO_2 metals

Préparation de monocristaux dopés de VO_2 et étude de leurs propriétés électriques
S. Koide, H. Takei, and S. Iida
Oyo Buturi, Japan, 37:815-820 (1968)

Vanadium dioxide coatings applied to alumina substrates
J. B. MacChesney, J. F. Potter, and H. J. Guggenheim
J. Am. Ceram. Soc., 51:176-177 (1968)

Priprava a vlastnosti jednoduchych a nekterych podvojnych kyslicniku vanadu
Emil Pollert
Chem. listy Svazek, 62:897-917 (1968)
Review, 52 refs.

Tri-arc furnace for Czochralski growth with a cold crucible
T. B. Reed and E. R. Pollard
J. Crystal Growth, 2:243-247 (1968)

Vapor phase crystallization of vanadium oxide by hydrolysis of vanadium oxychloride
H. Takei
Japan J. Appl. Phys., 7:827-836 (1968)

Über die Einkristallzüchtung von Vanadinpentoxid aus der Schmelze
W. Kleber, L. Ickert, K. Grasme, and G. Riedel
Kristall. Tech., 2:481-488 (1967)

Epitaxial growth of VO_2 single crystals and their anisotropic properties in electrical resistivities
S. Koide and H. Takei
J. Phys. Soc. Japan, 22:946 (1967)

Preparation and properties of crystalline and glassy vanadium pentoxide
T. N. Kennedy, R. Hakim, and J. D. Mackenzie
(Rensselaer Polytechnic Inst., Troy, N. Y.), Contract Nonr-491(21), TR-1; AD-630 013 (March 1966), 20 pp.

Optical absorption coefficients of vanadium pentoxide single crystals
N. Kenny, C. R. Kannewurf, and D. H. Whitmore
J. Phys. Chem. Solids, 27:1237 (1966)

Growth of vanadium dioxide whisker
I. Kitahiro, A. Watanabe, and H. Sasaki
J. Phys. Soc. Japan, 21:196 (1966)

Synthèse hydrothermale du bioxyde de vanadium
M. Kunitomi, T. Kunugi, K. Futaki, and N. Ameniya
J. Chem. Soc. Japan, Ind. Chem. Sect., 69:1896-1897 (1966)

Application of the isothermal flux evaporation method
Rustum Roy
Mat. Res. Bull., 1:299-302 (1966)

Polarization and recovery of single-crystal vanadium pentoxide electrodes
A. B. Scott, K. R. Newby, and C. Bryden
(Oregon State Univ., Dept. of Chemistry, Corvallis), Contract Nonr-1286(08), TR-2; AD-636 544 (July 1966), 12 pp.

Crystal growth of some vanadium oxides
L. E. Sobon and P. E. Greene
J. Am. Ceram. Soc., 49:196 (1966)
Includes V_6O_{13} or $V_{12}O_{26}$

Growth and electrical properties of vanadium-oxide single crystals by oxychloride decomposition method (VO_2, V_2O_3, V_2O_5)
H. Takei and S. Koide
J. Phys. Soc. Japan, 21:1010 (1966)

Production of monocrystals of high-melting, chemically active metals and various oxides by zone melting
N. A. Brilliantov, N. L. Mezentseva, and L. S. Starostina
Rost Kristallov. Vol. 6B (N. N. Sheftal' and A. V. Shubnikov, eds.) (1965), pp. 277-300
Growth of Crystals, Vol. 6B, Consultants Bureau, New York (1968), pp. 107-111

Electrical properties of crystalline, glassy and liquid vanadium oxide
R. Hakim, T. Kennedy, and J. D. Mackenzie
Am. Ceram. Soc. Bull., 44:303 (1965)

A new growing method for VO_2 single crystals
H. Sasaki and K. Watanabe
J. Phys. Soc. Japan, 19:1748 (1964)

Crystal growth of TiO_2 and V_2O_3 by flame fusion
A. R. Von Hippel et al.
(Mass. Inst. Tech.), Progress report No. 27, Contract AF33(616)-5920, Nonr-1841(10); AD-241 907 (July 1960), pp. 33-36

5.b.2.b. Niobium

Zur Darstellung und Kristallstruktur von $NbO_{2.454}$
R. Gruehn and R. Norin
Z. Anorg. Allgem. Chem., 367:209-218 (1969)

Growth and properties of NbO single crystals
E. R. Pollard and T. B. Reed
Lincoln Laboratory, MIT Solid State Research Quarterly Technical Summary-1 May through 31 July 1969, Contract AF19(628)-5167 (Peter E. Tannenwald, ed.), ESD-TR-69-211 (August 15, 1969), 76 pp.
Pulled from the melt in tri-arc furnace; crystal structure, physical, electrical, and magnetic properties; superconducting below 1.45°K

Growth and properties of Nb_2O_5 thin-film capacitors
H. R. Brunner, F. P. Emmenegger, M. L. A. Robinson, and H. Rotschi
J. Electrochem. Soc., 115:1287-1289 (1968)

Tri-arc furnace for Czochralski growth with a cold crucible
T. B. Reed and E. R. Pollard
J. Crystal Growth, 2:243-247 (1968)

Chemistry and niobium and tantalum
F. Fairbrother
Topics in Inorganic and General Chemistry. Monograph 10, American Elsevier Publishing Co., Inc., New York (1967), 250 pp.
Includes oxygen compounds

Preparation and crystal structure of $NbO_{2.480}$
R. Gruehn and R. Norin
Z. Anorg. Allgem. Chem., 355:176 (1967)

Resistance-heated crystal puller for operation at 2000°C
T. B. Reed and R. E. Fahey
Rev. Sci. Instr., 37:59-61 (1966)
NbO

The preparation of oxide films of protactinium, tantalum and niobium on platinum
M. G. Brown, S. J. Lyle, and G. R. Martin
Nucl. Instr. Methods, 35:353-354 (1965)

Single-crystal growth and polymorphy of Nb_2O_5 and Ta_2O_5
Rene Moser
Schweiz. Mineral. Petrogr. Mitt., 45:35-101 (1965)

Structure of niobium suboxide (Nb_4O) revealed by electron diffraction
Siegfried and Joerg Renner
Z. Metallk., 56:531-534 (Aug. 1965)

New phase of niobium pentoxide (zeta Nb_2O_5)
F. Laves, R. Moser, and W. Petter
Naturwissenschaften, 51:356-357 (1964)

The structure of anodic films — I. An electron diffraction examination of the products of anodic oxidation on tantalum, niobium, and zirconium
P. H. G. Draper and J. Harvey
Acta Met., 11:873 (1963)

A study of the thermal oxidation of tantalum and niobium by means of electron-diffraction and capacity measurements
J. Harvey and P. Y. G. Draper
(The work was carried out at the Imperial College of Science and Technology, London; Dr. Harvey is now at the Royal Aircraft Establishment, Farnborough, Hants)
J. Inst. Metals, 92:136 (1963-1964)

On the vacuum reduction process for the production of niobium metal
Hirozo Kimura and Yasuo Sasaki
Trans. Natl. Res. Inst. Metals (Tokyo), 5:213-218 (1963)

The crystal structure of $Nb_{12}O_{29}$
Rulf Norin
Acta Chem. Scand., 17:1391-1404 (1963)

Über die Darstellung der Niboxide und ihren Transport im Temperaturgefälle
Harold Schafer and Margarita Huesker
Z. Anorg. Allgem. Chem., 317:321 (1962)
NbO, NbO_2, Nb_2O_5

High-temperature oxidation of niobium
Per Kofstad, Hall Stein Kjollesdal, Joan Markali, and Nico Norman
Central Office for Industrial Research, Oslo-Blindern, Norway, SI Publ. No. 282; Technical (Scientific) Note No. 2, Contract AF61(052)-90 (April 1960)

Investigation of electrolytic processes for preparation of high-purity niobium metal
A. J. Kolk, Jr., M. E. Sibert, and M. A. Steinberg (Horizons, Inc., Cleveland), Contract AT(30-1)-1894, TID-6101 (1960), 26 pp.

Arc-melting furnace for Czochralski growth of single crystals (NbO)
T. B. Reed and E. R. Pollard
ESD-TR-67-167, p. 21

Precipitation of columbium oxide from a supersaturated columbium-oxygen solid solution
Lester Joseph Regitz
Dissertation Abstracts 63-3386, 108 pp.
Available from University Microfilms, Ann Arbor, Michigan

5.b.2.c. Tantalum

Studies in tantalum oxide crystal growth
G. M. Wolten and A. B. Chase
(Aerospace Corp., El Segundo, California), to be published in J. Solid State Chem. (Feb.-March 1970)

Growth, unit cells, and space groups of several Ta_2O_5 polymorphs
G. M. Wolten and A. B. Chase
Eighth International Congress of Crystallography, Buffalo, Stony Brook, and Upton, New York, August 7-24, 1969, American Institute of Physics, New York (1969)

A study of oxide plate formation in tantalum — I. Growth characteristics and morphology
J. Van Landuyt and C. M. Wayman
Acta Met., 16:803-814 (1968)

Chemistry of niobium and tantalum
F. Fairbrother
Topics in Inorganic and General Chemistry. Monograph 10, American Elsevier Publishing Co., Inc., New York (1967), 250 pp.
Includes oxygen compounds

Solid-state studies of oxygen-deficient tantalum pentoxide
Daniel R. Kudrak and M. J. Sienko
Inorganic Chem., 6:880-884 (1967)

Deposition of tantalum, tantalum oxide, and tantalum nitride with controlled electrical characteristics
E. Krikorian and R. J. Sneed
J. Appl. Phys., 37:3674-3681 (1966)

Vapor deposition of thin films of tantalum pentoxide
Donald J. Peacock
Electrochem. Tech., 4:443-444 (1966)

The preparation of oxide films of protactinium, tantalum, and niobium on platinum
M. G. Brown, S. J. Lyle, and G. R. Martin
Nucl. Instr. Methods, 35:354 (1965)

Single crystal growth and polymorphy of Nb_2O_5 and Ta_2O_5
Rene Moser
Schweiz. Mineral. Petrogr. Mitt., 45:35-101 (1965)

A study of the thermal oxidation of tantalum and niobium by means of electron-diffraction and capacity measurements
J. Harvey and P. Y. G. Draper

(The work was carried out at the Imperial College of Science and Technology, London; Dr. Harvey is now at the Royal Aircraft Establishment, Farnborough, Hants)
J. Inst. Metals, 92:136 (1963-1964)

Study of Electrical and Physical Characteristics of Secondary Emitting Surfaces
W. T. Peria, ed.
(Minnesota, Univ., Minneapolis), Contract AF33(657)-8040 (Jan. 23, 1963), 126 pp.

Growth and rectification of anodically formed tantalum oxide films
M. Krause
TT-66-13777
Technische Hochshule Stuttgart, thesis (1960)

5.b.3. Group VI

5.b.3.a. Chromium

Synthesis of CrO_2 by oxidation of $Cr(OH)_3$
Y. Shibasaki, F. Kanamaru, M. Koizumi, K. Ado, and S. Kume
Mater. Res. Bull., 5:1051-1058 (1970)

Preparation of ferromagnetic chromium dioxide
Joseph H. Balthis, Jr.
(E. I. du Pont de Nemours and Co.), US 3,423,320 (Jan. 21, 1969), 3 pp.

Preparation and structure of chromium (II) oxide
G. Brauer, H. Reuther, H. Walz, et al.
Z. Anorg. Allgem. Chem., 369(3/6):144-153 (1969)

Preparation of ferromagnetic CrO_2
R. Claude, G. Lorthioir, and C. Mazieres
Compt. Rend., Ser. C, Sci. Chim., 266:462-464 (1968)

Formation of chromium oxides in the Cr_2O_3 – CrO_3 region at elevated pressures up to 4 kilobar
Karl-Axel Wilhelmi
Acta Chem. Scand., 22:2565:2573 (1968)

Flux growth of single-crystal R_2O_3 oxides with the corundum structure
R. E. Barks
Dissertation Abstracts B 27, No. 9, 3055056 (1967)
Fe_2O_3, Cr_2O_3, and Al_2O_3

Crystal growth of CrO_2
B. L. Chamberland
Materials Res. Bull., 2:827-835 (1967)

Preparation of chromium (IV)-oxide from chromylchloride
K. J. de Vries
Naturwissenschaften, 21:563 (1967)

On the decomposition of CrO_2 in air
D. Rodbell and R. DeVries
Materials Res. Bull., 2:491-495 (1967)

Epitaxial growth of Cr_2O_3
R. H. Sailors, G. L. Liedl, and R. E. Grace
J. Appl. Phys., 38:4928 (1967)

Epitaxial growth of CrO_2
R. C. DeVries
Materials Res. Bull., 1:83-93 (1966)
General Electric, Schenectady, N. Y., 66-C-238 (Aug. 1966)

Growth of Cr_2O_3 monocrystals by Verneuil's method
A. A. Popova
Rost Kristallov, Vol. 4 (A. V. Shubnikov and N. N. Sheftal', eds.) (1964), pp. 148-150
Growth of Crystals, Vol. 4, (A. V. Shubnikov and N. N. Sheftal', eds.), Consultants Bureau, New York (1966)

Application of the isothermal flux evaporation method
Rustum Roy
Materials Res. Bull., 1:299-302 (1966)

Preparation and structural studies on chromium oxides and chromates
Karl-Axel Wilhelmi
Univ. of Stockholm, Inorganic Chemistry DIS No. 24 (1966)
Various Cr-O ratios

Etch pits in chromic oxide
W. S. Brower and E. N. Farabaugh
J. Appl. Phys., 36:1489-1490 (1965)

Preparation of especially pure chromic oxide by zone melting
M. V. Mokhosoev and T. T. Gotmanova
J. Appl. Chem. USSR 38:1559-1590 (1965)

Adsorption of polar molecules on metal oxide single crystals
Yung-Fang Yu Yao
J. Phys. Chem., 69:3930 (1965)

Growth of crystals by flame fusion
A. Linz, E. F. Farrell, M. J. Redman, A. Berkebile, and A. Vetrovs
MIT Technical Rept. 185, Dec. 1963, Contracts AF33(616)-8353 and AF19(628)-395

Preparation of high-purity chromium
A. Armington, G. Dillon, and M. Silcox
Ultrapurification of Semiconductor Materials (M. S. Brooks and J. K. Kennedy, eds.), Macmillan (1962)
Proceedings of Conference held in Boston, Mass. (1961)

Synthesis and properties of ferromagnetic chromium oxide
T. J. Swobada, Paul Arthur, Jr., N. L. Cox, J. N. Ingraham, A. L. Oppegard, and M. S. Sadler
J. Appl. Phys., 32 (Suppl.), 374 (1961)

Synthesis of fine particle size CrO_2
Paul Arthur, Jr.
Du Pont, U. S. Patent 2,956,955

Sintering and reactions of MgO and Cr_2O_3
Larry Leroy Hench
Dissertation Abstracts 65-1184, 214 pp.

5.b.3.b. Molybdenum

Electronic properties of inorganic solid solutions, final rept., Mar. 16, 1965-June 30, 1969
Lawrence E. Conroy
(Univ. Minnesota, Minneapolis), Contract DA-ARO(D)-31-124-G661, AD-695697; AROD-5532-3-C (Oct. 1, 1969), 5 pp.

Über Anomalien von MoO₃
Über Anomalien von MoO_3
F. A. Schroder and H. Felser
J. Less-Common Metals, 18:434–436 (1969)

Physical properties of a transition-metal oxide:
optical and photoelectric properties of single-
crystal and thin-film molybdenum trioxide
S. K. Deb
Proc. Roy. Soc., A, 304:211–231 (1968)

Process for purifying molybdenum trioxide
Clarence D. Vanderpool and Vincent Chiola
(Sylvania Electric Products, Inc.), U. S. Patent 3,393,971
(July 23, 1968)

The vapor growth of single crystals of WO_3 and
MoO_3
Barbara M. Wanklyn
J. Crystal Growth, 2:251–253 (1968)

Some optical and photoelectric properties of MoO_3
R. B. Dzhanelidze, I. M. Purtseladze, L. S. Khitarishvili,
R. I. Chikovani, and A. L. Shkol'nik
Fiz. Tverd. Tela, 7(8):2573–2575 (1965)
Sov. Phys. – Solid State, 7(8):2082–2083 (1966)

An improved technique for the electrolytic
growth of molybdenum (IV) oxide
D. S. Perloff and A. Wold
Crystal Growth (H. Steffen Peiser, ed.), Pergamon Press,
New York; Oxford; Edinburgh; London (1967), p. 361
Proceedings of an International Conf. on Crystal Growth,
Boston (June 20–24, 1966)

Growth of WO_3 and MoO_3 crystals from cryolite-
oxide melts
V. N. Vigdorovich and V. V. Marychev
Dokl. Akad. Nauk SSSR, 159:416 (1964)

Preparation and properties of sodium and potas-
sium molybdenum bronze crystals
A. Wold, W. Kunnmann, R. J. Arnott, and A. Ferretti
Inorg. Chem., 3:545 (1964)

Refining molybdenum trioxide by removing tung-
sten and certain other impurities
A. N. Zelikman, O. Ye. Kreyn, and N. N. Gorovits
Zh. Prikl. Khim., 34:679–682 (1961)

Research on high-temperature ferroelectric
storage media
C. F. Pulvari
AD-238 046 (April 1960), pp. 1–17, 24–31

Anomalous growth of MoO_3 crystals
Eiji Yoda
J. Phys. Soc. Japan, 15:821 (1960)

An improved technique for the electrolytic
growth of molybdenum (IV) oxide
David S. Perloff and Aaron Wold
CONF-660640-1; AED-CONF-66-187-2

5.b.3.c. Tungsten

Préparation de monocristaux d'oxyde tungstique
R. Le Bihan and Ch. Vacherand
Croissance de Composés Mineraux Monocristallins, No. 2
(1967–1968), pp. 147–157 (J. P. Suchet, ed.), Masson et
Cie, Editeurs, Paris-VIe (1969)

Electronic properties of inorganic solid solutions, final rept., March 16, 1965–June 30, 1969
Lawrence E. Conroy
(Univ. of Minnesota, Minneapolis), Contract DA-ARO(D)-31-
124-G661, AD-695697; AROD-5532-3-C (Oct. 1, 1969), 5 pp.

Der chemische Transport von Wolfram und Wolf-
ramdioxid
Jan Hendrik Dettingmeijer, Jurgen Tillack, und Harold
Schafer
Z. Anorg. Allgem. Chem., 369:161–177 (1969)

Thermodynamic analysis of the process of pro-
duction of refractory tungsten-based carbonyl
materials
V. G. Syrkin and A. A. Uzl'skii
Zh. Fiz. Khim., 43:2766–2770 (1969)
Reaction schemes for production of tungsten oxides and carbides

The vapor growth of single crystals of WO_3
and MoO_3
Barbara M. Wanklyn
J. Crystal Growth, 2:251–253 (1968)

WO_3 – crystal growth and crystallographic in-
vestigation
W. Kleber, M. Hahnert, and R. Muller
Z. Anorg. Allgem. Chem., 346(3/4):113–126 (1966)

$WO_{2.95}$, a new phase with shear structure in the
$W - O$ system
Pal Gado and Laszlo Imre
Acta Chim. Acad. Sci. Hung., 46(2):165–169 (1965)

Mass-spectrometric study of the oxidation of
tungsten
P. O. Schissel and O. C. Trulson
Union Carbide Research Institute Tech. Rept. No. C-26,
Contract DA-30-069-ORD-2787 (October 1964)

Growth of WO_3 and MoO_3 crystals from cryolite-
oxide melts
V. N. Vigdorovich and V. V. Marychev
Dokl. Akad. Nauk SSSR, 159:416 (1964)

Spectrochemical analysis of high-purity tung-
sten
R. W. Lewis, C. F. Earl, J. L. Potter, and J. R. Wells
(Bureau of Mines, Boulder City Metallurgy Research Lab.,
Nev. and Bureau of Mines, Reno Metallurgical Research
Center, Nev.), BM-RI-5814 (Jan. 1961), 14 pp.

Domain structure of WO_3 single crystals
S. Sawada and G. C. Danielson
Phys. Rev., 113:1005–1006 (1959)

Growth spiral on crystals of WO_3
S. Tanisaki
J. Phys. Soc. Japan, 11:620 (1956)

5.c. Structure, Phases, Thermal Properties

5.c.1. General

Electron-beam crystallization of anodic oxide
films
A. Aladjem, D. G. Brandon, J. Yahalom, and J. Zahavi
Electrochimica Acta, 15:663–671 (1970)
Al, Ti, Nb, Ta, and W

The use of electron microscopy in the study of
extended defects related to non-stoichometry
S. Amelinckx and J. Van Landuyt
The Chemistry of Extended Defects in Non-Metallic Solids,
North-Holland Publishing Co., Amsterdam (1970), pp. 295–
322

Contribution à l'étude des solutions solides titane α − oxygène et zirconium α − oxygène
A. Dubertret
Ph. D. thesis, Paris, C. N. R. S. No. 4115 (Feb. 1970), 138 pp.
Ti-O and Zr-O

Melting-point measurements in the tri-arc furnace
T. B. Reed
(Lincoln Lab., Massachusetts Institute of Tech.), Solid State Research Qtr. Tech. Rept. (1970:3); ESD-TR-70-234 (Aug. 1970), pp. 15-17
TiO, Ti_2O_3, $VO_{1.25}$, NbO

The role of oxygen pressure in the control and measurement of composition in 3d metal oxides
T. B. Reed
The Chemistry of Extended Defects in Non-Metallic Solids, North-Holland Publishing Company, Amsterdam (1970), pp. 21-35

Phase transformations in anodic oxide films
D. G. Brandon, J. Zahavi, A. Aladjem, and J. Yahalom
J. Vacuum Sci. Tech., 6(4):783-787 (1969)
Al_2O_3, ZrO_2, TiO_2, Ta_2O_5, β-Nb_2O_5

Phase relations in refractory metal-oxygen systems
L. L. Y. Chang and Bert Phillips
J. Am. Ceram. Soc., 52:527-533 (1969)
Phases and phase relations in the tungsten-oxygen, molybdenum-oxygen, tantalum-oxygen, and niobium-oxygen systems are reviewed

The Chemistry of Extended Defects in Non-Metallic Solids
LeRoy Eyring and Michael O'Keeffe
North-Holland Publishing Co., Amsterdam, London (1970)
Proceedings of the Institute for Advanced Study on the Chemistry of Extended Defects in Non-Metallic Solids, Casa Blanca Inn, Scottsdale, Arizona, April 16-26 (1969)
Crystallographic shear, the niobium oxides and oxide fluorides centered about Nb_2O_5 in composition, S. Andersson, pp. 164-166; The use of electron microscopy in the study of extended defects related to non-stoichiometry, S. Amelinckx and J. Van Landuyt, pp. 295-322 (WO_3 and TiO_2); Structural, electrical and magnetic properties of vacancy stabilized cubic TiO and VO, M. D. Banus and T. B. Reed, pp. 488-522; Oxygen stoichiometry of donor-doped $BaTiO_3$ and TiO_2, N. G. Eror and D. M. Smyth, pp. 62-74; High-temperature mass spectrometry and the thermodynamics of the titanium oxides, Paul W. Gilles, pp. 75-90; Point, line and planar defects in some non-stoichiometric compounds, B. G. Hyde and L. A. Bursill, pp. 347-378 (V_2O_5, MoO_3, TiO_2); Structural order and disorder in oxide of transition metals of the titanium, vanadium, and chromium groups, A. Magneli, pp. 148-163; The role of oxygen pressure in the control and measurement of composition in 3d metal oxides, T. B. Reed, pp. 21-35 (Sc through Zn); Effect of extended defects on the electrical properties of substoichiometric oxide single crystals, M. J. Sienko and J. M. Berak, pp. 541-554 (WO_3); Vacancy filling in titanium monoxide and similar semimetals, A. Taylor and N. J. Doyle, pp. 523-540 (TiO, VO, NbO); Electron-microscopic study on the structure of low-temperature modification of titanium monoxide phase, D. Watanabe, O. Terasaki, A. Jostsons, and J. R. Castles, pp. 238-258.

Changes of concentration in binary metal-gas solutions during simultaneous evaporation of

gas and metal. I. Theoretical considerations. II. Discussion of the changes of concentration in the systems V − O, Nb − O, Ta − O, V − N, Nb − N, and Ta − N
G. Horz
Z. Metallk., 60:115-126 (1969)

Determination of oxygen in refractory oxides
C. S. MacDougall, M. E. Smith, and G. R. Waterbury
Anal. Chem., 41:372-374 (1969)

Cohesion in cubic refractory monocarbides, mononitrides, and monoxides
S. P. Denker
J. Less-Common Metals, 14:1 (1968)

Calculation of the heats of formation of point defects in some transition-metal oxides
P. G. Dickens, R. Heckingbottom, and J. W. Linnett
Trans. Faraday Soc., 64:1489-1498 (1968)
TiO, VO, MnO, FeO, CoO, and NiO

Crystallographic distortion, electron-electron interaction and the metal-nonmetal transition
A. M. de Graaf and R. Luzzi
Helv. Phys. Acta, 41:764-766 (1968)

Self-diffusion in simple oxides (a bibliography)
P. J. Harrop
J. Materials Sci., 3:206-222 (1968)

Point defects and phase stability of transition-metal compounds
F. A. Kroger
J. Phys. Chem. Solids, 29:1889-1899 (1968)

Application of the electron microprobe to the study of refractory metal carbides, oxides, and silicides
S. H. Moll and G. W. Brun
Anisotropy in Single-Crystal Refractory Compounds, Vol. 1 (F. W. Vahldiek and S. A. Mersol, eds.), Plenum Press, New York (1968)

Improvement in the sensitivity of the spectrographic determination of trace impurities in refractory bases
T. M. Moroshkina
Tr. Kom. Anal. Khim. Akad. Nauk SSSR, Inst. Geokhim. Anal., 16:62-66 (1968)
Fe, Ti, Al, Si, Ca, Ni, Co, Mo, Nb, and Ta in tungsten and tantalum and their oxides

The description of non-stoichiometric transition-metal oxides. A logical extension of inorganic crystallography
Sten Andersson
Bull. Soc. Franc. Mineral. Crist., 90:522-527 (1967)

Aspects of the problems of synthesis and structure of some oxide or oxide-like compounds formed by the transition elements in the groups IV, V, and VI of the periodic table
Sten Andersson
Arkiv Kemi, 26:521 (1967)
Review: 103 refs.

Interstitial Alloys
H. J. Goldschmidt, ed.
Plenum Press, New York (1967)

Thermodynamics of non-stoichiometric oxide systems
L. E. J. Roberts and T. L. Markin
Proc. Brit. Ceram. Soc., 8:201-217 (1967)

Ionic diffusion in solid metal oxides, final report, Oct. 31, 1965-Oct. 31, 1967
W. W. Smeltzer
AFOSR-67-2633; AD-661579 (Oct. 31, 1967)

Propriétés thermodynamiques, physiques, et structurales des dérivés semi-métalliques
Orsay, 28 Sept.-1 Oct. 1965, No. 157, Colloq. Intern. du Centre National de la Recherches Scientifique, Editions du Centre National de la Recherches Scientifique, Paris (1967)

Engineering properties of Selected Ceramic Materials
J. F. Lynch, C. G. Ruderer, and W. H. Duckworth, comps. and eds.
American Ceramic Society, Inc., Columbus, Ohio (1966)

Phase Diagrams for Ceramists
E. M. Levin, C. R. Robbins, and H. F. McMurdie, comps. and Margie K. Reser, ed.
American Ceramic Society, Inc., Columbus, Ohio (1964)

Plenum Press Handbooks of High-Temperature Materials. No. 1. Materials Index
Peter T. B. Shaffer
Plenum Press, New York (1964)

Non-stoichiometric Compounds
L. Mandelcorn, ed.
Academic Press, New York (1963)

Compilation of the melting points of the metal oxides
S. J. Schneider
NBS Monograph 68 (Oct. 1963)

Thermodynamic properties of 65 elements — their oxides, halides, carbides, and nitrides
C. E. Wicks and F. E. Block
(Albany Metallurgy Research Center, Albany, Oregon), Bureau of Mines Bull. 605 (Dec. 1961)

Physical Properties and Phase Diagrams of Ten Refractory Oxides. Part II. Journal Literature: Selected Bibliography
Z. D. Lane, M. J. Tunis, E. A. Olmstead, J. H. Kennedy, and G. R. Maynard
UCRL-6262, Part 2

5.c.2. Group IV

5.c.2.a. Titanium

On the aggregation of Wadsley defects in slightly reduced rutile
L. A. Bursill and B. G. Hyde
Phil. Mag., 23:3-15 (1971)

Equilibria of intermediate oxides in the titanium-oxygen system
J. S. Anderson and A. S. Khan
J. Less-Common Metals, 22:219-223 (1970)

Volume compression of TiO_x
M. D. Banus and Mary C. Lavine
Solid State Research Quarterly Tech. Summary – 1 Feb. through 30 April 1970, pp. 42-44

Alan L. McWhorter (Lincoln Lab., Massachusetts Institute of Technology), MIT Solid State Research (1970:2); ESD-TR-70-148 (May 15, 1970), 92 pp.

New ordered phases of slightly reduced rutile and their sharp dielectric absorptions at low temperature
C. W. Chu
Phys. Rev., B1:4700-4708 (1970)

Reinterpretation and temperature dependence of epr in $TiO_2:Fe^{3+}$ (anatase)
M. Horn and C. F. Schwerdtfeger
Solid-State Commun., 8:1741-1743 (1970)

Phase relationships in titanium-oxygen alloys
A. Jostsons and P. McDougall
The Science Technology and Application of Titanium (R. Jaffee and N. Promisel, eds.), Pergamon Press, Oxford and New York (1970), pp. 745-763

Neutron-diffraction investigation of ordered structures in the titanium-oxygen system
I. I. Kornilov, V. V. Vavilova, L. E. Fykin, R. P. Ozerov, S. P. Soloviev, and V. P. Smirnov
Met. Trans., 1:2569-2571 (1970)

Oxygen x-ray emission band shifts applied to the characterization of transition-metal oxide surface layers
H. B. Krause, G. A. Savanick, and E. W. White
(Penn. State Univ., University Park, Pa.), Contract N00014-67-A-0385-0002, AD-699 544; TR-4 (Jan. 1970), 11 pp.

Shear structures in titanium oxide
J. van Landuyt and S. Amelinckx
Mat. Res. Bull., 5:267-274 (1970)

Diffraction effects due to shear structures: a new method for determining the shear vector
J. van Landuyt, R. de Ridder, R. Gevers, and S. Amelinckx
Mat. Res. Bull., 5:353-362 (1970)
TiO_2

Decorated dislocations in rutile
A. J. H. Mante and J. Volger
Phys. Letters, A, 32:204-205 (1970)

Determination of impurities in titanium and titanium dioxide by neutron activation analysis. Part V. Destructive and non-destructive determination of manganese, indium, and uranium
R. Neirinckx, F. Adams, and J. Hoste
Anal. Chim. Acta, 50:31-38 (1970)

Thermal expansion of rutile and anatase
K. V. Krishna Rao, S. V. Nagender Naidu, and Leela Iyengar
J. Am. Ceram. Soc., 53:124-126 (1970)

Possible topotaxy in the TiO_2 system
P. Y. Simons and Frank Dachille
Am. Mineralogist, 55:403-415 (1970)

Oxygen vacancies in rutile crystals doped with chromium
L. S. Sochava, I. I. Reshina, and D. N. Mirlin
Fiz. Tverd. Tela, 12:1214-1221 (1970)
Sov. Phys. — Solid State, 12(4):946-951 (1970)

Stabilization of zirconia by erbia
R. K. Stewart and O. Hunter, Jr.
J. Am. Ceram. Soc., 53:421-422 (1970)

Diffusion of titanium in single-crystal rutile
  D. A. Venkatu and L. E. Poteat
  Materials Sci. Engng. (Netherlands), 5(5):258-262 (May 1970)

Interstitial order disorder transformation in
the Ti − O solid solution. IV. A neutron-diffraction
tion study
  S. Yamaguchi, K. Hiraga, and M. Hirabayashi
  J. Phys. Soc. Japan, 28:1014 (1970)

Diffusion of defects in rutile during its par-
tial reduction in a vacuum
  V. I. Barbanel' and V. N. Bogomolov
  Fiz. Tverd. Tela, 11(9):2671 (1969)
  Sov. Phys. − Solid State, 11(9):2160-2161 (1970)

Nature of defects in partly reduced rutile crys-
tals
  V. I. Barbanel', V. N. Bogomolov, S. A. Borodin, and S. I.
    Budarina
  Fiz. Tverd. Tela, 11(2):534-536 (1969)
  Sov. Phys. − Solid State, 11(2):431-433 (1969)

Covalency of d^1 ions in rutile
  B. P. Berkovskii
  Fiz. Tverd. Tela, 11(9):2699-2700 (1969)
  Sov. Phys. − Solid State, 11(9):2187-2188 (1970)

New data on thermal conductivity, thermo emf
and electroconductivity of pure and doped rutile
single crystals
  V. N. Bogomolov, I. A. Smirnov, and E. V. Shadrichev
  Fiz. Tverd. Tela, 11(11):3214-3224 (1969)
  Sov. Phys. − Solid State, 11(11):2606-2613 (1970)

On a new family of titanium oxides and the na-
ture of slightly reduced rutile
  L. A. Bursill, B. G. Hyde, O. Terasaki, and D. Watanabe
  Phil. Mag., 20:347-359 (1969)

Contribution à l'étude, par résonance para-
magnétique électronique, de la nature et de la
réactivité des défauts créés par réduction du
bioxyde de titane
  M. Che
  Ph. D. thesis, Lyon, CNRS No. 3611 (Sept. 1969), 109 pp.

Pressure derivatives of the elastic properties of
polycrystalline quartz and rutile
  D. H. Chung and Gene Simmons
  (Mass. Inst. of Technology, Dept. of Earth and Planetary
    Sciences, Cambridge, Mass. 02139), to be published in
    Earth and Planetary Sci. Letters (1969)

Pressure-temperature studies of anatase,
brookite, rutile and $TiO_2(II)$: a reply
  Frank Dachille, P. Y. Simons, and Rustum Roy
  Am. Mineral., 54:1481-1482 (1969)

Dissociation energy of the molecule TiO(g) and
the thermodynamics of the system Ti − O
  J. Drowart, P. Coppens, and S. Smoes
  J. Chem. Phys., 50(2):1046-1048 (1969)

Dissociation energy of TiO(g) and the high-
temperature vaporization and thermodynamics
of titanium oxides
  V. P. W. Gilles and P. J. Hampson
  J. Chem. Phys., 50(2):1048-1049 (1969)

Pressure-temperature studies of anatase,
brookite, rutile, and $TiO_2(II)$: a discussion
  John C. Jamieson and Bart Olinger
  Am. Mineral., 54:1477-1481 (1969)

Configurational entropy of titanium monoxide
  A. Jostsons and A. E. Jenkins
  J. Phys. Chem., 73-749 (1969)

Lattice dynamics of crystals having rutile struc-
ture (FeF_2, MnF_2, MgF_2 and TiO_2)
  R. S. Katiyar and R. S. Krishnan
  J. Indian Inst. Sci., 51(2):121-133 (1969)

An experimental study of defect ordering and
transport phenomena in nonstoichiometric metal
oxides, annual summary report April 1, 1968-
March 31, 1969
  Clement J. Kevane
  (Arizona State University, Tempe), Contract N00014-68-A-
    0306, AD-687 737 (April 1969), 17 pp.

Structural study of Ti_5O_9
  L. K. Keys
  (Pennsylvania State Univ., Materials Research Lab.), in Na-
    tional Bureau of Standards Molecular Dynamics and Struc-
    ture of Solids (June 1969), pp. 273-276

Interstitial order-disorder transformation in the
Ti − O solid solution. II. A calorimetric study
  Masahiro Koiwa and Makoto Hirabayashi
  J. Phys. Soc. Japan, 27:801-806 (1969)

Interstitial order-disorder transformation in
the Ti − O solid solution. III. A statistical
theory
  Masahiro Koiwa and Makoto Hirabayashi
  J. Phys. Soc. Japan, 27:807-815 (1969)

Bond type in the dioxides of titanium, germani-
um and tin
  M. F. C. Ladd
  Acta Cryst., 25A:486-487 (1969)

Crystallography and defect chemistry of solid
solutions of vanadium and titanium oxides
  R. E. Loehman, C. N. R. Rao, and J. M. Honig
  J. Phys. Chem., 73:1781 (1969)

Elastic constants of single-crystal rutile under
pressures to 7.5 kilobars
  M. H. Manghnani
  J. Geophys. Res., 74:4317-4828 (1969)

Microwave sound absorption by anharmonic three-
phonon interactions in TiO_2
  R. Nava, R. Callarotti, H. Ceva, and A. Martinet
  Phys. Rev., 185:1177 (1969)

Determination of impurities in titanium and
titanium dioxide by neutron activation analysis.
Part II. Determination of 27 trace constituents
in titania powder
  R. Neirinckx, F. Adams, and J. Hoste
  Anal. Chim. Acta, 46:165-178 (1969)

Determination of impurities in titanium and
titanium dioxide by neutron-activation analysis.
Part IV. Determination of trace impurities in
titanium dioxide single crystals
  R. Neirinckx, F. Adams, and J. Hoste
  Anal. Chim. Acta, 48:1-11 (1969)

Madelung potentials and ordering energy in TiO
  M. O'Keeffe and M. Valigi
  J. Chem. Phys., 50:1490-1491 (1969)

Specific heat and paramagnetic susceptibility of stoichiometric and reduced rutile (TiO_2) from 0.3 to 20°K
T. R. Sandin and P. H. Keesom
Phys. Rev., 177(3):1370-1383 (1969)

Vibrational spectrum of crystals having the rutile structure and nonanalytic behavior of the limiting dipole frequencies
M. S. Shur and Yu. N. Tsarev
Fiz. Tverd. Tela, 10(10):2906-2908 (1968)
Sov. Phys. — Solid State, 10(10):2293-2298 (1969)

Anisotropy of diffusion in rutile
J. L. Steele and E. R. McCartney
Nature, 222:79-80 (1969)

Kinetics of the transition of titanium dioxide prepared by sulfate process and chloride process
A. Suzuki and R. Tukuda
Bull. Chem. Soc. Japan, 42(7):1853-1857 (1969)
Anatase-rutile

Rules for inelastic neutron scattering in rutile crystals
O. A. Usov and G. L. Bir
Fiz. Tekh. Poluprov., 3(8):1188-1191 (1969)
Sov. Phys. — Semicond., 3(8):998-1000 (1970)

Certain results of x-ray spectral investigation of the chemical band in variable composition ($Ti - TiO_{0.48}$) titanium oxides
E. E. Vainshtein and others
AD-695916; FTD-HT-23-989-68
Izv. Akad. Nauk SSSR, Otd. Sibir. Ser. Chim. Nauk (Moscow), No. 3, pp. 12-16 (1966) (May 12, 1969), 11 pp.

The structure of the low-temperature modification of titanium monoxide
A. W. Vere and R. E. Smallman
The Mechanism of Phase Transformations in Crystalline Solids, Proceedings of an International Symposium organized by the Institute of Metals and held in the University of Manchester from July 3-5, 1968, pp. 212-219, Monograph and Report Series No. 33, The Institute of Metals, London (1969)

The ordered structure of $TiO_{1.25}$
D. Watanabe, O. Terasaki, A. Jostsons, and J. R. Castles
The Mechanism of Phase Transformations in Crystalline Solids, Proceedings of an International Symposium organized by the Institute of Metals and held in the University of Manchester from July 3-5, 1968, pp. 220-221, Monograph and Report Series No. 33, The Institute of Metals, London (1969)

Interstitial order-disorder transformation in the Ti — O solid solution. I. Ordered arrangement of oxygen
Sadae Yamaguchi
J. Phys. Soc. Japan, 27:155 (1969)

Quenchable effects of high pressures and temperatures on the cubic monoxide of titanium
Mario D. Banus
Mat. Res. Bull., 3:723:734 (1968)

Defect state and the properties of some solids (TiO_2 and MgO)
J. O. Brittain

Anisotropy in single-crystal refractory compounds (F. W. Vahldiek and S. A. Mersol, eds.), Vol. 2, Plenum Press, New York (1968), p. 95

X-ray study of titanium monoxide in homogeneous region and titanium nitride
I. A. Brytov, M. A. Rumsh, and A. S. Parobets
Fiz. Tverd. Tela, 10(3):794-800 (1968)
Sov. Phys. — Solid State, 10(3):621-626 (1968)

Pressure-temperature studies of anatase, brookite, rutile and TiO_2-II
Frank Dachille, P. Y. Simons, and Rustum Roy
Am. Mineral., 53:1929-1939 (1968)

Neutron and x-ray diffraction studies on tapiolite and some synthetic substances of trirutile structure
Olov von Heidenstam
Arkiv for Kemi, 28:375-387 (1968)

Neue Phasen im System Titan-Sauerstoff
Ernst Hilti
Naturwissenschaften, 55:130-131 (1968)

Röntgenographische Untersuchung der Titanmonoxid-Tieftemperatur-Modifikation
E. Hilti and F. Laves
Naturwissenschaften, 55:131 (1968)

Force field, elastic constant and specific heat of rutile
K. Iiishi, Y. Shiro, H. Murata, and Y. Umegaki
J. Sci. Hiroshima Univ. A-II (Japan), 32:277-290 (1968)

The polymorphism of the oxide Ti_3O_5
H. Iwasaki, N. F. H. Bright, and J. F. Rowland
J. Less-Common Metals, 17:99-110 (1968)

On stoichiometric defects in rutile (TiO_{2-x})
W. Jakubowski
Acta Phys. Polon., 33:465-470 (1968)

High-pressure polymorphism of titanium dioxide
J. C. Janieson and B. Olinger
Science, 161:893-895 (1968)

The ordered structure of Ti_3O
A. Jostsons and A. S. Malin
Acta Cryst., B24:211 (1968)

Fault structures in Ti_2O
A. Jostsons and P. G. McDougall
Phys. Stat. Sol., 29:873 (1968)

Anion dipole contribution to the lattice energy of rutile
P. I. Kingsbury
Acta Cryst., 24A(5):578-579 (1968)

Defects in rutile. I. Electron paramagnetic resonance of interstitially doped n-type rutile. 2. Diffusions of interstitial ions. 3. Optical and electrical properties of impurities and charge carriers
P. I. Kingsbury, W. D. Ohlsen, and O. W. Johnson
Phys. Rev., 175:1091-1109 (1968)

Defects in rutile: II. Diffusion of interstitial ions
Paul I. Kingsbury, Jr., and W. D. Ohlsen
(Utah Univ., Salt Lake City), Contract AT(11-1)-1284, COO-1284-9 (April 30, 1968)

Hypersonic attenuation in rutile
J. N. Lange
Phys. Rev., 176:1030–1035 (Dec. 15, 1968)

Adsorcion de oxigeno por oxidos de titanio en la oscuridad y bajo radiacion ultravioleta. III. Estudio de la conductividad eléctrica y de las especies adsorbidas en muestras dopadas
G. Munuera and F. Gonzalez
R. Soc. esp Fis. Quim., An. Quim., 64:15–23 (1968)

Determination of impurities in titanium and titanium dioxide by neutron activation analysis. I. Simultaneous determination of 16 trace elements in titanium
R. Neirinckx, F. Adams, and J. Hoste
Anal. Chim. Acta, 43:369–380 (1968)

Crystal structure analysis of niobium-doped rutile single crystal
N. Niizeki
Advances in X-Ray Analysis, Vol. 11 (J. B. Newkirk, Q. R. Mallett, and H. G. Pfeiffer, eds.), Plenum Press, New York (1968), pp. 482–487

Crystallographic study of the transition in Ti_2O_3
C. N. R. Rao, R. E. Loehman, and J. M. Honig
Phys. Letters, 27A:271 (1968)

The specific heat and paramagnetic susceptibility of stoichiometric and reduced rutile (TiO_2) from 0.3 to 20°K
T. R. Sandin
Ph. D. thesis, Purdue Univ. (August, 1968)

The specific heat of vacuum reduced rutile (TiO_2) below 4.5°K
T. R. Sandin and P. H. Keesom
Phys. Letters, 26A:167 (1968)

The ordered structure of $TiO_{1.25}$
D. Watanabe, O. Terasaki, A. Jostsons, and J. R. Castles
J. Phys. Soc. Japan, 25:292 (1968)

Thermodynamic study of dilute solutions of defects in the rutile structure TiO_{2-x}, NbO_{2-x}, and $Ti_{0.5}Nb_{0.5}O_{2+x}$
C. B. Alcock, S. Zador, and B. C. H. Steele
Proc. Brit. Ceram. Soc., No. 8, pp. 231–245 (1967)

Studies of the defect structure of nonstoichiometric rutile, TiO_{2-x}
R. N. Blumenthal, J. Baukus, and W. M. Hirthe
J. Electrochem. Soc., 114:172 (1967)

Study of the nature of water in rutile by slow-neutron inelastic scattering
H. Boutin, H. Prask, and D. R. Lyengar
J. Catalysis, 9:309–312 (1967)

Some results of x-ray spectral studies of the higher oxides of titanium ($TiO_{1.75} - TiO_2$)
V. I. Chirkov and E. E. Vainshtein
Izv. Akad. Nauk SSSR, Neorg. Mater., 3(6):1022–1027 (1967)

Discrepancies in the thermodynamics of the titanium-oxygen system
Paul W. Gilles
J. Chem. Phys., 46:4987–4988 (1967)

High-temperature vaporization and thermodynamics of the titanium oxides: I. Vaporization characteristics of the crystalline phases
Paul W. Gilles, K. Douglas Carlson, Hugo F. Franzen, and Phillip G. Wahlbeck
J. Chem. Phys., 46:2461–2465 (1967)

Preparation of thin sections by ion bombardment for transmission electron microscopy (TiO_2)
W. M. Hirthe, A. T. Melville, and P. H. Wackman
Rev. Sci. Instr., 38:223–227 (1967)

EPR studies of titanium interstitials in n-type rutile
P. I. Kingsbury, W. D. Ohlsen, and O. W. Johnson
(Utah Univ., Salt Lake City), Contract AT(11-1)-1284
From American Inst. of Phys. Summer Mtg., Seattle, Wash.
CONF-670951-2 (Sept. 2, 1967), 6 pp.

Thermal expansion of rutile from 100 to 700°K
R. K. Kirby
J. Res. Nat. Bur. Stand., 71A:363–371 (1967)

Note on the defect structure of rutile (TiO_2)
P. Kofstad
J. Less-Common Metals, 13:635–638 (1967)

Enthalpy of the anatase-rutile transformation
A. Navrotsky and O. J. Kleppa
J. Am. Ceram. Soc., 50:626 (1967)

Variation du pouvoir réflecteur X du rutile sous l'influence d'un champ électrique
Rene Puget
Compt. Rend., 264:708 (1967)

Proton trapping in TiO_2 and simple oxides
R. Roy and R. T. Greer
Solid State Commun., 5(2):109–111 (1967)

The structure of TiO_2. II. A high-pressure phase of TiO_2
P. Y. Simons and F. Dachille
Acta Cryst., 23:334–336 (1967)

Specific heat and character of the chemical bond in Ti and V carbides, nitrides and oxides
V. V. Tarasov, A. F. Demidenko, and A. K. Maltsev
Izv. Akad. Nauk SSSR, Neorg. Mater., 3(6):957–962 (1967)
Inorg. Mater., 3(6):857–861 (1967)

Dissociation energy of TiO(g) and the high-temperature vaporization and thermodynamics of the titanium oxides: II, Tri-titanium pentoxide
Phillip G. Wahlbeck and Paul W. Gilles
J. Chem. Phys., 46:2465–2473 (1967)

The ordered structure of TiO
D. Watanabe, J. R. Castles, A. Jostsons, and A. S. Malin
Acta Cryst., 23:307 (1967)

Properties of titanium and rutile (TiO_2) thin films
Floyd Olaf Arntz
Tech. Rept. AFML-TR-66-122 (March, 1966)

Low-temperature internal friction peak in rutile
R. D. Carnahan and J. O. Brittain
J. Appl. Phys., 37:1808–1809 (1966)

Lattice parameters of nine oxides and sulfides as a function of pressure
  R. L. Clendenen and H. G. Drickamer
  J. Chem. Phys., 44:4223-4228 (1966)
  TiO₂

The structure and morphology of oxide films during the initial stages of titanium oxidation
  D. L. Douglass and J. van Landuyt
  Acta Met., 14:491 (1966)

The defect structure of rutile containing small additions of aluminum oxide
  Katrine Seip Forland
  Acta Chem. Scand., 20:2573-2578 (1966)

Densities, lattice parameters and defect chemistry of pure non-stoichiometric compounds (TiO$_x$)
  W. van Gool
  J. Mater. Sci., 1:261-268 (1966)

Isotopic mass dependence of Li diffusion in rutile
  O. W. Johnson and H. R. Krouse
  J. Appl. Phys., 37:668-670 (1966)

The fine structure of spots in electron diffraction resulting from the presence of planar interfaces and dislocations. II. Observations on crystals containing stacking faults
  J. van Landuyt, R. Gevers, and S. Amelinckx
  Phys. Stat. Sol., 18:363 (1966)
  TiO₂

Phase transition of titanium dioxide under various pressures
  F. W. Vahldiek
  J. Less-Common Metals, 11:99-110 (1966)

Internal friction in rutile containing Ni or Cr
  J. B. Wachtman, Jr., S. Spinner, W. S. Brower, T. Fridinger, and R. W. Dickson
  Phys. Rev., 148:811-816 (1966)

Ordered structure of titanium oxide
  D. Watanabe, J. R. Castles, A. Jostsons, and A. S. Malin
  Nature, 210:934-936 (1966)

Diffusion of transition-metal ions into rutile (TiO₂)
  James P. Wittke
  J. Electrochem. Soc., 113:193-194 (1966)

Dislocations and stacking faults in rutile crystals grown by flame-fusion methods
  D. J. Bareer and E. N. Farabaugh
  J. Appl. Phys., 36:2803-2806 (1965)

Propriétés du bioxyde de titanate (anatase) élaboré dans le réacteur à flamme
  J. Long, F. Fuillet, and S.-J. Teichner
  Rev. Hautes Temper. Refract., 2:163-172 (1965)

Investigation of catalyzed crystallization by means of Raman spectroscopy
  Ya. S. Bobovich
  FTD-TT-65-689; TT-66-60134; AD-625 786 (Sept. 1965), 11 pp.
  Stekloobraznoe Sostoyanie, No. 1, Katalizirovannaya Kristallizatsiya Stekla, Izd. Akad. Nauk SSSR, Leningrad (1963), pp. 87-90
  The Structure of Glass, Vol. 3, Catalyzed Crystallization of Glass, Consultants Bureau, New York (1964), pp. 93-95
  TiO₂

Anisotropy of oxygen and boron diffusion in rutile
  V. N. Bogomolov
  Fiz. Tverd. Tela, 5(7):2011 (1963)
  Sov. Phys. – Solid State, 5(7):1468 (1964)

Dipolar defects in reduced rutile
  Lawrence Adam Konrad Dominik
  Ph. D. thesis, Pennsylvania Univ., Philadelphia
  Available from Univ. of Microfilms, Ann Arbor, Michigan as No. 67-12741

Thermogravimetric studies of the defect structure of rutile (TiO₂)
  P. Kofstad
  J. Phys. Chem. Solids, 23:1579 (1962)

Studies on the brookite-rutile transformation
  C. N. R. Rao, S. R. Yoganarasimhan, and P. A. Faeth
  Trans. Faraday Soc., 57:504 (1961)

Effect of high pressure on oxides with defect-rocksalt structures (TiO$_x$)
  M. D. Banus
  Report ESD-TR-67-266, pp. 24-25

Studies in the titanium-oxygen system and the defect nature of rutile
  Vernon Ray Porter
  Dissertation Abstracts 66-4838, 107 pp.

The kinetics and mechanism of the anatase-rutile transformation
  Robert Day Shannon
  Dissertation Abstracts 64-13,094, 97 pp.

5.c.2.b. Zirconium

Melting point of pure zirconia
  R. E. Latta and E. C. Duderstadt
  (General Electric Co., Nuclear Systems Programs, Cincinnati, Ohio), GEMP-735 (April 1970), 5 pp.

Thermoelectron emission, electric conductivity, and thermogravimetry in study of point defects of zirconium oxide
  J. P. Loup and A. M. Anthony
  Phys. Stat. Sol., 38:499-512 (1970)

Lattice contraction of hot-pressed calcia-stabilized zirconia
  N. M. P. Low and A. C. D. Chaklader
  Mater. Res. Bull., 5:137-146 (1970)

Oxygen diffusion in monoclinic zirconia
  D. J. Poulton and W. W. Smeltzer
  J. Electrochem. Soc., 117:378-381 (1970)

Variations des paramètres cristallins de la solution solide α-zirconium – oxygène en fonction de la teneur en oxygène
  Pierre Boisot and Gerard Beranger
  Compt. Rend., 269C:587-590 (1969)

Interstitial ordering in α-zirconium – oxygen alloys
  M. Fehlmann, A. Jostsons, and J. G. Napier
  Z. Krist., 129:318-342 (1969)

Structural characteristics of cubic ZrO₂ stabilized by calcium
  S. K. Filatov and V. A. Frank-Kamenetskii
  Kristallografiya, 14(3):505-506 (1969)
  Sov. Phys. – Cryst., 14(3):414-415 (1969)

Anomalous thermal expansion of ZrO_2 and HfO_2 over the range 20-1200°C
S. K. Filatov and V. A. Frank-Kamenetskii
Kristallografiya, 14:804-808 (1969)
Sov. Phys. - Cryst., 14:696 (1970)

Effect of rare-earth oxide impurities on the polymorphism of zirconium dioxide
V. B. Glushkova and L. V. Sazonova
Chemistry of High-Temperature Materials (N. A. Toropov, ed.), Consultants Bureau, New York (1969), pp. 90-96

Dependence of the coefficient of thermal expansion of stabilized zirconium dioxide on the ratio of cubic and monoclinic phases
A. G. Karaulov, A. A. Grebenyuk, V. Ya. Belik, and I. N. Rudyak
Chemistry of High-Temperature Materials (N. A. Toropov, ed.), Consultants Bureau, New York (1969), pp. 48-52

Thermochemical investigation of the Zr — O system
A. N. Kornilov, I. M. Ushakov, S. M. Skuratov, and G. P. Shverkin
Dokl. Akad. Nauk SSSR, 186:831-834 (1969)
Dokl. - Chem., 186:439-442 (1969)

Microwave acoustic relaxation in reduced rutile
J. N. Lange
Phys. Rev., 179:860-862 (1969)

ZrO_2 — TiO_2 system: IV, liquidus curve measurement
Masao Mizuno, Tetsuo Noguchi, and Toyoaki Yamada
Nagoya Kogyo Gijutsu Shikensho Hokuku, 18:9-14 (1969)

Axial thermal expansion of ZrO_2 and HfO_2 in the range room temperature to 1400°C
R. N. Patil and E. C. Subbarao
J. Appl. Crystallog., Denm., 2(6):281-288 (1969)

Low-temperature cubic and tetragonal forms of zirconium dioxide
Yu. M. Polezhaev
FTD-HT-23-1066-68; AD-687 236 (Jan. 1969), 7 pp.
Zh. Fiz. Khim., 41:2958-2959 (1967)

Ermittlung der Struktur des Zirkoniumsuboxides Zr_2O mittels Elektronenbeugung
S. Steeb and A. Rickert
J. Less-Common Metals, 17:429-436 (1969)

Anisotropic thermal expansion of zirconia
M. Davies
(Morganite Research and Development Ltd., London), Special Ceramics 4 (P. Popper, ed.), Stoke-on-Trent, Eng., British Ceramic Research Association (1968), pp. 229-236
20-1100°C by x-ray diffraction

Ordered structures of Zr_6O and Zr_3O suboxides
L. M. Kovba, E. M. Kenina, I. I. Kornilov, and V. V. Glazova
Dokl. Akad. Nauk SSSR, 180:360-363 (1968)
Dokl. - Chem., 180:436-439 (1968)

Thermal expansion of ZrO_2 stabilized between 300 and 2600°K
D. M. Shakhtin, E. V. Levintovich, T. L. Pivovar, and G. G. Eliseeva
Izv. Akad. Nauk SSSR, Neorg. Mater., 4(9):1603-1604 (1968)
Inorg. Mater., 4(9):1401-1402 (1968)

Ordered arrangement of oxygen in the interstitial solid solution of zirconium-oxygen system
Sadae Yamaguchi
J. Phys. Soc. Japan, 24:855 (1968)

Influence of heating rate on the temperature characteristic of the nondiffusive transition of zirconium dioxide
A. M. Gavrish, B. Ya. Sukharevskii, and P. P. Krivoruchko
Dokl. Akad. Nauk SSSR, 177:886-889 (1967)
Dokl. - Chem. Tech., 177:176-179 (1967)

X-ray and spectrographic investigations of phase transitions in refractory oxides (ZrO_2 and rare-earth oxides)
V. B. Glushkova and E. K. Koehler
Mat. Res. Bull., 2:503-513 (1967)

Observations en microscopie électronique de films de zircone obtenus par oxydation du zirconium
G. Beranger, F. Duffaut, B. de Gelas, and P. Lacombe
J. Nucl. Mat., 19:290-300 (1966)

The atomic structure of amorphous ZrO_2
K. Doi
Bull. Soc. Franc. Mineral. Crist., 89:216-225 (1966)
CSIRO-TRANS-8339
Available from SLA as 68-10835-20B

Thermal expansions of zirconia and formation reaction of stabilized zirconia
Teruichiro Kubo, Masanori Kato, and Iwao Kohatsu
Kogyo Kagaku Zasshi, 69:2277-2280 (1966)

Freezing point measurement of metal oxides with a solar furnace
Tetsuo Noguchi, Masao Mizuno, and Takeshi Kozuka
Kogyo Kagaku Zasshi, 69:1705-1709 (1966)
ZrO_2, HfO_2, and Al_2O_3 are 2979° ± 20°, 3026° ± 20°, and 2322° ± 15°K, respectively
Zr and Hf

Phase transitions in ZrO_2 obtained by the thermal decomposition of $NaHZrSiO_5$; effect of preliminary pressure
Yu. M. Polezhaev and V. M. Ust'yantsev
Issledovaniya v oblasti khimii silikatov i okislov, Izdat. Nauka, Moscow (1965), pp. 217-220

Electrical resistivity and diffusionless phase transformations of zirconia at high temperatures and ultrahigh pressures
E. Dow Whitney
J. Electrochem. Soc., 112:91 (1965)

Compositional analysis of twenty-six zirconias
R. E. Heffelfinger, E. R. Blosser, and W. M. Henry
J. Am. Ceram. Soc., 47:646 (1964)

An annotated bibliography of research on the structural and physical properties of zirconia
Berthold C. Weber
ARL-64-205 (1964)

The structure of anodic films — I. An electron diffraction examination of the products of anodic oxidation of tantalum, niobium, and zirconium
P. H. G. Draper and J. Harvey
Acta Met., 11:873 (1963)

Inconsistencies in zirconia literature
B. C. Weber
J. Am. Ceram. Soc., 45:614 (1962)
The results of research in the field of zirconia systems have produced many discrepancies. This note is an attempt to point out and explain numerous erroneous statements reported on zirconium dioxide

The study of surfaces by low-energy electron
diffraction. Examination of the V_2O_3 (010)
surface by LEED
   L. Fiermans
   Verhandl. K. Vlaamse Acad. Wetensch. (Belgium), 31(111):9-
     149 (1969)

Particular LEED features on the V_2O_5 (010) surface and their relation to the LEED beam induced transition $V_2O_5 \rightarrow V_{12}O_{26}$
   L. Fiermans and J. Vennik
   Surface Si., 18:317-324 (1969)

Lattice polarization and Coulomb energies in
VO_2
   G. J. Hyland
   Phil. Mag., 20:837-841 (1969)

Thermal expansion anisotropy of oxides and
oxide solid solutions
   Henry P. Kirchner
   J. Am. Ceram. Soc., 52:379-386 (1969)
   VO_2

Crystallography and defect chemistry of solid
solutions of vanadium and titanium oxides
   R. E. Loehman, C. N. R. Rao, and J. M. Honig
   J. Phys. Chem., 73:1781 (1969)

Action ménagée de certains réducteurs sur V_2O_3;
mise en evidence de l'oxyde V_4O_9
   Francois Theobald, Robert Cabala, and Jean Bernard
   Compt. Rend., 269:Ser.C, 1209-1212 (1969)

Specific heat of cubic vanadium monoxide and
vanadium oxycarbide
   V. S. Chernyaev, E. N. Shchetnikov, and G. P. Shveikin
   Izv. Akad. Nauk SSSR, Neorg. Mater., 4(12):2117-2123 (1968)
   Inorg. Mater., 4(12):1840-1846 (1968)

Thermal capacity of cubic vanadium oxide and
oxycarbide
   V. S. Chernyaev, E. N. Shchetnikov, G. P. Shveikin, and P. V.
    Gel'd
   Izv. Akad. Nauk SSSR, Neorg. Mater., 4(12):2117-2123 (1968)
   Inorg. Mater., 4(12):1840-1846 (1968)

Kinetics of the oxygen up-take of group Va met-
als. II. Investigations in the $V-O$ system. III.
Investigations in the $Nb-O$ and $Ta-O$ systems
   G. Horz
   Z. Metallk., 59:180 and 283 (1968)

Priprava a vlastnosti jednoduchych a nekterych
podvojnych kyslicniku vanadu
   Emil Pollert
   Chem. listy Svazek, 62:897-917 (1968)
   Review, 52 refs.

Epitaxial growth of VO single crystals and their
electrical properties
   H. Takei and S. Koide
   J. Phys. Soc. Japan, 24:1394 (1968)
   Rock salt structure

Determination of tetravalent vanadium and iron
impurities in vanadium pentoxide
   S. Ya. Vinkovetskaya and T. I. Levitskaya
   Zavod. Lab., 34:278 (1968)

Electron-microscope study of oxygen interstitial
ordering in vanadium
   M. Cambini, M. Heerschap, and R. Gevers
   Luxembourg, Off. Centr. Vente Publ. Communautes europ.,
    EUR-3642e (1967), 24 pp.

Influence of the LEED beam on the structure
of the vanadium pentoxide (010) surface
   L. Fiermans and J. Vennik
   Phys. Letters, 25A:687-688 (1967)

Domain structure and twinning in crystals of
vanadium dioxide
   P. J. Fillingham
   J. Appl. Phys., 38:4823 (1967)

Structural diagram of the V_2O_5-KVO_3 system
   A. A. Fotiev, M. P. Glazyrin, and S. I. Alyamovskii
   Zh. Neorg. Khim., 12:1325-1329 (1967)

The phase diagram and phase transition of the
$V_2O_3 - V_2O_5$ system
   Koji Kosuge
   J. Phys. Chem. Solids, 28:1613-1621 (1967)

X-ray spectroscopic study of vanadium com-
pounds with metallic properties
   E. Z. Kurmaets, S. A. Nemnonov, A. Z. Men'shikov, and
    G. P. Shveikin
   Izv. Akad. Nauk SSSR, Ser. Fiz., 31:996-1001 (1967)

Etch pits indicating plane directions in VO_2
   A. Kuromatsu, I. Kitahiro, and A. Watanabe
   Japan. J. Appl. Phys., 6:1005 (1967)

On the phase transformations of VO_2
   T. Mitsuishi
   Japan. J. Appl. Phys., 6:1060 (1967)

Thermal expansion of tetragonal phase of VO_2
   K. V. Krishna Rao, S. V. Nagender Naidu, and Leela Lyengar
   J. Phys. Soc. Japan, 23:1380-1382 (1967)

Specific heat and character of the chemical bond
in Ti and V carbides, nitrides and oxides
   V. V. Tarasov, A. F. Demidenko, and A. K. Maltsev
   Izv. Akad. Nauk SSSR, Neorg. Mater., 3(6):957-962 (1967)
   Inorg. Mater., 3(6):857-861 (1967)

Etude cristallographique de la phase $VO_{2.33}$
(V_3O_7)
   D. Thomas, J. Tudo, and G. Tridot
   Compt. Rend., 265:183-184 (1967)

Hydrated vanadium pentoxide
   V. L. Zolotavin, L. K. Tolstov, A. A. Ivakin, A. A. Fotiev,
    V. V. Mochalov, and Yu. F. Bukreev
   Izv. Akad. Nauk SSSR, Neorg. Mater., 3(9):1601-1607 (1967)
   Inorg. Mater., 3(9):1393-1398 (1967)

Sur la non-stoichiometrie inhomogène dans le
pentoxyde de vanadium
   Elie Gillis and Germain Remaut
   Compt. Rend., 262:1215 (1966)

Thermodynamic properties of vanadium and its
compounds
   Alla D. Mah
   (Bureau of Mines), BM-RI-6727 (1966)
   Includes VO, V_2O_3, V_2O_4, V_6O_{13}, V_2O_5

Crystal chemistry studies
 Rustum Roy, S. Kachi, G. J. McCarthy, O. Muller, and W. B. White
 (Pennsylvania State Univ., Materials Research Lab.) Qtrly. Rept. No. 2, Aug. 13–Nov. 13, 1965, Contract DA-28-043-AMC-01304(E), Jan. 1966, 34 pp.
 Includes phase equilibrium studies and transitions in the system V_2O_3–V_2O_5

Anisotropic thermal expansion of V_2O_5
 I. Corvin and L. Cartz
 J. Am. Ceram. Soc., 48:328 (1965)

Thermodynamic properties of non-stoichiometric vanadium monoxide
 M. Hoch and D. Ramakrishnan
 J. Phys. Chem. Solids, 25:869–872 (1964)

Phase transition in VO_2
 T. Kawakubo and T. Nakagawa
 J. Phys. Soc. Japan, 19:517 (1964)

Phase relations in the $VO_{1.5}$– VO_2 system at high temperatures
 M. P. Morozova, M. V. Konopelko, and I. V. Pinchuk
 Vestn. Leningr. Univ., Ser. Fiz. Khim., 19:109–114 (1964)

Kinetics of vanadium trioxide oxidation
 V. V. Pechkovskii and A. G. Zvezdin
 Kinet. Kataliz., 5:424–429 (1964)

Infrared spectra-structure correlation study of vanadium-oxygen compounds
 L. D. Frederickson and D. M. Hausen
 Anal. Chem., 35:818–827 (1963)

Changes of the free energy of formation of oxides and nitrides with temperature. Construction of the diagrams. Examples of application
 M. Olette and M. I. Ancay-Moret
 Rev. Met., 60:569–582 (1963)

Lower oxides of vanadium
 S. Westman
 Acta Chem. Scand., 17:749–752 (1963)

Lattice structures of lower oxides of titanium and vanadium
 S. M. Ariya and Yu. G. Popov
 Zh. Obshch. Khim., 32:2077–2081 (1962)

Structure of vanadium dioxide
 R. Heckingbottom and J. W. Linnet
 Nature, 194:678 (1962)

Refinement of the α-Al_2O_3, Ti_2O_3, V_2O_3, and Cr_2O_3 structures
 R. E. Newnham and Y. M. de Haan
 Z. Krist., 117:235 (1962)

The crystal structure of vanadium pentoxide
 H. G. Bachmann, F. R. Ahmed, and W. H. Barnes
 Z. Krist., 115:110 (1961)

Bonding in the trigonal bipyramidal coordination polyhedra of V_2O_5 and of certain other structures containing pentavalent vanadium
 H. G. Bachmann and W. H. Barnes
 Z. Krist., 115:215 (1961)

Beta-delta-zeta phases in the vanadium-oxygen system
 P. V. Gel'd, S. I. Alyamovskii, and I. I. Matveenko
 Zh. Strukt. Khim., 2:301–307 (1961)

Lattice energies of compounds of the transition-type metals
 M. F. C. Ladd and W. H. Lee
 J. Inorg. Nucl. Chem., 23:199–205 (1961)

Heats and free energies of formation of oxides of vanadium
 A. D. Mah and K. K. Kelley
 U. S. Bur. Mines Rep. Invest. No. 5858 (1961), 11 pp.

Heat capacity of a few vanadium oxides
 M. S. Yakovleva and Z. L. Krasilova
 Vestn. Leningrad Univ., Ser. Fiz. Khim., 16:136–139 (1961)

Structural peculiarities of vanadium oxide
 P. V. Gel'd, S. I. Alyamovskii, and I. I. Matveenko
 Fiz. Metallov. Metalloved., 9:315–317 (1960)

Enthalpy of formation and gram-formula volumes of lower oxides of vanadium
 M. P. Morozova and G. Eger
 Zh. Obshch. Khim., 30:3514–3517 (1960)

Study of rhombohedral V_2O_3 by neutron diffraction
 A. Paoletti and S. J. Pickart
 J. Chem. Phys., 32:308–309 (1960)

Determination of the structures of refractory products
 K. K. Strelov
 Ogneupory, 25:269–275 (1960)

Low-temperature x-ray diffraction studies on vanadium sesquioxide
 E. P. Warekois
 J. Appl. Phys. Suppl., 31:346–347 (1960)

Phase analysis studies on the vanadium-oxygen system within the $VO_{0.25}$-$VO_{1.5}$ region at 800°
 S. Westman and C. Nordmark
 Acta Chem. Scand., 14:465–470 (1960)

Crystal chemistry of titanium, vanadium, and molybdenum oxides at elevated temperatures
 A. Magneli, S. Andersson, L. Kihlborg, S. Asbrink, S. Westman, B. Holmberg, and C. Nordmark
 NP-8054 (1959), 141 pp.

Kinetics of the reduction of higher to lower metal oxides. Part 1. Reduction of vanadium-4 oxide to vanadium-3 oxide with hydrogen
 O. Schmitz-Dumont and G. Woitas
 Z. Elektrochem., 63:122–129 (1959)

Mass-spectrometric study of the sublimation of some vanadium and niobium oxides
 S. A. Shchukarev, G. A. Semenov, and K. E. Frantseva
 Zh. Neorg. Khim., 4:2638 (1959)

Enthalpy of formation of vanadium oxides
 E. Volf and S. M. Ariya
 Zh. Obshch. Khim., 29:2470–2473 (1959)

X-ray study of the lower oxides of titanium and vanadium
 E. Volf, S. S. Tolkachev, and I. I. Kozhina
 Vestn. Leningrad Univ., Ser. Fiz. Khim., 14:87–92 (1959)

Exact and rapid procedure for the combined direct determination of the oxidation state of vanadium and the vanadium content of metal-vanadium-oxygen compounds

B. Reuter and J. Siewert
Z. Anal. Chem., 162:175-180 (1958)

Specific heat of vanadium sesquioxide between 70 and 300°
J. Jaffray and R. Lyand
Compt. Rend., 233:133-135 (1951)

Heat capacities of vanadium, vanadium trioxide, vanadium tetroxide and vanadium pentoxide at low temperatures
C. T. Anderson
J. Amer. Chem. Soc., 58:564-566 (1936)

Vaporization of vanadium oxides: thermodynamic and phase studies
Stanley Killingbeck
Dissertation Abstr., 65-7657, 194 pp.
Available from University Microfilm, Inc., Ann Arbor, Michigan

5.c.3.b. Niobium

Lattice parameters, thermal expansion coefficients, and densities of Nb, and of solid solutions Nb — O and Nb — N — O and their defect structure
M. E. Straumanis and S. Zyszczynski
J. Appl. Cryst., 3:1 (1970)

The investigation of oxygen compounds of niobium in region of $NbO_2 - Nb_2O_5$
A. Burdeze, E. V. Tkachenko, and F. Abbattista
Izv. Akad. Nauk SSSR, Neorg. Mater., 5(11):1957-1962 (1969)
Inorg. Mater., 5(11):1666-1670 (1969)

A new hexagonal-type oxide of niobium
T. Nakayama, T. Osaka, and A. Kitada
J. Less-Common Metals, 19:291-293 (1969)

Note on the phase transition in NbO_2
Kimiko Sakata
J. Phys. Soc. Japan, 26:582 (1969)

An oxide transformation phenomenon observed during the oxidation of Nb
J. S. Sheasby
Oxidation of Metals, 1:121-125 (1969)

Electron microscopy of high-temperature Nb_2O_5 and related phases
J. G. Allpress, J. V. Sanders, and A. D. Wadsley
Phys. Stat. Sol., 25:541 (1968)

Application de la microscopie électronique par transmission à l'étude des céramiques
P. Delavignette and S. Amelinckx
Bull. Soc. Franc. Ceram., No. 80, pp. 75-92 (1968)
Ta_2O_5, Nb_2O_5

On the phase diagram of the niobium — oxygen system
E. Fromm and H. Jehn
J. Less-Common Metals, 15:242-243 (1968)

Niobium physicochemical properties of its compounds and alloys. III. Crystal structure and densities
Hans Nowotny and Karl J. Seifert
At. Energy Rev., Spec. Issue No. 2, pp. 71-172 (1968)
NbO

Oxygen diffusion in alpha-niobium pentoxide
J. S. Sheasby and B. Cox
J. Less-Common Metals, 15:129-135 (1968)

Diffusion of oxygen in near-stoichiometric α-Nb_2O_4
W. K. Chen and R. A. Jackson
J. Chem. Phys., 47:1144-1148 (1967)

Study of phase transition in NbO_2
T. Sakata, K. Sakata, and I. Nishida
Phys. Stat. Sol., 20:K155 (1967)

On the superstructures of Ta_2O_5 and Nb_2O_5
J. Spyridelis, P. Delavignette, and S. Amelinckx
Phys. Stat. Sol., 19:683 (1967)

Direct observation of the superlattices of Ta_2O_5 and Nb_2O_5 in the electron microscope
J. Spyridelis, P. Delavignette, and S. Amelinckx
Mat. Res. Bull., 2:113-120 (1967)

The solid-solubility of oxygen in Nb and Nb-rich Nb — Hf, Nb — Mo, and Nb — W alloys. Part I: The Nb — O system
A. Taylor and N. J. Doyle
J. Less-Common Metals, 13:313-330 (1967)

The crystal structure of niobium monoxide
A. L. Bowman, T. C. Wallace, J. L. Yarnell, and R. G. Wenzel
Acta Cryst., 21:843 (1966)

Studies on the defect structure of α-Nb_2O_5
W. K. Chen and R. A. Swalin
J. Phys. Chem. Solids, 27:57-64 (1966)

Zur p-T-e-Darstellung der Systeme Niob-Sauerstoff und Tantal-Sauerstoff
Eckehard Fromm
Z. Metallk., 57:540-545 (1966)

Die niederen Oxide des Niobs
J. Niebuhr
J. Less-Common Metals, 11:191-203 (1966)

Crystal structure of $Nb_{12}O_{29}$ (mon)
Rolf Norin
Acta Chem. Scand., 20:871-880 (1966)

Das Bauprinzip der Kristallstruktur von η-Nb_2O_5
W. Petter and F. Laves
Naturwissenschaften, 52:617 (1965)

Structure of niobium suboxide (Nb_4O) revealed by electron diffraction
Siegfried Steeb and Joerg Renner
Z. Metallk., 56:531-534 (1965)

The crystal structure of the high-temperature form of niobium pentoxide
B. M. Gatehouse and A. D. Wadsley
Acta Cryst., 17:1545-1554 (1964)

A submicron sectioning technique for analyzing diffusion specimens of tantalum and niobium
R. E. Pawel and T. S. Lundy
J. Appl. Phys., 35:435 (1964)

Heat content, thermal expansion, and electrical conductivity of NbO_2
F. G. Kusenko and P. V. Gel'd
Izv. Vysshikh Uchebn. Zaveden., Tsvetn. Met., 3:102 (1960)

5.c.3.c. Tantalum

Electron-diffraction investigation of the phase transition in thin films of tantalum oxides having a perovskite-like structure
V. I. Khitrova and Z. G. Pinsker
Kristallografiya, 15(3):540-546 (1970)
Sov. Phys. - Cryst., 15(3):455-460 (1970)

Effect of oxide additions on the polymorphism of tantalum pentoxide. III. "Stabilization" of the low-temperature structure type
R. S. Roth and J. L. Waring
J. Res. Natl. Bur. Stand., A74:485-493 (1970)

Effect of oxide additions on the polymorphism of tantalum pentoxide. II. "Stabilization" of the high-temperature structure type
R. S. Roth, J. L. Waring, and W. S. Brower
J. Res. Natl. Bur. Stand., A74:477-484 (1970)

Oxides formed on sputtered Ta
D. G. Muth
J. Vacuum Sci. Technol., 6:749-752 (1969)

Ta_2O_5 and $\alpha-KPO_3$
G. M. Wolten and A. B. Chase
Z. Kristal., 129(5-6):365-368 (1969)

Application de la microscopie électronique par transmission à l'étude des céramiques
P. Delavignette and S. Amelinckx
Bull. Soc. Franc. Ceram., No. 80, 75-92 (1968)
Ta_2O_5, Nb_2O_5

Thermodynamic properties of tantalum oxides
V. G. Dneprova, T. N. Rezukhina, and Ya. I. Gerasimov
Zh. Fiz. Khim., 42:1532-1535 (1968)

An electron-diffraction study of the γ phase in the Ta − O system
V. V. Klechkovskaya and V. I. Khitrova
Kristallografiya, 13(3):523-526 (1968)
Sov. Phys. - Cryst., 13(3):428-430 (1968)

High-resolution electron-microscopic observations of the superstructure of tantalum pentoxide
J. Spyridelis, P. Delavignette, and S. Amelinckx
Mat. Res. Bull., 3:31-36 (1968)

Effect of oxide additions on the polymorphism of tantalum pentoxide (system $Ta_2O_5 - TiO_2$)
J. L. Waring and R. S. Roth
J. Res. Natl. Bur. Stand., 72A:175-186 (1968)

A study of oxide-plate formation in Ta. II. Crystallographie analysis
C. M. Wayman and J. Van Landutyt
Acta Met., 16:815-822 (1968)

The unit cell of beta-Ta_2O_5
G. M. Wolten
(Aerospace Corp., El Segundo Calif. Lab. Operations), Contract F04695-67-C-0158, Rept. No. TR-0158 (3250-10)-15
SAMSO-TR-68-254 (March 1968), 15 pp.

Electron diffraction study of tantalum dioxide in thin films
V. I. Khitrova, V. V. Klechkovskaya, and Z. G. Pinsker
Kristallografiya, 12(6):1044-1050 (1967)
Sov. Phys. - Cryst., 12(6):907-912 (1968)

Direct observation of the superlattices of Ta_2O_5 and Nb_2O_5 in the electron microscope
J. Spyridelis, P. Delavignette, and S. Amelinckx
Mat. Res. Bull., 2:113-120 (1967)

On the superstructures of Ta_2O_5 and Nb_2O_5
J. Spyridelis, P. Delavignette, and S. Amelinckx
Phys. Stat. Sol., 19:683 (1967)

Structure of Ta oxide
N. Terao
Japan. J. Appl. Phys., 6:21-34 (1967)

Zur p-T-e-Darstellung der Systeme Niob-Sauerstoff und Tantal-Sauerstoff
Eckehard Fromm
Z. Metallk., 57:540-545 (1966)

Electron diffraction by thin films of cubic tantalum oxides
V. I. Khitrova
Kristallografiya, 11(2):204-212 (1966)
Sov. Phys. - Cryst., 11(2):199-204 (1966)

Suboxides of tantalum
J. Niebuhr
J. Less-Common Metals, 10:312-322 (1966)
The literature of the suboxides of tantalum is critically evaluated

Indication of hardness of thin dielectric films
Aubrey J. Raffalovich
Rev. Sci. Instr., 37:368-369 (1966)

A submicron sectioning technique for analyzing diffusion specimens of tantalum and niobium
R. E. Pawel and T. S. Lundy
J. Appl. Phys., 35:435 (1964)

Crystallization of thin amorphous tantalum oxide films heated in air or vacuo, and the structure of the crystalline oxide
J. Harvey and H. Wilman
Acta Cryst., 14:1278 (1961)

The thermodynamics of the rhenium-oxygen and molybdenum-oxygen systems and the defect structure of alpha tantalum pentoxide
James Sheridan Foster
Dissertation Abstracts, 65-5636, 148 pp.

5.c.4. Group VI

5.c.4.a. Chromium

The crystal structure of CrO_3
J. S. Stephens and D. W. J. Cruickshank
Acta Cryst., 26B, Pt. 3, 222-226 (1970)

The design of tables of Shubnikov space groups of dichroic symmetry
R.P. Ozerov
Kristallografiya, 14(3):393-403 (1969)
Sov. Phys. - Cryst., 14(3):323-332 (1969)
Cr_2O_3

A mass-spectrometric study of the sublimation of chromium trioxide
  Charles A. Washburn
    (Lawrence Radiation Lab., Univ. of California, Berkeley),
      UCRL-18685, Feb. 1969, Contract W-7405-eng-48, 127 pp.
      (thesis)

Phase equilibrium in the system $CrO_2 - Cr_2O_3$
  Osamu Fukuunaga and Shinroku Saito
  J. Am. Ceram. Soc., 51:362-363 (1968)

Phase diagrams for the systems Si – O and Cr – O
  R. E. Johnson and Arnulf Muan
  J. Am. Ceram. Soc., 51:430-433 (1968)

Reduction of CrO_3 into CrO_2 and Cr_2O_3 under very high pressure and high temperature
  Nauto Kawai, Akira Sawaoka, Saburo Kikuchi, and Nobuaki Tamagawa
  Japan J. Appl. Phys., 6:1397-1399 (1967)

On the decomposition of CrO_2 in air
  D. Rodbell and R. DeVries
  Mat. Res. Bull., 2:491-495 (1967)

Film thickness determination by electron-probe microanalysis
  D.P. Whittle and G. C. Wood
  Corrosion Sci., 6:397-420 (1966)

Preparative and structural studies on chromium oxides and chromates
  Karl-Axel Wilhelmi
    (University of Stockholm, Inorganic Chemistry), DIS No. 24, 1966
    Various Cr–O ratios

Crystal structure of Cr_5O_{12}
  Karl A. Wilhelmi
  Acta Chem. Scand., 19:165-176 (1965)

Diffusion of chromium through its oxide
  I. M. Fedorchenko and Yu. B. Ermolovich
  AEC-tr-6927
  Ukr. Khim. Zh., 26:429-431 (1960), 5 pp.

The system $Cr - Cr_2O_3$
  Ya. I. Ol'shanskii and V. K. Shlepov
  Fulmer Research Institute Ltd., Translation No. 59
  Translated by S. Klemantaski from Dokl. Akad. Nauk SSSR, 91:561-564 (1953)

5.c.4.b. Molybdenum

Crystal structure of $MoO_3 \cdot 2H_2O$
  S. Asbrink, B. Brandt, and P. Kierkegaard
  Acata Chem. Scand., 23:2196 (1969)

Similarity between the tungsten bronzes and the dioxides of W and Mo
  L. Ben-Dor and L. E. Conroy
  Israel J. Chem., 7:713-715 (1969)
  Solid State Chem. in Israel

Free energies of formation of WO_2 and MoO_2
  Sven Berglund and Peter Kierkegaard
  Acta Chem. Scand., 23(1):329-330 (1969)

Crystallographic shear in molybdenum trioxide
  L. A. Bursill
  Proc. Roy. Soc. London, 311A:267-290 (1969)

Moire pattern produced by electron bombardment of a molybdenum trioxide single crystal
  P. K. Garg, J. Kumar, and D. L. Bhattacharya
  Acta Cryst., 25A, Pt. 4, p. 584 (1969)

Crystal structure of $Mo_4O_{10}(OH)_2$
  K. A. Wilhelmi
  Acta Chem. Scand., 23:419-428 (1969)

MoO_3-Whiskers mit gestörter Struktur
  H. Saalfeld and O. Jarchow
  Z. Krist., 126:241-243 (1968)

A refinement of the crystal structure of molybdenum dioxide
  Bjorn G. Brandt and A. C. Skapski
  Acta Chem. Scand., 21:661-672 (1967)

The crystal chemistry of molybdenum oxides
  Lars Kihlborg
  Nonstoichiometric Compounds, Symp. sponsored by Div. of Inorganic Chemistry, 141st Mtg., Am. Chemical Society, Washington, D. C., March 21-23, 1962, Roland Ward, Symp. Chairman (Am. Chemical Soc., Washington, D. C., 1963), Advances in Chemistry Series 39, p. 37

5.c.4.c. Tungsten

Observations of needle crystals of tungsten oxide by field emission and electron microscopy
  Fumio Okuyama
  J. Appl. Phys., 42:256-259 (1971)

The thermal expansion and orthorhombic-tetragonal transition of the tungsten trioxide phase
  R. J. Ackermann and Charles A. Sorrell
  High Temp. Sci., 2(2):119-120 (1970)

Disordered crystallographic shear in WO_{3-x} crystals
  J. G. Allpress and P. Gado
  Crystal Lattice Defects, 1:331-342 (1970)

The formation of shear structures in substoichiometric tungsten trioxide
  R. J. D. Tilley
  Mat. Res. Bull., 5:813-824 (1970)

Similarity between the tungsten bronzes and the dioxides of W and Mo
  L. Ben-Dor and L. E. Conroy
  Israel J. Chem., 7:713-715 (1969)
  Solid State Chem. in Israel

Free energies of formation of WO_2 and MoO_2
  Sven Berglund and Peter Kierkegaard
  Acta Chem. Scand., 23(1):329-330 (1969)

Further refinement of the structure of WO_3
  B. O. Loopstra and H. M. Rietveld
  Acta Cryst., 25B:1420-1421 (1969)

Thermionic emission properties of tungsten oxide
  E. Ya. Zandberg and U. Kh. Rasulev
  Zh. Tekhn. Fiz., 38:1793-1797 (1968)
  Sov. Phys. – Tech. Phys., 13:1446-1449 (1969)

About the symmetry of the WO_3 structure stable at room temperature, and its polymorphy
  P. Gado
  Hiki, 3(3-4):103-111 (1968)

Tungsten oxides
  G. D. Rieck
  Tungsten and Its Compounds, Pergamon Press, New York (1967), p. 93

Non-stoichiometry and interfacial dislocations in tungsten trioxide
J. Spyridelis, P. Delavignette, and S. Amelinckx
Mat. Res. Bull., 2:615-620 (1967)

Two higher phase transitions of WO_3
Haakon Broekken
J. Appl. Phys., 37:3635-3636 (1966)

Studies on structural defects in wolfram trioxide; influence of minor additions of tantala or niobia
Pal Gado and Arne Magneli
Mat. Res. Bull., 1:33-44 (1966)

Das Wolframtrioxid und seine Reaktion mit den Oxiden zweiwertiger Metalle. I. Phasenumwandlung und Zwillingsstruktur des Wolframtrioxids
V. Leute
Z. Phys. Chem.(Frankfurt), 48:307-318 (1966)

Neutron-diffraction investigation of WO_3
B. O. Loopstra and P. Boldrini
Acta Cryst., 21:158 (1966)

Nasschemische Spurenanalyse von MoO_3 and WO_3 und von reinem Molybdän- und Wolfram-metall
R. Puschel und E. Lassner
Mikrochim. Acta, 4:751 (1965)

Crystal structure of monoclinic tungsten trioxide at room temperature
Shigetoshi Tanisaki
J. Phys. Soc. Japan, 15:573-581 (1960)

A mass-spectrometric investigation of tungsten dioxide and tungsten trioxide
James Everett Battles
Dissertation Abstracts 65-3824, 71 pp.

5.d. Magnetic Properties

5.d.1. General

Bibliography of magnetic materials and tabulation of magnetic transition temperatures
T. F. Connolly and Emily D. Copenhaver
(Research Materials Information Center, Solid State Div., Oak Ridge National Lab., Oak Ridge, Tenn.), ORNL-RMIC-7 (Rev. 2), July 1970, 116 pp.

Superconductivity of ceramic compounds
K. M. Ralls, E. R. Parker, and V. F. Zackay
Ceramic Microstruct. Anal. Significance Product. Proc. 3rd International Materials Symp., Berkeley, Calif., 1966, John Wiley and Sons, New York; London; Sydney (1968), pp. 489-508

Transition-Metal Compounds: Transport and Magnetic Properties
E. R. Schatz, ed.
Gordon and Breach, New York (1964), 200 pp.

5.d.2. Group IV

5.d.2.a. Titanium

Cooperative magnetic transitions in the titanium-oxygen system: a new approach
L. N. Mulay and W. J. Danley
J. Appl. Phys., 41:877 (1970)

Properties of Ti_4O_7 single crystals
Koichi Nagasawa, Yasutoshi Kato, Yoshichika Bando, and Toshio Takada
J. Phys. Soc. Japan, 29:241 (1970)

Structural, electrical, and magnetic properties of vacancy-stabilized cubic TiO and VO
M. D. Banus, T. B. Reed, T. E. Stack, and R. E. Fahey
Solid State Research quarterly technical summary Feb. 1-April 30, 1969, Peter E. Tannenwald, Contract AF19(628)-5167, Solid State Research (1969:2), ESD-TR-69-110 (May, 1969), 60 pp.
Ti and V

Magnetic transitions in Ti_3O_5 and Ti_4O_7
W. J. Danley and L. N. Mulay
Bull. Am. Phys., Soc., 14:350 (1969)

On the magnetic nature of Ti_2O_3
G. J. Hyland
Phys. Stat. Sol., 35:K133-134 (1969)

Absence of antiferromagnetism in Ti_2O_3
R. M. Moon, T. Riste, W. C. Koehler, and S. C. Abrahams
(Oak Ridge National Lab., Oak Ridge, Tenn.), J. Appl. Phys., 40:1445-1447 (1969)

Magnetic susceptibility of titanium and of some Ti – O alloys
R. J. Wasilewski
J. Appl. Phys., 40:2677-2678 (1969)

Vacancies and superconductivity in titanium monoxide
N. J. Doyle, J. K. Hulm, C. K. Jones, R. C. Miller, and A. Taylor
Phys. Letters, 26A:604-605 (1968)

Neutron-diffraction data on Ti_2O_3 and V_2O_3
H. Kendrick, A. Arrott, and S. A. Werner
J. Appl. Phys., 39:585 (1968)

Anomalous magnetic properties of Ti_5O_9
L. K. Keys
J. Appl. Phys., 39:598 (1968)

Van Vleck paramagnetism in orthorhombic TiO_2 (brookite)
F. E. Senftle and A. N. Thorpe
Phys. Rev., 175:1144-1146 (1968)

Resistivity and magnetic order in Ti_2O_3
L. L. Van Zandt, J. M. Honig, and J. B. Goodenough
J. Appl. Phys., 39:594 (1968)

Fundamental studies in high-temperature materials phenomena
M. E. Bell, L. E. Cross, L. N. Mulay, and G. R. Barsch
(Penn. State Univ., University Park), Annual Progress Report, Contract AT(30-1)-2581, NYO-2581-7 (March 1967), 37 pp.

Etude cristallographique et magnétique d'un oxyde trirutile antiferromagnétique
Jean-Claude Bernier and Paul Poix
Compt. Rend., 265:1247 (1967)

Investigation of the x-ray emission spectra of titanium in the oxides Ti_2O_3 and Ti_3O_5 in connection with anomalies of their electrical and magnetic properties
V. I. Chirkov, E. E. Vainshtein, and Ya. V. Vasil'ev
Izv. Akad. Nauk SSSR, Neorg. Mater., 3(6):1017-1021 (1967)
Inorg. Mater., 3(6):906-909 (1967)

Magnetic studies of Ti_3O_5
L. K. Keys
Phys. Letters, 24A:628 (1967)

Magnetism of the titanium-oxygen system
L. K. Keys and L. N. Mulay
Japan. J. Appl. Phys., 6:122-123 (1967)

Magnetic-susceptibility studies on the Magneli phases of the titanium-oxygen system
L. K. Keys and L. N. Mulay
J. Appl. Phys., 38:1466 (1967)

Magnetic susceptibility measurements of rutile and the Magneli phases of the Ti−O system
L. K. Keys and L. N. Mulay
Phys. Rev., 154:453-456 (1967)

Magnetic behavior of the paraelectric susceptibility of the impurity system in reduced rutile at low temperatures
R. K. MacCrone
Bull. Am. Phys. Soc., 12:558 (1967)

Magnetic studies of the semiconductor to metal transitions in Ti_3O_5 and Ti_2O_3
L. K. Keys and L. N. Mulay
Appl. Phys. Letters, 9:248 (1966)

5.d.3. Group V

5.d.3.a. Vanadium

Magnetic anisotropy of V_2O_3 single crystals
Morris Greenwood, R. W. Mires, and A. R. Smith
J. Chem. Phys., 54:1417-1418 (1971)

Observation of nuclear specific heat in V_2O_3
K. Andres
Phys. Rev., B2:3768-3771 (1970)

High-pressure suppression of the magnetic state of V_2O_3: ^{51}V nuclear resonance at 4.2°K and 65 kbar
A. C. Gossard, D. B. McWhan, and J. P. Remeika
J. Appl. Phys., 41:864 (1970)

Measurement of the internal magnetic field in V_2O_3 using the inelastic spin-flip-scattering of neutrons
A. Heidemann
Z. Phys., 238(3):208-220 (1970)

The magnetic properties of $(V_{1-x}Cr_x)_2O_3$ system
A. Menth and J. P. Remeika
Bull. Am. Phys. Soc., 15:385 (1970)

Antiferromagnetism in V_2O_3
R. M. Moon
J. Appl. Phys., 41:883 (1970)

Growth and magnetic properties of V_8O_{15} single crystals
Koichi Nagasawa, Yoshichika Bando, Toshio Takada, Hiroyuki Horiuchi, Masayasu Tokonami, and Nobuo Morimoto
Japan. J. Appl. Phys., 9:841 (1970)

Onset of magnetism in vanadium oxides: ^{51}V NMR studies of VO
W. W. Warren, Jr., A. C. Gossard, and M. D. Banus
J. Appl. Phys., 41:881 (1970)

Structural, electrical, and magnetic properties of vacancy-stabilized cubic TiO and VO
M. D. Banus, T. B. Reed, T. E. Stack, and R. E. Fahey
Solid State Research quarterly technical summary, Feb. 1-April 30, 1969, Peter E. Tannenwald, Contract AF19(628)-5167, Solid State Research (1969:2): EST-TR-69-110 (May 1969), 60 pp.
Ti and V

Temperature dependence of the vanadium NMR frequency shift in V_2O_3
E. D. Jones
J. Phys. Soc. Japan, 27:1692-1693 (1969)

Contribution à l'étude des phases V_xO avec $x \cong 1$
P. Massard, J. C. Bernier, and A. Michel
Ann. Chim., 4:147-151 (1969)

On the magnetic properties of the V−O system
D. Mehandjiev and B. Anghelov
Compt. Rend. Acad. Bulgare Sci., 22(7):731-734 (1969)

Electrical and magnetic properties of NbO_2
Kimiko Sakata
J. Phys. Soc. Japan, 26:867 (1969)

Magnetic susceptibilities of metallic V_2O_3 single crystals
D. J. Arnold and R. W. Mires
J. Chem. Phys., 48:2231-2234 (1968)

Electrical and magnetic properties of vanadium dioxide
G. J. Hill and R. H. Martin
Phys. Letters, 27A:34-35 (1968)

Neutron-diffraction data on Ti_2O_3 and V_2O_3
H. Kendrick, A. Arrott, and S. A. Werner
J. Appl. Phys., 39:585 (1968)

Magnetic susceptibility of V_2O_5
Fam Zoan Khan, L. N. Blinov, and L. P. Strakhov
Fiz. Tekhn. Poluprovod., 2(4):457-462 (1968)
Sov. Phys. − Semicond., 2(4):377-381 (1968)

The phase transition in VO_2
Koji Kosuge
J. Phys. Soc. Japan, 22:551 (1967)

Electric and magnetic properties of "VO"
S. Kawano, K. Kosuge, and S. Kachi
J. Phys. Soc. Japan, 21:2744 (1966)

Crystal chemistry studies
R. Roy, S. Kachi, G. J. McCarthy, O. Muller, and W. B. White (Pennsylvania State Univ., University Park, Pa.) Qtrly. Rept., Aug. 13, 1965-Nov. 13, 1965, Contract DA 28-043 AMC-01304 (E), Jan. 1966

Antiferromagnetism of V_2O_3 observed by the Mössbauer effect
T. Shinjo, K. Kosuge, S. Kachi, H. Takaki, M. Shiga, and Y. Nakamura
J. Phys. Soc. Japan, 21:193 (1966)

Contributions to the vanadium-51 nuclear magnetic resonance frequency shift and susceptibility in vanadium sesquioxide
E. D. Jones
Phys. Rev., 137:978-982 (1965)

d. Magnetic Properties

The colour problem of vanadium pentoxide. II. Temperature dependence of magnetic susceptibility of vanadium pentoxide
  A. R. Tourky, T. M. Salen, Z. Hanafi, and K. Al Zewel
  Z. Phys. Chem.(Leipzig), 230:184-188 (1965)

Electrical conductivity of vanadium oxides
  S. Kachi, T. Takada, and K. Kosuge
  J. Phys. Soc. Japan, 18:1839 (1963)

Phase diagram and magnetism of the $V_2O_3 - V_2O_5$ system
  K. Kosuge, T. Takada, and S. Kachi
  J. Phys. Soc. Japan, 18:318-319 (1963)

American Institute of Physics Handbook, 2nd ed.
  McGraw-Hill Book Co., Inc., New York (1963), pp. 5-200

Magnetic susceptibility of vanadium at 80-370°K
  N. I. Bogdanova and G. M. Loginov
  Fiz. Tverd. Tela, 4(1):236-238 (1962)
  Sov. Phys. — Solid State, 4(1):167-169 (1962)

Electrical conductivity anomaly in vanadium sesquioxide
  G. Goodman
  Phys. Rev. Letters, 9:305 (1962)

Electric and magnetic properties of V_2O_3 and related sesquioxides
  A. J. MacMillan
  AD-291 459 (1962), 32 pp.

Magnetic susceptibility of solid solutions of VO_2 and TiO_2
  C. M. Arya and G. Grossman
  Fiz. Tverd. Tela, 2(6):1283-1286 (1960)
  Sov. Phys. — Solid State, 2(6):1166-1169 (1960)

Magnetic transitions in titanium-sesquioxide and vanadium-trioxide
  P. H. Carr and S. Foner
  J. Appl. Phys. Suppl., 31:344-345 (1960)

Direct cation — cation interactions in primarily ionic solids
  J. B. Goodenough
  J. Appl. Phys., 31:359S (1960) (Suppl.)

Study of rhombohedral V_2O_2 by neutron diffraction
  A. Paoletti and S. J. Pickart
  J. Chem. Phys., 32:308 (1960)

Measurement of the susceptibilities of different oxides of vanadium and their variation as a function of temperature
  J. Roch
  Compt. Rend., 250:2167-2169 (1960)

Low-temperature x-ray diffraction studies on vanadium sesquioxide
  E. P. Warekois
  J. Appl. Phys., 31:346S (1960) (Suppl.)

Analysis of ferromagnetic and antiferromagnetic second-order transitions
  J. A. Hofman, A. Paskin, K. J. Tauer, and R. J. Weiss
  Phys. Chem. Solids, 1:45-60 (1956)

5.d.4. Group VI

5.d.4.a. Chromium

Rotation du plan de polarisation d'une onde ultra sonore se propageánt dans un cristal anti-ferromagnétique
  M. Boiteux, P. Doussineau, B. Ferry, J. Joffrin, and A. Levelut
  Solid-State Commun., 8:1609-1613 (1970)
  Cr_2O_3

Magnetic properties of $(Cr_{1-x}Fe_x)O_2$
  E. Hirata, T. Mihara, T. Kawamata, and M. Asanuma
  Japan J. Appl. Phys., 9:647-651 (1970)

Contribution à l'étude des oxydes simples et mixtes de fer, de chrome et d'aluminium
  A. Rousset
  Ph.D. thesis, Lyon, C.N.R.S., No. 4468 (May 1970), 235 pp.

Inelastic neutron scattering investigation of spin waves and magnetic interactions in Cr_2O_3
  E. J. Samuelsen, M. T. Hutchings, and G. Shirane
  Physica, 48:13-42 (1970)

Ultrasonic behavior near the spin-flop transition of Cr_2O_3
  Y. Shapira
  Phys. Rev., 187:734-736 (1970)

Domain switching measurements in an antiferromagnet
  C. A. Brown and T. H. O'Dell
  IEEE Trans. Magnet., 5(4):964-967 (1969)
  Cr_2O_3 single crystals

Temperature effects of spin waves in Cr_2O_3 studied by means of inelastic neutron scattering
  E. J. Samuelsen
  Physica, 45:12 (1969)

Inelastic neutron scattering investigation of spin waves and magnetic interactions in Cr_2O_3
  E. J. Samuelsen, M. T. Hutchings, and G. Shirane
  Solid State Commun., 7:1043-1045 (1969)

Magnetic and magneto-optic properties of FeRh and CrO_2
  A. M. Stoffel
  J. Appl. Phys., 40:1238 (1969)

Pressure dependence of the Néel temperature of Cr_2O_3
  R. M. Brugger and R. B. Bennion
  J. Phys. Chem. Solids, 29:435-438 (1968)

Effect of particle size on the Néel temperature of powdered chromic oxide
  W. Gunsser, W. Hille, and A. Knappwost
  Z. Naturforsch., 23A:781-783 (1968)

Materials
  Nuclear Technology Branch Yearly Progress Report for period ending June 30, 1968 (R. L. Heath, ed.), pp. 141-179
  (Idaho Nuclear Corp., Idaho Falls), Contract AT(10-1)-1230, IN-1218 (Dec. 1968), 357 pp.
  Includes magnetic structure of chromium oxides and pressure dependence on Néel temperature.

Neutron investigation of the spin system dynamics in Cr_2O_3
  S. Krasnicki, H. Rzany, A. Wanic, A. Kowalska, R. A. Alikhanov, Z. Dimitrijevic, and J. Todorovic
  (Inst. of Nuclear Research, Cracow, Poland), INP-661 (Oct. 1968)

Susceptibility derived sublattice magnetization in Cr_2O_3
  H. Shaked and S. Shtrikman
  Solid State Commun., 6:425-426 (1968)

Magnetic critical-point behavior of CrO_2
  J. S. Kouvel and D. S. Rodbell
  J. Appl. Phys., 38:979-981 (1967)

Magnetoelectric effect in Cr_2O_3 single crystal as studied by dielectric-constant method
  H. B. Lal, R. Srivastava, and K. G. Srivastava
  Phys. Rev., 154:505-507 (1967)

Effect of domain structure on the magnetic properties of chromium oxide
  D. N. Astrov and B. I. Alshin
  Zh. Éksperim. i Teor. Fiz., 51:28-31 (1966)
  Sov. Phys. – JETP, 24(1):18-20 (1967)

Antiferromagnetic domain switching in Cr_2O_3
  T. J. Martin and J. C. Anderson
  IEEE Trans. Magnetics, MAG-2, 446-49 (1966)

Pulse measurements of the magnetoelectric effect in chromium oxide
  T. H. O'Dell
  IEEE Trans. Magnetics, MAG-2, 449-452 (1966)

Magnetocrystalline anisotropy of single crystal CrO_2
  D. S. Rodbell
  J. Phys. Soc. Japan, 21:1224-1225 (1966)

Survey of magnetoelectric materials
  R. P. Santoro and R. E. Newnham
  (Lab. for Insulation Research, Mass. Inst. Tech., Cambridge, Mass.), Contract AF 33(615)-2199, Report AFML-TR-66-327 (Sept. 1966), 24 pp.

Magnetoelectrics – a new class of materials
  T. H. O'Dell
  Electronics and Power (Aug. 1965), p. 266

Observation of the magneto-electric effect in polycrystalline chromium oxide
  T. H. O'Dell
  Phil. Mag., 10:899 (1964)

Susceptibility of iron group compounds in high magnetic fields
  R. Stevenson
  Canadian J. Phys., 40:1385-1393 (1962)
  Cr_2O_3

5.e. Electrical and Electronic Properties

5.e.1. General

Electrical and optical properties of transition-metal oxides
  D. Adler
  Radiat. Eff., 4:123-131 (1970)
  International Conference on Nonmetallic Crystals, New Delhi, India, Jan. 1969

Metallic and nonmetallic behavior in transition metal oxides
  I. G. Austin and N. F. Mott
  Science, 168(3927):71-78 (1970)
  Electron correlation effects in narrow d bands and polarons

Molecular-orbital interpretation of the soft x-ray $L_{II,III}$ emission and absorption spectra from some titanium and vanadium compounds
  D. W. Fischer
  J. Appl. Phys., 41(9):3561-3569 (1970)

Semiconductor $\rightleftharpoons$ metal phase transitions
  G. J. Hyland
  J. Solid State Chem., 2:318-331 (1970)
  V_2O_3, VO_2, and Ti_2O_3

X-ray diffraction studies of the metal insulator transitions in Ti_4O_7, V_4O_7, and VO_2
  M. Marezio, P. D. Dernier, D. B. McWhan, and J. P. Remeika
  Mat. Res. Bull., 5:1015-1024 (1970)

Electrical conduction in metal oxides
  C. N. R. Rao and G. V. Subba Rao
  Phys. Stat. Sol., 1A:597-652 (1970)
  Review, 319 refs.

A study of the electronic structure of the first-row transition-metal compounds
  T. M. Wilson
  Intern. J. Quantum Chem., Symp. 3, Part 2 (1970), 757-774

Metal-nonmetal transitions in narrow-band materials; crystal structure versus correlation
  Z. Zinamon and N. F. Mott
  Phil. Mag., 21:881 (1970)
  Ti_2O_3, VO_2, V_2O_3

Electronic properties of refractory monoxides having intrinsic lattice vacancy concentrations, final report
  S. P. Denker
  (Columbia Univ., Dept. of Electrical Engineering, New York), CU-3553-16 (June 1969), 11 pp.

The insulator – metal transition
  S. Doniach
  Adv. Phys., 18(76):819-848 (1969)

Band approach to the transition-metal oxides
  J. Feinleib
  (Lincoln Lab., Mass. Inst. of Tech., Lexington, Mass.)
  Electronic Structures in Solids, pp. 231-258, 1969
  Also AD-694 133 (1969), 29 pp.

Electronic conduction in transition-metal oxide films
  W. J. Frey, T. N. Kennedy, and J. D. Mackenzie
  Am. Ceram. Soc. Bull., 48(4):425 (1969)
  Electronic switches, active semiconducting devices or magnetic memories

Crystal-field splitting of levels and x-ray spectra of transition-metal monoxides
  A. Z. Menshikov, I. A. Brytov, and E. Z. Kurmaev
  Phys. Stat. Sol., 35:89-93 (1969)

Metallic and non-metallic behaviour in compounds of transition metals
  N. F. Mott
  Phil. Mag., 20:1-21 (1969)
  Titanates, vanadium and titanium oxides, and Fe_3O_4

Study of the electronic structure and interatomic bonds in some compounds and binary alloys by the method of x-ray spectroscopy
  S. A. Nemnonov, A. Z. Menshikov, K. M. Kolobova, E. Z. Kurmayev, and V. A. Trapeznikov

Trans. Met. Soc. AIME, 245:1191-1198 (1969)
TiO, VO, Cr_2O_3

Thin-film dielectric properties of r.f. sputtered oxides
I. H. Pratt
Solid State Tech., 12(12):49-57 (1969)
Includes Ta and Hf

Dielectric characteristics of r.f. sputtered oxide films
I. H. Pratt and J. J. McCarthy
(Army Electronics Command, Ft. Monmouth, N. J.), ECOM-3173; AD-695 629 (Sept. 1969), 42 pp.
Ta−O and Hf−O

Insulating and metallic states in transition-metal oxides
David Adler
Solid State Physics Advances in Research and Applications, Vol. 21 (Frederick Seitz, David Turnbull, and Henry Ehrenreich, eds.), Academic Press, New York (1968), pp. 1-115

Mechanisms for metal−nometal transitions in transition-metal oxides and sulfides
David Adler
Rev. Modern Phys., 40:714-736 (1968)
181 refs.; it now appears likely that band theory can be adapted to describe most of the materials that exhibit metal−nometal transitions

Electrical properties of transition-metal compounds
David Adler and Harvey Brooks
Comments Solid State Phys., 1(5):145-150 (1968)
Oxides, sulfides, and selenides

Electronic properties of refractory monoxide having intrinsic lattice vacancy concentrations
S. P. Denker
(Columbia University, New York), Annual Progress Report, 1 Jan. 1967 to 31 Jan. 1968, CU-3553-9 (Feb. 1968)

Possible anomalies at a semimetal−semiconductor transition
B. I. Halperin and T. M. Rice
Rev. Modern Phys., 40:755 (1968)

Lattice instability associated with metal−semiconductor transitions
E. Hanamura
Rev. Modern Phys., 40:744 (1968)

Model for the metal−nonmetal transition in impure semiconductors
D. Jerome
Rev. Modern Phys., 40:833 (1968)

Suboxide theory of the oxidation of transition metals
I. I. Kornilov
Dokl. Akad. Nauk SSSR, 183(5):1087-1096 (1968)

The metal−insulator transition, session II-electrons in narrow bands
N. F. Mott
Rev. Modern Phys., 40:677 (1968)

The electrical and optical properties of refractory oxides
W. C. Tripp and E. T. Rodine
ARL-68-0221 (Dec. 1968)

Theory of semiconductor-to-metal transitions
David Adler and Harvey Brooks
Phys. Rev., 155:826 (1967)

Transition Metal Compounds: Transport and Magnetic Properties
E. R. Schatz, ed.
Gordon and Breach, New York (1964), 200 pp.

Semiconductors
N. B. Hannay
Academic Press, New York and London (1963), pp. 409-423
The state of the art, transition-metal oxides; theory; classification into metals, semiconductors, and insulators

5.e.2. Group IV

5.e.2.a. Titanium

A reference list on semiconducting and nonstoichiometric titanium oxides
John T. Milek
Electronic Properties Information Center, Hughes Aircraft Co., Culver City, Calif., Jan. 1971
74 refs.

Semiconductor-metal transition in Ti_3O_5
C. N. R. Rao, S. Ramdas, R. E. Loehman, and J. M. Honig
J. Solid State Chem., 3:83-88 (1971)

Electrical and thermal properties of nonstoichiometric rutile at low temperatures
Gary Allen Baum
Dissertation, Oklahoma State University, 1970
Available from University Microfilms, Inc., Ann Arbor, Mich., Order No. 70-21,343

Direct-current electrical breakdown of thin titania films in high vacuum
K. G. Bouchard
J. Vacuum Sci. Tech., 7:531-533 (1970)

Nickel-doped rutile as a spin-echo material
D. A. Bozanic, D. Mergerion, and R. W. Minarik
J. Appl. Phys., 41:5041-5042 (1970)

The electrical transition in V-doped Ti_2O_3
G. V. Chandrashekhar, Q. Won Choi, J. Moyo, and J. M. Honig
Mat. Res. Bull., 5:999-1008 (1970)

New ordered phases of slightly reduced rutile and their sharp dielectric absorptions at low temperature
C. W. Chu
Bull. Am. Phys. Soc., 15:381 (1970)

Electronic band structure and the K and L x-ray spectra from TiO, TiN, and TiC
D. W. Fischer
J. Appl. Phys., 41:3922-3926 (1970)

A current instability of TiO_2 thin film
T. Hada, S. Hayakawa, and K. Wasa
Japan. J. Appl. Phys., 9(9):1078-1084 (1970)

Electrical conductivity of slightly reduced rutile between 2° and 370°K
R. R. Hasiguti, E. Yagi, and M. Aono
Radiation Eff., 4:137-140 (1970)
International Conference on Nonmetallic Crystals, New Delhi, India, Jan. 1969

Magnetic field dependence of the dielectric behavior of bound polarons in rutile (TiO_2)
  H. S. Matis and R. K. MacCrone
  Phys. Rev. Letters, 25:1715–1718 (1970)

New electronic transitions in TiO
  C. M. Pathak and H. B. Palmer
  J. Mol. Spectros., 33:137–146 (1970)

Electrical conduction in TiO_2
  T. P. Pearsall
  J. Phys. D: Appl. Phys., 3:1837–1848 (1970)

Model of the insulator–metal transition in Ti_2O_3
  P. M. Raccah and H. J. Zeiger
  Solid State Research Quarterly Technical Summary–1 Feb. through 30 April 1970, Alan L. McWhorter (Lincoln Lab., Massachusetts Institute of Tech.), MIT Report Solid State Research (1970:2), pp. 49–50; ESD–TR–70–148 (May 15, 1970), 92 pp.

Structure in the two-photon absorption spectrum of TiO_2 (rutile)
  H. S. Waff and K. Park
  Phys. Letters, 32A:109–110 (1970)

Photoelectronic properties of stoichiometric flux-grown rutile (TiO_2) crystals
  F. G. Wakim
  Bull. Am. Phys. Soc., 15:385 (1970)

Photoelectronic effects in TiO_2 single crystals and powder layers. The role of chemisorbed gas
  R. R. Addiss, Jr., and F. G. Wakim
  Phot. Sci. Eng., 13(3):111–119 (1969)

Electrical properties of some titanium oxides
  Robert F. Bartholomew and D. R. Frankl
  Phys. Rev., 187:828–833 (1969)

Hall mobility of reduced rutile in the temperature range 300–1250°K
  I. Bransky and D. S. Tannhauser
  Solid State Commun., 7:245:248 (1969)

Dielectric measurements on flux-grown crystals of rutile (TiO_2) without contacting electrodes
  L. E. Cross and C. F. Groner
  J. Appl. Phys., 40:126 (1969)

Physical properties and stability of TiO from the APW–VCA energy bands
  S. P. Denker and J. M. Schoen
  Bull. Am. Phys. Soc., 14:431 (1969)

Hyperfine and nuclear quadrople interactions in copper-doped TiO_2
  T. C. Ensign, Te-Tse Chang, and A. H. Kahn
  Phys. Rev., 188:703–709 (1969)

Investigation of the effect of adsorption of water vapors on the work function and the electrical conductivity of titanium dioxide
  E. N. Figurovskaya
  Kinet. i Katliz., SSSR, 10(2):453–456 (1969)
  Kinetics and Catalysis, 10(2):374–377 (1969)

Photoelectronic processes in rutile
  Amal K. Ghosh, F. G. Wakim, and R. R. Addiss, Jr.
  Phys. Rev., 184:979–988 (1969)

Electrical properties of titanium dioxide deposited by chemical-vapor transport
  D. R. Harbison and H. L. Taylor
  Thin Film Dielectrics, Montreal, Canada, 7–11 Oct. 1968, Electrochemical Soc., New York (1969), pp. 254–278

Negative-magnetoresistance effects in Ti_2O_3
  J. M. Honig, L. L. Van Zandt, T. B. Reed, et al.
  Phys. Rev., 182:863–871 (1969)

Photoconductivity of the titanium dioxide crystal
  T. Iida and H. Nozaki
  Bull. Chem. Soc. Japan, 42:929–933 (1969)

Thermoelectric power of titanium monoxides
  Shinji Kawano
  Japan. J. Appl. Phys., 8:1264 (1969)

Electrical properties of specially reduced rutile (TiO_2) single crystals
  H. B. Lal
  Indian J. Pure and Appl. Phys., 7:72–73 (1969)

Dielectric properties of normal, reduced, and specially reduced rutile (TiO_2) single crystals at room temperature
  H. B. Lal and K. G. Srivastava
  Can. J. Phys., 47:3–6 (1969)

Mise en évidence de phénomènes de polarisation interfaciale dans des poudres d'oxyde de tita
  Lisette Lavielle and Guy Perny
  Compt. Rend., 269B:556–559 (1969)

Thermal emission of TiO_2 surfaces
  A. T. Nicolau, C. Topliceanu, and V. Ruxandra
  Rev. Roumaine Phys., 14:99–108 (1969)

X-ray study of inner level shifts and band structure of TiC and related compounds
  Lars Ramqvist, Borje Ekstig, Elisabeth Kallne, Erik Noreland, and Rolf Manne
  J. Phys. Chem. Solids, 30:1849–1860 (1969)

Charge transfer in transition-metal carbides and related compounds studied by ESCA
  Lars Ramqvist, Kjell Hamrin, Gunilla Johansson, Anders Fahlman, and Carl Nordling
  J. Phys. Chem. Solids, 30:1835–1847 (1969)

Adaptation of the augmented plane wave method to random, three-dimensional, substitutional alloys: application to titanium monoxide
  Joel Mark Schoen
  Ph.D. thesis, Columbia Univ. (1969)

Band structure, physical properties and stability of TiO by the APW–VCA
  Joel M. Schoen and Stephen P. Denker
  (Columbia Univ., Dept. of Electrical Eng., New York), Contract AT(30–1)3553, CU-3553–14 (March 14, 1969)
  Bull. Am. Phys. Soc., 14:431 (1969)

Résistivité électrique et coeff. de Hall de la phase non-stoechiométrique de TiO
  S. Takeuchi and K. Suzuki
  J. Jap. Inst. Metals, 33(3):284–290 (1969)

Surface photovoltage effects in titanium dioxide films
  P. Vohl
  Phot. Sci. Eng., 13(3):120–126 (1969)

Switching phenomena in titanium oxide thin films
F. Argall
Solid State Electron., 11:535-541 (1968)

Capacitance and resistance measurements of TiO_2 rectifying barriers
F. L. English
Solid State Electronics, 11:473-479 (1968)

Band structure and the titanium $L_{II,III}$ x-ray emission and absorption spectra from pure metal, oxides, nitride, carbide, and boride
D. W. Fischer and W. L. Baun
J. Appl. Phys., 39:4757-4776 (1968)

High-frequency conductivity of reduced rutile
T. Goto and T. Okada
J. Phys. Soc. Japan, 25:289 (1968)

The effect of hydrogen on the electrical properties of rutile
G. J. Hill
Brit. J. Appl. Phys., Ser. 2, Vol. 1, pp. 1141-1162 (1968)

Research on defect controlled electrical properties of rutile
W. M. Hirthe
ARL-68-0065 (April 1968)

Electrical properties of Ti_2O_3 single crystals
J. M. Honig and T. B. Read
Phys. Rev., 174:1020-1026 (1968)

Nature of the electrical transition in Ti_2O_3
J. M. Honig
Rev. Modern Phys., 40:748 (1968)

Field-enhanced conductivity in TiO_2 (rutile)
O. W. Johnson
Appl. Phys. Letters, 13:338 (1968)

Defects in rutile. III. Optical and electrical properties of impurities and charge carriers
O. W. Johnson, W. D. Ohlsen, and P. I. Kingsbury, Jr.
Phys. Rev., 175:1102 (1968)

Semiconductor-to-semiconductor transition in the pseudobinary system $TiO_2 - VO_2$
I. K. Kristensen
J. Appl. Phys., 39:5341 (1968)

Dielectric properties of thoroughly reduced rutile (TiO_2) single crystals
H. B. Lal
Indian J. Pure Appl. Phys., 6:328 (1968)

Explanation of photocurrents in rutile (TiO_2) single crystal based on the field-emission concept
H. B. Lal and K. G. Srivastava
Indian J. Pure Appl. Phys., 6:598-599 (1968)

Time-dependent electrical conduction in rutile single crystals
L. J. Van Ruyven and J. D. Chase
Appl. Phys. Letters, 12:214 (1968)

Study of the electron and atomic structure of titanium carbide and titanium oxide
M. P. Arbuzov, E. T. Kachkovskaya, and B. V. Khaenko
Khimicheskaya Svyaz' v Poluprovodnikakh i Termodinamika
(Chemical Bonds in Semiconductors and Thermodynamics), 1966, Nauka i Tekhnika, Minsk (Sept. 1967), pp. 64-71

The nature of the electrical conduction transient behavior of rutile

R. N. Blumenthal, W. N. Hirthe, and B. A. Pinz
Anisotropy in Single-Crystal Refractory Compounds, Vol. 1, (Fred W. Vahldiek and Stanley A. Mersol, eds.), Plenum Press, New York (1968), pp. 139-151
Proceedings of an International Symposium held in Dayton, Ohio (June 13-15, 1967)

Electronic mobility in rutile (TiO_2) at high temperatures
R. N. Blumenthal, J. C. Kirk, Jr., and W. M. Hirthe
J. Phys. Chem. Solids, 28:1077-1079 (1967)

Electrical properties of rutile containing dissolved niobium
N. P. Bogoroditskii, V. Kristya, and Ya. I. Panova
Fiz. Tverd. Tela, 9(1):253-256 (1967)
Sov. Phys. − Solid State, 9(1):187-189 (1967)

Dielectric relaxation of hopping electrons in reduced rutile, TiO_2
L. A. K. Dominik and R. K. MacCrone
Phys. Rev., 156:910-913 (1967)

Reversible and irreversible changes in the work function and electric conductivity of titanium dioxide in response to the chemisorption of oxygen
E. N. Figurovskaya and V. F. Kiselev
Dokl. Akad. Nauk SSSR, 175:1336 (1967)

Influence of impurity doping on photoconductivity of TiO_2 (rutile)
T. Iida and H. Nozaki
Kogyo Kagaku Zasshi, 70:1285-1287 (1967)

The energy levels of transition-metal ions (Fe^{3+}, Co^{2+}) in TiO_2
T. Iida and H. Nozaki
Kogyo Kagaku Zasshi, 70:1624-1627 (1967)

Hall effect and thermoelectric power in semiconductive TiO_2
M. Itakura, N. Niizeki, H. Toyoda, and H. Iwasaki
Japan. J. Appl. Phys., 6:311 (1967)

Ti_3O_5-effective mass
L. K. Keys
J. Phys. Soc. Japan, 23:478-479 (1967)

Conduction through titanium dioxide thin films with large ionic space charge
J. Maserjian and C. A. Mead
J. Phys. Chem. Solids, 28:1971-1983 (1967)

Electron mobility and the donor centers in reduced and Li-doped rutile (TiO_2)
G. A. Acket and J. Volger
Physica, 32:1680-1692 (1966)

Electrical conductivity of nonstoichiometric rutile single crystals from 1000 to 1500°C
R. N. Blumenthal, J. Coburn, J. Baukus, and W. M. Hirthe
J. Phys. Chem. Solids, 27:643-654 (1966)

Research on high-temperature dielectric materials
F. Chernow, W. B. Westphal, and R. E. Newnham
(Mass. Inst. Tech., Cambridge, Mass.), Summary Tech. Rep. No. 2, Nov. 1965 to Nov. 1966, Contract AF33(615)-2199, AFML-TR-67-27 (Nov. 1966), 38 pp.

Electronic properties of titanium monoxide
Stephen Paul Denker
J. Appl. Phys., 37:142-149 (1966)

Conduction through thin titanium dioxide films
J. Maserjian
(Jet Propulsion Lab., Calif. Institute of Technology, Pasadena, Calif.), Tech. Report No. 32-976 (Oct. 1966)

Energy-level diagram of titanium oxide
G. Perny and R. Lorang
J. Chim. Phys., 63:833-837 (1966)
Lockheed Missiles and Space Co., Palo Alto, Calif., translation N67-19165, 1966, 8 pp.

Recherches sur les propriétés diélectriques de l'oxyde de titane pigmentaire
P. Bombenger
Ph.D. thesis, Strasburg (1965)

Relation between stoichiometry and electrical resistivity of titanium monoxide
Stephen Paul Denker
Propriétés Thermodynamiques Physiques et Structurales des Dérives Semimetalliques, Colloq. Intern. du C.N.R.S., No. 157, Orsay, 28 Sept.-1 Oct. 1965 (publ. 1967)

Variation with time of the electrical conductivity of rutile
M. G. Harwood
Brit. J. Appl. Phys., 16:1493 (1965)

Electrical conductivity of rutile. III. Ionic effects in pure material. IV. Influence of added impurities
M. G. Harwood
Spec. Ceram. 1964 Proc. Symp. Stoke-on-Trent, Academic Press, New York, London (1965), pp. 221-242

Thermal conductivity and thermoelectric power of rutile (TiO_2)
W. R. Thurber and A. J. H. Mante
Phys. Rev., 139:A1655-1665 (1965)

Conduction phenomena in rutile single crystals
John A. Van Raalte
J. Appl. Phys., 3365-3369 (1965)

Photoconductivity of rutile TiO_2 single crystals
F. G. Wakim
Bull. Am. Phys. Soc., 18:1092 (1965)

TiO_2 rectifying barriers
F. English and B. Gossick
Solid State Electron., 7:193-204 (1964)

Conductivity of reduced rutile at low temperatures
R. R. Hasiguti, N. Kawamiya, and E. Yagi
Proc. International Congress Phys. Semiconducteurs, Paris (1964), p. 92

Activation energies of low-temperature conductivity of reduced rutile
R. R. Hasiguti, N. Kawamiya, and E. Yagi
J. Phys. Soc. Japan, 19:573 (1964)

Electron energy bands in $SrTiO_3$ and TiO_2 (theory and experiment)
A. K. Kahn, H. P. R. Frederikse, and J. H. Beckerd
(Informal Proc. Bull., Intern. Conf. Materials, Pittsburgh, Pa., 1963), in: Translation-Metal Compounds (E. R. Schatz, ed.), Gordon and Breach Sci. Publ., New York (1964), pp. 53-64

Electrical studies of TiO_2 films formed on Ti by diffusion
N. M. Davidovich and V. A. Darin
Phys. Metals Metallogr., 16(2):96-99 (1963)

Frequency and temperature behavior of oxide semiconductors
H. Oettel
Hermsdorfer Tech. Mitt., 3:98-105 (1962)

Transfer of electric charges through rutile single crystals
K. G. Srivastava
Phys. Rev., 119:520 (1960)

5.e.2.b. Zirconium

Electrical resistivity of h.c.p. zirconium-oxygen alloys
T. M. Giam and F. Claisse
J. Nucl. Mater., 34(3):325-331 (1970)
Zr_6O, Zr_4O, Zr_3O, and $Zr_{24}O_{10}$

Determination of the (anionic) O_2 semiconductivity in ZrO_2 solid electrolytes from permeability measurements
R. Hartung and H. H. Mobius
Z. Phys. Chem.(Leipzig), 243(1/2):133-138 (1970)

Emission thermoélectrique de l'oxyde de zirconium en présence d'oxygène
Jean-Pierre Loup and Anne-Marie Anthony
Compt. Rend., 268C:772-775 (1969)

Comments on the paper "The dielectric constant of zirconia"
B. Cox
Brit. J. Appl. Phys., 1:671-672 (1968)

The dielectric constant of zirconia
P. J. Harrop and J. N. Wanklyn
Brit. J. Appl. Phys., 18:739 (1967)

Mesure de la conductibilité électrique d'un cristal de zircone quadratique
Anne-Marie Anthony, Andre Quillot, and Pompiliu Nicolau
Compt. Rend., 262:896-899 (1966)

Temperaturabhängigkeit des Photoeffekts von anodisch hergestellten ZrO_2-Schichten
M. Hartl and K. Achtziger
Phys. Stat. Sol., 14:355-362 (1966)

Electrical conductivity studies of tetragonal zirconia
L. A. McClaine and C.P. Coppel
J. Electrochem. Soc., 113:80-85 (1966)

The electrical conductivity of zirconium dioxide films at intermediate temperatures
D. K. Dawson and R. H. Creamer
Brit. J. Appl. Phys., 16:1643 (1965)

Electrical properties and defect structure of zirconia: II. Tetragonal phase and inversion
R. W. Vest and N. M. Tallan
(Systems Research Labs., Inc., Dayton, Ohio), Contract AF33(615)-2765, ARL-65-272; AD-637 078 (March 1965), 8 pp.

The electrical behavior of refractory oxides. III
R. W. Vest, J. A. Crawford, W. C. Tripp
(Systems Research Labs., Inc., Dayton, Ohio), Contract AF33(657)-10815, ARL-65-259 (Dec. 1965)

Some electrical properties of ZrO_2 films
M. L. Young, P. J. Harrop, N. J. M. Wilkins, and J. N. Wanklyn
AERE-R 4957 (1965), 7 pp.

5.e.2.c. Hafnium

Structure and electrical properties of sputtered films of hafnium and hafnium compounds
F. T. J. Smith
J. Appl. Phys., 41:4227–4231 (1970)
Hafnium, hafnium nitrides, and hafnium dioxide

A comprehensive bibliography of literature published before 1969 pertaining to dielectric properties of Al_2O_3, BeO, HfO_2, MgO, SiO_2, and ThO_2
S. C. Keeton, R. R. Schemmel, and J. L. Bates
(Battelle Memorial Institute, Pacific Northwest Labs., Washington), BNWL–1180, BNWL–1181 (October, 1969)

5.e.3. Group V

5.e.3.a. Vanadium

The electron structure and the chemical bonds in vanadium pentoxide
A. A. Abdullaev and L. D. Kislovskii
Sov. Phys. — Cryst., 15(5):860–862 (1971)

Electrical and optical properties of VO_2 near the semiconductor — semimetal transition point
B. S. Borisov, S. T. Koretskaya, V. G. Mokerov, A. V. Rakov, and S. Solov'ev
Sov. Phys. — Solid State, 12(8):1763–1769 (1971)

Structure and electrical properties of nonstoichiometric single crystals of vanadium sesquioxide V_2O_3
V. N. Novikov, B. A. Tallerchik, E. I. Gindin, and V. G. Prokhvatilov
Sov. Phys. — Solid State, 12(9):2061–2064 (1971)

Influence of an electric field on a phase transition and switching effect in V_2O_3
V. N. Andreev, A. G. Aronov, and F. A. Chudnovskii
Fiz. Tverd. Tela, 12(5):1557–1559 (1970)
Sov. Phys. — Solid State, 12(5):1230–1231 (1970)

Semiconducting and semimetallic behavior in VO
M. D. Banus, T. B. Reed, and A. J. Strauss
(Lincoln Lab., Massachusetts Institute of Tech.), Solid State Research Rept. (1970:3), pp. 21–23
ESD-TR-70-234 (Aug. 1970)

Optical and electrical properties of V_2O_3 and Cr doped V_2O_3
A. S. Barker, Jr., H. C. Montgomery, and J. P. Remeika
Bull. Am. Phys. Soc., 15:386 (1970)

The orientation dependence of line parameters of the characteristic x-ray spectrum of the atoms of elements in monocrystals. Basic parameters of K- and L-series V line in V_2O_5 monocrystal
I. B. Borovskii and V. I. Matyskin
Dokl. Akad. Nauk SSSR, 192(1):63–66 (1970)

Symmetry considerations and the vanadium dioxide phase transition
J. R. Brews
Phys. Rev., B1:2557–2568 (1970)

Characteristic energy loss studies of V_2O_5 and vanadium
L. Fiermans and J. Vennik
Phys. Stat. Sol., 41:621–629 (1970)

The conductivity anomaly in vanadium dioxide
M. Guntersdorfer
Solid-State Electron., 13(3):355–367 (1970)

Schalteffekte in VO_2
M. Guntersdorfer
Solid-State Electron., 13:369–379 (1970)

Electronic structure and anomalous thermal expansion in FeF_2 and VO_2
Y. Hazony and H. K. Perkins
J. Appl. Phys., 41:5130 (1970)

Contact potential difference measurements on (010) surfaces of vanadium pentoxide
I. Hevesi, A. Suli, and J. Gyulai
Acta Phys. Acad. Sci. Hung., 29:79–83 (1970)
Pt electrodes on a freshly cleaved surface

Critical behavior of the Mott transition in Cr-doped V_2O_3
A. Jayaraman and J. P. Remeika
Bull. Am. Phys. Soc., 15:386 (1970)

Magnetic field dependence of the metal-insulator "Mott transition": a study of the compound VO_2
A. P. Klein
J. Phys. C. Proc. Phys. Soc. (Solid State Phys.), Ser. 3, No. 3, pp. 166–171 (1970)

Energy bands of VO
T. E. Norwood and J. L. Fry
Phys. Rev., 2:472–481 (1970)

Electrical properties of V_8O_{15} single crystal
H. Okinaka, K. Kosuge, S. Kachi, K. Nagasawa, Y. Bando, and T. Takada
Phys. Letters, 33A:370–371 (1970)
Metal-to-semiconductor phase transition at 70°K

Electrical properties of the V_4O_7 single crystals
H. Okinaka, K. Nagasawa, K. Kosuge, Y. Bando, S. Kachi, and T. Takada
J. Phys. Soc. Japan, 28:798–799 (1970)

Electrical properties of the V_8O_9 single crystals
H. Okinaka, K. Nagasawa, K. Kosuge, Y. Bando, S. Kachi, and T. Takada
J. Phys. Soc. Japan, 28:803 (1970)

Electrical properties of V_6O_{11} and V_7O_{13} single crystals
Hideyuki Okinaka, Koichi Nagasawa, Koji Kosuge, Yoshichika Bando, Sukeji Kachi, and Toshio Takada
J. Phys. Soc. Japan, 29:245–246 (1970)

The present position of theory and experiment for VO_2
William Paul
Mater. Res. Bull., 5:691–702 (1970)

Metal-insulator transition in transition-metal oxides (V_2O_3)
T. M. Rice and D. B. McWhan
IBM J. Res. Develop., 14:251-257 (1970)

Hall effect in VO_2 as a function of temperature near the transition temperature
William H. Rosevear and William Paul
Bull. Am. Phys. Soc., 15:316 (1970)

Metal-insulator transition in V_2O_3 in a strong electric field
K. A. Valiev, Yu. V. Kipaev, V. G. Mokerov, A. V. Rakov, and S. G. Solov'ev
JETP Letters, 12:12-14 (1970)

Two switching devices utilizing VO_2
R. H. Walden
IEEE Trans. Electron Devices, ED-17, 603-11 (1970)

Mössbauer-effect study of metal-insulator transition in V_2O_3
G. K. Wertheim, J. P. Remeika, H. J. Guggenheim, and D. N. E. Buchanan
Phys. Rev. Letters, 25:94-96 (1970)

Energy band structure of NiO and VO
Timothy M. Wilson
Bull. Am. Phys. Soc., 15:254 (1970)

The nature of the metallic state in V_2O_3 and related oxides
I. G. Austin and C. E. Turner
Phil. Mag., 19:939-949 (1969)

Hydrostatic-pressure dependence of the electronic properties of VO_2 near the semiconductor-metal transition temperature
C. N. Berglund and A. Jayaraman
Phys. Rev., 185:1034-1039 (1969)

Electronic properties of VO_2 near the semiconductor-metal transition
C. N. Berglund and H. J. Guggenheim
Phys. Rev., 185:1022 (1969)

Vanadium $L_{II,III}$ x-ray emission and absorption spectra from metal, oxides, nitride, carbide, and boride
D. W. Fischer
J. Appl. Phys., 40:4151-4163 (1969)

Oscillation phenomena and switching phenomena of V_2O_4 semiconductors (CTR)
H. Futaki
Hitachi Rev. (Japan), 18(11):427-430 (1969)

Effects of various doping elements on the transition temperature of vanadium oxide semiconductors
Hisao Futaki and Minoru Aoki
Japan. J. Appl. Phys., 8:1008-1013 (1969)

Semiconductor-to-metal transition and positron annihilation in V_2O_3
A. Gainotti, C. Ghezzi, and M. Manfredi
Nuovo Cimento, Serie X, 62B:121-129 (1969)

Electrical properties of V_2O_5
Johan Haemers
Thesis, State University of Ghent (1969)
In Dutch

Thermoelectric power of vanadium monoxides
Shinji Kawano, Koji Kosuge, and Sukeji Kachi
J. Phys. Soc. Japan, 27:1076 (1969)

A vanadium oxide film-switching element
T. N. Kennedy and F. M. Collins
AD-683368 (Feb. 1969), 27 pp.

Suppression of the semiconductor-metal transition in vanadium oxides
T. N. Kennedy and J. D. Mackenzie
J. Non-cryst. Solids, 1:326-330 (1969)

Growth and electrical properties of vanadium dioxide single crystals containing selected impurity ions
J. B. MacChesney and H. J. Guggenheim
J. Phys. Chem. Solids, 30:225-234 (1969)

Critical pressure for the metal-semiconductor transition in V_2O_3
D. B. McWhan and T. M. Rice
Phys. Rev. Letters, 22:887-890 (1969)

Optical properties and band structure of vanadium dixode and pentoxide single crystals
V. G. Mokerov and A. V. Rakov
Fiz. Tverd. Tela, 11(1):197-199 (1969)
Sov. Phys.—Solid State, 11(1):150-152 (1969)

Band structure and lattice distortion in V_2O_3
I. Nebenzahl and M. Weger
Phys. Rev., 184:936 (1969)

A D.T.A. study of the semiconductor-metallic transition temperature in $V_{1-x}V_xO_2$, $0 \leq x \leq 0.067$
M. Nygren and M. Israelsson
Mater. Res. Bull., 4:881-886 (1969)

Electrical properties of the V_3O_5 single crystals
Hideyuki Okinaka, Koichi Nagasawa, Koji Kosuge, Yoshichika Bando, Toshio Takada, and Sukeji Kachi
J. Phys. Soc. Japan, 27:1366-1367 (1969)

Photoemission from VO_2
R. J. Powell, C. N. Berglund, and W. E. Spicer
Phys. Rev., 178:1410-1415 (1969)

Field effect relaxation of contact potential difference between stabilized vanadium pentoxide single crystal and platinum surfaces
A. Suli, I. Hevesi, and J. Gyulai
Acta Phys. Chem. Szeged. (Hungary), 15(3-4):99-102 (1969)

The colour problem of vanadium pentoxide. IV. Hall effect and dependence of conductivity on oxygen pressure
A. R. Tourky, Z. Hanofi, and K. Al Zewel
Z. Phys. Chem. Leipzig (Germany), 242(5-6):305-311 (1969)

Conduction mechanism in vanadium pentoxide
D. S. Volzhenskii and M. V. Pashkovskii
Fiz. Tverd. Tela, 11(5):1168-1172 (1969)
Sov. Phys.—Solid State, 11(5):950-953 (1969)

High-speed thermal switches based on vanadium dioxide
R. G. Cope and A. W. Penn
J. Phys. D. (British J. Appl. Phys.), 1:161-168 (1968)

Anisotropy in the electrical resistivity of vanadium dioxide single crystals
C. R. Everhart and J. B. MacChesney
J. Appl. Phys., 39:2872 (1968)

Transport properties of sputtered vanadium dioxide thin films
D. H. Hensler
J. Appl. Phys., 39:2354 (1968)

Zum Widerstandssprung bei der Phasenumwandlung in VO$_2$
W. Heywang and M. Guntersdorfer
Helv. Phys. Acta, 41:908-913 (1968)
Semiconductor to metal transition depends on the free carrier density influencing the band gap in the semiconducting state

On the electronic phase transitions in the lower oxides of vanadium
G. J. Hyland
J. Phys., 1:189 (1968)

Some remarks on electronic phase transitions and on the nature of the metallic state in VO$_2$
G.J. Hyland
Rev. Modern Phys., 40:739 (1968)

ESR evidence of the state of carriers in V$_2$O$_5$
V. A. Ioffe and I. B. Patrina
Fiz. Tver. Tela, 10(3):815-821 (1968)
Sov. Phys. − Solid State, 10(3):639-644 (1968)

ESR of VO in argon matrix at 4°K; establishment of its electronic ground state
Paul H. Kasai
J. Chem. Phys., 49:4979-4984 (1968)

Suppression of the semiconductor − metal transition in vanadium oxides
T. N. Kennedy and J. D. Mackenzie
(Rensselaer Polytechnic Inst., Troy, New York), Contract N00014-68-A-0117, TR-1, AD 681 874 (Dec. 1968), 13 pp.

Electrical properties of vanadium pentoxide
John C. McCulloch
(Oregon State Univ., Corvallis), Rept. No. TR-3, Contract Nonr-1286(08), May 1968, 53 pp.
AD-670 560

Semiconductor properties and deviation from the stoichiometry of molten V$_2$O$_5$
E. A. Pastukhov, O. A. Esin, and N. A. Vatolin
Izv. Akad. Nauk SSSR, Neorg. Mater., 4(11):1960-1965 (1968)
Inorg. Mater., 4(11):1704-1708 (1968)

Effect of pressure on the electrical properties of semiconducting oxides
W. Paul
Am. Ceram. Soc. Bull., 47:383 (1968)
V$_2$O$_3$

Photoemission from VO$_2$
R. J. Powell, C. N. Berglund, and W. E. Spicer
Rev. Modern Phys., 40:737 (1968)

Preparation of thin films of vanadium (di-, sesqui-, and pent-) oxide
G. A. Rozgonyi and W. J. Polito
J. Electrochem. Soc., 115:56-57 (1968)

Structural and electrical properties of vanadium dioxide thin films
G. A. Rozgonyi and D. H. Hensler
J. Vacuum Sci. Tech., 5:194-199 (1968)

Epitaxial growth of VO single crystals and their electrical properties
H. Takei and S. Koide
J. Phys. Soc. Japan, 24:1394 (1968)
No semiconductor-to-metal transition at 126°K

Metal −semiconductor transition of VO$_2$: A study of magnetic resonance
J. J. Umeda, K. Narita, and H. Kusumoto
Hitachi Rev. (Japan), 17:204-211 (1968)

The Hall effect in V$_2$O$_3$ single crystals in the metallic conductivity region
V. P. Zhuze, A. A. Andreev, and A. I. Shelykh
Fiz. Tverd. Tela, 10:3674-3677 (1968)
Sov. Phys. − Solid State, 10:2914 (1969)

Semiconductor to metal transitions in transition-metal compounds
David Adler, Julius Feinleib, Harvey Brooks, and William Paul
Phys. Rev., 155:851 (1967)

Electrical conductivity of vanadium oxide at low temperatures
S. M. Ariya, B. Ya Brach, and V. A. Vladimirova
Vestn. Leningr. Univ. Ser. Fiz. i Khim (Leningrad), No. 22, pp. 157-159 (1967)
N68-17385, Lockheed Missiles and Space Co., Sunnyvale, Calif.

Semiconductor-to-metal transition in V$_2$O$_3$
Julius Feinleib and William Paul
Phys. Rev., 155:841-850 (1967)

Reactively sputtered vanadium dioxide thin films
E. N. Fuls, D. H. Hensler, and A. R. Ross
Appl. Phys. Letters, 10:199 (1967)

Electronic structure and mutual solubility of TiO, VO, and NbO
P. V. Gel'd, G. P. Shveikin, and S. I. Alyamovskii
Zh. Neorg. Khim., 12:2001-2007 (1967)

High-frequency conductivity of VO$_2$
S. Kabashima, Y. Tsuchiya, and T. Kawakubo
J. Phys. Soc. Japan, 22:932 (1967)

Electrical transition in V$_6$O$_{13}$
K. K. Kanazawa
(Univ. of California, Riverside), Bull. Am. Phys. Soc., 12:1120 (1967)

Shift of transition temperature of vanadium dioxide crystals
I. Kitahiro and A. Watanabe
Japan. J. Appl. Phys., 6:1023 (1967)
RMIC Cart. 105

Electrical conductivity and magnetic susceptibility of oxide semiconductors based on vanadium oxides
A. Yu. Kudzin and I. M. Chernenko
Fiz. Tverd. Tela, 9(6):1822-1823 (1967)
Sov. Phys. − Solid State, 9(6):1430-1431 (1967)

Thin-film switching elements of VO$_2$
K. van Steensel, F. van de Burg, and C. Kooy
Philips. Res. Repts., 22:170-177 (1967)

Electric and magnetic properties of "VO"
S. Kawano, K. Kosuge, and S. Kachi
J. Phys. Soc. Japan, 21:2744 (1966)

Hall effect of vanadium dioxide powder
I. Kitachiro, T. Ohashi, and A. Watanabe
J. Phys. Soc. Japan, 21:2422 (1966)

Thermoelectric power of vanadium dioxide whisker
  I. Kitachiro and A. Watanabe
  J. Phys. Soc. Japan, 21:2423 (1966)

A new phase appearing in metal—semiconductor transition in VO₂
  J. Umeda, S. Ashida, H. Kusumoto, and K. Narita
  J. Phys. Soc. Japan, 21:1461 (1966)

On the conduction mechanism of some transition-metal oxides
  Gerard A. Acket
  Ph. D. Thesis, Univ. of Utrecht, Nov. 15, 1965
  J. Am. Ceram. Soc., 48:365 (1965)

Anisotropy of the electrical conductivity of VO₂ single crystals
  P. F. Bongers
  Solid State Commun., 3:275 (1965)

Electrical properties of crystalline, glassy and liquid vanadium oxide
  R. Hakim, T. Kennedy, and J. D. MacKenzie
  Amer. Ceram. Soc. Bull., 44:303 (1965)

Pressure dependence of the resistance of VO₂
  C. H. Neuman, A. W. Lawson, and R. F. Brown
  J. Chem. Phys., 41:1591 (1964)

Semiconductor-to-metal transition in V₂O₃
  David Adler and Julius Feinleib
  Phys. Rev. Letters, 12:700 (1964)

Thermoelectric forces and electrical conductivity of vanadium and niobium pentoxides
  A. I. Manakov, O. A. Esin, and B. M. Lepinskikh
  Fiziko-khimicheskie osnovy proizvodstva stali. Trudy VI konferentsii, Moskva, 1961, Nauka, Moscow (1964), pp. 148-153

Effect of pressure on the metal-to-insulator transition in V₂O₄ and V₂O₃
  S. Minomura and H. Nagasaki
  J. Phys. Soc. Japan, 19:131-132 (1964)

Electrical conduction in glasses containing vanadium pentoxide
  H. Nestor
  ScD. thesis,, Dept. of Metallurgy, MIT, (January 1964)

Electrical properties of vanadium pentoxide
  I. B. Patrina and V. A. Ioffe
  Fiz. Tverd. Tela, 6(11):3227-3234 (1964)
  Sov. Phys. — Solid State, 6(11):2581-2585 (1965)

Stress effects on insulator-to-metal transitions. Part I. Vanadium(2)-oxygen (3). Part 2. Vanadium spinels
  J. Feinleib
  AD-431 609 (1963), 255 pp.

Electrical conductivity of vanadium oxides
  S. Kachi, T. Takada, and K. Kosuge
  J. Phys. Soc. Japan, 18:1839 (1963)

Effect of pressure on the electrical resistance of some transition-metal oxides and sulfides
  S. Minomura and H. G. Drickamer
  J. Appl. Phys., 34:3043-3048 (1963)

Zur Frage des Leitungsmechanismus oxidischer Halbleiter bei höheren Temperaturen
  J. Rudolph
  Tech.-Wissensch. Abh. Osram. Gesellsch., Dtsch., 8:86-99 (1963)

Effect of pressure on the metal-to-insulator transition in vanadium trioxide
  I. G. Austin
  Phil. Mag., 7:961-967 (1962)

Electrical conductivity anomaly in vanadium sesquioxide
  Gilbert Goodman
  Phys. Rev. Letters, 9:305 (1962)

Electric and magnetic properties of V₂O₃ and related sesquioxides
  A. J. MacMillan
  MIT Lab. for Insulation Research, Tech. Report No. 172; AD-291459 (Oct. 1962), 32 pp.

Electric conductivity of the lower vanadium oxides
  N. I. Bogdanova and S. M. Ariya
  Vestn. Leningrad Univ., Ser. Fiz. Khim., 16:143-147 (1961)

Energy spectrum of charge carriers in ferro- and antiferromagnetic substances
  E. K. Kudinov and A. G. Samoilovich
  Izv. Akad. Nauk SSSR, Ser. Fiz., 25:1339-1342 (1961)

Microwave dielectric properties of the group of 3d-transition-metal oxides
  A. A. Samokhvalov
  Fiz. Tverd. Tela, 3(12):3593-3601 (1961)
  Sov. Phys. — Solid State, 3(12):2613-2618 (1961)

Forms of the higher oxides of vanadium from data on their conductivity
  N. I. Bogdanova and S. M. Ariya
  Zh. Obshch. Khim., 30:3-7 (1960)

Electrical properties of certain semiconducting oxide glasses
  V. A. Ioffe, I. V. Patrina, and S. V. Poberovskaya
  Fiz. Tverd. Tela, 2(4):656-663 (1960)
  Sov. Phys. — Solid State, 2(4):609-614 (1960)

Electrical conductivity of high-vanadium phosphate glass
  M. Munakata
  Solid State Electronics, 1:159 (1960)

Vanadium pentoxide glass with a low electrical resistance
  J. E. Stanworth, E. P. Denton, and H. Rawson
  Chem. Abstr., 54:17831i (1960)

Electrical conductivity of certain titanium and vanadium oxides
  S. M. Ariya and N. I. Bogdanova
  Fiz. Tverd. Tela, 1(7):1022-1026 (1959)
  Sov. Phys. — Solid State, 1(7):936-939 (1960)

Oxides that show a metal-to-insulator transition at the Néel temperature
  F. J. Morin
  Phys. Rev. Letters, 3:34-36 (1959)

The effect of low-valency vanadium on the electrical conductivity of high-vanadium phosphoric acid glasses
  M. Munakata, S. Karvamura, J. Ashara, and M. Iwamoto
  Yogyo Kyokai Shi, 67:344 (1959)

5.e.3.b. Niobium

X-ray K-absorption edge of niobium in niobium metal and its oxides
V. G. Bhide and M. K. Bahl
J. Chem. Phys., 52:3093-3096 (1970)

Electrode effects and bistable switching of amorphous Nb_2O_5 diodes
T. W. Hickmott and W. R. Hiatt
Solid State Electronics, 13:1033-1047 (1970)

Electrical properties of NbO
J. M. Honig, W. Wahnsiedler, and P. Eklund
In: Massachusetts Institute of Technology, Francis Bitter National Magnet Laboratory's Quarterly Technical Status Report Covering the Period from 1 July 1970 to 30 Sept. 1970, pp. 26-27, QTSR-41 (Sept. 1970)
Resistivity, Hall coefficient, and magnetoresistance measurements

Photoconduction, trapping and chemisorption effects in sputtered niobium oxide films
Frank G. Ullman, Carl L. Deertz, and Lucas W. Smith
Bull. Am. Phys. Soc., 15:385 (1970)
J. Appl. Phys., 41(9):3872 (1970)

Preparation, optical and dielectric properties of vapor-deposited niobium oxide thin films
M. T. Duffy, C. C. Wang, A. Waxman, and K. H. Zaininger
J. Electrochem. Soc., 116:234-239 (1969)

Electronic effects on bistable switching and dielectric breakdown of Nb_2O_5 diodes
T. W. Hickmott
J. Vac. Sci. Tech., 6:601 (1969)

Electroluminescence, bistable switching, and dielectric breakdown of Nb_2O_5 diodes
T. W. Hickmott
J. Vac. Sci. Tech., 6:828-833 (1969)

Photoconductivity, electroluminescence and hole injection in thin Nb_2O_5 diodes
T. W. Hickmott
Thin Solid Films, 3:85-107 (1969)

Seebeck effect in α-Nb_2O_5 near stoichiometry
Kenneth E. Jesse
J. Appl. Phys., 40(8):3386 (1969)

Electrical properties of NbO and NbO_2
J. A. Roberson and R. A. Rapp
J. Phys. Chem. Solids, 30:1119-1124 (1969)

Semiconducting anodic Nb_2O_5 on Nb
D. Stuttzle and K. E. Heusler
Phys. Chem., 65:201-215 (1969)

Voltage controlled negative resistance (VCNR) in reactively sputtered thin Nb_2O_5 films
R. Pinto and B. M. Shaha
Japan. J. Appl. Phys., 7:1542-1543 (1968)

Electrical conductivity and thermoelectric power of niobium dioxide
R. F. Janninck and D. H. Whitmore
J. Phys. Chem. Solids, 27:1183 (1966)

The melting point and other properties of the lower oxides of niobium
O. P. Kolchin and N. V. Sumarokova
Atomnaya Energiya, 10(2):168-169 (1961)
Soviet Atomic Energy, 10(2):167-169 (1961)
NbO and NbO_2

[Title Not Given]
F. G. Kusenko and P. V. Gel'd
Izv. Vysshikh Uchebn. Zavendenii, Tsvetn. Met., 3:102 (1960)
Heat content, thermal expansion, and electrical conductivity of NbO_2

Measurements with the help of liquid helium. XXII. The resistance of metals, alloys and compounds
W. Meissner, H. Franz, and H. Westerhoff
Ann. Phys., 17:593 (1933)
Low-temperature conductivity, NbO

5.e.3.c. Tantalum

Schottky currents in dielectric films
J. C. Schug, A. C. Lilly, and D. A. Lowitz
Phys. Rev. B, Solid State, 1:4811-4818 (1970)
Mylar, silica, tantalum pentoxide

The influence of the electrolyte on the composition of "anodic oxide films" on tantalum
G. Amsel, C. Cherko, G. Feuillode, and J. P. Nadai
J. Phys. Chem. Solids, 30:2117-2134 (1969)
The film properties (rate of chemical attack, dielectric constant, formation field, molar volume, etc.) depend on the nature of the electrolyte used for formation

Asymmetric conduction in thin-film tantalum-tantalum oxide-metal structures: interstitial and substitutional impurity effects and direct detection of flaw breakdown
N. N. Axearod and N. Schwartz
J. Electrochem. Soc., 116:460-465 (April 1969)

Negative Photoleifähigkeit in Ta_2O_5-Schichten
M. Hartl and W. Schwarz
Z. Naturforsch., 24a:296-297 (1969)

Dielectric properties of Ta_2O_5 thin films
D. L. Pulfrey, P. S. Wilcox, and L. Young
J. Appl. Phys., 40:3891-3898 (1969)

Dielectric properties of Ta_2O_5 films in electrolytes
M. Hartl and E. H. Pauli
TC-3115 (Nov. 22, 1966), 12 pp.
Electrochimica Acta, 11:1417-1424 (1966)

Electrical conduction through thin films of tantalum oxide
S. Horiuchi, T. Sasami, and J. Yamaguchi
Electron. Eng. Japan, 86:48-56 (1966)

Ionic conductivity, dielectric constant, and optical properties of anodic oxide films on two types of sputtered tantalum films
D. Mills, L. Young, and F. G. R. Zobel
J. Appl. Phys., 37:1821-1824 (1966)

5.e.4. Group VI

5.e.4.a. General

X-ray $K\beta$ emission spectra and energy levels of compounds of 3d-transition metals – I. Oxides
A. S. Koster and H. Mendel
J. Phys. Chem. Solids, 31:2511-2522 (1970)
Cr

The effect of chemisorption of atomic oxygen on conductivities of semiconducting oxides. I. Pyrolysis and photolysis of oxygen.
G. V. Malinova and I. A. Myasnikov
Kinet. i Kataliz, 10(2):328-335 (1969)
Kinetics and Catalysis, 10(2):263-268 (1969)
ZnO, MoO_3, and WO_3

5.e.4.b. Chromium

Transport processes in amorphous Cr_2O_3 films
D. F. Barbe and D. S. Herman
J. Appl. Phys., 41:3136-3120 (1970)

Disorder model for a semiconductor with characteristic defect centres, taking chromium sesquioxide as an example
K. Hauffe et al.
NLL-CE-Trans-5211-(9022.09)
Z. Phys. Chem. (Frankfurt), 198(516):232-247 (1951) (Feb. 24, 1970), 16 pp.

Electron tunneling $Cr - Cr_2O_3 -$ metal junctions
G. I. Rochlin and P. K. Hansma
Phys. Rev., B2(6):1460-1463 (1970)

Electrical properties of nonstoichiometric Cr_2O_3
L. N. Cojocaru
J. Phys. Chem., 64:255-262 (1969)

Electrical properties of nonstoichiometric oxides of $CrO_3 - Cr_2O_3$ systems
L. N. Cojocaru, T. Costea, and I. Negoescu
Z. Phys. Chem. (Frankfurt), 60:152-158 (1968)

The electrical conductivity of chromic oxide in the range 500-1300°C
H. E. N. Stone
Materials Sci. Eng., 2:348-349 (1968)

Electric and catalytic properties of chromia doped with magnesium and titanium
J. Deren and J. Haber
Bull. Acad. Polon. Sci., 15:103 (1967)

Magnetoelectric effect in Cr_2O_3 single crystal as studied by dielectric-constant method
H. B. Lal, R. Srivastava, and K. G. Srivastava
Phys. Rev., 154:505-507 (1967)

The K-series emission lines of the x-ray spectrum of chromium and chromium oxide
V. V. Nemoshkalenko and M. A. Mindlina
In: Investigating the Electronic Properties of Metals and Alloys, Izd. Akad. Nauk SSSR, Kiev (1967), pp. 78-91

Electrical resistivity and electron-microscopic observation of sintered chromium dioxide
Tomozo Nishikawa, Nobuo Nakayama, and Eiichi Hirota
Yogyo Kyokai Shi, 74(8):256-261 (1966)

Pulse measurements of the magnetoelectric effect in chromium oxide
T. H. O'Dell
IEEE Trans. Magnetics, MAG-2, 449-452 (1966)

Sur le mécanisme de la conduction électrique dans le sesquioxyde de chrome pur
J. Roche
Ann. Fac. Sci. Univ., Clermont, 32:105-109 (1966)

Electrical resistivity of single-crystal CrO_2
D. S. Rodbell, J. M. Lommel, and R. C. DeVries
J. Phys. Soc. Japan, 21:2430 (1966)

Survey of magnetoelectric materials
R. P. Santoro and R. E. Newnham
(Lab. for Insulation Research, Mass. Inst. Tech., Cambridge, Mass.), AFML-TR-66-327 (Sept. 1966), 24 pp.

Magnetoelectrics — a new class of materials
T. H. O'Dell
Electronics and Power (Aug. 1965), p. 266

Observation of the magneto-electric effect in polycrystalline chromium oxide
T. H. O'Dell
Phil. Mag., 10:899 (1964)

Dielectric constant of Cr_2O_3 crystals
P. H. Fang and W. S. Brower
Phys. Rev., 129:1951 (1963)

The electrical conductivity of single-crystal Cr_2O_3
Julian Anthony Crawford
Dissertation Abstracts 64-7499, 78 pp.

5.e.4.c. Molybdenum

Electrical properties of evaporated molybdenum oxide films
G. S. Nadkarni and J. G. Simmons
J. Appl. Phys., 41:545-551 (1970)

Charge compensation of impurities and electrical properties of MoO_3
V. A. Ioffe, I. B. Patrina, E. V. Zelenetskaya, et al.
Phys. Stat. Sol., 35:535-542 (1969)

Stabilization of the tunnel structure of Mo_5O_{14} by partial metal-atom substitution
L. Kihlborg
Acta Chem. Scand., 23:1834-1835 (1969)

On the photosensitization of molybdenum trioxide single crystals
A. L. Shkolynik
Soobshch. Akad. Nauk Gruz. SSR, 42:563-566 (1966)

Zur Frage des Leitungsmechanismus oxidischer Halbleiter bei höheren Temperaturen
J. Rudolph
Tech.-Wissensch. Abh. Osram. Gesellsch., Dtsch., 8:86-99 (1963)

5.e.4.d. Tungsten

Raman spectroscopy of WO_3
D. M. Hannon
Bull. Am. Phys. Soc., 15:297 (1970)

Contribution à l'étude du mode de conduction dans l'oxyde de tungstène WO_3 monocristallin
A. M. Baticle, P. Lemasson, F. Perdu, P. Vennereau, and J. Vernieres
Compt. Rend., 2688:1203-1206 (1969)

Raman scattering from the "soft" ferroelectric mode in tungstic oxide
J. F. Scott
Bull. Am. Phys. Soc., 14:738 (1969)

Properties of WO_3 modified by the substitution of Mo and Cr
A. B. Knox
Trans. British Ceram. Soc., 66:85-91 (1967)

Zur Frage des Leitungsmechanismus oxidischer Halbleiter bei höheren Temperaturen
J. Rudolph
Tech.-Wissensch. Abh. Osram Gesellsch., Dtsch., 8:86-99 (1963)

Electrical transport properties of tungstic oxide and analysis of electrical properties of tungsten-bronze systems
  Billy Lee Crowder
  Dissertation Abstracts 64-1022, 107 pp.

5.f. Optical Properties

5.f.1. General

Plasma resonance in TiO, VO, and NbO
  C. N. R. Rao, W. E. Wahnsiedler, and J. M. Honig
  J. Solid State Chem., 2:315-317 (1970)

Polarons in crystalline and non-crystalline materials
  I. G. Austin and N. F. Mott
  Adv. Phys., 18:41-102 (1969)

Infrared absorption spectra of carbides, and V, Nb, and Ta lower oxides
  S. I. Alyamovskii, G. P. Shveikin, and P. V. Gel'd
  Zh. Neorg. Khim., 12:1738-1742 (1967)

Infrared spectra of various metal oxides in the region of 2 to 26 microns
  D. W. Sheibley and M. H. Fowler
  NASA TN D-3750 (Dec. 1966)

Infrared absorption study of metal oxides in the low frequency region (700-240 cm^{-1})
  N. T. McDevitt and W. L. Baun
  Spectrochim. Acta, 20:799-808 (1964)
  Characteristic frequencies of oxides of 52 metals

Reflectance spectra of non-stoichiometric titanium oxide, niobium oxide, and vanadium oxide
  F. Vratny and F. Micale
  Trans. Faraday Soc., 59:2739 (1963)

Infrared spectra of titanium and vanadium oxides in the crystalline state
  S. M. Ariya and M. V. Golomolzina
  Fiz. Tverd. Tela, 4(10):2921-2924 (1962)
  Sov. Phys. – Solid State, 4(10):2142-2144 (1963)

5.f.2. Group IV

5.f.2.a. Titanium

Optical absorption and photochromism in iron-doped rutile
  W. Clark and P. Broadhead
  J. Phys. C: Solid State Phys., 3:1047-1054 (1970)

Red-orange photoluminescence in TiO_2:W
  R. B. Lauer and Amal K. Ghosh
  Appl. Phys. Letters, 16(9):341-346 (1970)

Optical absorption, EPR, and electrical conductivity in chromium-doped rutile crystals
  D. N. Mirlin, I. I. Reshina, and L. S. Sochava
  Fiz. Tverd. Tela, 11(9):2471-2480 (1969)
  Sov. Phys. – Solid State, 11(9):1995-2001 (1970)

Electronic band structure. Optical reflectance spectrum of single crystal Ti_2O_3
  P. M. Raccah and W. J. Scouler
  (Lincoln Lab., Massachusetts Institute of Technology), pp. 23-24, ESD-TR-70-33 (May 1970)

Investigation of optical absorption in rutile via two-photon absorption
  Harve S. Waff
  Bull. Am. Phys. Soc., 15:387 (1970)

Structure in the two-photon absorption spectrum of TiO_2 (rutile)
  H. S. Waff and K. Park
  Phys. Letters, 32A:109-110 (1970)

Experimental observation of small polarons in rutile (TiO_2)
  V. N. Bogomolov, Yu. A. Firsov, E. K. Kudinov, et al.
  Phys. Stat. Sol., 35:555-558 (1969)

Frequency dependence of the polaron absorption in conducting TiO_2 crystals
  E. K. Kudinov, D. N. Mirlin, and Yu. A. Firsov
  Fiz. Tverd. Tela, 11:2789-2801 (1969)
  Sov. Phys. – Solid State, 11:2257 (1970)

Interpretation of the infrared absorption spectra of stannic oxide and titanium dioxide (rutile) powders
  J. T. Luxon and R. Summitt
  J. Chem. Phys., 50:1366-1370 (1969)

Thermally stimulated currents and luminescence in rutile (TiO_2)
  R. R. Addiss, A. K. Ghosh, and F. G. Wakim
  Appl. Phys. Letters, 12:397-400 (1968)
  Luminescence at 850 $m\mu$; 83 to 373°K; slightly reduced crystals.

Crystal structure and optical properties of thin organogenic titanium oxide layers on glass substrates
  H. Bach and H. Schroeder
  Thin Solid Films, 1(4):254-276 (1968)

Optical absorption by polarons in rutile (TiO_2) single crystals
  V. N. Bogomolov and D. N. Mirlin
  Phys. Stat. Sol., 27:443-453 (1968)

Caractéristiques optiques de couches minces de TiO_2, préparées par évaporation réactive de Ti_2O_3
  S. Katsube and Y. Katsube
  Bull. Govt. Ind. Res. Inst., Japan, 19:203-208 (1968)

Interband Faraday rotation in some perovskite oxides and rutile
  W. S. Baer
  J. Phys. Chem. Solids, 28:677-687 (1967)

Infrared absorption in conducting rutile crystals
  V. N. Bogomolov and D. N. Mirlin
  Zh. Éksper. Teor. Fiz. Pis'ma (USSR), 5:293-296 (1967)
  JETP Letters, 5:241-243 (1967)

Electromodulation of the optical constants of rutile in the UV
  A. Frova, P. J. Boddy, and Y. S. Chen
  Phys. Rev., 157:700-708 (1967)

Application of infrared spectroscopy to order-disorder problems in simple ionic solids
  W. B. White
  Mat. Res. Bull., 2:381-394 (1967)
  TiO_2, rutile and anatase

Infrared study of the surface of rutile
K. E. Lewis and G. D. Parfitt
Trans. Faraday Soc., 62:204 (1966)

Electrical and optical properties of titanium dioxide thin films
T. S. Travina and Yu. A. Mukhin
Izv. Vyssh. Uchebn. Zaved., Fiz., 6:74-80 (1966)

Optical and structural properties of oxidized titanium films
F. Arntz and T. Chernow
J. Vacuum Sci. Tech., 2:20-23 (1965)

Optical properties and band structure of wurtzite-type semiconductors and rutile
M. Cardona and G. Harbeke
International Congr. Phys. Semiconducteurs, Paris (1964), p. 122

Optische Konstanten und Energie-Band Struktur von TiO_2 und Halbleiter mit Wurzitstruktur
G. Harbeke and M. Cardona
Phys. Verh. D.P.G., 8:280 (1964)

Anodic oxidation of metallic titanium. Part 6. Specific gravity, crystal structure, refractive index, and dielectric constant of the anodic film on titanium
M. Koyama
Rigagaku Kenkyusho Kokoku, 39:21-26 (1963)

Electrical and optical investigation of absorption centers in rutile single crystals
K. G. Srivastava
Phys. Rev., 119:516 (1960)

Studies of the optical absorption spectrum of rutile single crystals
B. H. Soffer
(Mass. Inst. Tech., Cambridge), Contracts Nonr-184(10) and AT(30-1) 1937, Technical Rept. 140; AD-227 945 (Aug, 1959), 28 pp.

Point defect relaxation, optical absorption and dislocation damping in rutile
Robert D. Carnahan
Dissertation Abstracts, 64-2460, 167 pp.

5.f.2.b. Zirconium

Thermoluminescence of anodic and thermal zirconium oxide layers
J. Mochniak
Acta Phys. Polon., A37:909-918 (1970)

Thermoluminescence of ZrO_2
C. Bettinali, G. Ferraresso, and J. W. Manconi
J. Chem. Phys., 50:3957-3961 (1969)

Determination of the refractive index and thickness of oxide films on anodized zirconium from transmission interference measurements
J. C. Banter
Measurement Techniques for Thin Films (B. Schwartz and N. Schwartz, eds.), Electrochemical Society, Inc., New York (1967), pp. 273-280

Electroluminescence of zirconium oxide films
J. Mochniak
Acta Phys. Polon., 30:559-566 (1966)

Preparation and luminescent properties of Ti-activated zirconia
J. F. Sarver
J. Electrochem. Soc., 113:124-128 (1966)

Optical energy gaps in the monoclinic oxides of hafnium and zirconium and their solid solutions
J. G. Bendoraitis and R. E. Salomon
J. Phys. Chem., 69:3666-3667 (1965)

Réflexion et transmission dans l'infrarouge de la zircone monoclinique
Bernard Piriou and Jean Tsakiris
Compt. Rend., 261:3079-3081 (1965)

5.f.2.c. Hafnium

Optical energy gaps in the monoclinic oxides of hafnium and zirconium and their solid solutions
J. G. Bendoraitis and R. E. Salomon
J. Phys. Chem., 69:3666-3667 (1965)

5.f.3. Group V

5.f.3.a. Vanadium

Optical and electrical properties of V_2O_3 and Cr-doped V_2O_3
A. S. Barker, Jr., H. C. Montgomery, and J. P. Remeika
Bull. Am. Phys. Soc., 15:386 (1970)

Optical properties of V_2O_3 doped with chromium
A. S. Barker, Jr., and J. P. Remeika
Solid State Commun., 8:1521-1524 (1970)

Optical characteristics of vanadium pentoxide single crystals
A. A. Abdullaev, L. D. Kislovskii, and L. M. Belyaev
Opt. i Spektroskopiya, 26(6):1043-1044 (1969)
Opt. Spectrosc., 26:566 (1969)

Optical and transport properties of high-quality crystals of V_2O_4 near the metallic transition temperature
Larry A. Ladd and William Paul
Solid State Commun., 7:425-428 (1969)

Absorption spectrum of vanadium dioxide below the semiconductor—metal transition point
D. N. Mirlin
Fiz. Tverd. Tela, 10(12):3697-3698 (1968)
Sov. Phys.—Solid State, 10(12):2938-2939 (1969)

The colour problem of vanadium pentoxide. III. Conductivity and infrared absorption
A. R. Tourky, Z. Hanafi, and K. Al Zewel
Z. Phys. Chem. Leipzig (Germany), 242(5-6):298-304 (1969)

Investigation of the reflection spectra of vanadium dioxide single crystals at the semiconductor—metal phase transition
V. G. Mokerov and A. V. Rakov
Fiz. Tverd. Tela, 10(5):1556-1557 (1968)
Sov. Phys.—Solid State, 10(5):1231-1232 (1968)

Optical properties of VO_2 between 0.25 and 5 eV
Hans W. Verleur, A. S. Barker, Jr., and C. N. Berglund
Phys. Rev., 172:788-798 (1968)

Optical properties of VO$_2$ above and below the transition temperature
A. S. Barker, H. W. Verleur, and H. J. Guggenheim
Bull. Am. Phys. Soc., 12:60 (1967)

Optical absorption near the absorption edge in V$_2$O$_5$ single crystals
Z. Bodo and I. Hevesi
Phys. Stat. Sol., 20:K45 (1967)

On the optical properties of vanadium pentoxide single crystals
I. Hevesi
Acta Phys. Hungary, 23:415-424 (1967)

On diffuse reflection spectra of V$_2$O$_5$
I. Hevesi
Acta Phys. Chem. Szeged (Hungary), 13:39-42 (1967)

Determination of optical constants and thickness of anisotropic crystal plates from transmission measurements
I. Hevesi
Acta Phys. Acad. Sci. Hung., 23:75-85 (1967)
V$_2$O$_5$

Optical properties of vanadium pentoxide
R. T. Jacobsen and Milton Kerker
J. Opt. Soc. Am., 57:751 (1967)

Infrared optical properties of vanadium dioxide above and below the transition temperature
A. S. Barker, Jr., H. W. Verleur, and H. J. Guggenheim
Phys. Rev. Letters, 17:1286 (1966)

Optical-absorption coefficients of vanadium pentoxide single crystals
N. Kenny, C. R. Kannewurf, and D. H. Whitmore
J. Phys. Chem. Solids, 27:1237 (1966)

The colour problem of vanadium pentoxide, I. X-ray studies
A. R. Tourky, M. S. Farag, T. M. Salem, Z. Hanafi, and Al Zewel
Z. Phys. Chem. (Leipzig), 230:179-183 (1965)

The optical absorption of vanadium pentoxide single crystals
Neal S. Kenny
Dissertation Abstr., 65-12, 114, 127 pp.
University Microfilms, Inc., Ann Arbor, Mich.

5.f.3.b. Niobium

Nonlinear optical polarizability of the niobium-oxygen bond
C. R. Jeggo and G. D. Boyd
J. Appl. Phys., 41:2741-2742 (1970)

5.f.3.c. Tantalum

Electro-optic effects in thin films
A. Frova and P. Migliorato
Atta Frequenza (Italy), 39(6):562-565 (1970)
Tantalum oxide

Electro-optic effect and modulated interference in tantalum oxide films
A. Frova and P. Migliorato
Appl. Phys. Letters, 15:406-408 (1969)

Ellipsometer measurements of the optical constants of tantalum oxide films at high absorption
F. G. Ullman, C. D. Spivey, R. W. Laws, and B. J. Holden
Bull. Am. Phys. Soc., 14:432 (1969)

5.f.4. Group VI

5.f.4.a. Chromium

Magnetic Davydov splittings in the optical absorption spectrum of Cr$_2$O$_3$
J. W. Allen, R. M. MacFarlane, and R. L. White
Phys. Rev., 179:523-541 (1969)

Die Schwingungsspektren von CrO$_3$ und K$_2$Cr$_4$O$_{13}$
Rainer Mattes
Z. Naturforschung, 24b:772-773 (1969)

Optical spectrum of Cr$_2$O$_3$
J. W. Allen, R. M. MacFarlane, and R. L. White
Bull. Am. Phys. Soc., 13(3), paper CK15 (March 1968)

On the infrared spectra of CrO$_2$
R. M. Chrenko and D. S. Rodbell
Phys. Letters, 24A:211 (1967)

Optical properties of reactively evaporated chromium oxide films
R. I. Frank and W. L. Moberg
J. Vacuum Sci. Tech., 4:133-134 (1967)

Optical spectrum of antiferromagnetic Cr$_2$O$_3$
J. P. Van der Ziel
Phys. Rev., 161:483-492 (1967)

Infrared lattice spectra of α-Al$_2$O$_3$ and Cr$_2$O$_3$
R. Marshall and S. S. Mitra
J. Chem. Phys., 43:2893-2894 (1965)

5.f.4.b. Molybdenum

Die Ramanspektren von MoO$_3$ und WO$_3$
W. Krasser
Naturwissenschaften, 56:213-214 (1969)

Physical properties of a transition-metal oxide: optical and photoelectric properties of single-crystal and thin-film molybdenum trioxide
S. K. Deb
Proc. Roy. Soc., A, 304:211-231 (1968)

Optical properties of MoO$_3$
A. L. Shkol'nik
Izv. Akad. Nauk SSSR, Ser. Fiz., 31:2050-2051 (1967)
Bull. Acad. Sci. USSR, Phys. Ser., 31:2093-2095 (1967)

Optical properties and color-center formation in thin films of molybdenum trioxide
S. K. Deb and J. A. Chopoorian
J. Appl. Phys., 37:4818-4825 (1966)

Some optical and photoelectric properties of MoO$_3$
R. B. Dzhanelidze, I. M. Purtseladze, L. S. Khitarishvili, R. I. Chikovani, and A. L. Shkol'nik
Fiz. Tverd. Tela, 7(8):2573-2575 (1965)
Sov. Phys. – Solid State, 7(8):2082-2083 (1966)

5. Binary Transition Metal Oxides

5.f.4.c. Tungsten

Lattice vibrations of δ-uranium and tungsten trioxides
  K. Ohwada
  Spectrochimica Acta, 26A:1035-1044 (1970)

Die Ramanspektren von MoO$_3$ und WO$_3$
  W. Krasser
  Naturwissenschaften, 56:213-214 (1969)

Spectre de vibration de l'oxyde de tungstène WO$_3$
  Claude Rocchiccioli-Deltcheff
  Compt. Rend., Ser. B; 268:45-47 (1969)

6. Binary Transition Metal Chalcogenides

6.a. General

In vacuo synthesis, structure, and thermal stability of some transition-metal diselenides
E. P. Gladchenko, A. G. Duksina, and V. L. Kalikhman, V. A. Obolonchik, and L. M. Prokoshina
Porosh. Met., 10(3):76–83 (1970)
Ti, Nb, W, Mo

Thermal conductivity of some binary and tertiary selenides of transition elements
V. A. Ivanova, D. Sh. Abdinov, and G. M. Aliev
Izv. Akad. Nauk SSSR, Neorg. Mater., 6(3):566–568 (1970)
Inorg. Mater., 6(3):500–502 (1970)

Growing of heavy-metal chalcogenide single crystals by chemical transport with aluminum chloride
Heinz D. Lutz, Cs. Lovasz, K. H. Bertram, M. Sreckovic, and U. Brinker
Monatsh. Chem., 101(2):519–524 (1970)
Cr_2S_3, Cr_3S_4, Cr_5S_6, Cr_7S_8, CrS, Cr_2Se_3, Cr_7Se_8, CoS_2, CoS, NiS_2, NiS, $ZnCr_2S_4$, $CdCr_2S_4$, $MnCr_2S_4$, $ZnCr_2S_4$, $CdCr_2Se_4$, $HgCr_2S_4$, $CuCr_2Se_4$, $NiCr_2S_4$, $Ni_{0.95}Cr_{0.05}S$, and $Ni_{0.75}Cr_{0.25}$

The preparation of high-purity, metal–sulphur alloys
S. Marich
J. Sci. Instr. (J. Phys. E), 3(4):317 (April 1970)

The synthesis and x-ray study of tungsten and rhenium tellurides
A. A. Opalovski, V. I. Fedorov, E. U. Lobkov, B. G. Erenburg, and L. N. Senchenko
Izv. Akad. Nauk SSSR, Neorg. Mater., 6(3):561–563 (1970)
Inorg. Mater., 6(3):495–497 (1970)

Electrical conductivity of thin single crystals of the IVB-VIB dichalcogenides
T. J. Wieting
J. Phys. Chem. Solids, 31:2148–2151 (1970)
MoS_2, $MoSe_2$, α-$MoTe_2$, WS_2, WSe_2, $NbSe_2$, HfS_2: vapor transport crystal growth

Infrared absorption spectra of sulfides and their analogs
A. I. Boldyrev, L. N. Egorova, and A. S. Povarennykh
Konst. Svoistva Miner., 3:5–16 (1969)

Group IV and V transition-metal rich chalcogenide x-ray powder diffraction data
H. F. Franzen, J. G. Smeggil, and B. R. Conard
(Ames Lab., Ames, Iowa), IS-2183 (1969), 18 pp.
Ti_2S, Ti_2Se, Zr_2S, Zr_2Se, Hf_2S, Hf_2Se, HfSe, $Nb_{31}S_8$, Nb_2Se, Ta_2S, and Ta_6S

Monocristallisation par réaction de transport en phase gazeuse
Pierre Gibart and Gaston Collin
Croissance de Composés Minéraux Monocristallins, No. 2 (J. P. Suchet, ed.), pp. 127–145, Masson et Cie, Paris (1969)
$CoCr_2Te_4$, Fe_2Te_3, Cr_3Te_4, $Cr_{0.67}Te$, $Co_{0.67}Te$

Electron energy loss studies of the transition-metal dichalcogenides
W. Y. Liang and S. L. Cundy
Phil. Mag., 19:1031–1043 (1969)

New x-ray data regarding tungsten and rhenium selenides
A. A. Opalovsky, V. E. Federov, B. G. Erenburg, E. U. Lobkov, and L. N. Senchenko
Izv. Sib. Otedel. Akad. Nauk SSSR, (Khim.), 5(12):62–65 (Sept. 1969)

The transition-metal dichalcogenides, discussion and interpretation of the observed optical, electrical, and structural properties
J. A. Wilson and A. D. Yoffe
Advan. Phys., 18(73):193–335 (1969)
435 refs.

Mechanisms for metal –nonmetal transitions in transition-metal oxides and sulfides
David Adler
Rev. Modern Phys., 40:714–736 (1968)
181 refs.

Electrical properties of transition-metal compounds
David Adler and Harvey Brooks
Comments Solid State Phys., 1(5):145-150 (1968)
Oxides, sulfides, 29 refs.

Selenium and selenides
D. M. Chizhikov and V. P. Shchastlivyi
(Translated from the Russian by Eugene M. Elkin), Collet's
Publishers Ltd., London and Wellingborough (1968)

Electrical properties of the group IV disulfides
TiS_2, ZrS_2, HfS_2, and SnS_2
L. E. Conroy and Kyu Chang Park
Inorg. Chem., 7(3):459-463 (1968)

Band model for transition-metal chalcogenides
having layer structures with occupied trigonal-
bipyramidal sites
John B. Goodenough
Mater. Res. Bull., 3(5):409-415 (1968)

Crystal-structure determination of transition-
metal compounds with octahedral configurations
D. K. Hohnke
Thesis, Univ. of Pennsylvania, Philadelphia, 1968, 183 pp.
Available from University Microfilms, Ann Arbor, Mich.,
Order No. 69-15,066

Crystal chemistry of the chalcogenides and
pnictides of the transition elements
F. Hullinger
Structure and Bonding, Vol. 4 (C. K. Jorgensen et al., eds.),
Springer-Verlag, Berlin (1968), pp. 83-229
532 refs.

Conductance of transition-element compounds
with ordered vacancies
V. A. Inova, D. Sh. Abdinov, and G. M. Aliev
Dokl. Akad. Nauk SSSR, 182(5):1111-1113 (1968)
Beta-CrSe, Cr_7Se_8, Cr_3Se_4, Cr_2Se_3, $NiCr_2Se_4$, VCr_2Se_4, and
$FeCr_2Se_4$

Vapor pressure and dissociation of metal sul-
fides
R. A. Isakova
Izd. Nauka, Alma-Ata (1968), 230 pp.
Zn, Cd, Pb, Cu, Ni, Co, Ge, In, Ga, Ta, Mo, Re, Ag, Sn, Sb,
As, Bi, Hg and Fe and complex sulfides such as FeAsS,
FeS_2, CuFeS, and Cu_5FeS_4; 274 refs.

Structure and Bonding, Vol. 4
C. K. Jorgensen, J. B. Neilands, Ronald S. Nyholm, D. Reinen,
and R. J. P. Williams, eds.
Springer-Verlag, Inc., New York (1968)

Magnetic semiconductors
Siegfried Methfessel and Daniel C. Mattis
Encyclopedia of Physics, Vol. 18/1 (H. P. J. Wijn, ed.,
S. Flugge, chief editor), Springer-Verlag, New York (1968),
pp. 389-562

Molybdenum telluride ($MoTe_2$), lanthanum tel-
luride (La_2Te_3) and vanadium telluride (V_3Te)
E. Montignie
Z. Anorg. Allg. Chem., 362(5-6):329-330 (1968)

Synthesis of molybdenum, tungsten, and niobium
diselenides
V. A Obolonchik and L. M. Prokoshina
Izv. Akad. Nauk SSSR, Neorg. Mater., 4:1654 (1968)

New transition metal dichalcogenides formed at
high pressure
Tom A. Bither
In: Proceedings of the Second International Conference on Solid
Compounds of Transition Elements, June 12-16, 1967, pp.
26-28

Non-stoechiométrie et ordre lacunaire dans les
sulfures, séléniures et tellurures des métaux
transition
M. Chevreton
Bull. Soc. Mineral. Cristallogr., 90:592-597 (1967)

Description of transition-metal compounds:
application to several sulfides
John B. Goodenough
Propriétés thermodynamiques physiques et structurales des
dérives semi-métalliques, No. 157, Colloq. Intern. Centre
National de la Récherche Scientifique, Orsay, 28 Sep-
tembre-1 Octobre 1965, Editions du Centre National de la
Récherche Scientifique, Paris (1967), p. 264

Progrès récents dans l'étude cristallochimique
des sulfures, séléniures et tellurures des métaux
de transition
Y. Jeannin
Bull. Soc. Fr. Mineral. Cristallogr., 90:528-536 (1967)

The preparation and electrical properties of
niobium selenide and tungsten selenide
R. Kershaw, M. Vlasse, and A. Wold
Inorg. Chem., 6:1599 (1967)

Some physical properties of transition-metal
sulfides
V. Kh. Oganesyan
Khal'kogenidy (Svoitstva Metody Poluch. Primen.), Mater.
Seminara, 1st, Kiev, 1965 (published 1967), pp. 160-166
Sulfide phases of Ti, Zr, Nb, Ta, Cr, Mo, and Fe

Transition-element chalcogenides (magnetic
properties)
M. Schieber
Experimental Magnetochemistry, Selected Topics in Solid
State Physics, North Holland (1967), pp. 432-459

First transition series tellurides with the $B8_1$
structure
J. P. Suchet and R. Druilhe
Colloq. Intern. Centre Nat. Rech. Sci., No. 157:307-322, discussion 322-323 (1967)

Electrolytic growth and preparation of transi-
tion-metal compound single crystals
Aaron Wold
AFML-TR-67-239, April 30, 1967
($NbSe_2$, WSe_2) AD 660615

2nd International Conference on Solid Compounds of Transition Elements
Enschede, Netherlands (June 12-16, 1967), 140 pp.

Thermal expansion of some sulfides of the
transition metals
E. M. Dudnik and V. Kh. Oganesyan
Soviet Power Met. Metal Ceram. (English Transl.), No. 2,
pp. 125-127 (1966)
Ti_2S_3, ZrS_2, Nb_2S_3, alpha-TaS_2, Cr_2S_3, and MoS_2

Thermal Constants of Compounds, Issue 2
V. P. Glushko
Academy of Sciences, USSR, Moscow (1966)
Translation PB-175628 available through CFSTI

Phase transitions in transition-metal chalcogenides
  C. Haas
  Solid State Commun., 4:419-421 (1966)

Contributions to the chemistry of the binary compounds of the transition elements
  Haakon Haraldsen
  Angew. Chem. Intern. Edit., 5(1):58 (1966)
  Phase and structural relationships of the sulfur, selenium, and tellurium compounds of the 4d and 5d transition elements of groups IV to VII of the periodic system are discussed

Nature of the chemical bond in semiconducting sulfides of metals belonging to the fourth and fifth groups of the periodic system
  G. V. Samsonov and V. Kh. Oganesyan
  Izv. Akad. Nauk SSSR, Neorg. Mater., 2(10):1757-1762 (1966)
  TiS, Ti_2S_3, ZrS_2, Nb_2S_3, TaS_2, Cr_2S_3, MoS_2, and FeS

On the bonding of S, Se, and Te in transition-metal monochalcogenides
  Hugo F. Franzen
  IS-1108 (March 3, 1965)

Thermodynamic properties of the metal bromides, iodides, sulfides, sulfates, chromates, metaphosphates, orthophosphates, molybdates, tungstates, borates, and hydroxides, final report
  W. J. Cooper and D. A. Scarpiello
  (Callery Chemical Co.), SC-RR-64-67, January 1964

Preparation and properties of the single crystalline AB_2-type selenides and tellurides of niobium, tantalum, molybdenum, and tungsten
  L. H. Brixner
  J. Inorg. Nucl. Chem., 24:257-263 (1962)

Molybdenum and niobium sulfides
  F. Jellinek, C. Brauer, and H. Mueller
  Nature, 185:376 (1960)

Properties of sulfide semiconductors — NiS, WS_2, PbTe, MoS_2, Ag_2Te, As_2Te_3, Sb_2Se_3
  J. W. Buttrey
  (Armour Res. Found.), U.S. Govt. Res. Rept., 32:765(A) (1959)

6.b. Group IV

The zirconium – sulfur system at high temperatures
  B. R. Conrad and H. F. Franzen
  High Temp. Sci., 3:49-55 (1971)
  Synthesis, crystal growth, structures

The crystal structures of ZrS and $Zr_{0.77}S$
  B. R. Conrad and H. F. Franzen
  The Chemistry of Extended Defects in Non-Metallic Solids, Proc. Inst. for Advanced Study on the Chemistry of Extended Defects in Non-Metallic Solids, Casa Blanca Inn, Scottsdale, Arizona, April 16-26, 1969 (LeRoy Eyring and Michael O'Keeffe, eds.), North-Holland Publishing Co., Amsterdam, London (1970), pp. 207-219

Spectres de pertes d'énergie dans films minces monocristallins de cristaux lamellaires, TiS_2, $TiSe_2$, $TiTe_2$
  Roger Vilanove
  Compt. Rend., 271B:1101-1113 (1970)

The system titanium-sulfur. II. The structures of Ti_3S_4 and Ti_4S_5
  G. A. Wiegers and F. Jellinek
  J. Solid State Chem., 1:519-525 (1970)

Electronic properties of inorganic solid solutions, final report
  Lawrence E. Conroy
  (Dept. of Chemistry, Minnesota Univ., Minneapolis), AROD-5532-3-C; AD-695697 (Oct. 1969), 5 pp.
  Crystal growth of metal disulfides (TiS_2, ZrS_2, HfS_2, SnS_2)

Etude du système nickel-ZrS_2. Structure et surstructure $Ni_{0.50}ZrS_2$
  Luc Trichet and Jean Rouxel
  Compt. Rend., 269C:1040-1043 (1969)

Thermodynamics of TiS_3
  E. Tronc and M. Huber
  Compt. Rend., 268C:1771-1774 (1969)

Etude structurale du composé Ti_5Te_8
  Simmone Brunie and Maurice Chevreton
  Mat. Res. Bull., 3:309-314 (1968)

Electrical properties of the group IV disulfides TiS_2, HfS_2, and SnS_2
  Lawrence E. Conrow and Kyu Chang Park
  Inorg. Chem., 7:459-463 (1968)

The crystal structure of Zr_2Se
  H. F. Franzen and L. J. Norrby
  Acta Cryst., 24B:601 (1968)

Contribution à l'étude des sulfures et des sulfo-séléniures de titane
  Y. Jacquin
  Thèse Doct. Sci. Phys., Paris (1968)
  Arch. orig. Centre Document. C.N.R.S., No. 2229 (Mar 14, 1968)

Preparation of zirconium sulfides
  G. N. Dubrovskaya and G. V. Samsonov
  Zh. Prikl. Khim., 40(1):7-10 (1967)

Method for synthesizing titanium chalcogenides
  V. N. Eremenko and V. E. Listovnichii
  Khal'kogenidy (Svoistva Methody Poluch. Primen.), Mater. Seminara, 1st, Kiev, 1965, pp. 66-68 (published 1967)

The Group IV di-transition-metal sulfides and selenides
  H. F. Franzen, J. Smeggil, and B. R. Conrad
  Materials Res. Bull., 2:1087-1092 (1967)
  Ti_2Se, Zr_2Se, Hf_2Se, Ti_2S, Hf_2S

The crystal structure of Ti_2S
  J. P. Owens, B. R. Conrad, and H. F. Franzen
  Acta Cryst., 23:77 (1967)

Optical and electrical properties of some group IV disulfides
  Kyu Chang Park
  Ph.D. thesis, University of Minnesota, 1967
  University Microfilms, Ann. Arbor, Michigan, Order No. 68-1558
  Large single crystals of TiS_2, ZrS_2, HfS_2, and SnS_2 prepared by chemical-vapor transport

Densities, lattice parameters, and defect chemistry of pure non-stoichiometric compounds
  W. van Gool
  J. Materials Sci., 1:261-268 (1966)
  TiS_2, Ti_2S_3

Sur une loi de régularité des propriétés électriques des chalcogénures des métaux de transition
V. Kh. Oganesyan
Izv. Akad. Nauk Arm. SSR, Ser. Tekh. Nauk, 19(1):30–34 (1966)
Sulfides of Ti, Zr, Hf

Crystal structure of dihafnium sulfide
H. F. Franzen and J. Graham
IS-1165 (May 25, 1965)

Preparation and optical properties of group $IVB-VI_2$ chalcogenides having the CdI_2 structure
D. L. Greenaway and R. Nitsche
p. 447, Propriétés thermodynamiques physiques et structurales des dérives semi-métalliques No. 157, Colloq. Intern. Centre National de la Récherche Scientifique, Orsay, 28 Septembre–1 Octobre, 1965, Editions du Centre National de la Récherche Scientifique, Paris, 1967
Single crystals of TiS_2, $TiSe_2$, $TiTe_2$, ZrS_2, $ZrSe_2$, HfS_2, $HfSe_2$ prepared by vapor transport with iodine

Electrical and optical properties of some chalcogenides of elements of the fourth subgroup
H. G. Grimmeiss, A. Rabenau, H. Hahn, and P. Ness
Z. Elektrochem., 65(9):776–783 (1961)
TiS_3, ZrS_3, $ZrSe_3$, HfS_3, TiS, TiSe, TiTe, Ti_3S_4, TiS_2, $TiSe_2$, $TiTe_2$

X-ray study of titanium selenides
Fredrik Gronvold and F. J. Langmyhr
Acta Chem. Scand., 15(10):1949–1962 (1961)

The crystal structure of Ti_5Te_4
F. Gronvold, A. Kjekshus, and F. Raaum
Acta Cryst., 14, Pt. 9, 930–934 (Sept. 10, 1961)

Subchalcogenide phases of titanium
Harry Hahn and Peter Ness
Z. Anorg. Allgem. Chem., 302(1–2):27–36 (Nov. 1959)
Ti_3S_2, Ti_2Se_2, Ti_2Te, $ZrTe_3$, Zr_4Te_3, Zr_3Te_2, ZrTe

Concerning the zirconium-selenium system
Harry Hahn and Peter Ness
Z. Anorg. Allgem. Chem., 302(1–2):37–49 (Nov. 1959)

The zirconium-tellurium system
Harry Hahn and Peter Ness
Z. Anorg. Allgem. Chem., 302(3–4):136–154 (Nov. 1959)

6.c. Group V

On the superconductivity of lamellar structures. Investigation of solid solutions in the $NbSe_2 - NbTe_2$ system
E. A. Antonova, K. V. Kiselva, and S. A. Medvedeev
Zh. Éksper. Teor. Fiz., 59(1):54–58 (1970)
Sov. Phys. – JETP, 32(1):34–36 (1971)

Optical properties and superconductivity of $NbSe_2$
R. Bachmann, H. C. Kirsch, and T. H. Geballe
Solid State Commun., 9:57–60 (1971)

Construction and properties of weak-link detectors using superconducting layer structures
F. Consadori, A. A. Fife, R. F. Frindt, and S. Gygar
Appl. Phys. Letters, 18:233–235 (1971)
$NbSe_2$

Absence of antiferromagnetism in $NbSe_2$ and $TaSe_2$
E. Ehrenfreund, A. C. Gossard, F. R. Gamble, and T. H. Geballe
J. Appl. Phys., 42:1491–1493 (1971)

On the polymorphism of niobium diselenide
F. Kaduk and F. Jellinek
J. Less-Common Metals, 23:437–441 (1971)

Anisotropy of the superconducting properties of niobium diselenide and correlation with crystal structure and composition
E. A. Antonova, S. A. Medvedev, and I. Yu. Shebalin
Zh. Éksper. Teor. Fiz., 57(2):329–345 (1969)
Sov. Phys. – JETP, 30(2):181–189 (1970)

The crystal structure of Ta_6S
H. F. Franzen and J. G. Smeggil
Acta Cryst., B26:125–129 (1970)

The non-stoichiometric phases $Nb_{1+x}Se_1$ and $Ta_{1+x}Se_2$
R. Huisman, F. Kadijk, and F. Jellinek
J. Less-Common Metals, 21:187–193 (1970)

Electron-microscope study of the substructure of niobium ditelluride single crystals
J. Van Landuyt, G. Remaut, and S. Amelinckx
Mater. Res. Bull., 5:731–742 (1970)

Electron-microscopic study of the structure defects and their interactions in niobium-ditelluride
J. Van Landuyt, G. Remaut, and S. Amelinckx
Phys. Stat. Sol., 41:271–289 (1970)

The low-temperature electrical and magnetic properties of $TaSe_2$ and $NbSe_2$
H. N. S. Lee, M. Garcia, H. McKinzie, and A. Wold
J. Solid State Chem., 1:190–194 (1970)

Optical density and thickness of $NbSe_2$ and $Nb_{1.04}Se_2$ lamellae
G. E. Myers and G. L. Montet
Thin Solid Films, 6:R7–R8 (1970)

Optical properties of single crystals of $NbSe_2$ and $Nb_{1.04}Se_2$
G. E. Myers and G. L. Montet
J. Appl. Phys., 41:4642 (1970)

Production of niobium telluride and its properties
K. R. Allakhverdiev, E. A. Antonova, and G. A. Kalyuzhnaya
Izv. Akad. Nauk SSSR, Neorg. Mater., 5(9):1653–1654 (1969)
Inorg. Mater., 5(9):1401–1402 (1969)

Superconductivity in layered structures of the niobium diselenide type
E. A. Antonova, K. V. Kiseleva, and S. A. Medvedev
Fiz. Metallov Metalloved., 27(3):441–445 (1969)

Superconductivity of niobium diselenide
E. A. Antonova, K. V. Kiseleva, G. A. Kalyuzhnaya, and S. A. Medvedev
Fiz.-Khim., Metalloved. Metallofiz. Sverkhprovodnikov, Tr. Vses. Soveshch., 4th 1968 (E. M. Savitskii, ed.), Izd. "Nauka," Moscow, USSR (1969), pp. 23–28

Reflection spectra of niobium diselenide
E. A. Antonova, V. G. Vorob'ev, G. A. Kalyuzhnaya, and V. V. Sobolev
Fiz. i Tekhn. Poluprovod., 3:922–923 (1969)
Sov. Phys. – Semicond., 3:777–778 (1969)

Crystal structure of diniobium monoselenide
B. R. Conrad, L. J. Norrby, and Hugo F. Franzen
Acta Cryst., 25B(9):1729–1736 (1969)

Studies of solid compounds of group IV and V transition metals formed at high temperature. I. The crystal structures of Ti_2S, $Nb_{21}S_8$, and Nb_2Se. II. The metal-rich region of the zirconium-sulfur system
Bruce Randolph Conrad
(Ames Lab., Iowa), IS-T-307 (Aug. 1969), 291 pp.

The crystal structure of Ta_2S
H. F. Franzen and J. G. Smeggil
IS-1881, July 18, 1968
Acta Cryst., 25:1736 (1969)

Two new subsulfides of tantalum (preparation and structure)
H. F. Franzen and J. G. Smeggil
J. Am. Chem. Soc., 91:2814–2815 (1969)

Crystal chemistry of transition metal-chalcogen systems
H. F. Franzen, J. G. Smeggil, B. R. Conard, L. J. Norrby, and D. Strachan
Annual Summary Research Report Ceramic and Mechanical Engineering, Chemical Engineering, Chemistry, Mathematics, and Computer Science, Metallurgy, Physics, and Reactor Divisions, July 1, 1968–June 30, 1969, Ames Laboratory, Ames, Iowa, Contract W-7405-eng-82, IS-2100 (July 1969)
Ta_2S, Ta_6S, and Nb_2Se

X-ray and magnetic study of vanadium sulfides in the range V_5S_4 to V_5S_8
F. Gronvold, H. Haraldsen, B. Pedersen, and T. Rufte
Rev. Chim. Miner., 6(1):215–240 (1969)

The system $Nb-S$
D. Hodouin
Compt. Rend., 269C:1943–1944 (1969)

On the polymorphism of tantalum diselenide
R. Huisman and F. Jellinek
J. Less-Common Metals, 17:111–117 (1969)

The system niobium — sulfur
F. Kadijk and F. Jellinek
J. Less-Common Metals, 19:421–430 (1969)

The low-temperature transport properties of $NbSe_2$
H. N. S. Lee, H. McKinzie, D. S. Tannhauser, and A. Wold
J. Appl. Phys., 40:602 (1969)

Energy band model of Nb- and Ta-dichalcogenide superconductors
M. H. Van Maaren and H. B. Harland
Phys. Letters, 29A:571–573 (1969)

Étude structurale de V_5Te_8
Simone Brunie and Maurice Chevreton
Bull. Soc. Fr. Mineral. Crist., 91:422–427 (1968)

A phase analysis of vanadium monosulfide
Thomas John Burger
IS-T-239 (May 1968)

Vanadium chalcogenides and the nickel-arsenide structure
F. M. A. Carpay
Philips Res. Repts., Suppl. No. 10 (1968), 99 pp.

The crystal structure of Nb_2Se
B. R. Conrad and L. J. Norrby
IS-1872 (July 2, 1968)

A new subsulfide of tantalum
H. F. Franzen
IS-1846, 1968

Second-order phase transition in the VS one-phase region
H. F. Franzen and T. J. Burger
J. Chem. Phys., 49:2268–2272 (1968)

The crystal structure of $Nb_{21}S_8$
H. F. Franzen, T. A. Beineke, and B. R. Conrad
IS-1521, Feb. 10, 1967
Acta Cryst., B24:412 (1968)

The crystal structures of niobium (III) selenide and tantalum (III) selenide
F. Kadijk, R. Huisman, and F. Jellinek
Acta Cryst., B24: 1102–1106 (1968)

Niobium physicochemical properties of its compounds and alloys. III. Crystal structures and densities
Hans Nowotny and Karl J. Seifert
At. Energy Rev., Spec. Issue, No. 2, pp. 71–172 (1968)
NbSe

The production and the physicochemical properties of niobium sulfide
V. Kh. Oganesyan, V. F. Bukhanevich, and S. V. Radzikovskaya
Armyanskii Khim. Zh., 19:161–166 (1966)
(Wright-Patterson AFB, Ohio), FTD-HT-23-330-68, AD-685 142 (July 19, 1968), 10 pp.

Anisotropic superconductor
Eugene Revolinsky, Donald, J. Beerntsen, and Glen A. Spiering
(Allis-Chalmers Manufg. Co.), U.S. Patent 3,406,362, Oct. 15, 1968
$NbSe_2$

Preparation, structure and properties of Nb_3S_4
A. F. J. Ruysink, F. Kadijk, A. J. Wagner, and others
Acta Cryst., B24:1614–1619 (1968)

On the structural properties of the Ta_1Se_2 phase
Einar Bjerkelund and Arne Kjekshus
Acta Chem. Scand., 21:513–526 (1967)

Chemistry of niobium and tantalum
F. Fairbrother
Topics in inorganic and general chemistry, Monograph 10, American Elsevier Publishing Co., New York (1967)
Includes sulfur, selenium, and tellurium systems

Some new niobium and tantalum sulfides and selenides
R. Huisman, F. Kadijk, and F. Jellinek
J. Less-Common Metals, 12:423–424 (1967)
$Nb_{1.05}Se_2 - Nb_{1.10}Se_2$, $Ta_{1.10}Se_2$

Some new superconducting group VA dichalcogenides
M. H. van Maaren and G. M. Schaeffer
Phys. Letters, 24A:645 (1967)
$NbTe_2$, TaS_2, $TaSe_2$

The crystal structures of $NbTe_2$ and $TaTe_2$
Bruce E. Brown
Acta Cryst., 20:264 (1966)

Preparation and properties of tantalum disulfide
  V. F. Bukhanevich and S. V. Radzikovskaya
  FTD-HT-23-1120-67; AD-677229
  Ukr. Khim. Zh., 32(9):926-929 (1966)

Superconductivity in group VA dichalcogenides
  M. H. van Maaren and G. M. Schaeffer
  Phys. Letters, 20:131 (1966)
  $NbSe_2$, NbS_2

Preparation and physicochemical properties of Nb sulfide
  V. Kh. Oganesyan, V. F. Bukhanevich, and S. V. Radzikovskaya
  Arm. Khim. Zh., 19(3):161-166 (1966)

The magnetic susceptibility of tantalum diselenide
  Rod. K. Quinn, Robert Simmons, and John J. Banewicz
  J. Phys. Chem., 70:230 (1966)

Intermediate phases in the systems niobium — selenium, niobium — tellurium, tantalum — selenium, and tantalum — tellurium
  K. Selte, E. Bjerkelund, and A. Kjekshus
  J. Less-Common Metals, 11:14-30 (1966)

Critical currents in superconducting single-crystal niobium diselenide
  G. A. Spiering, E. Revolinsky, and D. J. Beerntsen
  J. Phys. Chem. Solids, 27:535-541 (1966)

Superconductive anomaly in specific heats of some niobium and vanadium compounds
  K. Ukei and E. Kanda
  Ann. Acad. Sci. Fennicae A VI (Finland), No. 210, pp. 104-107 (1966)
  Proc. Low Temp. Calorimetry Conf., Otaniemi (1966)
  NbS_2

On the crystal structure of $TaSe_3$
  Einar Bjerkelund and Arne Kjekshus
  Acta Chem. Scand., 19:701-710 (1965)

Layer structure polytypism among niobium and tantalum selenides
  B. E. Brown and D. J. Beerntsen
  Acta Cryst., 18:31-36 (1965)

Study of niobium and tantalum sulfides
  V. A. Obolonchik, S. V. Rodzikovskaya, and V. F. Bukhanevich
  Porosh. Met., Akad. Nauk Ukr. SSSR, 11:9-14 (1965)

Superconductivity in the niobium-selenium system
  E. Revolinsky, G. A. Spiering, and D. J. Beerntsen
  J. Phys. Chem. Solids, 26:1029 (1965)

The selenide and telluride systems of niobium and tantalum
  E. Revolinsky, B. E. Brown, D. J. Beerntsen, and C. H. Armitage
  J. Less-Common Metals, 8:63-72 (1965)

Report on the chemistry of the elements niobium and tantalum. XLIII. Compounds of niobium with P, As, Sb, S, Se, Te. Synthesis and chemical transport
  Harald Schaefer and Werner Fuhr
  J. Less-Common Metals, 8:375-387 (June 1965)

Magnetic properties of niobium selenides and tellurides
  Karl Selte and Arne Kjekshus
  Acta Chem. Scand., 19(1):258-260 (1965)
  Nb_5Se_4, Nb_3Se_4, $Nb_{1+x}Se_2$, $NbSe_4$, Nb_5Te_4, Nb_3Te_4, $NbTe_2$, and $NbTe_4$

On the crystal structure of $TaTe_4$
  E. Bjerkelund and A. Kjekshus
  J. Less-Common Metals, 7:231-234 (1964)

On the vanadium selenides
  E. Rost and L. Gjertsen
  Z. Anorg. Allgem. Chem., 328:299-308 (1964)
  Structure, antiferromagnetic properties

Über die Vanadintelluride und ihre magnetischen Eigenschaften
  Erling Rost, Liv Gjertsen, and Haakon Haraldsen
  Z. Anorg. Allgem. Chem., 333:301 (1964)
  V_5Te_4

The crystal structures of Nb_3Se_4 and Nb_3Te_4
  Kari Selte and Arne Kjekshus
  Acta Cryst., 17:1568 (1964)

On the crystal structure of $NbTe_4$
  Kari Selte and Arne Kjekshus
  Acta Chem. Scand., 18:690-696 (1964)

Tantalum diselenide and triselenide
  L. A. Aslanov, Yu. M. Ukrainskiy, and Yu. P. Simanov
  Zh. Neorg. Khim., 8(8):1801-1805 (1963)

Magnetic susceptibility of vanadium sulphides at elevated temperatures
  G. M. Loginov
  Russian J. Inorg. Chem., 6(2):133-137 (1961)

Lower vanadium sulphides ($VS-V_2S_3$)
  G. M. Loginov
  Russian J. Inorg. Chem., 5(1):105 (1960)

Investigation of the system vanadium — tellurium
  Yu. M. Ukrainskiy, A. V. Novoselova, and Yu. P. Simanov
  Nauchnye Dokl. Vysshei Shkoly, Khimiia i Khimicheskaia Tekhnologiia, No. 1, pp. 62-66 (1959)
  Metallic character of the bond in the vanadium tellurides may be presumed

Vanadium tellurides
  Fredrik Gronvold, Olaf Hagberg, and Haakon Haraldsen
  Acta Chem. Scand., 12:971-982 (1958)

The chemistry of compounds of variable composition. IV. The tantalum — selenium system
  S. M. Ariya, A. I. Zaslavsky, and I. I. Matveeva
  Zh. Obshch. Khim., 9:2373-2375 (1956)
  Journal of Gen. Chem., 9:2651-2654 (1956)

6.d. Group VI

Spin structure and magnetic anisotropy of Cr_5S_6 and rhombohedral Cr_2S_3
  T. J. A. Popma, C. Haas, and B. Van Laar
  J. Phys. Chem. Solids, 32:581-590 (1971)

The magnetic structure of Cr_2Te_3, Cr_3Te_4, and Cr_5Te_6
  A. F. Andresen
  Acta Chem. Scand., 24:3495-3509 (1970)

X-ray crystallographic study of Mo_3Se_4 and Mo_3Te_4
O. Bars, D. Grandjean, A. Meerschant, and M. Spiesser
Bull. Soc. Fr. Mineral. Crist., 93(4):498-499 (1970)

A semi-empirical tight-binding calculation of the band structure of MoS_2
R. A. Bromley
Phys. Letters, 33A:242-243 (1970)

Examination of molybdenum disulfide with LEED and Auger emission spectroscopy
D. H. Buckley
Lewis Research Center, NASA, Cleveland, Ohio, NASA-TN-D-7010; E-5855 (Dec. 1970)

Magneto-optic properties of CrTe films prepared by sequential evaporation
R. L. Comstock and P. H. Lissberger
J. Appl. Phys., 41:1397 (1970)

Study of the spin-reordering transition in Cr_5S_6
K. Dwight, N. Menyuk, and J. A. Kafalas
Phys. Rev., B2(9):3630-3638 (1970)

Derivation of optical constants from transmission measurements alone — applied to $MoSe_2$
R. A. Hazelwood
Thin Solid Films, 6:329-341 (1970)

Thermal conductivity of binary and ternary selenides of transition elements
V. A. Ivanova, D. Sh. Abdinov, and G. M. Aliev
Izv. Akad. Nauk SSSR, Neorg. Mater., 6(3):566-568 (1970)
Beta-CrSe, Cr_7Se_8, Cr_2Se_3, $NiCr_2Se_4$, VCr_2Se_4, and $FeCr_2Se_4$

Magnetic properties of Cr_5S_6
N. Menyuk, K. Dwight, and J. A. Kafalas
(Lincoln Lab., Masschusetts Institute of Technology, Lexington, Mass.), pp. 38-40, Solid State Research Report 4; ESD-TR-69-336 (1969 ; issued Jan. 1970)

A fluorine bomb calorimetric study of molybdenum disulfide. The standard enthalpies of formation of the di- and sesquisulfides of molybdenum
P. A. G. O'Hare, Edward Benn, F. Yu. Chemg, and George Kuzmycz
J. Chem. Thermodynamics, 2:797-804 (1970)

Determination of rhenium and other trace elements in molybdenites by means of intrumental neutron activation analysis
Z. Randa, J. Benada, J. Kuncir, and M. Vobecky
Radiochem. Radioanal. Letters, 3/4:227-237 (1970)

Lattice mode degeneracy in MoS_2 and other layer compounds
J. L. Verble and T. J. Wieting
Phys. Rev. Letters, 25:362-365 (1970)

Preparation, structure, and properties of chromium selenides. Crystal growth with selenium vapor as a novel transport agent
F. H. Wehmeier, E. T. Keve, and S. C. Abrahams
Inorg. Chem., 9:2125-2131 (1970)

Excitons and photoconductivity in transition-metal dichalcogenides
T. J. Wieting and A. D. Yoffe
Phys. Stat. Sol., 37:353-366 (1970)
MoS_2, $MoSe_2$, and $MoTe_2$

Anomalous dependence of the spin-echo intensity on the external magnetic field in the ferromagnetic Cr_7Te_8
Masuhiro Yamaguchi and Takasu Hashimoto
J. Phys. Soc. Japan, 29:238 (1970)

Effect of pressure on the Curie point of the pseudo-NiAs-type Cr_3Te_4
K. Ozawa, S. Yanagisawa, T. Yoshimi, M. Ogawa, and S. Anzai
Phys. Stat. Sol., 38:385-391 (1970)

Electron energy losses and optical anisotropy of MoS_2 single crystals
K. Zeppenfeld
Optics Communications, 1:377-378 (1970)

The thermal stability and friction of the disulfides, diselenides, and ditellurides of molybdenum and tungsten in vacuum (10^{-9} to 10^{-6} torr)
William A. Brainard
(NASA, Lewis Research Center), NASA-TN-D-5141 (April 1969)

A new mineral: brezinaite, Cr_3S_4 and the Tucson meteorite
T. E. Bunch and L. H. Fuchs
Am. Mineral., 54:1509-1518 (1969)

Effects of pressure and temperature on exciton absorption and band structure of layer crystals: molybdenum disulfide
G. A. N. Connell, J. A. Wilson, and A. D. Yoffe
J. Phys. Chem. Solids, 30:287-296 (1969)

Low-energy electron observation of graphite and molybdenite crystals. Application to the study of graphite oxidation
Gerard David
(Commissariat a l'Energie Atomique, Centre d'Études Nucléaires, Saclay, France), CEA-R-3841 (Dec. 1969), 57 pp.

Spontaneous magnetostriction of chromium telluride
V. A. Gordienko, V. V. Zubenko, and V. I. Nikolaev
Zh. Éksper. i Teor. Fiz. 57(5):1597-1600 (1969)
Sov. Phys. – JETP, 30(5):864-865 (1970)

Magnetic properties of Cr_7Te_8
Takasu Hashimoto and Masuhiro Yamaguchi
J. Phys. Soc. Japan, 27:1121-1126 (1969)

Photovoltage in single crystals of α-$MoTe_2$
G. P. Kekelidze and B. L. Evans
British J. Appl. Phys., 2:855-861 (1969)

On the thermal decomposition rate of molybdenum sesquisulphide Mo_2S_3 in vacuum
Yoichi Maru, Hiroshi Yoshida, and Yoshio Kondo
Trans. Japan Inst. Metals, 10:8-11 (1969)

Transition-metal chalcogen compounds. Formation and decomposition of transition-metal chalcogen anions in aqueous solution
A. Mueller, O. Glemser, E. Diemann, and H. Hofmeister
Z. Anorg. Allgem. Chem., 371(1-2):74-80 (1969)
Mo and W

The pressure dependence of the lattice parameters of CrTe and CrSb
H. Nagasaki, I. Wakabayashi, and S. Minomura
J. Phys. Chem. Solids, 30:2405-2408 (1969)

Structural and magnetic phase transitions of chromium sulfides $Cr_{1-x}S$ with $0 \leq x \leq 0.12$
  T. J. A. Popma and C. F. van Bruggen
  J. Inorg. Nucl. Chem., 31:73–80 (1969)

Electron-microscope observations in WSe_2
  M. C. Richman
  Metallography, 1:227–232 (1969)

Influence of the magnetic transition of the heat capacity behavior of chromium telluride
  S. G. Sankar, V. G. Gunjikar, and A. B. Biswas
  (Indian Institute Tech., Bombay, India), Proceedings Nuclear Physics Solid State Physics, 13th Symposium (1968), pp. 179–182 (publ. 1969)

Chromium chalcogenides prepared at high pressure and the crystal growth of chromium sesquisulfide
  Arthur W. Sleight and Tom A. Bither
  Inorg. Chem., 8:566–569 (1969)

The $Mo^{VI}-Se^{IV}$ system
  P. Souchay, M. Cadiot, and C. Volfovskii
  Compt. Rend., C269(15):826–829 (1969)

Characterization and physicochemical study of nonstoichiometric molybdenum selenides
  M. Spiesser, Jean Rouxe, Kerriou, and Guy Goureaux
  Bull. Soc. Chim., 5:1427–1431 (1969)

Electrical properties of molybdenite
  S. R. Guha Thakurta
  Indian J. Phys., 43:169–172 (1969)
  MoS_2

Structural transformations in synthetic MoS_2
  A. N. Zelikman, G. V. Indenbaum, M. V. Teslitskaya, and V. P. Shalankova
  Kristallografiya, 14(5):795–799 (1969)
  Sov. Phys. – Cryst., 14(5):687 (1970)

Identification and some properties of point defects and non-basal dislocations in molybdenite surfaces
  O. P. Bahl, E. L. Evans, and J. M. Thomas
  Proc. Roy. Soc. London, Ser. A, 306-53–65 (1968)

Impact synthesis of divalent chromium chalcogenides, 2
  S. S. Batsanov and E. S. Zolotova
  Dokl. Akad. Nauk SSSR, 180(1):93–94 (1968)
  Dokl. – Chem., 180(1):383–384 (1968)

Etude de Cr_2S_3 rhomboédrique par diffraction neutronique et mesures magnétiques
  E. F. Bertaut, J. Cohen, B. Lambert-Andron, and P. Mollard
  J. Phys., 29:813–824 (1968)

Etude structurale de nouveaux séléniures de chrome Cr_2Se_3
  Maurice Chevreton and Bernard Dumont
  Compt. Rend., 267:884–887 (1968)

The optical absorption properties of synthetic MoS_2
  A. Clark and R. H. Williams
  Brit. J. Appl. Phys., 1:1222 (1968)

Gold-decoration of topographical features at surfaces of synthetic molybdenum disulphide
  E. L. Evans, O. P. Bahl, and J. M. Thomas
  Trans. Faraday Soc., 64:3354–3357 (Dec. 1968)

The photovoltage in thin crystals of MoS_2
  B. L. Evans and K. T. Thompson
  Brit. J. Appl. Phys., 1:1619–1623 (1968)

An experimental study on the anomalous transmission of electrons through crystals. Measurements with molybdenite films at 200 and 500 kV
  Ayahiko Ichimiya
  Japan. J. Appl. Phys., 7:1425 (1968)
  MoS_2

Magnetic and electric properties of lower chalcogenides of molybdenum
  A. A. Opalovskii and V. E. Fedorov
  Izv. Akad. Nauk SSSR, Neorg. Mater., 4(2):293–294 (1968)
  Mo_3Se_4, Mo_3Te_4, MoS_2, and $MoTe_2$

Synthesis and structure of nonstoichiometric Mo/Se phases
  M. Spiesser, C. Marchal, and J. Rouxel
  Compt. Rend., 266(22):1583–1586 (1968)

Lattice constants, thermal expansion coefficients, and densities of molybdenum and the solubility of sulfur, selenium, and tellurium in it at 1100°
  Martin E. Straumanis and Ramesh P. Shodhan
  Z. Metallk., 59(6):492–495 (1968)

Intensity anomaly in Kikuchi lines of molybdenite
  Hidewo Takahashi
  Japan. J. Appl. Phys., 7:1310 (1968)
  MoS_2

Semiconducting properties of single crystals of n- and p-type tungsten diselenide (WSe_2)
  L. C. Upadhyayula, J. J. Loferski, A. Wold, W. Giriat, and R. Kershaw
  J. Appl. Phys., 39:4736–4740 (1968)

The observation of thick specimens by high-voltage electron microscopy. II. Experiment with molybdenite films at 50–1200 kV
  Ryozi Uyeda and Minoru Nonoyama
  Japan. J. Appl. Phys., 7:200 (1968)

Lattice parameter measurements on molybdenum disulphide
  P. A. Young
  Brit. J. Appl. Phys. (J. Phys. D), Ser. 2, Vol. 1, pp. 936–938 (1968)

Structure and properties of the chromium sulfides
  C. F. van Bruggen and F. Jellinek
  Propriétés thermodynamiques physiques et structurales des dérives semimétalliques, No. 157, Colloq. Intern. Centre National de la Récherche Scientifique, Orsay, 28 Septembre-1 Octobre 1965, Editions du Centre National de la Récherche Scientifique, Paris (1967), p. 31

Structure et conductibilité électrique des tellurures de chrome
  M. Chevreton, M. Murat, and E. F. Bertaut

Propriétés thermodynamiques physiques et structurales des dérives semi-métalliques, No. 157, Colloq. Intern. Centre National de la Récherche Scientifique, Orsay, 28 Septembre-1 Octobre, 1965, Editions du Centre National de la Recherche Scientifique, Paris (1967), p. 49

Synthesis and x-ray phase study of chromium sulfoselenide and telluroselenide
I. M. Doronina, V. S. Filatkina, and S. S. Batsanov
Izv. Akad. Nauk SSSR, Neorg. Mater., 3(9):1696-1697 (1967)

Exciton spectra in thin crystals (MoS_2)
B. L. Evans and P. A. Young
Proc. Roy. Soc. London, Ser. A, 298:74-97 (1967)

Electrical properties of single crystals of tungstenite (WS_2)
S. R. Guhathakurta
Indian J. Phys., 41:99-107 (1967)

Effect of pressure on the Curie temperature of CrTe and MnSb compounds of the nickel arsenide type
Hideaki Ido, Takejiro Kaneko, and Kazuo Kamigaki
J. Phys. Soc. Japan, 22:1418 (1967)

On some characteristics of chromium selenides
V. A. Ivanova, D. Sh. Abdinov, and G. M. Aliev
Phys. Stat. Sol., 24:K145 (1967)

Electrical properties of the chromium selenides
V. A. Ivanova and G. M. Aliev
Izv. Akad. Nauk SSSR, Neorg. Mater., 3(8):1490-1491 (Aug. 1967)

Magnetic structure of trigonal Cr_2S_3
B. van Laar
Phys. Letters, 25A:27-29 (1967)

Ferrimagnetic and antiferromagnetic structures of Cr_5S_6
B. van Laar
Phys. Rev., 156:654 (1967)

Reflection spectra of molybdenum disulphide crystals
W. U. Liang
Phys. Letters, 24A:573 (1967)

The detection of vacancies in molybdenite
G. L. Montet
Appl. Physics. Letters, 11:223 (1967)

New methods for synthesizing simple and mixed molybdenum chalcogenides
A. A. Opalovskii and V. E. Fedorov
Khal'kogenidy (Svoistva Metody Poluch. Primen.), Mater. Seminara, 1st, Kiev 1965 (published 1967), pp. 79-85

Temperature variation of magnetic susceptibility of molybdenite single crystal
D. Paul
Indian J. Phys., 41(12):943-944 (1967)

Tungsten and Its Compounds
G. D. Rieck
Pergamon Press, New York (1967)

Transport phenomena of $Cr_{1-x}Te_{1+x}$ alloys in paramagnetic state
Masaaki Sekinobu and Minoru Nogami
Japan. J. Appl. Phys., 6:1405 (1967)

Méthode de mesure de l'effet Hall ordinaire des cristaux magnétiques et application au système chrome-tellure
J. Serre and J. P. Suchet
Compt. Rend., 264:1412-1415 (1967)

Non-stoichiometric Mo telluride phases
M. Spiesser and J. Rouxel
Compt. Rend., 265:92-95 (1967)

Certain thermistor characteristics of single crystals of tungstenite (WS_2)
S. R. G. Thakurta
Indian J. Phys., 41:618-621 (1967)

Thermoelectric power of tungstenite (WS_2) single crystals
S. R. G. Thakurta
Indian J. Phys., 41:940-942 (1967)

Relaxation of the spin of iron in the ferromagnetic compounds, manganese antimonide, and chromium telluride, below the Curie point
S. S. Yakimov, V. Ya. Gamlitskii, V. I. Nikolaev, and S. R. Rodin
Dokl. Akad. Nauk SSSR, 177(6):1313-1315 (1967)
Dokl. – Sov. Phys., 12(12):1153-1154 (1968)

The crystal structures of WTe_2 and high-temperature $MoTe_2$
Bruce E. Brown
Acta Cryst., 20:268 (1966)

Transport phenomena in layered structures
R. Fivaz
Helv. Phys. Acta, 39(3):247-261 (1966)
MoS_2, $MoSe_2$, and WSe_2

Single crystals of MoS_2 several molecular layers thick
R. F. Frindt
J. Appl. Phys., 37:1928-1929 (1966)

Phase transitions in transition-metal chalcogenides
C. Haas
Solid State Commun., 4:419-421 (1966)

Pressure effect on magnetic transitions in $CrS_{1.17}$
Takahiko Kamigaichi, Tetsuhiko Okamoto, Nobuo Iwata, and Eiji Tatsumoto
J. Phys. Soc. Japan, 21:2730 (1966)

Hall effect in chromium telluride
Minoru Nogami
Japan. J. Appl. Phys., 5:134 (1966)

Molybdenum chalcogenides
A. A. Opalovskii and V. E. Fedorova
Uspekhi Khimii, 35(3):427-459 (March 1966)

Preparation and physicochemical properties of Cr_2S_3
S. V. Radzikovskaya and V. Kh. Oganesyan
Armyansk. Khim. Zh., 19(11):844-848 (1966)

Electrical properties of alpha- and beta-$MoTe_2$ as affected by stoichiometry and preparation temperature
E. Revolinsky and D. J. Beerntsen
J. Phys. Chem. Solids, 27:523-526 (1966)

Rhombohedral molybdenum ditelluride
　Meyer S. Silverman
　(Pennsalt Chemicals Corp.), U. S. Patent 3,385,667, May 28, 1968, Appl. Feb. 11, 1966

Ultra-high pressure synthesis of rhombohedral MoS_2 and WS_2
　M. S. Silverman
　AD 655307 (1966)

Molybdenum diselenide: rhombohedral high pressure – high temperature polymorph
　L. C. Towle, R. E. Stajdohar, V. Oberbeck, and B. E. Brown
　Science, 154:895–896 (1966)

Optical absorption and dispersion in molybdenum disulphide
　B. L. Evans and P. A. Young
　Proc. Roy. Soc., A254:402–422 (1965)

Optical absorption of a few unit-cell layers of MoS_2
　R. F. Frindt
　Phys. Rev., 140:A536–539 (1965)

Propriétés semiconductrices du ditellurure de molybdène
　A. Lepetit
　J. Phys. (Fr.), 26:175 (1965)

Molybdenum diselenide
　M. S. Silverman
　(Pennsalt Chemicals Corp.), U. S. Patent 3,472,621 (Oct. 14, 1969, Appl. July 9, 1965), 3 pp.

Molybdenum diselenide having a rhombohedral crystal structure and method of preparing same
　M. S. Silverman
　(Pennsalt Chemicals Corp.) U. S. Patent 347221 (July 9, 1965)

Heat capacities of Cr_5Te_6, Cr_3Te_4, and Cr_2Te_3 from 5 to 350°K
　F. Gronvold and E. F. Westrum, Jr.
　Z. Anorg. Allgem. Chem., 328:272–282 (1964)

Semiconducting behavior of substituted tungsten diselenide and its analogues
　W. T. Hicks
　J. Electrochem. Soc., 111:1058 (1964)

Sulfides of tungsten
　E. Ya. Rode and B. A. Lebedev
　Zh. Neorgan. Khim., 9(9):2068–2075 (1964)

Conductibilité électrique des cristaux à structure de nickeline
　Jacques Suchet, Daniel Calecki, and Kieu Van Con
　Compt. Rend., 258:4486–4489 (1964)
　CrSe, CrTe, MnTe

Detailed crystal structure of rhombohedral MoS_2 and systematic deduction of possible polytypes of molybdenite
　Y. Takeuchi and W. Nowacki
　Schweiz. Mineralogische und Petrographische Mitteilungen, 44:105 (1964)

Origin of ferrimagnetism in compounds Cr_5S_6 and Cr_2S_3
　Motoyoshi Yuzuri and Yoji Nakamura
　J. Phys. Soc. Japan, 19(8):1350–1354 (1964)

A neutron-diffraction investigation of Cr_2Te_3 and Cr_5Te_6
　Arne F. Andresen
　Acta Chem. Scand., 17:1335–1342 (1963)

Structure and electric conductivity of the ordered vacancy compounds of the chromium-selenium system
　M. Chevreton, M. Murat, and C. Eyraund
　J. Phys. (Fr.), 24:443–446 (1963)

The optical properties of single crystals of WSe_2 and $MoTe_2$
　R. F. Frindt
　J. Phys. Chem. Solids, 24:1107–1112 (1963)

Preparation and properties of the single crystalline AB_2-type selenides and tellurides of niobium, tantalum, molybdenum, and tungsten
　L. H. Brixner
　J. Inorg. Nucl. Chem., 24:257 (1962)
　WSe_2

Magnetic properties of Cr_5S_6 in chromium sulfides
　K. Dwight, R. W. Germann, N. Menyuk, and A. Wold
　J. Appl. Phys., 33, Suppl., 1341 (1962)

Anomalies in electrical conductivity and magnectic susceptibility of chromium selenides
　Kan-ichi Masumoto, Tadamiki Hihara, and Takahiko Kamigaichi
　J. Phys. Soc. Japan, 17:1209–1210 (1962)

Magnetic structure of chromium selenide
　L. M. Corliss, N. Elliott, J. M. Hastings, and R. L. Sass
　Phys. Rev., 122:1402–1406 (1961)

Chalcogenides of the transition elements. III. Molybdenum ditelluride
　Osvald Knop and Roderick D. MacDonald
　Can. J. Chem., 39:897–904 (1961)

On crystal structure of rhombohedral MoS_2
　S. A. Semiletov
　Kristallografiya, 6(4):536–540 (1961)

Diselenides of molybdenum and rhenium
　Yu. M. Ukrainskiy and A. V. Novoselova
　Dokl. Akad. Nauk SSSR, 139(5):1136–1137 (1961)
　Dokl. – Chem., 11(5):828–829 (1961)

The problem of the phase composition of the chromium-tellurium system
　L. G. Gaydukov, V. N. Novogrudskiy, and I. G. Fakidov
　Fiz. Metallov Metalloved., 9(1):152–154 (1960)

The magnetic properties of single crystals of chromium selenide
　Ichiro Tsubokawa
　J. Phys. Soc. Japan, 15:2243–2247 (1960)

Melting point and sublimation of molybdenum disulfide
　Peter Cannon
　Nature, 183:1612–1613 (1959)

The supposed molybdenum sesquisulfide
　V. Montoro
　NASA-TT-F-11051, June 1967, translated from Atti Pontif. Acad. delle Sci. (Rome), 9(6):331–337 (1929)

6.e. Group VII

The dichalcogenides of technetium and rhenium
J. C. Wildervanck and F. Jellinek
J. Less-Common Metals, 24:73-81 (1971)
Single crystals by chem. trans.; crystal structures

Synthesis of manganese disulfide
M. Avinor and G. De Pasquali
J. Inorg. Nucl. Chem., 32:3403-3404 (1970)

Synthesis of manganese diselenide
M. Avinor and G. De Pasquali
J. Inorg. Nucl. Chem., 32:3959 (1970)

Covalency parameters in MnSe and MnSe$_2$
A. J. Jacobson and B. E. F. Fender
J. Chem. Phys., 52:4563-4566 (1970)

Electrical properties of manganese sulfide and manganese selenide single crystals
A. G. Rustamov, I. G. Kerimov, L. M. Valiev, and S. Kh. Babaev
Izv. Akad. Nauk SSSR, Neorg. Mater., 6(7):1339-1340 (1970)
Inorg. Mater., 6(7):1176-1177 (1970)

Physicochemical study of manganese telluride
V. G. Vanyarkho, V. P. Zlomanov, and A. V. Novoselova
Izv. Akad. Nauk SSSR, Neorg. Mater., 6(7):1257-1259 (1970)
Inorg. Mater., 6(7):1102-1104 (1970)

[55]Mn-nuclear acoustic resonance in MnTe
K. Walther
Phys. Letters, 32A:201-202 (1970)

Manganese disulfide (hauerite) and manganese ditelluride. Thermal properties from 5 to 350°K and antiferromagnetic transitions
E. F. Westrum, Jr., and F. Gronvold
J. Chem. Phys., 52:3820-3826 (1970)

Developments and problems in applied magnetism
H. Bosma, Th. Holtwijk, and A. R. Miedema
Ned. Tijdschr. Natuurk., 35:52-66 (1969)
MnTe

Optical properties of alpha-MnSe
Donald L. Decker
thesis, Univ. of Calif., Riverside, Calif.
University Microfilms, Order No. 69-19,124, Ann Arbor, Mich.
Diss. Abstr. Int. B 30, No. 5, 2369 (1969)

Interactions magnétoélastiques dans le sulfure de manganèse
Roland Georges
Compt. Rend., Ser. B:268, 16-19 (1969)

Chaleur spécifique à basse température du sulfure et de l'oxyde de manganèse
Roland Georges, Eberhard Gmelin, David Landau, and Jean-Claude Lasjaunias
Compt. Rend., 269B:827-830 (1969)

Total intensities of some crystal field transitions in MnO and MnS related to the antiferromagnetism
Donald R. Huffman
J. Appl. Phys., 40:1334 (1969)

Fine structure in the absorption spectrum of alpha-MnS
Hiroo Komura
J. Phys. Soc. Japan, 26:1446-1451 (1969)

Preparation and properties of Ga$_2$Te$_3$, Ga$_2$Te, ReTe$_2$, and Re$_2$Te
E. Montiginie
Z. Anorg. Allg. Chem., 366:111-112 (1969)

The pressure dependence of the lattice parameters of MnSb and MnTe
Hiroshi Nagasaki, Ippei Wakabayashi, and Shigeru Minomura
J. Phys. Chem. Solids, 30:329-337 (1969)

Low temperature heat capacity measurements on the transition metal oxides RuO$_2$ and IrO$_2$ and on the insulating magnetic materials MnS, EuS and EuTe
Burr Charles Passenheim
Dissertation, Univ. of California, Riverside, 1969, 156 pp.
Available from University Microfilms, Inc., Ann Arbor, Mich., Order No. 69-19128

Magnetization near an iodine impurity in antiferromagnetic MnTe$_2$
M. Pasternak
Phys. Rev., 184:523-527 (1969)

Tellurium-125 Moessbauer effect in paramagnetic and antiferromagnetic manganese telluride
M. Pasternak and A. L. Spijkervet
Phys. Rev., 181(2):574-579 (1969)

Slip behavior and hardness indentations in MnSe and MnSe - MnS solid solutions
Paul G. Riewald and Lawrence H. van Vlack
J. Am. Ceram. Soc., 52:370-375 (1969)

Electrical transport phenomena in MnTe, an antiferromagnetic semiconductor
Jan Douwe Wasscher
Ph.D. thesis, Technical University, Eindhoven, Netherlands, Philips Research Reports Suppl., No. 8 (1969)

Phase studies of the systems Mn - S, Mn - Se, and MnS - MnSe
Heribert Wiedemeier and A. Gary Sigai
High Temp. Sci., 1:18-25 (1969)

Spin-polarized energy band structure of MnO and alpha-MnS
Timothy M. Wilson
Bull. Am. Phys. Soc., 14:431 (1969)

Thermal conductivity near a continuous-phase transition
V. P. Zhuze, O. N. Novruzov, and A. I. Shelykh
Fiz. Tverd. Tela, 11:1287-1296 (1969)
Sov. Phys. - Solid State, 11:1044-1051 (1969)
CoO, NiO, MnTe, BaTiO$_3$, and NaNO$_3$

Electrical conductivity of the antiferromagnetic semiconductor MnTe below its Néel temperature
A. I. Zvyagin, L. V. Povstyani, and R. M. Aref'eva
Fiz. Tverd. Tela, 11(11):3392-3394 (1969)
Sov. Phys. - Solid State, 11(11):2759 (1970)

Manganese-tellurium system
N. Kh. Abrikosov, K. A. Dyul'dina, and V. V. Zhadanova
Izv. Akad. Nauk SSSR, Neorg. Mater., 4:1878-1884 (1968)
Inorg. Mater., 4:1638-1642 (1968)

Synthesis and some chemical properties of rhenium diselenide
M. I. Ermolaev and Yu. A. Gukova
Tr. Voronezh. Tekhnol. Inst., 17(1):208–210 (1968)

Magnetic susceptibility and magnetostriction of MnSe
I. G. Kerimov and T. A. Mamedov
Fiz. Metallov Metalloved., 26:188–191 (July 1968)

Magnetic susceptibility of manganese selenide and manganese sulfide in the magnetic transformation region
I. G. Kerimov, T. A. Mamedov, and N. G. Aliev
Izv. Akad. Nauk Azerb. SSR, Ser. Fiz.-Tekh. Mat. Nauk, No. 4, pp. 3–6 (1968)

Antiferromagnetic transitions in MnS_2 and $MnTe_2$
M. S. Lin and H. Hacker, Jr.
Solid State Commun., 6:687–689 (1968)

Structure and some physical properties of manganese selenide and manganese telluride and their mutual solutions
G. I. Makovetskii
Vestsi Akad. Navuk Belarus. SSR, Ser. Fiz.-Mat. Navuk, No. 5, pp. 91–97 (1968)

Thermal expansion and magnetostriction of manganese chalcogenides
T. A. Mamedov, I. G. Kerimov, and N. G. Aliev
Dokl. Akad. Nauk Azerb. SSR, 24(10):15–20 (1968)

X-ray spectroscopic study of chemical bonding in manganese diselenide and cobalt diselenide
C. Mande and A. S. Nigavekar
Proc. Indian Acad. Sci., 67A(3):168–174 (March 1968)

Electrical conductivity and thermal emf of $Li_xMn_{1-x}Se$
A. G. Rustamov, I. G. Kerimov, and L. M. Valiev
Izv. Akad. Nauk Azerb. SSR, Ser. Fiz.-Tekh. Mat. Nauk, No. 1, pp. 28–32 (1968)

The free energy of formation of ReS_2
Juan Sodi and John F. Elliott
Trans. Met. Soc. AIME, 242:2143–2145 (1968)

Optical properties of alpha-MnS
Donald R. Huffman and Robert L. Wild
Phys. Rev., 156:989–997 (1967)

The growth of gamma-MnS crystals in sodium silicate gels
Abraham Schwartz, Arthur Tauber, and Joel R. Shappirio
Mat. Res. Bull., 2:375–380 (1967)

The electrical resistivity of MnTe at elevated temperatures
A. M. J. H. Seuter
Propriétés thermodynamiques physiques et structurales des dérives semi-métalliques, No. 157, Colloq. Intern. Centre National de la Récherche Scientifique, Orsay (28 Septembre-1 Octobre 1965), Editions du Centre National de la Récherche Scientifique, Paris (1967), p. 459

Antiferromagnetic transformations in some manganese compounds
N. N. Sirota and G. I. Makovetskii
Tr. Mezhdunar. Konf. Fiz. Nizkikh Temp., 10th, No. 4, pp. 176–180 (held in 1966, published 1967), M. P. Malkov, eds. (VINITI: Moscow, USSR)

Electron density and potential distribution in the manganese selenide lattice
N. N. Sirota, N. M. Olekhnovich, and G. I. Makovetskii
Khimicheskaya Svyaz' v Poluprovodnikakh i Termodinamika (Chemical bonds in semiconductors and thermodynamics), 1966, Nauka i Tekhnika, Minsk (Sept. 1967), pp. 59–63

Ultrasonic relaxation at the Néel temperature and nuclear acoustic resonance in MnTe
K. Walther
Solid State Commun., 5:399–403 (1967)

Weak ferromagnetism in the NiAs structure and galvanomagnetic measurements on MnTe
J. D. Wasscher
Propriétés thermodynamiques physiques et structurales des dérives semi-métalliques, No. 157, Colloq. Intern. Centre National de la Récherche Scientifique, Orsay (28 Septembre-1 Octobre 1965), Editions du Centre National de la Récherche Scientifique, Paris (1967), p. 465

Optical measurements on the antiferromagnetic semiconductor MnTe
G. Zanmarchi
J. Phys. Chem. Solids, 28:2123–2130 (1967)

Specific heat of MnS through the Néel temperature
Donald R. Huffman and Robert L. Wild
Phys. Rev., 148:526–527 (1966)

Optical properties of alpha-MnS in the fundamental absorption region
D. R. Huffman, R. L. Wild, and J. Callaway
J. Phys. Soc. Japan, 21, Suppl., 623 (1966)
Proc. Intern. Conf. Physics Semiconductors, Kyoto (1966)

Specific heat of MnS at liquid-helium temperatures
D. C. McCollum
Bull. Am. Phys. Soc., 11:35(A) (1966)

Effect of pressure on the magnetic transition point of manganese telluride
K. Ozawa, S. Anzai, and Y. Hamaguchi
Phys. Letters, 20:132 (1966)

Effect of pressure on the Néel temperature of $MnTe_2$
Akira Sawaoka, Syohei Miyahara, and Shigeru Minomura
J. Phys. Soc. Japan, 21:1017–1018 (1966)

Neutron-diffraction study of the antiferromagnetic transformation in manganese telluride
N. N. Sirota and G. I. Makovetskii
Dokl. Akad. Nauk SSSR, 170(6):1300–1302 (1966)
Sov. Phys. – Dokl., 11(6):888–890 (1967)

On the properties of alpha-MnS and MnS_2
S. Furuseth and A. Kjekshus
Acta Chem. Scand., 19:1405–1410 (1965)

Evidence of weak ferromagnetism in MnTe from galvanomagnetic measurements
J. D. Wasscher
Solid State Commun., 3:169–171 (1965)

Neutron-diffraction study on manganese telluride
Nobuhiko Kunitomi, Yoshikazu Hamaguchi, and Shuichiro Anzai
J. Phys., 25:568–574 (1964)

Spin-disorder scattering, anomalous behavior of the Hall coefficient and magnon-drag in anti-ferromagnetic $\overline{Mn}Te$
J. D. Wasscher, A. M. J. H. Seuter, and C. Haas
Proc. 7th Intern. Conf. on Physics of Semiconductors, Paris (1964), p. 1269

Magnetic properties of manganese telluride single crystals
Takemi Komatsubara, Miyuki Murakami, and Eiji Harahara
J. Phys. Soc. Japan, 18:356 (1963)

The heats of combustion of ReS_2 and Re_2S_7 and the thermodynamic functions for transition metal sulfides
J. E. McDonald and J. W. Cobble
J. Phys. Chem., 66:791 (1962)
ΔH_1 for ReS_2

High-temperature magnetic susceptibilities of MnO, MnSe, and MnTe
John J. Banewicz, Robert F. Heidelberg, and Allan H. Luxem
J. Phys. Chem., 65:615-617 (1961)

Diselenides of molybdenum and rhenium
Yu. M. Ukrainskiy and A. V. Novoselova
Dokl. Akad. Nauk SSSR, 139(5):1136-1137 (1961)
Dokl. – Chem., 139(5):828-829 (1961)

Antiferromagnetic structures of MnS_2, $MnSe_2$, and $MnTe_2$
J. M. Hastings, N. Elliott, and L. M. Corliss
Phys. Rev., 115:13-17 (1959)

The stoichiometry of MnTe
H. A. Johansen
J. Inorg. Nucl. Chem., 6:344-345 (1958)

Magnetic structures of the polymorphic forms of manganous sulfide
Lester Corliss, Norman Elliott, and Julius Hastings
Phys. Rev., 104:924-928 (1956)

Magnetic and electric properties of manganese telluride
Enji Uchida, Hisamoto Kondoh, and Nobuo Fukuoka
J. Phys. Soc. Japan, 11:27-32 (1956)

Electrical resistivity of antiferromagnet $MnTe_2$
Akira Sawaoka and Shohei Miyahara
J. Phys. Soc. Japan, 20:2087 (1955)

6.f. Group VIII

Anomalous exchangestriction in ferromagnetic pyrite and chromium chalcogenide spinel compounds
W. Bindloss
J. Appl. Phys., 42:1474-1475 (1971)

Conceptual phase diagram and its application to the spontaneous magnetism of several pyrites
J. B. Goodenough
J. Solid State Chem., 3:26-38 (1971)
FeS_2, CoS_2, NiS_2, and ternaries

Epitaxial growth and phase change in monoselenide of cobalt
A. Goswami and P. S. Nikam
J. Cryst. Growth, 8:247-251 (1971)

Electrical and magnetic properties of single-crystal hexagonal nickel sulfide
J. L. Horwood, L. G. Ripley, M. G. Townsend, and R. J. Tremblay
J. Appl. Phys., 42:1476-1477 (1971)

Magnetic susceptibility of single-crystal hexagonal nickel sulphide
J. L. Horwood and M. G. Townsend
Solid-State Commun., 9:41-43 (1971)

The magnetic structure of Fe_3Se_4
A. F. Andresen and B. van Laar
Acta Chem. Scand., 24:2435-2439 (1970)

Ferroelectricity in the system $Fe_{1-x}S$
C. B. Van den Berg
Phys. Stat. Sol., 40:K65-K68 (1970)

Préparation et étude par diffraction X de nouveaux sulfures de fer dans le domaine de composition $Fe_{0.85}S - FeS$
M. Chevreton, B. Petit, S. Brunie, and J. Kaufmann
Compt. Rend., 270C:426-429 (1970)

The magnetic structure of the spinel Fe_3S_4
J. M. D. Coey, M. R. Spender, and A. H. Morrish
Solid-State Commun., 8:1605-1608 (1970)

Ordered moment of NiS_2
L. M. Corliss and J. M. Hastings
Bull. Am. Phys. Soc., 15:154 (1970)

Etude des systèmes Co – S et Ni – S fonction de la température, de la pression de soufre et de la composition chimique
Jean-Pierre Delmaire, Henri Le Brusq, and Fernand Marion
Compt. Rend., 271C:1449-1451 (1970)

Thermodynamic properties of nickel-tellurium alloys
M. Ettenberg, K. L. Komarek, and E. Miller
J. Solid State Chem., 1:583-592 (1970)

Study of iron sulfides by the Mössbauer effect
G. N. Goncharov, Yu. M. Ostanevich, and S. B. Tomilov
Izv. Akad. Nauk SSSR, Ser. Geol., 8:79-88 (1970)

Mössbauer effect in FeS_{1+x} system
G. N. Goncharov, Yu. M. Ostanevick, S. B. Tomilov, and L. Cser
Phys. Stat. Sol., 37:141-150 (1970)

Conceptual phase diagram and its application to the spontaneous magnetism of pyrites
J. B. Goodenough
Solid State Research (Alan L. McWhorter, ed.), ESD-TR-70-234 (Aug. 15, 1970), pp. 23-26
FeS_2

Heat capacities and thermodynamic properties of the $Ni_{1-x}Se$-phase from 298 to 1050°K
F. Gronvold
Acta Chem. Scand., 24:1036-1050 (1970)

Ordered moment of NiS_2
J. M. Hastings and L. M. Corliss
IBM J. Res. Develop., 14:227-228 (1970)

The variation of the electrical conductivity of FeS with sulfur pressure at temperatures between 670 and 900°C
  H. I. Kaplan and W. L. Worrell
  The Chemistry of Extended Defects in Non-Metallic Solids, Proc. Inst. for Advanced Study on the Chemistry of Extended Defects in Non-Metallic Solids, Casa Blanca Inn, Scottsdale, Arizona, April 16–26, 1969 (LeRoy Eyring and Michael O'Keeffe, eds.), North-Holland Publishing Co., Amsterdam, London (1970), pp. 561–574

Neutron-diffraction study of Fe_7Se_8. II
  Masaru Kawaminami and Atsushi Okazaki
  J. Phys. Soc. Japan, 29:649–655 (1970)

The preparation of high-purity, metal-sulphur alloys
  S. Marich
  J. Phys. E. Sci. Instr., 3:317 (1970)

Cellular structure in the iron-iron sulphide eutectic
  S. Marich
  Met. Trans., 1:1060–1061 (1970)

Cubic FeS, a metastable iron sulfide
  Rinaldo de Medicis
  Science, 170:1191–1192 (1970)

Une nouvelle forme de sulfure de fer: FeS cubique
  Rinaldo de Medicis
  Rev. Chim. Minerale, 7:723–728 (1970)

Thermoelectric power of $Ni_{1-x}S$
  T. Ohtani, K. Kosuge, and S. Kachi
  J. Phys. Soc. Japan, 29:521 (1970)

Hall effect of NiAs-type $Ni_{1-x}S$
  Tsukio Ohtani, Koji Kosuge, and Sukeji Kachi
  J. Phys. Soc. Japan, 28:1588 (1970)

The ordered vacancy arrangement in pyrite defect structures
  E. Parthe and D. Hohnke
  The Chemistry of Extended Defects in Non-Metallic Solids, Proc. Inst. for Advanced Study on the Chemistry of Extended Defects in Non-Metallic Solids, Casa Blanca Inn, Scottsdale, Arizona, April 16–26, 1969 (LeRoy Eyring and Michael O'Keeffe, eds.), North-Holland Publishing Co., Amsterdam, London (1970), pp. 220–222

Interstitials and vacancies in FeS
  R. C. Thiel
  Phys. Stat. Sol. (Letters Sect.), 40:17–20 (1970)

X-ray diffraction measurements on metallic and semiconducting hexagonal NiS
  Jeffrey Trahan and R. G. Goodrich
  Phys. Rev. B: Solid State, 2:2859 (1970)

X-ray diffraction measurements on hexagonal NiS
  J. F. Trahan and R. G. Goodrich
  Bull. Am. Phys. Soc., 15:393 (1970)

Energy bands of hexagonal NiS
  J. M. Tyler and J. L. Fry
  Phys. Rev., B, 1(12):4604–4616 (1970)

Magnetic anisotropy of Fe_3S_4 as revealed by electron diffraction
  S. Yamaguchi and H. Wada
  J. Appl. Phys., 40:1873–1874 (1970)

Magnetism of pyrite-type compounds
  Kengo Adachi
  Nippon Butsuri Gakkaishi, 24(8):518–528 (1969)
  $Co(S_xSe_{1-x})_2$

Magnetization curves of Fe_7S_8 and Fe_7Se_8 under high magnetic field
  Kengo Adachi and Kiyoo Sato
  J. Phys. Soc. Japan, 26:581 (1969)

Magnetic properties of cobalt and nickel dichalcogenide compounds with pyrite structure
  Kengo Adachi, Kiyoo Sato, and Motohiko Takeda
  J. Phys. Soc. Japan, 26(3):631 (1969)
  CoS_2, $CoSe_2$, NiS_2, $NiSe_2$, and the system $Co(S_xSe_{1-x})_2$

Preparation of very pure Fe_xS
  C. B. Van Den Berg and R. C. Thiel
  Z. Anorg. Allgem. Chem., 368(1/2):106–112 (1969)

α-Transition in FeS — a ferroelectric transition
  C. B. Van Den Berg, J. E. van Delden, and J. Bouman
  Phys. Stat. Sol., 36:K89–K93 (1969)

Redetermined crystal structure of FeS_2 (pyrite)
  G. Brostigen and A. Kjekshus
  Acta Chem. Scand., 23(6):2186–2188 (1969)

Thermal decomposition of iron, cobalt, and nickel dichalcogenides
  Livio Cambi, Mario Elli, and E. Guidici
  Chim. Ind. (Milan), 51(8):795–810 (1969)

Production of iron and germanium sulfides
  D. M. Chizhikov, L. V. Nikiforov, and Yu. A. Lainer
  Zh. Neorg. Khim. SSSR, 14(9):2299–2302 (1969)

The insulator — metal transition
  S. Doniach
  Advan. in Phys., 18(76):819–848 (1969)
  NiO, NiS, and V_2O_3; 40 refs.

On the magnetic properties of $CoSe_2$, NiS_2, and $NiSe_2$
  S. Furuseth and A. Kjekshus
  Acta Chem. Scand., 23(7):2325–2334 (1969)

Préparation de monocristaux de sulfure de nickel
  J.-P. Gamondes and M. Laffitte
  Rev. Chim. miner., Fr., 6(4):755–763 (1969)

Thermodynamic properties of $CoTe_2$
  V. A. Geiderikh, Ya. I. Gerasimov, and O. B. Matlasevich
  Khim. Svyaz Krist. (N. N. Sirota, ed.), Izd. "Nauka Tekhnika," Minsk, USSR (1969), pp. 292–295

Préparation de monocristaux de tellurures de cobalt
  Pierre Gibart and Chantal Vacherand
  J. Cryst. Growth, 5:111–114 (1969)

Mössbauer effect in the FeS_{1+x} system
  G. N. Goncharov, Yu. M. Ostanevich, S. B. Tomilov, and L. Cher
  (Joint Inst. for Nuclear Research, Dubna, USSR, Lab. of Neutron Phys.), JINR-P14-4312 (1969), 21 pp.

Phase relations in the system nickel-sulfur-selenium at 500°
  K. Haugsten and E. Roest
  Acta Chem. Scand., 23(10):3599–3600 (1969)

The treatment of cobalt selenide under selenium pressure
  K. Igaki and Y. Noda

Nippon Kinzoku Gakkai-Si, 33(2):371-375 (1969)
Electrical, magnetic, and structural properties of cobalt selenide

Fe—Se system
Gunnar Kullerud
(Geophys. Lab., Washington, D. C.), Carnegie Inst. Wash., Yearb. 1967-1968, Vol. 67, pp. 175-177 (Publ. 1969)

Cubic-hexagonal inversions in some M_3S_4-type sulfides
Gunnar Kullerud
(Geophys. Lab., Washington, D. C.), Carnegie Inst. Wash., Yearb., 1967-1968, Vol. 67, pp. 179-182 (Publ. 1969)

Thermodynamics of the cobalt—selenium system
M. Laffitte and O. Cerclier
High Temperatures-High Pressures, 1(4):449-455 (1969)

Etude par diffraction neutronique de Fe_3Se_4
B. Lambert-Andron and G. Berodias
Solid State Commun., 7:623-629 (1969)

Mössbauer spectroscopic study of magnetic properties of Fe_3S_4
E. F. Makarov, A. S. Marfunin, A. R. Mkrtchyan, G. N. Nadzharyan, V. A. Povitskii, and R. A. Stukan
Fiz. Tverd. Tela, 11:495-497 (Feb. 1969)
Sov. Phys. – Solid State, 11:391 (1969)

X-ray spectroscopic study of chemical bonding in NiSe and $NiSe_2$ (semiconductivity)
C. Mande and A. S. Nigavekar
Proc. Indian Acad. Sci., 69A:316-323 (1969)

The structure of iron-iron sulphide eutectic alloys
S. Marich and G. Brinson
J. Australian Inst. Metals, 14(4):283-291 (1969)

Magnetic properties of $CoS_{2-x}Se_x$
N. Menyuk, K..Dwight, J. A. Kafalas, and A. Wold
Solid State Research, quarterly technical summary Feb. 1-April 30, 1969 (Peter E. Tannenwald, ed.), Contract No. AF19(628)-5167, Solid State Research (1969:2)
ESD-TR-69-110 (May 1969), 60 pp.

Mössbauer studies of iron sulphides
J. A. Morice, L. V. C. Rees, and D. T. Rickard
J. Inorg. Nuclear Chem., 31(12):3797-3802 (1969)

Pressure effect of Curie point of CoS_2 and ferromagnetic properties of $Co(S_xSe_{1-x})_2$
Kiyoo Sato, Kengo Adachi, Tetsuhiko Okamoto, and Eiji Tatsumoto
J. Phys. Soc. Japan, 26(3):639 (1969)

Contribution à l'étude des propriétés électriques et magnétiques des séléniures de fer
J. Serre
Dissertation, University of Paris, 1969, 44 pp.

Propriétés physiques de Fe_7Se_8
Jacques Serre, Pierre Gibart, and Jacques Bonnerot
J. de Phys., 30:93-96 (1969)

Effet Hall et magnétorésistance transverse dans Fe_7Se_8
Jacques Serre and Gerard Villers
Compt. Rend., 268:162-165 (1969)

Neutron-diffraction study of NiS under pressure
F. A. Smith and J. T. Sparks
J. Appl. Phys., 40:1332 (1969)

Phase transitions in nickel and copper selenides and tellurides
A. L. N. Stevels
Thesis, University of Groningen, Philips Res. Rept. Suppl., No. 9 (1969)

Exoelectron emission during phase transformations of magnetic pyrites
B. Sujak
Acta Phys. Polon., 35:475-478 (1969)

Mössbauer measurements on $Fe_{1.11}Te$
J. Suwalski, J. Piekoszewski, and J. Leciejewicz
J. Phys. Soc. Japan, 26:1546 (1969)

Energy band calculations for NiS
J. M. Tyler, T. E. Norwood, and J. L. Fry
Bull. Am. Phys. Soc., 14:431 (1969)

Infrared studies of lattice vibrations in iron pyrite
J. L. Verble and R. F. Wallis
Phys. Rev., 182:783-789 (1969)

X-ray K absorption edges of iron and selenium in iron selenide
A. N. Vishnoi, B. K. Agarwal
Indian J. Pure Appl. Phys., 7:812-813 (1969)

Preparation and properties of new magnetic transition-metal chalcogenides, final report February 15, 1966-January 31, 1969
Aaron Wold
(Brown University), Contract No. DA-31-124-ARO-D-433, AD-682017, AROD 6079.9-MC (Jan. 31, 1969), 4 pp.
The pyrites CoS_2 and $CoSe_2$ have been prepared as single crystals

Technik zur Magnetoanalyse von Pulvern durch Elektronenbeugung
S. Yamaguchi and H. Wada
Exper. Tech. Phys., Dtsch., 17(5-6):527-529 (1969)
Fe_3S_4

Hydrothermal synthesis of iron thiospinel
S. Yamaguchi
Kristallografiya, 14(4):752-753 (1969)
Sov. Phys. – Cryst., 14(4):653 (1970)
Fe_3S_4

The Chemistry of Cobalt
Cobalt Monograph, C.I.C., Brussels (1969), Cobalt Information Center

Magnetic properties of some compounds with pyrite structure
K. Adachi, K. Sato, and M. Takeda
J. Appl. Phys., 39:900 (1968)
CoS_2, $CoSe_2$, NiS_2, $NiSe_2$, $Co(S_xSe_{1-x})$

Neutron-diffraction investigation of Fe_3Se_4
A. F. Andresen
Acta Chem. Scand., 22(3):827-835 (1968)

Transition-metal pyrite dichalcogenides. High-pressure synthesis and correlation of properties
T. A. Bither, R. J. Bouchard, W. H. Cloud, and others
Inorg. Chem., 7:2208-2220 (1968)

Gamma-resonant Mössbauer spectroscopy of iron sulfides
B. V. Borshagovskii, A. S. Marfunin, A. R. Mkrtchyan, and others
Izv. Akad. Nauk SSSR, Ser. Khim., 33:1267-1271 (1968)

The preparation of single crystals of FeS_2, CoS_2, and NiS_2 pyrites by chlorine transport
    R. J. Bouchard
    J. Crystal Growth, 2:40-44 (1968)

Contribution à l'étude du système fer-soufre limité du domaine du monosulfure de fer pyrrhotine
    W. Burgmann, Jr., G. Urbain, and M. G. Frohberg
    Mem. Sci. Rev. Metal, Fr., 65:567-578 (1968)

Phase studies on the iron-selenium system
    J. E. Dutrizac, M. B. I. Janjua, and J. M. Toguri
    Can. J. Chem., 46:1171 (1968)

Heat capacities and thermodynamic properties of the iron selenides $Fe_{1.04}Se$, Fe_7Se_8, and Fe_3S_4 from 298 to 1050°K
    Fredrik Groenvold
    Acta Chem. Scand., 22(4):1219-1240 (1968)

Mössbauer study of the magnetic structure of Fe_7S_8
    Lionel M. Levinson and D. Treves
    J. Phys. Chem. Solids, 29:2227-2231 (1968)

An investigation of the crystal growth of the heavy metal sulfides in supercritical hydrogen sulfide
    Leroy Crawford Lewis
    Ph.D. Thesis from Oregon State University, Corvallis (1968), 172 pp
    Sulfides, iron

X-ray spectroscopic study of chemical bonding in manganese diselenide and cobalt diselenide
    C. Mande and A. S. Nigavekar
    Proc. Indian Acad. Sci., 67A(3):168-174 (March 1968)

Magnetic properties of FeS_2 and CoS_2
    Syohei Miyahara and Teruo Teranishi
    J. Appl. Phys., 39:896 (1968)

Thermodynamics of pyrrhotite ($Fe_{1-x}S$) at the temperature range of 800-1100°C
    M. Nagamori and M. Kameda
    Trans. Japan. Inst. Metals, 9(3):187-194 (1968)

Synthesis of smythite-rhombohedral Fe_3S_4
    D. T. Rickard
    Nature, 218-356-357 (1968)

The influence of certain elements (aluminum, manganese, chromium, and titanium) on the chemical composition of iron sulphides
    Jiri Skala and Richard Riman
    Sbornik V. S. B. Ostrave (Hutnicka), 14(3):115-122 (1968)

Metal-to-semiconductor transition in hexagonal NiS
    Joseph T. Sparks and Ted Komoto
    Rev. Modern. Phys., 40:752 (1968)

Crystal structures of ruthenium diselenide, osmium diselenide, platinum diarsenide, and α-nickel diarsenide
    W. N. Stassen and R. D. Heyding
    Can. J. Chem., 46(12):2159-2163 (1968)

Phase transitions in the nickel selenide Ni_3Se_2 and related phases
    A. L. N. Stevels, J. Bouwma, G. A. Wiegers, and F. Jellinek
    Recueil des Traveaux Chimiques des Pays-Bas, 87(7):705-708 (1968)

Temperature dependence of hyperfine interactions in near-stoichiometric FeS. I. Experiment
    R. C. Thiel and C. B. van den Berg
    Phys. Stat. Sol., 29:837-846 (1968)

The structure of tetragonal FeS
    M. Uda
    Z. Anorg. Allg. Chem., 361-94-98 (1968)

Synthesis of magnetic Fe_3S_4
    Masayuki Uda
    Scientific Papers of the Institute of Physical and Chemical Research (Tokyo, Japan), 62(14):14-23 (March 1968)

The miscibility gap between FeS and $Fe_{1-x}S$
    Richard A. Yund and Henry T. Hall
    Mater. Res. Bull., 3:779-784 (1968)

Phase transition in Fe_xS (x = 0.90-1.00) studied by neutron diffraction
    A. F. Andresen and P. Torbo
    Acta Chem. Scand., Danem., 21(10):2841-2848 (1967)

Observation en diffraction électronique de la sulfuration de films minces de cobalt
    Marc Brieu, Renée Calsou, and Lucien Lafourcade
    Compt. Rend., 264:1300-1302 (1967)

Effet Mössbauer et nature des liaisons dans quelques composés semi-métalliques du fer
    A. Gerard
    Propriétés thermodynamiques physiques et structurales des dérives semi-métalliques, No. 157, Colloq. Intern. Centre National de la Récherche Scientifique, Orsay (28 Septembre-1 Octobre 1965), Editions du Centre National de la Récherche Scientifique, Paris (1967), p. 55

On the magnetocrystalline anisotropy of iron selenide Fe_7Se_8
    Takashi Kamimura, Kazuo Kamigaki, Tokutaro Hirone, and Kiyoo Sato
    J. Phys. Soc. Japan, 22:1235 (1967)

Neutron-diffraction study of Fe_7Se_8
    Masaru Kawaninami and Atsushi Okazaki
    J. Phys. Soc. Japan, 22:924 (1967)

Anomalous electrical resistivity of Fe_7Se_8
    Masaru Kawaminami and Atsushi Okazaki
    J. Phys. Soc. Japan, 22:925 (1967)

Preparation and magnetic properties of cobalt disulfide
    B. Morris, V. Johnson, and A. Wold
    J. Phys. Chem. Solids, 28:1565-1567 (1967)

Self-diffusion of iron in ferrous sulphide
    S. Mrowec
    Bull. Acad. Polon. Sci., Ser. Sci. Chim., 15(11):521-525 (1967)

Electric and magnetic properties of nickel tellurides
    G. Saut
    In: Proceedings of Second International Conference on Solid Compounds of Transition Elements, June 12-16, 1967, at the Technical University of Twente, Netherlands, pp. 87-89

Propriétés électriques, optiques et magnétiques des tellurures de cobalt et de nickel (phase de structure NiAs)
    Georges Saut
    These from Faculté des Sciences de l'Université de Paris (1967)

The structure of synthetic Fe_3S_4 and the nature of transition to FeS
  Masayuki Uda
  Z. Anorg. Allgem. Chem., 350:105 (1967)

Kristallisation von Disulfiden aus Schmelzlösungen
  K.-Th. Wilke, D. Schultze, and K. Topfer
  J. Crystal Growth, 1:41-44 (1967)

Propriétés physiques de sulfures de type pyrite et de type spinelle
  Mme. M. Wintenberger and J. Bonnerot
  Propriétés thermodynam. phys. struct. Dérives semi-métal. Coll. internation. Orsay, 1965, Paris, C.N.R.S. (1967), pp. 369-372
  NiS_2

Structural characteristics of the Fe—FeS eutectic
  D. L. Albright and R. W. Kraft
  Trans. AIME, 236:998 (1966)

On the tellurides of nickel
  J. Barstad, F. Gronvold, E. Rost, and E. Vestersjo
  Acta Chem. Scand., 20:2865 (1966)

New transition-metal dichalcogenides formed at high pressure
  T. A. Bither et al.
  Solid State Commun., 4:533-535 (1966)
  Synthesis and properties of the Cu dichalcogenide series CuS_2, $CuSSe$, $CuSe_2$, $CuSeTe$, and $CuTe_2$ as well as new MX_2 compounds of Fe, Co, and Ni

Zur Kenntnis des Eisen (III)-Sulfids
  H. P. Boehm and E. Flaig
  Angew. Chem., Dtsch., 78(21):987 (1966)
  Fe_2S_3 and Fe_3S_4

High-temperature phases $Ni_{3\pm x}Se_2$ and Ni_6Se_5
  F. Gronvold, R. Mollerud, and E. Rost
  Acta Chem. Scand., 20:1997-1998 (1966)

The magnetic susceptibilities of the cobalt-sulfur system
  R. F. Heidelberg, A. H. Luxem, S. Talhouk, and J. J. Banewicz
  Inorg. Chem., 5(2):194-197 (1966)

Méthode d'étude du spectre d'absorption des cristaux à forte densité de porteurs libres
  P. Manca and G. Saut
  Compt. Rend., 262:1621-1623 (1966)
  Applic. aux tellurures de fer et de cobalt

Magnetic and x-ray spectroscopic study of CoSe
  A. S. Nigavokar and C. Mande
  Proc. Nucl. Phys. and Solid State Phys. Symp. (1966), pp. 83-87, Solid State Phys., Vol. 2, Bombay, Physics Committee, Dept. of Atomic Energy

Sur un nouveau séléniure de nickel Ni_3Se_2
  Jean-Pierre Rouche and Pierre Lecocq
  Compt. Rend., 262:555-556 (1966)

Propriétés électriques et magnétiques des tellures de nickel
  Georges Saut
  Compt. Rend., 263:1174 (1966)

Etude métallurgique et propriétés électriques des séléniures de fer de structure NiAs
  Jacques Serre and Raymond Druilhe
  Compt. Rend., 262:639-641 (1966)
  $FeSe$, Fe_7Se_8, Fe_3Se_4

Preparation of copper, cadmium, cobalt, and nickel selenides by selenite reduction in aqueous solutions
  V. M. Shul'man and V. L. Varand
  Izv. Akad. Nauk SSSR, Ser. Khim., No. 5, pp. 934-935 (1966)

A study of the S $K\beta_1$ line from some sulfide crystals
  E. Suoninen, V. Lantto, and V. Polvi
  Ann. Acad. Sci. Fenn., Ser. A, VI. Physica, 296 (1966)
  FeS, FeS_2, NiS

Magnetic structures recently studied at CEN-Grenoble
  E. F. Bertaut
  Bull. Soc. Sci. Bretagne, Vol. 39, "Fascicule Hors Serie," 67-68 (1964)
  Colloq. Assoc. Francais de Cristallographie Changements de Phase dans les Solides Inorganiques, Rennes (1965), Soc. Sci. Bretagne
  FeS

Some remarks on vacancy order
  E. F. Bertaut
  Bull. Soc. Sci. Bretagne, Vol. 39, Fascicule Hors Serie, 73-74 (1964)
  CEA-TP-6862, Colloq. Assoc. Franc. Cristallographie, Rennes (April 1-3, 1965)
  Fe_7S_8

Thermal expansion of iron pyrites
  R. S. B. Chrystall
  Trans. Faraday Soc., 61, Pt. 8, 1811-1815 (1965)

Optics of hexagonal pyrrhotine ($\sim Fe_9S_{10}$)
  Kurt von Gehlen
  Mineralogical Mag., 35:335-346 (1965)

The system nickel—tellurium. I. Structure and some superstructures of the $Ni_{3\pm q}Te_2$ phase
  R. B. Kok, G. A. Wiegers, and F. Jellinek
  Recueil des Travaux Chimiques des Pays-Bas, 84:1585 (1965)

Preparation of large nickel and cobalt monosulfide crystals
  V. I. Rybnikov and V. I. Smirnov
  Issledovaniya v oblasti Khimii i tekhnologii mineral'nykh solei i okislov, Moskva; Leningrad, Izdat. Nauka (1965), pp. 91-96

Electric and magnetic properties of cobalt telluride (gamma phase)
  G. Saut
  Compt. Rend., 261:3339-3342 (1965)

Crystal structure and electrical properties of some cobalt-group chalcogenides
  F. Hulliger
  Nature, 204:644-646 (1964)
  $Co_{1/2}Rh_{1/2}S_2$, $RhSSe$, $RhSe_2$, $RhSeTe$, $RhTe_3$, $Rh_{1/2}Ir_{1/2}Se_2$, IrS_3, $IrSSe$, $IrSe_2$, $IrSe_{1/2}Te_{1/2}$, $IrTe_2$, $Co_{1/2}Rh_{1/2}S_{\sim3}$, $RhS_{\sim3}$, $IrS_{\sim3}$, $IrSe_{\sim3}$, $IrTe_{\sim3}$, Rh_2S_3

Magnetocristalline anisotropy of pyrrhotite
  Kiyoo Sato, Motohiko Yamada, and Tokutaro Hirone
  J. Phys. Soc. Japan, 19:1592 (1964)

Etudes de diffraction neutronique à Livermore
  J. T. Sparks and T. Komoto
  J. Phys., 25:567 (1964)

Propriétés semi-conductrices du tellurure de fer (gamma phase)
Jacques Suchet
Compt. Rend., 259:3219-3222 (1964)

A neutron-diffraction study of magnetic ordering in iron telluride
Janusz Leciejewicz
Acta Chem. Scand., 17:2593-2599 (1963)

Neutron-diffraction study of NiS
Joseph T. Sparks and Ted Komoto
J. Appl. Phys., 34:1191-1192 (1963)

Mössbauer study of hyperfine field, quadrupole interaction, and isomer shift of ^{57}Fe in $FeS_{1.00}$, $FeS_{1.05}$, and $FeS_{1.07}$
K. Ono, A. Ito, and E. Hirahara
J. Phys. Soc. Japan, 17(10):1615-1620 (1962)

Low-temperature heat capacities and thermodynamic properties of the $Ni_{1-x}Se$ phase
Fredrik Gronvold, Torkild Thurmann-Moe, Edgar F. Westrum, Jr., and Norman E. Levitin
Acta Chem. Scand., 14:634-640 (1960)

The system nickel — selenium
Joh E. Hiller and W. Wegener
Neues Jahrb. Mineral. Abhandl., 94:1147-1159 (1960)

Magnetic and electrical anomalies of iron telluride single crystals
Rokuroh Naya, Miyuki Murakami, and Eiji Hirahara
J. Phys. Soc. Japan, 15:360-361 (1960)

Investigation of the cobalt-tellurium system
L. D. Dudkin and K. A. Diul'dina
Zh. Neorg. Khim., 4(10):2313-2319 (1959)

The magnetic susceptibility of iron ditelluride. II
D. M. Finlayson, J. P. Llewellyn, and T. Smith
Proc. Phys. Soc. (London), 74, Part 1, 75-80 (1959)

The magnetic-susceptibility of iron ditelluride. I
J. P. Llewelly and T. Smith
Proc. Phys. Soc. (London), 74, Pt. 1, 65-74 (1959)

Thermodynamic properties of nonstoichiometric nickel tellurides and of tellurium
Robert Engel Machol
(Univ. of Michigan), Dissertation Abstracts, 20(4):1193-1194 (Oct. 1959)

On the magnetic property of iron telluride
Ichiro Tsubokawa and Shoku Chiba
J. Phys. Soc. Japan, 14:1120 (1959)

Heat capacities and thermodynamic properties of the iron tellurides $Fe_{1.11}Te$ and $FeTe_2$ from 5 to 350°K
Edgar F. Westrum, Jr., Chien Chou, and F. Gronvold
J. Chem. Phys., 30(3):761-764 (1959)

Magnetic and electrical anisotropies of iron sulfide single crystals
Eiji Hirahara and Miyuki Murakami
J. Phys. Chem. Solids, 7:281 (1958)

Magnetic properties of the substances Fe_xSe_{1-x}
I. Maxim
Acad. rep. polulare Romine. Inst. fiz. atomica so Inst. fiz., Studii cercetari fiz., Vol. 9, pp. 323-329 (1958)

Crystal structure of nickel selenide, Ni_3Se_2
R. P. Agarwala and A. P. B. Sinha
Z. Anorg. Allgem. Chem., 289(1/4):203-206 (Feb. 1957)

The thermal expansion of iron ditelluride
J. P. Llewellyn and T. Smith
Proc. Phys. Soc. (London), 70B:1113-1122 (1957)

Temperature dependence of the lattice constants in the mixed crystal series $NiTe - NiTe_2$
A. Schneider and K. H. Imhagen
Naturwissenschaften, 44(11):324-325 (1957)

Some electrical characteristics of single-crystal iron monotelluride
D. M. Finlayson, D. Greig, J. P. Llewellyn, and T. Smith
Proc. Phys. Soc. (London), 69B:860-862 (1956)

The magnetic properties of $FeSe_x$ with NiAs structure
Tokutaro Hirone and Shu Chiba
J. Phys. Soc. Japan, 11:666-670 (1956)

Structural study of iron selenides, $FeSe_x$. I. Ordered arrangement of defects of Fe atoms
Atsushi Okazaki and Kinshiro Hirakawa
J. Phys. Soc. Japan, 11:930-936 (1956)

Magnetic properties of nickel telluride
Enji Uchida and Hisamoto Kondoh
J. Phys. Soc. Japan, 11:21-27 (1956)

Note on the ferromagnetism of CoTe
Enji Uchida
J. Phys. Soc. Japan, 11(4):465-466 (1956)

The magnetic properties and phase diagram of the iron-tellurium system
Shoku Chiba
J. Phys. Soc. Japan, 10:837-842 (1955)

Magnetic properties of cobalt telluride
Enji Uchida
J. Phys. Soc. Japan, 10(7):517-522 (1955)

6.g. Group VIII (others)

The palladium telluride $Pd_{4-x}Te$ crystal structure
V. S. Khar'kin, R. M. Imamov, and S. A. Semiletov
Kristallografiya, 14(5):907-910 (1969)
Sov. Phys. — Cryst., 14(5):779 (1970)

Mixed platinum chalcogenides
E. D. Ruchkin, L. A. Vostrikova, and S. S. Batsanov
Izv. Akad. Nauk SSSR, Neorg. Mater., 6(2):252-256 (1970)
Inorg. Mater., 6(2):220-223 (1970)

Konstitution einiger Mischungen des Platins mit B-Elementen (B = Sn, Sb, Te)
S. Bhan, T. Godecke, und K. Schubert
J. Less-Common Metals, 19:121-140 (1969)

Tetrapalladium sulfide and tetrapalladium selenide: heat capacities and thermodynamic properties from 5 to 350°K
Fredrik Groenvold, Edgar F. Westrum, and Ray Radebaugh
J. Chem. Eng. Data, 14(2):205-207 (1969)

A high-pressure form of palladium disulfide
R. A. Munson and J. S. Kasper
Inorg. Chem., 8(5):1198-1199 (1969)

Crystal structure of pyrite-related Rh_3Se_8
(rhodium-selenium)
    D. Hohnke and E. Parthé
    Z. Kristallogr., 127:164-172 (1968)

A new structure type with octahedron pairs for
Rh_2S_3, Rh_2Se_3, and Ir_2S_3
    E. Parthé, D. Hohnke, and F. Hulliger
    Acta Cryst., 23:832-840 (1967)

Chalcogenides of the transition elements. V.
Crystal structure of the disulfides and ditellurides of ruthenium and osmium
    Sutarno, Osvald Knop, and K. I. G. Reid
    Can. J. Chem., 45:1391 (1967)

The crystal structure of Rh_2Te_3
    W. H. Zachariasen
    Acta Cryst., 20:334 (1966)

Redetermined crystal structures of $NiTe_2$, $PdTe_2$,
PtS_2, $PtSe_2$, and $PtTe_2$
    Sigrid Furuseth, Kari Selte, and Arne Kjekshus
    Acta Chem. Scand., 19:257-258 (1965)

Constitution and magnetic and electrical properties of palladium tellurides ($PdTe - PdTe_2$)
    A. Kjekshus and W. B. Pearson
    Can. J. Phys., 43:438 (1965)

The occurrence of superconductivity in sulfides,
selenides, tellurides of Pt-group metals
    Ch. J. Raub, V. B. Compton, T. H. Geballe, B. T. Matthias,
      J. P. Maita, and G. W. Hull, Jr.
    J. Phys. Chem. Solids, 26:2051-2057 (1965)
    PtTe, $IrTe_3$, $PdTe_2$, $PdTe_{1.04}$

The crystal structures of Pd_4Se and Pd_4S
    Fredrik Gronvold and Erling Rost
    Acta Cryst., 15:11-13 (1962)

On the sulfides, selenides, and tellurides of
platinum
    Fredrik Gronvold, Haakon Haraldsen, and Arne Kjekshus
    Acta Chem. Scand., 14:1879-1893 (1960)

Coprecipitation of palladium with tellurium.
Formation of palladium tellurides
    Marcelle Segui-Cros
    Bull. Soc. Chim. France, No. 3, pp. 451-452 (March 1960)

High-temperature x-ray study of the thermal
expansion of PtS_2, $PtSe_2$, $PtTe_2$, and $PdTe_2$
    Arne Kjekshus and Fredrik Gronvold
    Acta Chem. Scand., 13:1767-1774 (1959)

The crystal structure of iridium diselenide
    Luisa Brahde Barricelli
    Acta Cryst., 11:75-79 (1958)

Superconductivity in the Pd - Se system
    B. T. Matthias and S. Geller
    J. Phys. Chem. Solids, 4:318-319 (1958)

Sulfides, selenides, and tellurides of palladium
    Fredrik Gronvold and Erling Rost
    Acta Chem. Scand., 10:1620-1634 (1956)

The crystal structures of RhTe and $RhTe_2$
    S. Geller
    J. Am. Chem. Soc., 77:2641-2644 (1955)

The crystal structure of $RhSe_2$
    S. Geller and B. B. Cetlin
    Acta Cryst., 8:272-274 (1955)